AF587892

BINOMIUM CHITIN-CHITINASE: RECENT ISSUES

BINOMIUM CHITIN-CHITINASE: RECENT ISSUES

SALVATORE MUSUMECI
AND
MAURIZIO G. PAOLETTI
EDITORS

Nova Biomedical Books
New York

For permission to use material from this book please contact us:
Telephone 631-231-7269; Fax 631-231-8175
Web Site: http://www.novapublishers.com

Library of Congress Cataloging-in-Publication Data

Binomium chitin-chitinase : recent issues / editors, Salvatore Musumeci and Maurizio G. Paoletti.

p. ; cm.

Includes bibliographical references and index.

ISBN 978-1-60692-339-9 (hardcover)

1. Chitin. 2. Chitinase. I. Musumeci, Salvatore. II. Paoletti, M. G.

[DNLM: 1. Chitin. 2. Chitinase. QU 83 B614 2009]

QP702.C5B56 2009

573.7'74--dc22

2008050192

Published by Nova Science Publishers, Inc. ✦ New York

Contents

Preface

The binomial chitin/chitinase arena arises from the observation that chitin is, after cellulose, the widest spread bio-polymer in nature. It has been estimated that approximately 200 billions tons of chitin (at least 100 billions in the oceans) are produced every year by invertebrates including nematodes, mollusks, arthropods, crustaceans, insects, but also fungi, some algae and yeasts.

Its huge presence and its potential impact on the environment has led to the development in nature of chitinases with the goals of hydrolyzing chitin either as a potential source of energy or as a constitutive protection toward chitin-containing pathogens. Chitinases are also needed for molding in the metamorphosis of organisms such as most invertebrates and arthropods in particular.

While chitin structure is strictly conserved, chitinases, particularly the pre-mammalian ones, have been evolved through impressive structural and functional adaptations to the different evolutionary contexts. This evolution has been differently traced, according to the biological scale, but also taking into account the new bioinformatic procedures to support of those previously used.

Obviously, the increasing body of knowledge on chitin and chitinase biology have rapidly spread from the basic and transitional applications such as in designing insecticides up to gain a relevant role in clinical application. It is only two decades since chitinases have been isolated in humans and some have been found in the last decade.

Therefore, we are facing new and exciting interests on a previously neglected subject and an increasing pace of discoveries is likely coming.

Gaucher disease has been the cradle of chitotriosidase (Chit) produced by macrophages and this enzyme has been considered a marker of activity in this disease and a prognostic parameter for the enzymatic substitutive therapy.

Chitinases expression in humans has rapidly gain a role from a mere tool of macrophage activation to a useful clinical marker in the diagnosis of conditions in which the innate immunity could have a role as trigger or effector. In fact, chitotriosidase, now represents a valid marker of vascular wall alteration and atherosclerosis. Various contributors recently appeared in the literature and, certainly, other surprises are in gestation on the role of chitotriosidase in the modeling of atherosclerotic plaques. Also, Chit is acquiring a new and promising role in the pathogenesis of multiple sclerosis, perhaps in linkage with its macrophage/microglia-derived origin, and as a prognostic factor of the disease evolution. Chit has taken a position as a clinical marker not only in demyelinating diseases but also in the degenerative disease of the grey matter, Alzheimer's disease. Recent researches opened new horizons which help in the interpretation of Chit role during the course of the disease.

Recently, several studies have been published on the role of another chitinase, an acidic mammallian chitinase (AMCase) in the pathogenesis of asthma in humans and mice, in the mechanism of allergic conjunctivitis and in dry-eye pathology.

The ability of AMCase contained in human gastric juice to hydrolyze chitin has opened new possible implication for the use of chitin-containing organisms for feeding purposes. This subject is fascinating several researchers. An ability to eventually digest chitin, will be dramatically important not only for nutritional aims but also for the prevention of allergies to chitin-containing parasites like mites and as a weapon to face parasites of the digestive tracts.

Moreover a positive relationship between Chit1 expression level in antral gastric mucosa and both flogosis and *Helicobacter pylori* infection was also found.

Genetics of chitinases in humans is still in its infancy. Studies on population genetics have not reached an agreement on whether chitinases had protective function against chitin-containing pathogens in underdeveloped countries and on whether new functions could be selected in environments in which parasitic diseases are less frequent.

Currently, the role of chitinases and particularly of Chit in the immune response mainly derives from experimental observations. The transition from basic to clinical studies certainly will rapidly solve doubts about its immuno-regulatory function, which remains unsolved to date.

The hydrolytic ability of Chit should not be seen as its only function; interestingly, other chitinases are described in mammals and humans to retain the ability of binding chitin, but also of mediating relationships between tissue cells and macrophages. This unexpected ability open a new scenario of a relationship between dys-functional chitinases and cancer.

Moreover chitinases produced by bacteria have been seen to play a lytic role on certain types of cancer cells in culture.

The possibility to use chitinases in the diagnosis and treatment of certain fungal infections has been proposed with convincing evidences. Their ability in triggering immune response through cytokines such as IL-13, and their inactivation through inhibitors such as allosamidine or xantine also seem very promising, and open new ways in controlling inflammatory responses in mice and humans with allergic asthma, allergic conjunctivitis and dry-eye syndrome.

In conclusion the binomial chitin/chitinase which arises from the observation that chitin is, after cellulose, the widest spread bio-polymer in nature and that chitinase have been developed in syntony to chitin, maintains its interest expecially in the modeling of fungi, algae and yeast. However in mammals, where the chitin synthase is not present, a determinant role in the relation with the external environment can be placed by chitinases. This consideration open a new prospective in the role of chitinase, primarialy developed for the defence against chitin containing parasites. Moreover chitinases became new mediator of immune innate response adapted to regulate the relation among chitin largely diffused in nature and the chitinase producing organisms. The enormous presence of chitin in nature produced by invertebrates including nematodes, mollusks, crustaceans, insects, but also fungi, some algae and yeasts, promote an adaptative response on vertebrates and on mammals, followed by evolutive growth of chitinase functions.

This book offers a collection of articles on the binomial chitin/chitinase arena from authors who have all personally contributed to the development and increase of this topic. The book will hopefully represent a milestone for future researchers by taking into account the known archaic function of chitinases and the new description of chitin and chitinases roles in the innate immunity. This recently discovered role has likely been vital in the process of genetic selection and the broadening of researches in this field will pursuit others possible, previously neglected, functions.

Chapter I. - In living tissues both chitin polymorphs occur in covalent combination with either proteins or glucan and are often cross-linked following the quinone tanning process. Most of the chitin, however, is highly ordered as crystallites called nanofibrils that can be recovered as aqueous suspensions for nanotechnological applications. In spite of the inherent insolubility of chitin in water, chitin aqueous systems include ethers obtained from alkali chitin such as O-carboxymethyl chitin and glycol chitin; 6-oxychitin and partially reacetylated chitin are also easy to prepare, while chitin oligomers are *per se* water-soluble. Several solvents for chitin are also available such as the dimethylacetamide-LiCl mixture. Chitinases are enzymes involved in growth, defense, aggression and feeding secreted by animals, fungi and bacteria; they are finding applications in agriculture, particularly after the genomes of plants such as rice *Oryza sativa* and pests such as *Tribolium castaneum* were fully elucidated. Transgenic rice plants are endowed with novel and powerful chitinases that are promptly activated in case of infection by rice pathogens. Phytoparasitism has been put to profit by engaging *Trichoderma harzianum* in protecting plants against pathogens such as *Rhizoctonia solani*. The importance of lysozyme is highlighted insofar as chitinases retain an ancient structural motif of lysozyme, and actually lysozyme is an enzyme ubiquitously present in the human body for defense against microbes and parasites. Unspecific enzymes such as cellulase, hemicellulase and lipase are currently used to prepare chitin oligomers to be used in the biomedical field, particularly in medication and drug delivery.

Chapter II. - Chitinases hydrolyze the β1-4 linkages of chitin, an unbranched polymer of β1-4 linked N-acetyl-D-glucosamine (GlcNAc). Chitin is the second most abundant polymer in nature and many organisms including prokaryotes, vertebrates, plants, fungi and insects produce chitinases. The roles of chitinases in these organisms are diverse. For example, in bacteria, chitinases play a role in nutrition and parasitism. Occurrence of multiple chitinases helps bacteria to utilize various chitinous substrates. Chitinases play a critical role in viral pathogenicity. It is suggested that viral chitinases along with cathepsin are associated with loss of integrity of host tissues permitting mature polyhedra to escape into the environment and promoting horizontal virus transmission. In fungi, the biosynthesis and hydrolysis of chitin plays an important role in formation of a functional cell wall. Chitinases are thought to have autolytic, nutritional, and morphogenetic roles as they contribute to breakage and reforming of bonds within and between polymers, leading to re-modeling of the cell wall during growth and morphogenesis. In insects, chitin functions as scaffold material so insect growth and morphogenesis are strictly dependent on the coordination of chitin synthesis and its degradation which requires strict control of the participating enzymes during development. In addition chitinases are associated with the need for partial degradation of old cuticle in crustaceans and insects. Similar to insects, crustaceans and fungi, chitinases play important role in the life cycle of several protozoan and metazoan parasites that infect humans. Some pathogens use chitinase to invade or exploit the chitin containing structures of their host to establish successful infection or transmission to another host via insect vectors. In plants, chitinases have been implicated in plant resistance against fungal pathogens. Moreover, by reducing the defense reaction of the plant, chitinases allow symbiotic interaction with nitrogen-fixing bacteria or mycorrhizal fungi. They are also involved in numerous physiological events. In vertebrates, chitinases are usually part of the digestive tract and recently chitinases have been found to be implicated in various human diseases such as asthma, arthritis, multiple sclerosis, Gaucher disease, Alzheimer's disease, Fabry storage diseases etc. It is also suggested that chitinases expressed in human tissues may confer protection against fungi in a manner analogous to the protection provided by lysozyme against bacteria. The complexity and functional diversity of the chitinases has made them important candidate for study. The present chapter is focused on the diverse roles of chitinases in these organisms.

Chapter III. - In this contribution we reconsider the phylogeny of mammalian proteins homologous to the glycosyl hydrolase 18 family: chitinases and chitinase-like proteins. This problem has been recently dealt with in two important papers (Bussink *et al.* 2007; Funkhouser and Aronson, 2007). A clear scheme emerges from these analyses, in which chitinase-like proteins are specialized, tissue-specific, mammalian proteins that have lost the chitinolytic function and have acquired a wealth of possible new functions, mainly related to inflammatory processes. We present here preliminary results from different methods of sequence analysis based on: i) multiple alignments; ii) compression algorithms; iii) statistical over(under)-representation of short k-grams. From our preliminary exploration we formulate and discuss the hypothesis that, chitinase-like proteins are the ancestor group, present as pre-chitinase activators in an ancestral unicellular world from which active chitinases originated as a response to the emergence of chitin synthesis. Chitinase-like proteins in mammals could play a role, in inflammation and in cancer development, similar to the ancient role of activator or signalling molecules in unicellular organisms.

Chapter IV. - Chitinases are ubiquitous chitin fragmenting enzymes identified in several organisms. Two distinct chitinases have recently been identified in humans, chitotriosidase expressed in phagocytes and an acidic mammalian chitinase (AMCase) expressed in the gastrointestinal tract and to a lesser extent in lung.

A role for human chitotriosidase in innate immunity is suggested by several findings. *In vitro* and *in vivo* evidence link chitotriosidase overexpression by macrophages and its release from polymorphonuclear neutrophils (PMNs), via exocytosis of specific granules, to the immune response elicited in microbial infections. Initial, *in vitro* studies showing its chitinolytic activity towards the cell wall chitin of *Candida albicans* have been strengthend by later findings showing that it causes growth inhibition, hyphal tip bursting and prevention of hyphal switch in chitin containing fungi. Furthermore, administration of human recombinant chitotriosidase improved the survival of neutropenic mouse models of systemic candidiasis and aspergillosis.

Increased chitotriosidase plasma and tissue activity has been found in guinea pigs infected by *Aspergillus fumigatus*. Recently, increases in chitotriosidase activity, that run in parallel to their clinical outcome, were observed in neonates not only with systemic candidiasis and aspergillosis but also with bacterial infections. It is of interest that the highest chitotriosidase levels were observed in the neonates that succumbed to their fungal infection. Approximately 6% of the general population in Caucasians cannot synthesize an enzymatically active chitotriosidase and genetic variants in chitotriosidase were shown to be associated with gram-negative bacteremia in leukemic patients. On the other hand, no conclusive corresponding evidence exist regarding succeptibility and survival in fungal infections.

The role of AMCase in innate immunity is not well studied. Given its chitinolytic activity towards fungal cell wall chitin, it has been suggested that it might partly compensate for the above mentioned deficiency of chitotriosidase, however relevant data are missing, and the ones available link AMCase to allergic reactions rather than defence mechanisms.

Clearly more studies are required in order to fully understand the role of chitinases in fungal and bacterial infections and to assess their possible value as therapeutic agents in these infections.

Chapter V. - Chitin, the polymer of b-1,4 linked b-N-Acetyl-glucosamine (GlcNAc), is the most abundant biopolymer in marine environments and the second most abundant in nature, after cellulose. The degradation of chitin is mediated by chitinolytic hydrolases such as chitinases (EC.3.2.1.14) and b-hexosaminidases (EC.3.2.1.52). Based on sequence homologies chitinases fall into two groups: families 18 and 19 of glycosyl hydrolases.

Though mammals lack endogenous chitin, chitinases and chi-lectins, highly homologous proteins lacking enzymatic activity due to catalytic amino acid substitutions, are found in a wide variety of mammalian species. All belong to the family 18 of glycosyl hydrolases (GH18).

Despite the wealth of structural information that is available regarding the mammalian chitinase protein family, insight into their exact physiological role(s) remains limited.

This review gives an overview of all mammalian family members with special emphasis on their occurrence and expression in humans.

Recent molecular phylogenetic analyses suggest that both active mammalian chitinases (chitotriosidase and AMCase) result from an early gene duplication. Further gene duplication events, followed by loss of function mutations, allowed the evolution of the chi-lectins. The homologous genes coding family 18 glycosyl hydrolases are clustered in two distinct chromosomal loci.

The phylogenetic analyses suggest that the evolution of this gene family is in accordance with a form of multigene family evolution referred to as "birth-and-death evolution under strong purifying selection". Finally, several chitinase family members are present only in certain lineages of mammals and their tissue specific expression patterns differ profoundly between species, exemplifying recent evolutionary adaptations in the chitinase protein family.

Chapter VI. - CHIT1 has been the first human gene encoding a chitinolytic enzyme to be discovered. CHIT1 gene product, designated as chitotriosidase (Chit) is a member of the chitinase family and it synthesized by activated macrophages. Sequence homology studies indicate that CHIT1 gene is conserved across the evolutionary scale and consequently has an important biological role. Recently, a genetic polymorphism (a 24 bp duplication in exon 10) was found to be responsible for the common deficiency in Chit activity, frequently encountered in different populations. The presence of the duplication in individuals from various ethnic groups suggests that this mutation is relatively old.

Here we discuss the analysis of the CHIT1 gene in some ethnic groups from the Mediterranean, African to Asian areas, to evaluate whether the CHIT1 gene polymorphism H correlates with the changes in environmental features. From a population point of view, the understanding of the variability of the CHIT1 variants improve the knowledge on origin and diffusion of the gene from an original population to other people living in different world areas.

We can also use the study of CHIT1 variants to perform a correlation between the mean Chit enzyme activity with a particular genotype and the origin ancestry of population. The median enzyme activity in wild-type subjects was significantly higher in subjects of European ancestry, than subjects of African and Asian ancestry. Moreover, genomic analysis of individuals heterozygous or wild type for the H polymorphism with little or absence of enzyme activity allows to identified several polymorphisms related to the Chit activity. The presence of mutations e/o polymorphisms, as the G354R and the A442V, occurring predominantly in subjects of African ancestry directly influence the Chit activity.

Chapter VII. - Gaucher disease (GD) has been the cradle of the human phagocyte chitinase, also known as chitotriosidase (CHIT1). GD is caused by deficiency of glucocerebrosidase, the enzyme responsible for the lysosomal breakdown of the lipid glucosylceramide. The disease is characterized by the accumulation in various tissues of pathological, lipid laden macrophages, so-called Gaucher cells. The search for suitable markers of Gaucher cells resulted in the identification of a thousand-fold increased chitinase activity in plasma from symptomatic Gaucher patients. Biochemical investigations identified a single responsible enzyme, named chitotriosidase based on its ability to hydrolyze 4-methylumbelliferyl-chitotrioside. Next, the properties of the chitotriosidase protein and gene were characterized. In the wake of the identification of chitotriosidase, the existence in mammals of another chitinase (AMCase) was discovered. This review focuses on the current knowledge on the features of the chitotriosidase protein and gene, the potential function of the enzyme in innate immunity and its value as disease marker in conditions involving macrophages. Attention is also paid to the biology of the Gaucher cell, the lipid-laden macrophage that so massively overproduces chitotriosidase.

Chapter VIII. Atherosclerosis is an inflammatory disease in which macrophages play a very important role in its pathogenesis. Chitotriosidase is one of the proteins highly secreted by activated macrophages. Moreover, chitotriosidase is highly expressed by macrophages within the vascular atherosclerosis plaques, suggesting that this enzyme could be involved in the inflammatory process associated with modified LDL particles in the arteries. Several groups have recently demonstrated that serum chitotriosidase activity is related to the extension of atherosclerosis, and predicts the risk of new cardiovascular events with a predictive value similar to CRP, and, when combined with CRP, the risk prediction of new cardiovascular events and the identification of a lower risk group seem to improve. The mechanism of these associations is not fully understood but could be related, as occurs with other chitinases, through the contribution of chitotriosidase to a T helper 2 immune response to oxidized LDL.

Chapter IX. - Juvenile idiopathic arthritis (JIA) is an inflammatory joint disease of unknown aetiology. The pathogenesis is driven by T and B-cells. The role of macrophages remains unclear. Sarcoidosis is a chronic granulomatous inflammation. The clinical spectrum in childhood is heterogeneous. Angiotensin converting enzyme (ACE) activity is used as a marker for disease activity. An unknown agent activates resident T-cells and macrophages, which subsequently release cytokines and chemokines which prime and activate neighbouring cells and are chemotactic for mononuclear cells. Human chitotriosidase is produced in macrophages. Chitotriosidase belongs to the chitinase protein family and is secreted by activated macrophages. The chitinases are able to catalyze the hydrolysis of chitin or chitin-like substrates such as 4-methylumbelliferyl chitotrioside. Serum chitotriosidase levels could represent the activity of macrophages in the synovial fluid in JIA. Serum chitotriosidase concentrations may be a useful marker for monitoring disease activity in sarcoidosis.

Chapter X. - Chitotriosidase (Chit) is a member of mammalian chitinase family with structural homology to chitinases from other species. Chit has yet unexplored roles in the immune network occurring in ischemic, inflammatory and degenerative neurological diseases, in which the macrophage-microglia activation is known to be pathogenic. Its prominent archaic hydrolytic function on chitin may only be a windscreen beyond which new functions can be discovered to support its clinical importance.

Chit is synthesized and secreted by activated macrophages and immature neutrophils and its natural substrate, chitin, is a N-acetylglucosamine polymer of fungi cell wall and several human parasites. In principle, as Chit plays a major role in defence mechanisms against chitin-containing pathogens, the clinical monitoring of its activity may be relevant in human infectious diseases. Contrary to this theoretical assumption, plasma Chit activity has been shown to have a positive correlation with normal ageing and to have application, as a lipid-laden macrophage marker, in the monitoring of non pathogen-mediated diseases such as Gaucher and Fabry storage diseases. Our study group have recently suggested that Chit elevation represents an useful marker of other, non-infectious, neurological diseases such as stroke, Alzheimer's disease (AD) and multiple sclerosis (MS). Peripheral and intrathecal Chit activity in MS have been also found to strongly correlate with MS severity. These findings are reviewed along with new unpublished data.

Chaper XI. - Alzheimer's disease is the most common cause of dementia, affecting over four million patients in the Unites States and fifteen million worldwide. As the average life expectancy increases in the United States and worldwide, the number of patients will proportionally increase, affecting over thirteen million individuals in the United States by 2050. The diagnosis of Alzheimer's disease carries significance, because the life span of these patients is halved when compared with healthy population controls. The therapeutic efforts are directed towards eradicating brain lesions, without the accurate knowledge of whether these are actually pathogenic. Therefore, currently, our understanding of this neurodegenerative disease precludes us from attaining a more elusive cure. Several drugs are currently utilized in the treatment of Alzheimer's disease, and when started early, the progression might be momentarily halted. No significant information has emerged from clinical trials involving immune therapy. Authors in our group have demonstrated that Amyloid β deposition, a histopathological landmark in Alzheimer's disease brains, may confer a protective effect against oxidative damage induced by reactive oxygen species. Chitin confers antioxidant properties of comparable strength to super oxide dismutase.

For several decades the hypothesis of Amyloid β mediated pathogenesis has perhaps diverted us from the real issues in this disease's etiology, hence, devoting millions of dollars and research hours into the Amyloid cascade hypothesis. Some investigators have shifted their efforts from the Amyloid β dogma into other possible pathophysiological processes such as oxidative stress. As a matter of fact, oxidative stress has taken a significant role in the study of several neurodegenerative diseases such as Creutzfeldt Jakob, Pick's disease, diffuse Lewy body dementia and Cerebrotendinous Xanthomatosis. Decreased glucose metabolism, deficiency in antioxidant metals such as zinc, and mitochondrial abnormalities in the electron transport chain mediate the generation of toxic reactive oxygen species that coupled with redox active metals leads to free radical damage. Antioxidant vitamin supplementation with Vitamin E and C has been included in the treatment of Alzheimer's patients. Furthermore, some investigators have demonstrated that in fact oxidative stress is an early process in neurodegeneration and that it precedes Amyloid β deposition.

Deranged glucose utilization and subsequent hyperglycemia in Alzheimer's disease patients is mediated by diminished numbers of cellular glucose transporters and down regulation of genes involved in the oxidative phosphorylation of glucose. This in turn leads to the activation of the hexosamine pathway and thus increases the synthesis of glucosamine polymers as described for other diseases such as Diabetes Mellitus. In this pathway there is synthesis of O linked glycoproteins from glucose by means of fructose and fructose-6 phosphate. These glucosamine polymers are the building blocks of chitin.

The neuropathological examination of Alzheimer's disease brains involves the quantification and location of histopathological landmarks such as Amyloid plaques, neurofibrillary tangles, and Amyloid angiopathy. Studies in familial and sporadic Alzheimer's disease patients have localized chitin and chitin-like polysaccharides in both the Amyloid plaque as well as in the neurofibrillary tangles by utilization of Calcofluor; a fluorochrome that is notable for identifying chitin *In vivo* by interacting with the β 1-4 bonds that make up the chitin polymer. Amyloid is a highly insoluble molecule which stains with Congo Red and displays apple green birefringence when exposed to polarized light.

These properties, which were initially attributed to the Amyloid protein conformation, are shared by commercial chitin and could perhaps represent that chitin imparts these biochemical features to the cerebral Amyloid deposition. The role of chitin in the Amyloid plaque is unknown. However, chitin might function as a primer of Amyloid deposition as inferred by the staining profiles with Calcofluor and with Amyloid β immunohistochemistry. Therefore chitin might be a protective accumulation against oxidative stress.

Chapter XII. - Cancer is a serious disease of human beings. So far satisfactory treatment and method of prevention are lacking. Many papers about the action of chitin, chitosan, and chitinase against cancer have been published. In this chapter, we are trying to collect related information from different areas, and the emphasis will be on the potential roles of chitin and chitinase in anticancer therapy. Much of the date reported in those papers is preliminary, many of the studies even had been done before the chitinase in human and mammalian animals being proven, and some of the discussions do need to have further direct evidences. Indeed, for further development in this important area, systematic studies are urgently needed. However, the results so far obtained are suggesting that development of low toxic anticancer treatment and preventive method from the study of chitin and chitinase is possible.

Chapter XIII. - YKL-40 (also named Chitinase-3-like-1, CHI3L1) is a 40 kDa heparin-, chitin- and collagen-binding glycoprotein without chitinase activity and a member of "mammalian chitinase-like proteins". The YKL-40 gene is located on chromosome 1q32.1, has a size of 7948 base pairs and contains 10 exons. The crystallographic structure for YKL-40 is known, but cellular receptors are not identified. High YKL-40 mRNA and protein expressions are found in human embryonic and fetal cells, macrophages during late state of differentiation, macrophages in inflammed synovial membrane, atheromatous plaques, arteritic vessels, alveolar macrophages in inflamed lung tissue, microglia/macrophages from central nervous system, and in tumor-associated macrophages, neutrophils, mast cells, arthritic chondrocytes, differentiated vascular smooth muscle cells, fibroblast-like synovial cells, endothelial cells and by several types of cancer cells. The YKL-40 gene and protein are overexpressed compared to normal tissues in glioblastoma, melanoma, squamous cell carcinoma and many types of adenocarcinoma.

The exact biological functions of YKL-40 are unknown. YKL-40 is a growth factor for fibroblasts and chondrocytes, modulates the rate of type I collagen fibril formation, acts synergistically with IGF-1, is regulated by TNFα and IL-6, requires sustained activation of NF-kappaB, initiates MAP kinase and PI-3K signalling cascades leading to the phosphorylation of ERK-1/2 MAP kinase and protein kinase B (AKT)-mediated signalling cascades, which are associated with control of mitogenesis. YKL-40 may play a role in inflammation and the innate immune response, enhances bacterial adhesion to colonic epithelial cells and has a role in cancer cell proliferation, differentiation, metastasis potential, protects the cells from undergoing apoptosis, stimulates angiogenesis, and has an effect on extracellular tissue remodeling surrounding the tumour, although *in vivo* proof of this is yet to be obtained.

Plasma levels of YKL-40 are elevated compared to healthy subjects in patients with acute inflammation (e.g. pneumonia, endotoxaemia, hepatitis) or chronic inflammation (e.g. rheumatoid arthritis, inflammatory bowel disease, asthma, sarcoidosis, type II diabetes, coronary artery disease) and in patients with liver fibrosis. Plasma YKL-40 levels are also elevated in some patients with primary or metastatic cancer and may be useful as an independent "prognosticator" of survival, a predictor of treatment response, and in monitoring cancer recurrence/progression after treatment. Unfortunately, most of these studies are small and retrospective. Recently, two large studies suggest that plasma YKL-40 may have a value in screening for colorectal cancer.

In the future, more research on the function of YKL-40 is needed and large prospective, longitudinal clinical studies should be performed to determine if plasma YKL-40 levels have a clinical value as a biomarker in patients with inflammation, tissue remodeling, fibrosis and cancer.

Chapter XIV. - Chitotriosidase, a functional chitinase secreted by activated macrophages, is extremely increased in plasma of patients with Gaucher disease (GD) or beta-glucocerebrosidase deficiency. GD is a lysosomal storage disorder characterized by blocked catabolism of glucosylceramide (GC), a metabolic intermediate derived from the cellular turnover of membrane gangliosides and globosides. The primary cell type affected in GD is the macrophage (Gaucher cell) where the presence of GC and other sphingolipids at non-physiological concentrations is thought to interfere with other biochemical pathways outside the lysosome, leading to cell dysfunction including reticuloendothelial expansion and macrophage activation. In 1999, we investigated if plasma chitotriosidase levels are increased in patients with beta-thalassemia, an haematological disorder characterized by the genetic defect of beta-globin chains synthesis and resulting in unproductive erythropoiesis and enormous expansion of the reticuloendothelial system. We found that plasma chitotriosidase activity was increased to a variable extent in a group of patients with beta-thalassemia major including those treated with the intense transfusion regimen and iron chelation therapy. We suggested that the increased chitotriosidase production in beta-thalassemia might reflect macrophage activation probably related to the intracellular iron overload and storage of erythrocytes membrane break-down products. After this initial description, chitotriosidase evaluation in patients with beta-thalassemia has been the object of fürther clinical work up. We will review here the present knowledge on chitotriosidase in beta-thalassemia with the aim to describe the role and significance of human chitinase increase in this haematological disorder.

Chapter XV. - High levels of plasma chitotriosidase (Chit) are a marker of macrophage activation in several infectious pathologies and, in particular, in human malaria. *Plasmodium falciparum (P. falciparum)*, during its maturative cycle in the midgut of the *Anopheles* mosquito, secretes a specific chitinase enabling it to cross chitin-containing peritrophic membrane (PM) which surrounds the blood meal. This represents a necessary step in the migration of the parasite from the midgut to the salivary glands of malaria's vector. The cooperation between human Chit and the chitinase produced by *P. falciparum* in attacking the peritrophic membranes in the Anopheles midgut has been recently demonstrated by *in vivo* experiments, and seems to favour the trasmissibility of human malaria in African sub-Saharan regions. Optical microscopy (OP) showed that the formation of the PM was completed after 16 h in the posterior midgut of *Anopheles* already fed with healthy donor bloods. In contrast, PM formation was partly conserved after 16 h, when mosquitoes were fed with malaria and Gaucher patient blood, but the PM appeared clearly damaged at 20 and 24 hours. In addition, the PM formation was almost completely inhibited in the midgut of *Anopheles* fed with *P. falciparum* chitinase enriched blood. These alterations in the PM formation were confirmed by Transmission Electronic Microscopy (TEM). This functional homology between human Chit and *P. falciparum* chitinase was confirmed also by computational methods. A simple sequence analysis method, potentially useful to assess fine textual closeness in families of homologous proteins, was applied to a set of chitinases from mammals and *plasmodia*. This analysis confirmed the clustering and the phylogenetic relationships obtained with well known alignment methods, but also showed that the sequences of chitinases from different malaria hosts and from different malaria parasites are strictly correlated. This correlation confirms a functional homology among chitinases, which is seen as a condition for the spreading of the different forms of malaria. From this perspective, one can get insight into the origins of malaria, and its genetic or pharmacological control.

Chapter XVI. - Chitin is abundant in the structural coatings of fungi, insects, and parasitic nematodes, but it is not produced in mammals. The host defense against chitin-containing pathogens includes production of chitinases. An acidic mammalian chitinase

(AMCase) is produced in human epithelial cells of lower airways and conjunctiva via a Th2-specific, IL-13-dependent pathway and seems to be associated with asthma and allergic ocular pathologies. The understanding of the role of AMCase in allergic disease is only at its beginning and many issues open new possibilities for its control using specific inhibitors of AMCase activity or modulating its expression. In patients with vernal keratoconjunctivitis (VKC) and with seasonal allergic conjunctivitis (SAC) the level of AMCase activity in the tears was found significantly elevated when compare to healthy controls and the highest levels were found in VKC. When RNA was extracted by conjunctival epithelial cells of these patients, quantitative Real Time PCR measurement confirmed that mRNA expression correlates with tear AMCase activity and the expression was significantly higher in VKC and SAC.

Also Receiver Operating Characteristic (ROC) analysis demonstrated that the sensitivity and specificity of AMCase measurement were 100 %, addressing the use of AMCase assay in the biochemical diagnosis of VKC and SAC.

Recent studies in rabbits, where a reactive uveitis was induced by LPS injection into the eye's anterior chamber, confirmed that increased AMCase activity was measurable in tears and that epithelial cells of conjunctiva express specific mRNA. A well as it was previously demonstrated in experimental model of mouse asthma, the inflammatory reaction induced by LPS was controlled by the chitinase inhibitor and steroid, instilled at 3 hr interval in conjunctival sacs.

In dry eye, another non allergic ocular pathology, an increased AMCase activity was documented and the specific mRNA expressed by epithelial conjunctival cells. In this pathology the eye inflammation can be ascribed to a common mechanism mediated by AMCase, via a Th2 specific, IL-13 dependent way. In synthesis, AMCase may be considered an important mediator in the pathogenesis of Th2 inflammation eye's diseases, suggesting its potential diagnostic and therapeutic utility.

Chapter XVII. - The mammalian family 18 chitinase members include different enzymes. The true enzymes which hydrolyze chitin are Chitotriosidase (Chit) and AMCase. The YKL-40, YKL-39, SI-CLP, oviductin and murine Ym1/2 are chitinase like proteins which have lost the hydrolytic activity. Several studies, in the last years, demonstrated the role of chitinases in the immunological response. The first human observation was that in Gaucher disease the lipid-laden macrophages are able to produce very high level of Chit in response to the presence of glucosylceramide and ceramide. Moreover clinical data showed also that Chit is higher in patients with *Plasmodium falciparum* malaria, expression of macrophage activation. Recent findings support the hypothesis that chitinases have a role in the innate immunity. Our research of some years ago demonstrated that the INF-gamma, TNF-alpha, LPS and Prolactin stimulate monocyte-derived macrophages to produce Chit, conditioning immune function. These results open a new view on the function of innate immunity, in the modulation of adaptive immune response and in allergic diseases. In fact AMCase has been found to be implicated in the Th2-mediated inflammations such as asthma, inflammatory bowel disease, chronic rhinosinusitis and eye pathologies. A recent study in mice suggested that the presence of chitin determines the accumulation of innate immune cells in tissues with allergy and that this mechanism could be abrogated by AMCase, concluding that chitinase may also have a regulatory mechanism in mounting the immune response. Moreover Chit has been associated to neurodegenerative diseases as shown by the studies of multiple sclerosis and Alzheimer disease, which make of Chit measurement in blood and in cephalorachidian liquid the most sensible parameters for follow up. The group of chitinase like proteins, expressed in several tissue and cells, YKL-40 and YKL-39 are implicated in autoimmune diseases as rheumatoid arthritis, where they are also involved in immune regulatory mechanisms. In this chapter we will try to illustrate the known mechanisms implicated in immunological response.

Chapter XVIII. - Asthma is a disease characterized by chronic inflammation of the airway, thought to result from inappropriate activation of the Th2 immune response. A master regulator of Th2 inflammation is the cytokine IL-13, which stimulates the expression of many effectors responsible for the airway hyperresponsiveness and eosinophilic inflammation that is characteristic of asthma. Recent work has shown that both chitinase and chi-lectin proteins are strongly upregulated by IL-13 expression. Inhibition of the acidic mammalian chitinase (AMCase), blocked the inflammation and hyperresponsiveness observed in a mouse model of asthma. Furthermore, chitin, a widespread polymer of *N*-acetyl-b-D-glucosamine and substrate of chitinases, is found in many organisms including those for which exposure is linked to asthma such as, dust mites, fungi and cockroaches. Interestingly, in a mouse model of asthma, investigators have demonstrated that chitin can induce the recruitment of immune cells associated with allergic asthma to the lung. Moreover, they find that AMCase enzymatic activity negatively regulates this process. Herein, these seminal studies on the role of chitinases and chitin in Th2 inflammation are highlighted in the context of other human and murine data on chitinases.

Chapter XIX. - Family of human Glyco_18-domain-containing proteins comprises catalytically active chitinases and chitinase-like proteins. Human chitinase-like proteins include YKL-39, YKL-40 and SI-CLP (stabilin-1 interacting chitinase-like protein). In addition, chitinase-like proteins YM1 and YM2 were identified in rodents, but their human homologues do not exist. In contrast to true chitinases, YKL-39, YKL-40 and SI-CLP are enzymatically inactive due to the lack of critical catalytic aminoacids in the enzymatic site within their Glyco_18 domain. While true chitinases bind chitin via C-terminal chitin-binding domain, YKL-39, YKL-40 and SI-CLP do not posses chitin-binding domain, and contain solely Glyco_18-domain. However the Glyco_18 domain of YKL-40 mediates binding to heparin, hyaluronan, and chitin. All three human chitinase-like proteins are secreted into the extracellular space. Elevated levels of YKL-40 are associated with several chronic inflammatory disorders and cancers. Biological activities of YKL-40 include regulation of cell proliferation, adhesion, migration and activation. YKL-40 promotes growth of human synovial cells, chondrocytes, skin and foetal lung fibroblasts. Two biological activities of YKL-39 are suggested to contribute to progression of osteoarthritis. One is the induction of autoimmune response, and second is participation in tissue remodeling. Biological activity of SI-CLP is currently under investigation in our laboratory. Both YKL-40 and SI-CLP are expressed by several cell types including tumour cells and macrophages. We found antagonistic regulation of expression of YKL-40 and SI-CLP in human macrophages. YKL-40 is strongly induced by IFNgamma, Th1 cytokine which initiates classical inflammation. Th2 cytokine IL-4 as well as glucocortiocid dexamethasone suppress YKL-40 expression. In contrast, IL-4 and dexamethasone synergistically activate SI-CLP, while IFNgamma abrogates this effect. YKL-39 was identified as an abundantly secreted protein in primary culture of human articular chondrocytes. YKL-39 is currently recognized as a biomarker for the activation of chondrocytes and the progress of the osteoarthritis in human. Recently we found that mRNA of YKL-39 is dramatically upregulated in human macrophages by IL-4 and TGF-beta, a crucial growth factor regulating tumour growth and atherosclerosis. Thus, all three mammalian chitinase-like proteins YKL-39, YKL-40 and SI-CLP are indicators of macrophage activation modes found in distinct pathological situations. Macrophages utilise several tightly regulated pathways for secretion of soluble mediators. The mechanism which regulates intracellular sorting for human chitinase-like proteins and their commitment for secretion was not elucidated. We investigated the mechanism of SI-CLP sorting and secretion in human alternatively activated macrophages. Using biochemical and cell biology approaches we showed, that SI-CLP directly interacts with multifunctional macrophage receptor stabilin-1. Stabilin-1 recognises newly synthesised SI-CLP in late Golgi compartment and targets it to the secretory lysosomes. However SI-CLP can be delivered into lysosomes also in the absence for stabilin-1, indicating that more than one intracellular receptor is involved in its sorting. We also showed, that glucocorticoid dexamethasone in therapeutic concentrations despite inducing expression of SI-CLP, blocks its secretion leading to the intracellular accumulation of high levels of SI-CLP. Significance of these findings for understanding of macrophage-mediated pathologies is discussed.

Chapter XX. - Chitin-containing food is an interesting but underestimated source of locally available, in most cases sustainable, food although chitin digestion by humans has generally been questioned or denied. Only in recent times chitinases have been found in several human tissues and their role has been associated with defence against parasite infections as well as with some allergic conditions. We reflected that crustaceans, and to some extent molluscs, mushrooms and most arthropods containing chitin, are sometime a consistent part of food regimes for local communities. Finally, we demonstrated that AMCase is present in gastric juices and it is associated with chitin digestion. In most tropical and some temperate countries, such as Japan and Korea, a significant number of adult insects and larvae are consumed raw, or cooked along with diverse local specialities. At present, up to 2,000 species of insects and other terrestrial arthropods have been listed as edible in Africa, Asia, Central and South America, Australia and Europe. Both insects and crustaceans are covered by chitin teguments and mushrooms contain some chitin. In most cases, the hard covering of polysaccharide chitin on insects accounts for 5-20% of their dry weight. In general, chitinases can digest chitin and reduce it to simple compounds such as *N*-acetyl-glucosamine. Western society does not consider insects an important food, however: crustaceans, such as lobsters and crabs, are commonly eaten after discarding the hardened chitin-rich tegument, with the exception of small shrimps, which are generally eaten fried. Therefore, Western nutrition does not seem to depend on chitinases. These and other considerations, including the absence of chitin as a human body component, have led us to ask whether humans are capable of chitin digestion.

To assess chitinases' function as tools to digest chitin, we have examined 48 patient's gastric juices, obtained during gastroscopy, at Padova University Hospital. We found that 14.6% of total samples studied showed AMCase activity from 36.270 to 3.540 nmol/ml/h. The majority of involved subjects (75%) had lower values, from 2.800 to 0.178 nmol/ml/h; while in 10.4% of subjects the chitinolitic activity varied from 0.086 to 0.013 nmol/ml/h, and could be considered absent. We reported superficial digestion of fly forewings, utilizing gastric juice of a patient with an AMCase activity of 19.410 nmol/ml/h.

If AMCase enzyme, present in gastric juice, is truly involved in chitin digestion, we should expect a higher presence of expressed AMCase in populations currently accustomed to eating mushrooms and/or invertebrates bearing chitin.

We also found a positive relationship between *CHIT* expression level in antral gastric mucosa and both flogosis and *Helicobacter pylori* infection.

Chapter XXI. - Inflammatory bowel diseases (IBD), including Crohn's disease (CD) and ulcerative colitis (UC), are a group of chronic inflammatory disorders that affect individuals throughout life. The etiology and pathogenesis of these two major forms of IBD is largely unknown. Several studies have indicated that dysregulated host/enteric microbial interactions are required for the development of IBD. Both the colonic epithelial cells (CECs) that form a barrier between the luminal contents (including microorganisms and other antigens) and the underlying immune cells play important roles in maintaining adequate host/microbial interactions. In fact, CECs actively participate in the induction of both innate and adaptive immune responses to the luminal contents by inducing several specific molecules. By utilizing DNA microarray screening technology, our group has unexpectedly identified a novel intestinal inflammation-associated molecule, Chitinase 3-like-1 (CHI3L1, YKL-40 or HC-gp39), which is produced mainly by CECs and macrophages only under inflammatory conditions. We have also provided novel insight into the pathophysiological role of CHI3L1 for enhancing bacterial adhesion and invasion on/into CECs. CHI3L1 is characterized by a strong binding affinity to chitin without enzymatic activity. The ability of a host to produce chitinases, which have enzymatic activity, could be a critical factor in the regulation of the initial immune response against pathogen (e.g., fungi, parasites)-derived chitin. In contrast, exaggerated production of mammalian chitinases may cause harmful and pathogenic effects in mucosal regions. Although bacteria do not possess chitin as a structural component, some strains of bacteria can express chitin-binding proteins (CBPs) upon exposure to chitin. In fact, our recent experimental results suggest that over-expression of CHI3L1 on CECs and CBP on bacteria can significantly enhance the bacterial adhesion on CECs. Therefore, bacterial CBPs may form an important bridge directly or indirectly in facilitating the binding of luminal bacteria to CHI3L1 on the colonic epithelial surface. Interestingly, many pathogenic and potentially pathogenic bacteria are presumably able to express CBPs. In this chapter, we will discuss about the physiological function of mammalian chitinases and bacterial CBPs in the intestine and their potentially pathogenic role during the development of human IBD.

Chapter XXII. - Chitinases occur widely in nature with various physiological roles dictated by the producing organisms. Chitinase inhibitors are useful to investigate a physiological role of chitinase of each organism and have a potential as useful drugs, mainly as insecticides or anti-asthmatic agents. Among chitinase inhibitors, allosamidin, a *Streptomyces* metabolite, has been used in basic research most frequently. Its structure and effects on a variety of organisms including insects, yeasts, parasites and mammals have provided clues to elucidate chitinase enzymology and physiological roles.

Chapter XXIII. - Chitin is a widespread carbohydrate polymer with unique biomechanical properties. In its crystalline form and many occurring additional modifications, chitin is a tough, resilient compound which is difficult to degrade, even by specialised enzymes. It is totally insoluble in water and plays an important part in the carbon and nitrogen cycles and as an energy source, particularly in the marine biosphere. Chitinolytic bacteria are endowed with a complex machinery enabling them to detect the presence of chitin, move chemotactically towards it following a gradient, attach to its surface via specialised pili, to release enzymes and accessory proteins for its degradation, and finally to import and metabolise its fragments. Binding of bacteria to chitin also has implications for human health, as pathogenic bacteria such as *Vibrio cholerae* are found predominantly in association with chitin-bearing copepods or other organisms, and thus can be filtered from water easily. Modified chitin-like substances (nod factors) released by Rhizobia (nitrogen fixing bacteria) are involved in the symbiotic relationship with legumes. For most plants, however, chitin detection signals an impending danger from a fungal pathogen. Considerable advances in the understanding of chitin sensing in plants have been achieved in the past few years, leading to the identification and cloning of several receptors containing extracellular LysM and intracytoplasmic Ser/Thr kinase domains involved either in recognition of nod factors or in the detection of chitin fragments, the latter eliciting a complex immune response by the infected plant.

Very little is known regarding the interactions of chitin with the immune system of animals. Mammals do not contain chitin thus, similarly to what is seen in plants, chitin could constitute a 'danger' signal. Indeed, recent work has suggested that chitin is recognised by the mammalian immune system. This work has led the authors to postulate receptors for chitin e.g. on macrophages. Because chitin is very insoluble in aqueous solutions, sensing of chitin in most systems studied to date is mediated via recognition of its soluble degradation fragments $GlcNAc_n$. Size discrimination by the known receptors allows distinction of chitin-derived chito-oligomers ($GlcNAc_n$ with $n \geq 2$) from GlcNAc monomers, which could also be derived from the degradation of glycoproteins or glycolipids, and thus do not constitute a danger signal. Taken together, the examples seen in bacteria and plants point to the possibility that chitin-sensing pattern recognition receptors could also be found in higher animals such as mammals, and that chitin recognition could be mediated via interaction of chitin-oligosaccharides (of variable size) with these receptors.

Salvatore Musumeci and Maurizio G. Paoletti

Aknowledgements

Many colleagues and friends have stimulated us in undertaking this edited book with formal and informal talks and discussions. We are especially thankful for all chapter authors that have accepted to read and improve chapters of other authors in this book. In addition sections or full chapters have been seen by Maria Luisa Mostacciuolo, Tullio Pozzan, Mila Tommaseo Ponzetta, Silvio Tosatto, Livio Trainotti.

In: Binomium Chitin-Chitinase: Recent Issues
Editor: Salvatore Musumeci and Maurizio G. Paoletti ISBN 978-1-60692-339-9

Chapter I

New Aspects of Chitin Chemistry and Enzymology

Riccardo A.A. Muzzarelli[1]
Institute of Biochemistry, University of Ancona,
IT-60100 Ancona, Italy

Abstract

In living tissues both chitin polymorphs occur in covalent combination with either proteins or glucan and are often cross-linked following the quinone tanning process. Most of the chitin, however, is highly ordered as crystallites called nanofibrils that can be recovered as aqueous suspensions for nanotechnological applications. In spite of the inherent insolubility of chitin in water, chitin aqueous systems include ethers obtained from alkali chitin such as O-carboxymethyl chitin and glycol chitin; 6-oxychitin and partially reacetylated chitin are also easy to prepare, while chitin oligomers are *per se* water-soluble. Several solvents for chitin are also available such as the dimethylacetamide-LiCl mixture. Chitinases are enzymes involved in growth, defense, aggression and feeding secreted by animals, fungi and bacteria; they are finding applications in agriculture, particularly after the genomes of plants such as rice *Oryza sativa* and pests such as *Tribolium castaneum* were fully elucidated. Transgenic rice plants are endowed with novel and powerful chitinases that are promptly activated in case of infection by rice pathogens. Phytoparasitism has been put to profit by engaging *Trichoderma harzianum* in protecting plants against pathogens such as *Rhizoctonia solani*. The importance of lysozyme is highlighted insofar as chitinases retain an ancient structural motif of lysozyme, and actually lysozyme is an enzyme ubiquitously present in the human body for defense against microbes and parasites. Unspecific enzymes such as cellulase, hemicellulase and lipase are currently used to prepare chitin oligomers to be used in the biomedical field, particularly in medication and drug delivery.

1 Riccardoposta@excite.it.

1. Chitin Nanofibrils

Crustaceans and insects protect themselves from predators and pathogens by secreting an exoskeleton that provides mechanical support to the body, armor against predators, and defense against pathogens while permitting mobility through the formation of joints and attachment sites for muscles. The matrix, made of four superimposed layers and composed of chitin associated with proteins, is produced by an underlying monolayer of epidermal cells that also secrete modest quantities of lipids in the epicuticle, and carothenoids in the pigmented layer. Enzymes, for example carbonic anhydrase, are also synthesized in relation with the control of calcification, which occurs in the two middle layers: maximum activity is attained during the initial stages of calcification in producing carbonate ions.

Chitin fibers in crustacean shells are associated with carbonate that diffuse and precipitate after the fibrous component has been excreted and stabilized; the same occurs with collagen fibers and calcium phosphate in bones. In these tissues, the supporting organic component is made of preformed nanometer to micrometer-size elongated particles arranged into supramolecular structures with geometries analogous to those of some liquid crystals. In compact bones, arthropod cuticles and plant cell walls, these structures exhibit the macroscopic features of a cholesteric phase, except fluidity. In most cases, collagen, chitin and cellulose can be extracted from the biological tissues and dispersed in aqueous media to form colloidal suspensions. At appropriate concentrations, liquid crystalline phases can be identified, indicating that rod-like or spindle-like particles tend to align cooperatively in these systems. The particles are rigid and their shape is constant throughout the phase diagram. This helps understand the influence of various parameters, such as concentration, pH, and ionic strength on the behavior of the suspensions (Li *et al.* 1996, 1997; Nair and Dufresne, 2003).

The presence of crystalline chitin fibrils in the integuments was described several decades ago: early reports are those by Richards (1951), Weis-Fogh (1970) and Rudall (1967). The subject has been dealt with in a chapter of the first book devoted to chitin (Muzzarelli, 1977), in the books by Hepburn (1976), Neville (1975, 1993), Jollès and Muzzarelli (1999) and Muzzarelli (1993, 1996, 2001). Recent chapters and reviews include those by Giraud-Guille *et al.* 2004 and Kumar *et al.* 2004).

In order to grow, the animals must replace their old exoskeleton periodically by a new one in a process termed molting. Before the old cuticle is shed, a new, thin and not yet mineralized cuticle is secreted by the epidermal cells. After the molt the animals expand and the new soft cuticle is completed and mineralized.

As can be seen in Figure 1, the smallest sub-units in the structural hierarchy of the cuticle of the lobster *Homarus americanus* are the chitin macromolecules. Their chains are arranged in an antiparallel fashion forming α-chitin that prevails in the exoskeleton of large crustaceans; 18–25 of these chains together form nanofibrils of diameter ca. 2–5 nm and length ca. 300 nm. These nanofibrils cluster to form long chitin–protein fibers with diameters between 50 and 350 nm. The fibers assemble in planar honeycomb shaped arrays. These are stacked along their normal direction forming a twisted plywood-type structure. A stack that has been rotated from one plane to another by 180° about its normal is referred to as a Bouligand or plywood layer (Figure 1).

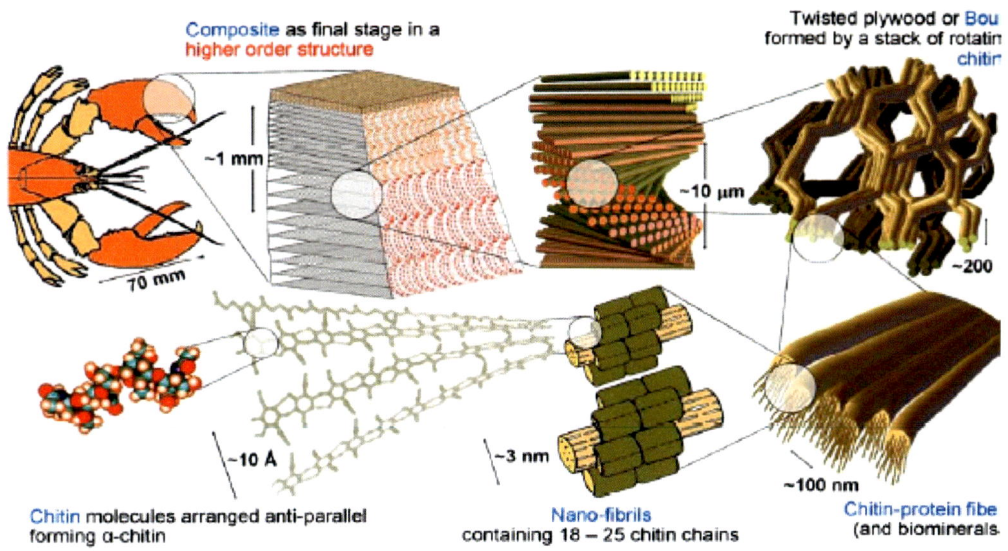

Figure 1. Hierarchical microstructure of the cuticle of the lobster *Homarus americanus*. Reprinted from *Acta Materialia* 53. Raabe D, Romano P, Sachs C. The crustacean exoskeleton as an example of a structurally and mechanically graded biological nanocomposite material. Pages 4281-4292. Copyright (2005), with permission from Elsevier.

Characteristic for the lobster cuticle is the presence of a well-developed pore canal system with many such canals penetrating the plywood structure. The pore canals contain long soft tubes. The fibers of each chitin–protein plane are arranged around the lenticellate cavities of the pore canals, building a structure that resembles a twisted honeycomb. In the hard parts of the lobster, the exo- and endo-cuticles are mineralized with calcium carbonate in the form of small crystallites a few nanometers in diameter (Raabe *et al.* 2007).

Ultra-thin sections of the organic matrix, in crab carapaces and in compact bone osteons as well, reveal typical arced patterns that however do not result from authentic curved filaments. In an ideal representation, the molecular directions are drawn as parallel and equidistant straight lines on a series of rectangles and from one card to the next, the lines turn by a small and constant angle. Series of nested arcs appear on oblique sides of the model, just as they appear in microscopy after sectioning of the material. Another consequence of the twisted plywood arrangement is the presence of periodic extinctions when the sections are viewed in polarized light microscopy, with a planar disposition observed in the crab cuticle (Giraud-Guille *et al.* 2004). This helical arrangement is revealed by the fingerprint patterns typical of cholesteric liquid crystals. The distance between two dark bands corresponds to a 180° rotation of the molecular orientations and corresponds to the half-cholesteric pitch.

Aqueous suspensions of nanocrystals can be prepared by acid hydrolysis of the purified polysaccharide. The effect of this treatment is to dissolve the chitin regions of low lateral order so that the insoluble, highly crystalline residue may be converted into a stable suspension by subsequent vigorous mechanical shearing action. For cellulose and chitin, these monocrystals appear as rod-like nanofibrils whose dimensions depend on the biological source of the substrate. In the case of starch they consist of platelet-like nanoparticles.

α-Chitin from shrimp shells was subjected to extensive hydrolysis in boiling 3 M hydrochloric acid. X-ray diffraction data indicated an increase of chitin crystallinity after hydrolysis, as the less-ordered chitin domains were digested. Line broadening data were used to measure crystallite size and particle size in the chitin nanocrystals. Congo Red adsorption was used to measure the specific surface area of the chitin nanocrystals, which was found to be 347 m^2/g, compared to 124 for chitin fibres, 249 for pulp cellulose nanocrystals, 272 for bacterial cellullose nanocrystals, and 88 for pulp fibres (Goodrich *et al.* 2007). The nanofibrils are slightly cationic (1 g is titrated with 0.16 mmol NaOH). In water, the protonated amino groups and their counter-ions form an electrical double layer around the crystallites; perturbation of the layer by solvents or electrolytes promotes reversible aggregation.

The ATR-FTIR spectrum in Figure 2, one of the most clearly resolved ever recorded on arthropod chitin (Muzzarelli *et al.* 2007), indicates that the material is pure α-chitin with split absorbance peaks at 1659 and 1625 cm^{-1} for the amide I region and a peak at 1563 cm^{-1} for the amide II region. No protein signal is detectable at 1540 cm^{-1}, where proteins would normally give rise to absorption. ^{13}C CP-MAS NMR spectroscopy was used to estimate the average degree of N-acetylation for the chitin nanocrystals. The degree of acetylation was determined by the ratio of the integration values of the methyl carbon to the anomeric carbon signal (Figure 3), to be 0.90 for the chitin nanocrystals after hydrolysis, and 0.89 prior to hydrolysis, indicating that the hydrolysis treatment had no effect. It is also evident that the isolated chitin nanocrystals are pure and residual proteins and minerals are absent.

The proposed crystal structures of α- and β-chitin are represented in Figures 4 and 5. In both structures, the chitin chains are organized in sheets where they are tightly held by a number of intra-sheet hydrogen bonds.

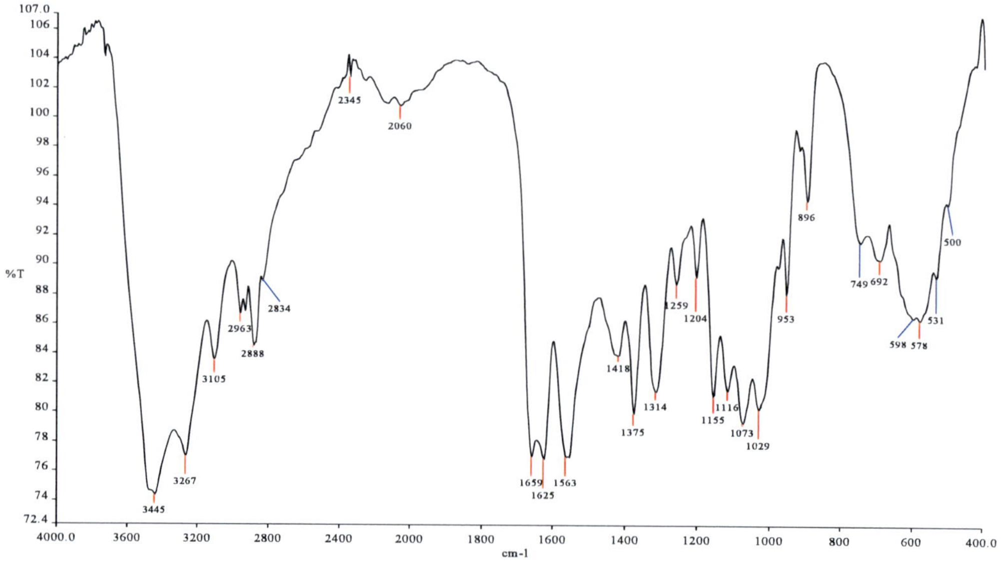

Figure 2. FTIR spectrum of chitin nanocrystals (Muzzarelli RAA, 2005; original results).

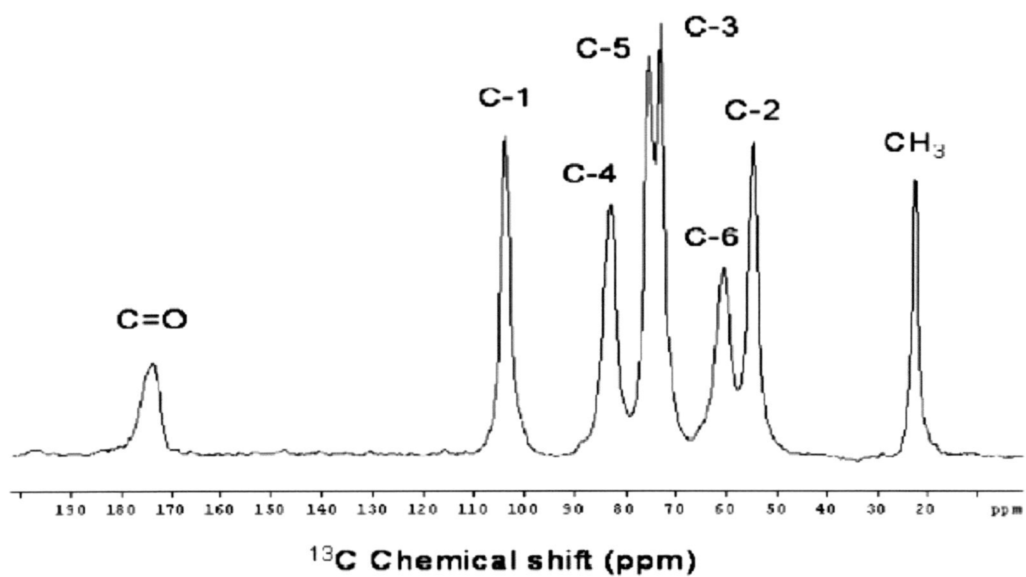

Figure 3. Typical 13C-CP-MAS spectrum of chitin nanocrystals.

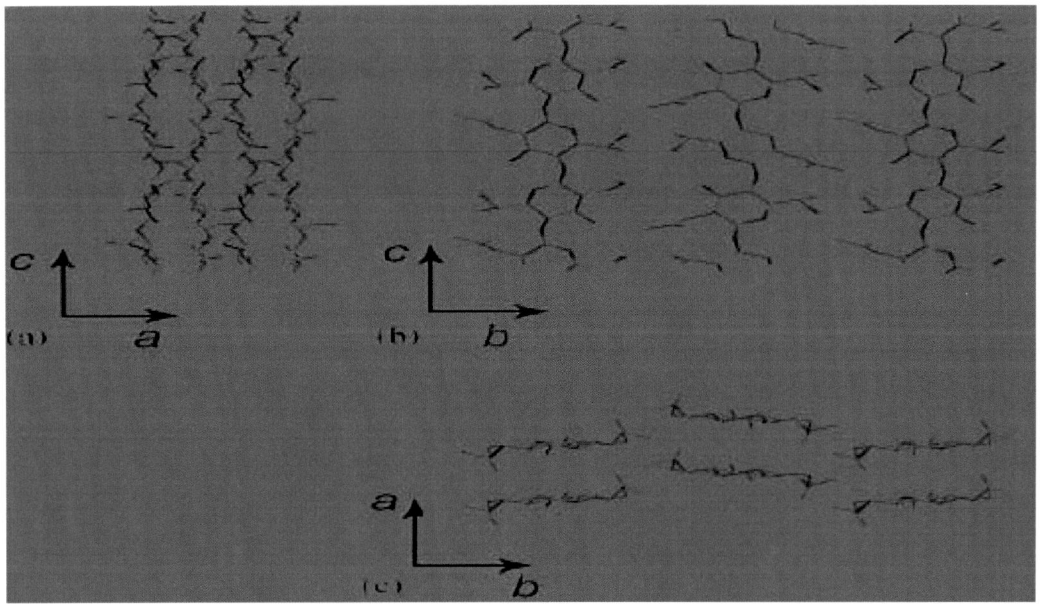

Figure 4. Structure of α-chitin: (a) ac projection; (b) bc projection; (c) ab projection. The structure contains a statistical mixture of 2 conformations of the $–CH_2OH$ groups. Reprinted from *Progress in Polymer Science* 31. Rinaudo M. Chitin and chitosan: properties and applications. Pages 603–632. Copyright (2006), with permission from Elsevier.

This tight network, dominated by the rather strong C–O…NH hydrogen bonds, maintains the chains at the distance of about 0.47 nm along the a parameter of the unit cell.

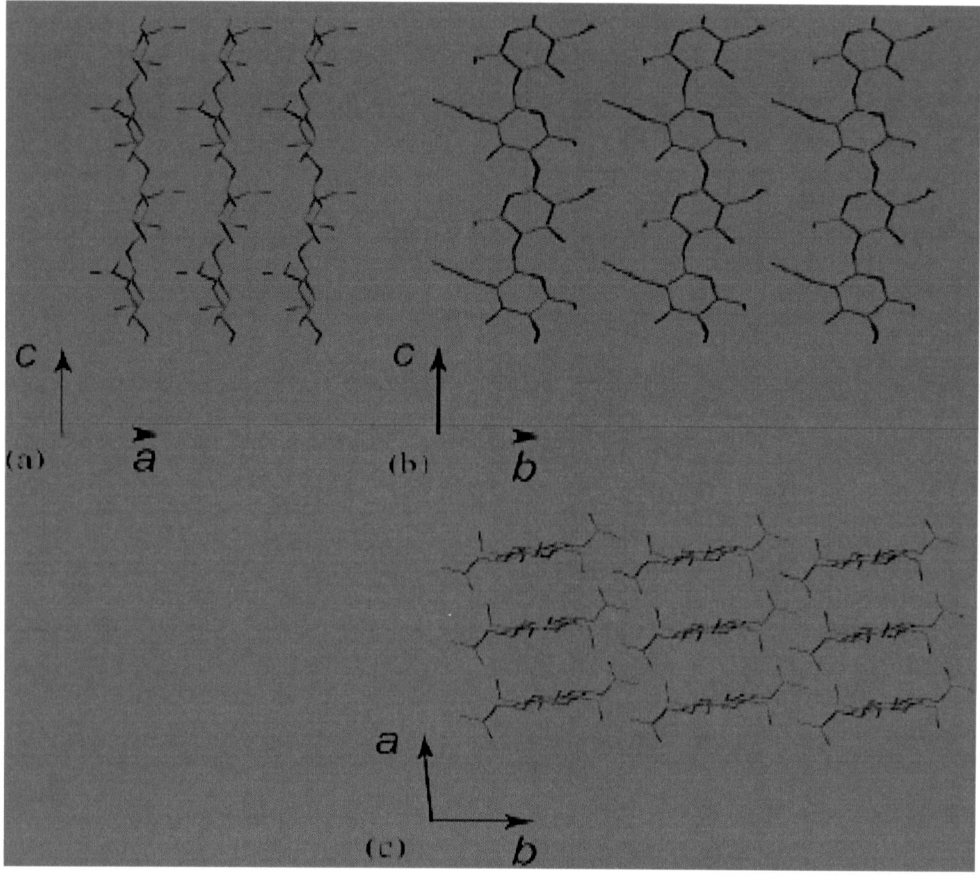

Figure 5. Structure of anhydrous β-chitin: (a) ac projection; (b) bc projection; (c) ab projection. A major point of difference from α-chitin is the absence of hydrogen bonds in the b direction Reprinted from *Progress in Polymer Science* 31. Rinaudo M. Chitin and chitosan: properties and applications. Pages 603–632. Copyright (2006), with permission from Elsevier.

It is important to note that in the α-chitin there are also some inter-sheet hydrogen bonds along the b parameter of the unit cell, involving the hydroxymethyl groups of adjacent chains. This peculiar feature is not found in the structure of β-chitin, which is therefore more susceptible than α-chitin to intra-crystalline swelling (Rinaudo, 2006). In fact, inclusion complexes in β-chitin having well defined guest-host (amine-chitobiose) ratios have been reported for aliphatic mono- and di-amines up to heptamethylenediamine by Noishiki *et al.* (2003).

1.1 α-Chitin

Chitin is a common constituent not only of the crustacean exoskeleton, but also of the arthropod cuticle in general, including insects, chelicerates, and myriapods. It also occurs in mollusk shells and fungal cell walls. All chitins are made of chitin nanofibrils (crystallites) embedded into a less crystalline chitin. α-Chitin is the most abundant polymorph; it occurs in fungal and yeast cell walls, in the crustacean exoskeltons, as well as in the insect cuticle. Studies on the crystallographic texture of the crystalline α-chitin matrix in the biological composite material forming the exoskeleton of the lobster *Homarus americanus* have shown that everywhere in the carapace the texture is optimized in such a way that the same crystallographic axis of the chitin matrix is parallel to the normal to the local tangent plane of the carapace. Notable differences in the texture are observed between hard mineralized parts and soft membranous parts (Raabe *et al.* 2005, 2006).

The hard chitinous tissues found in some invertebrate marine organisms are paradigms for robust, lightweight materials. Well illustrated examples are the oral grasping spines of *Sagitta* and Chaetognaths (Saito *et al.* 1995; Bone *et al.* 1983), the granular chitin in the epidermis of nudibranch mollusks (Martin *et al.* 2007) and the filaments of the seaweed *Phaeocystis* (Chretiennot-Dinet *et al.* 1997). These uncommon α-chitins have proved particularly interesting for structural studies since they present remarkably high crystallinity and high purity i.e. absence of pigments, proteins and calcite.

The main constituents of the beak of the jumbo squid (*Dosidicus gigas*) are chitin fibers (15-20 wt.%) and histidine- and glycine-rich proteins (40-45%). Notably absent are mineral phases, metals and halogens. Despite being fully organic, beak hardness and stiffness are at least twice those of the most competitive synthetic organic materials (notably engineering polymers) and comparable to those of *Glycera* and *Nereis* jaws. Furthermore, the combination of hardness and stiffness makes the beaks more resistant to plastic deformation than virtually all metals and polymers. In fact the closure forces exerted by the mandibular muscles of some species are large enough to crush the shells of gastropods. Moreover, the presence of intact beaks in the stomachs of squid predators indicates a high resistance to proteolysis.

The 3,4-dihydroxy-L-phenylalanine and abundant histidine content in the beak proteins as well as the pigmented hydrolysis-resistant residue testify cross-linking via quinone tanning. A high cross-linking density between the proteins and chitin may be the most important determinant of hardness and stiffness in the beak. Even after prolonged hydrolysis, some aminoacids remain in the chitin; while this is a general situation at the aminoacid trace level, the data for the *Dosidicus gigas* chitin indicate the presence of substantial amounts of 15 aminoacids with prevalence of glycine, alanine and histidine (Miserez *et al.* 2007).

In insects the tanning of cuticles is particularly important for the mechanical properties of the wings. The native insect cuticle was studied at several different stages throughout its tanning process (also known as sclerotization). As tanning proceeds, the catechols react with the proteins via an enzymatic process catalyzed by the enzyme laccase. The catechols are hydrophobic components and hence the water content of the cuticle decreases as tanning proceeds. Changes in mechanical properties take place as a function of tanning state. Elytra from the beetles *Tribolium castaneum* (red flour beetle) and *Tenebrio molitor* (yellow mealworm) were tested by dynamic mechanical analysis. In *Tribolium,* an economically important agricultural pest, it was possible to use RNA interference techniques to selectively suppress laccase gene expression during sclerotization in order to test the role of laccase in cuticle tanning.

In a series of experiments, the fracture stress of the fully tanned elytra increased to 45 ± 12 MPa; stiffness increased dramatically as shown by the Young's modulus of 1674 ± 383

MPa and the elastic modulus of 4860 ± 1840 MPa. Laccase silencing accompanied by water loss resulted in cuticles with both poor fracture stress and strain, proving quantitatively that laccase plays a major role in tanning as well as showing that water loss alone is not responsible for the superior mechanical properties of fully tanned insect cuticle.

Enzymes belonging to the β-N-acetylglucosaminidase family cleave chitin oligomers produced by the action of chitinases on chitin, into the N-acetylglucosamine monomer. Four genes encoding putative N-acetylglucosaminidases in *Tribolium castaneum*, and three other related hexosaminidases were identified by searching the recently completed genome. Full-length cDNAs for all four N-acetylglucosaminidases were cloned and sequenced, and the exon-intron organization of the corresponding genes was determined. RNA interference was most effective in interrupting all three types of molts, larval-larval, larval-pupal, and pupal-adult: treated insects died after failing to completely shed their old cuticles (Arakane *et al.* 2007).

1.2 β-Chitin

The less abundant β-chitin is found in association with proteins in squid pens and in the tubes synthesized by pogonophoran and vestimetiferan worms. It occurs also in aphrodite chaetae as well as in the lorica of protozoa like *Eufolliculina uhligi*, and certain seaweeds. A particularly pure form of β-chitin is found in the monocrystalline spines of the diatom *Thalassiosira fluviatilis,* currently exploited in the biomedical field. An exhaustive search of the crystal structure of β-chitin recently made by Yui *et al.* (2007) confirmed the original structure proposed by Gardner and Blackwell (1971). A strong x-ray diffraction ring, quoted as the α-chitin signature is found at 0.338 nm whereas a similar ring occurs at 0.324 nm in β-chitin; an inner ring at 0.918 nm in β-chitin is sensitive to hydration, moving to 1.16 nm in the presence of liquid water, whereas a similar strong inner ring at 0.943 nm in α-chitin is insensitive to hydration (Rinaudo, 2006).

The pens from the squids *Loligo sanpaulensis* and *Loligo plei* have become available in considerable amounts from the fisheries in Brazil, for the extraction of β-chitin. Due to the low content of inorganic compounds the demineralization step is skipped and a two-step alkaline treatment was deemed adequate to produce β-chitin with low ash contents (< 0.7%). Indeed, the inorganic contents were particularly low: Ca < 10.4 ppm, Mg < 2.5 ppm, Mn < 3.1 ppm and Fe < 1.8 ppm (Lavall *et al.* 2007).

Similarly, β-chitin and the corresponding chitosan have been isolated from the pens of *Loligo lessoniana* and *Loligo formosana*; they have been chemically characterized to qualify a potential chitin source. Also in this case, thanks to negligible ash content, the demineralization step was omitted and only deproteinization was carried out in the chitin isolation, with the yield of 35-38%, without significant difference either between the two species or the collection seasons. Mild alkaline deacetylation with various time lengths was employed in the chitosan preparation. The nitrogen contents indicated the effectiveness of the deproteinization method used. The samples showed moderate hygroscopicity. Trace elements present in the pens could not be removed in full, however (Chandumpai *et al.* 2004).

2. Nanotechnology

2.1 Technological Mimics of Quinone Tanning

The enzymatic reaction between chitosan and phenols promoted by tyrosinase or analogous enzymes such as laccase and horseradish peroxidase occurs in two steps: the oxidation of phenols to quinones with atmospheric oxygen, immediately followed by the chemical reaction of the quinones with the amino groups of chitosan. The kinetics were studied by Payne and coworkers (Yi *et al.* 2005) and by Muzzarelli *et al.* (1994, 2002). Biologically significant quinones, such as menadione (vitamin K), plumbagin, ubiquinone (CoQ_{10}) and CoQ_3 were examined along with 1,2-naphthoquinone and 1,4-naphthoquinone for their capacity to react with five chitosans in freeze-dried or film form. CoQ_{10} and CoQ_3 did not react with the chitosans, whilst menadione and 1,4-naphthoquinone were reactive and yielded deeply colored modified chitosans. The maximum capacity of chitosans for 1,4-naphthoquinone corresponded to an amine/quinone molar ratio close to 2: developments in the pharmaceutical and cosmetic areas were foreseen (Muzzarelli *et al.* 2003).

The use of biological materials for fabrication based on the mimic of quinone tanning (biofabrication) was reviewed by Yi *et al.* (2005) who focused on three specific approaches: directed assembly, where localized external stimuli are employed to guide assembly; enzymatic assembly, where selective biocatalysts are enlisted to build macromolecular structure; and self-assembly, where information internal to the biological material guides its own assembly. Chitosan offers a combination of properties uniquely suited for biofabrication because it can be directed to assemble in response to locally applied electrical signals, and its backbone provides sites that can be employed for the assembly of proteins, nucleic acids, and virus particles.

2.2 Reinforced Rubber and Poly(Caprolactone)

High reinforcing capability was reported resulting from the intrinsic chemical nature of chitins and from their hierarchical structure. During the last decade, many works have been devoted to mimic biocomposites by blending cellulose nanofibrils from different sources with polymer matrices (Dufresne, 2006). Reinforced natural rubber nanocomposites were developed from colloidal suspension of chitin nanofibrils and latex of un-vulcanized and pre-vulcanized natural rubber. The chitin nanofibrils consisted of slender parallelepiped rods with an average length around 240 nm and an aspect ratio close to 16. After the aqueous suspensions of chitin nanofibrils and rubber were mixed and stirred, solid composite films were obtained by casting and evaporating methods. For un-vulcanized systems a freeze-drying and subsequent hot-pressing processing technique was also used. All the results lead to the conclusion that the processing technique plays a major role in the properties of final composites developed. The chitin nanofibrils form a three-dimensional rigid network only in the evaporated samples, and it is assumed to be governed by a percolation mechanism. The critical volume fraction of chitin nanofibrils at the percolation threshold was found to be 4.4 vol % (around 6.4 wt %). The preparation of the latex requires the use of poloxamer 407

BASF Lutrol F127, a surfactant, in order to obtain a stable suspension (Morin *et al.* 2002). It is a bloc copolymer of ca. 70 wt % poly(ethylene oxide) and 30 wt % poly(propylene oxide).

Poly(caprolactone) is a biodegradable, semi-crystalline and thermoplastic polymer used for instance to manufacture suture threads; there is much interest in improving its mechanical properties and biochemical significance. Chitin nanofibrils (much longer than those of animal origin: L = 0.5-10 μm, d = 18 nm) were obtained from tubes secreted by *Riftia*, a vestimentiferan worm. The results showed that at high temperature and above 5 % nanofibrils, the chitin network is allowed to restore thus stabilizing the mechanical properties of the composite (Morin *et al.* 2002).

2.3 Composites with Chitosan or with Poly(Vinyl Alcohol)

α-Chitin nanofibril-reinforced poly(vinyl alcohol) composite films were prepared by solution-casting technique. The as-prepared nanofibrils exhibited the length in the range of 150-800 nm and the width in the range of 5-70 nm, with the average length and width being about 417 and 33 nm, respectively. The thermal stability of the nanocomposite films was improved compared to those of the pure PVA film with increasing nanofibril content. The presence of the nanofibrils did not have any effect on the crystallinity of the PVA matrix. The tensile strength of α-chitin nanofibril-reinforced PVA films increased, at the expense of the percentage of elongation at break, from that of the pure PVA film with initial increase in the nanofibril content and leveled off when the nanofibril content was greater than 2.96 % (Sriupayo, *et al.* 2005 a).

Similar preparations were made with α-chitin nanofibrils dispersed in chitosan by solution-casting, thanks to the high filmogenicity of chitosan. The tensile strength of α-chitin nanofibril-reinforced chitosan films increased with increasing nanofibril content to reach a maximum at the nanofibril content of 2.96 % and decreased gradually with further increase in the nanofibril content, while the percentage of elongation at break decreased with increasing nanofibril content and leveled off when the nanofibril content was greater than 2.96 %. As in the case of chitin nanofibril composites with PVA, both the addition of α-chitin nanofibrils and heat treatment helped improve water resistance, leading to decreased percentage of weight loss and percentage degree of swelling of the nanocomposite films (Sriupayo *et al.* 2005 b).

3. Chemical Aspects

Chitin is known for its scarce solubility that is justified by the structural characteristics illustrated above. Nevertheless, chitin nanofibril permanent suspensions are as easy to handle as solutions. In addition chitin can be brought into solution in a variety of ways. Austin *et al.* (1981) calculated solubility parameters for chitin in various solvents and experimentally obtained the chitin / LiCl complex soluble in dimethylacetamide and in N-methyl-2-pyrrolidone. Later on, the same solvents, especially the LiCl / DMAc mixture, were found to dissolve also cellulose. For a certain time the most widely used solvent for chitin was a

DMAc / LiCl mixture, so that Yoshimura *et al.* (2005) could esterify chitin with succinic anhydride in the presence of 4-dimethylamino pyridine as a catalyst; they obtained a superabsorbent hydrogel that absorbed water about 300 times of its dry weight, comparable to polyacrylate.

Saturated solutions of lithium thiocyanate were also developed. Calcium chloride dihydrate-saturated methanol was also employed by Tamura *et al.* (2006) for the preparation of chitin hydrogels under mild conditions: they refluxed $CaCl_2.2H_2O$ (850 g, 30 min) in methanol (1 l) and filtered after overnight standing. This solvent was used for either 20 g of α-chitin or 10 g of β-chitin powder under reflux for several hours. After extensive dialysis, a 3-6 % hydrogel was made available for the preparation of chitin sheets that exhibited cationic character, and were developed into biomedical items. Phosphoric, formic, dichloroacetic and trichloroacetic acids were found suitable for dissolution of chitins at room temperature. Other solvents are hexafluoroisopropanol and hexafluoracetone sesquihydrate.

3.1 Alkali Chitin

Etherification is one of the most important modifications used to prepare water soluble chitin derivatives, including carboxymethyl chitin, hydroxyethyl chitin, and hydroxypropyl chitin (Hirano,1988; Qin *et al.* 2006; Sini *et al.* 2005). Etherification of chitin is performed by treating alkali chitin with alkyl halides. The treatment with NaOH swells and decrystallizes the chitin structure and improves the access of alkyl halides or other reactants into chitin; otherwise the products have poor solubility due to dyshomogeneity. Unlike alkali treatment of cellulose however, the research work on alkali chitin has been occasional and far from being systematic (Thor and Henderson, 1940; Heuser, 1944). Martin-Gil *et al.* (1992) reported that cooling reduced the hydrogen bonds between the chitin sheets, and that the alkali enhanced decrystallization. Similarly, Feng *et al.* (2004) reported that crystallinity of chitin decreased by one half after alkali-freezing treatment in 50% NaOH solution for 3 days. Noishiki *et al.* (2003) studied the crystal conversion from β-chitin to α-chitin by alkali treatment, and reported that the minimum NaOH concentration for chitin swelling was between 25 wt% and 30 wt%. Many authors (Morita and Jinno, 1988; Somorin *et al.* 1979; Wan *et al.* 2004) have reported on ethers of chitin or chitosan. When soaked in 20 % NaOH solution, the gelation of chitin occurred, and the total absorbency reached 800 g per 100 g chitin maximally. The NaOH and water combined by hydrogen bonds with chitin could not be removed by centrifugation: water absorbency made the major contribution to the swelling. When frozen, the volume of water expanded thus enabling more alkali solution to penetrate into chitin particles and form a gel. When the NaOH concentration reached 30 %, the total absorbency began to decrease and stabilized at 200 g per 100 g chitin.

The addition of minor amounts of ethanol increased the absorbency of chitin. X-ray diffraction patterns and infrared spectra measurements showed that chitin was decrystallized and hydrogen bonds in chitin were weakened by alkali treatment. The measurement of molecular weight and degree of acetylation suggested that higher temperature and prolonged treatment time should be avoided during alkali treatment of chitin (Liu *et al.* 2008).

The chaotropic effect of urea has been found advantageous in promoting chitin dissolution thus urea has been used with a view at reducing the amounts of alkali. In fact the mixture was effective in dissolving chitin over a 5-day period at -20 °C when it contained as little as 8 % NaOH with 4 % urea. It was observed that this solvent had no detrimental effect on chitin structure and the urea was of benefit to the stability of the chitin solution. This alkali chitin aqueous solution transformed into a gel as a consequence of temperature increases (Hu *et al.* 2007).

3.2 6-Oxychitin

Crustacean chitins in water suspension were submitted to regiospecific oxidation at C-6 with the aid of 4 % w/v aqueous NaOCl (24 ml) in the presence of the stable nitroxyl radical 2,2,6,6-tetramethyl-1-piperidinyloxy (Tempo®) and NaBr at 25°C at room temperature. The resulting oxychitins have anionic character and are fully soluble over the pH range 3 - 12; their degree of substitution is 1.0 in all cases; the molecular weight is in the range 5-10 kDa, the final yield is 36 % and the degree of acetylation remains the original one (Muzzarelli *et al.* 1999).

In order to favour the regiospecific oxidation, one should take advantage of the possibility to change the crystal structure from alpha to beta or to put the chitin in the amorphous state: in both cases the penetration of the reagents would be favoured. In fact, the same preparation carried out with a chitin kept in warm water at 50°C for 4 hr, with a reacetylated chitosan, or with a colloidal chitin reprecipitated from sulfuric acid solution permitted to improve the yield from 36% to 97% while keeping the molecular weight at higher values (40 kDa). The moisture absorption and retention abilities of these types of compounds were superior to those of sodium hyaluronan and carboxymethyl chitosan (Sun *et al.* 2006).

6-Oxychitins lend themselves to metal chelation, polyelectrolyte complex formation with a number of biopolymers including chitosan, and to microsphere and bead formation. Oxychitin sodium salt coagulates a number of proteins, including papain, lysozyme and other hydrolases (Muzzarelli *et al.* 1999).

Incidentally, chitin nanocrystals (340 x 8 nm) dispersed in water were prepared following the same approach. Since the oxidized chitin had a crystallinity as high as that of the original α-chitin, and deacetylation did not occur, the C6 carboxylate groups were regarded as being present only on the chitin crystallite surfaces. (Fan *et al.* 2008).

Similarly treated biomasses of *Aspergillus niger, Trichoderma reesei* and *Saprolegnia parasitica* yielded polyuronans in the sodium salt form, fully soluble in water over the pH range 3 - 12. Yields were much higher than for the chitosan extraction. The polyuronans characterized by ^{1}H-NMR and FTIR spectrometry contain 20 % and > 75 % oxychitin, for *A. niger* and *T. reesei*, respectively. Since the fungi examined are representative of the three major types of cell walls, and are of industrial importance, it is concluded that the process is of wide applicability. The process allows upgrading the spent biomasses and the exploitation of their polysaccharides for industrial applications: examples are given where the polyuronans are reacted with glycerol or with poly(ethyleneglycol) to provide water soluble products with enhanced viscosity (Muzzarelli *et al.* 2002).

Oxychitin keeps the regenerative properties of chitin and chitosan; in a model study, surgical lesions in rat condylus were treated with N,N-dicarboxymethyl chitosan and 6-oxychitin sodium salt. Morphological data indicated that the best osteoarchitectural reconstruction was promoted by 6-oxychitin, even though healing was slightly slower compared to with N,N-dicarboxymethyl chitosan.

On Ti-6Al-4V alloy plates, plasma-sprayed with hydroxyapatite or with bioactive glass, a chitosan film was deposited to be reacted with 6-oxychitin and to form a polyelectrolyte complex. The latter was optionally contacted with 1-ethyl-3-(3-dimethylaminopropyl) carbodiimide at 4°C for 2 hr to form amide bonds between the two polysaccharides. Flat surfaces exempt from fractures were visible at the electron microscope. The results were used for the preparation of prosthetic articles possessing an external organic coating capable to promote colonization by cells, osteogenesis and osteointegration (Mattioli-Belmonte *et al.* 1999).

Oxychitin was reacted with chitosan to produce microspheres by polyelectrolyte coacervation. The most suitable chitosan salts were chitosan glutamate and chitosan chloride; the preferred chitosan molecular weight was > 150 kDa; the coacervation process took place at chitosan pH < 5 and in a pH range between 2 and 10 for oxychitin. The best reaction condition are as follows: chitosan solution at low pH (preferably 2), oxychitin solution at high pH (preferably 10); molar ratio should be in favour of oxychitin. The preparation of microspheres is made in a short time at 20°C in aqueous system.

6-Oxychitins in the form of free acids, salts and esters were proposed for use as surrogates of hyaluronans and of bacterial antigens in medical and health care products.

4. Chitin Depolymerization *in Vitro*

4.1 With Chitinases

The enzymatic depolymerization of chitin in vitro has been explored in consideration of the technical disadvantages of the chemical hydrolysis, namely partial deacetylation of the resulting oligomers, difficulty of controlling the depolymerization rate, low yield, cumbersome isolation of oligomer mixture, and heterogeneous conditions. Preparation of chitooligomers by enzymatic depolymerization does not produce deacetylation. It would seem that the most obvious approach to the production of chitooligomers is the use of chitinases, but the comparative evaluation of the data so far available points to a different conclusion as discussed below.

A recommended procedure with a chitinase is the following (Aiba, 1994; Muzzarelli *et al.* 1999). Chitin is swollen by treatment with acid to make it more easily digestible by chitinases: crab chitin (10 g) is mixed with concentrated HCl (100 ml) and stirred slowly at 4°C for 24 hr. The viscous material is diluted with cold water (1 liter). The swollen chitin is recovered by filtration on a Büchner funnel with a glass microfibre filter. The white amorphous solid is neutralized with 2 N NaOH and washed extensively with water. The washed chitin is kept at 4°C. According to Zhu and Laine (1997), the swollen chitin suspension (10 ml) is centrifuged at 5,000 rpm for 10 min. The white slurry (2 g) at the top layer is mixed with phosphate buffer, pH 6.0 (15 ml), containing 5 units chitinase from *Streptomyces griseus,* or recombinant chitinase from *Vibrio parahemolyticus* (Laine *et al.* 1994). Enzymatic reaction is considered complete when the cloudy chitin suspension becomes a clear solution. After incubation, the mixture is heated at 100°C for 3 min to inactivate the enzyme. The reaction time, temperature and products are shown in Table 1: actually only the lower oligomers are produced in all cases. The overall yield of oligomers from the 2-g chitin slurry is 20-25 mg.

The enzymatic degradation of chitin, that typically yields a mixture of chito-oligomers of various sizes, can be followed by measuring the production of educing ends. The widely used Schales' procedure does not permit accurate determination of reducing ends since the spectrophotometric signal increases with the length of the chito-oligosaccharides. In contrast, the 3-methyl-2-benzothiazolinone hydrazone method gives the same absorbance per reducing end, irrespective of the length of the chito-oligomers. The new method is the most sensitive reducing end assay for the quantification of chito-oligomers, detecting concentrations down to 5 microM (Horn and Eijsink, 2004).

It should be added that chitinases possess a specific domain that binds tightly to the insoluble substrate. In fact, the depolymerization of chitin is quite fast in most cases, notwithstanding the highly crystalline and insoluble nature of the chitin nanofibrils described above. Analogous domains are present in amylases, cellulases and xilanases.

4.2 With Lysozyme

The hydrolytic activity of lysozyme towards chitins and chitosans is genrally considered unspecific, lysozyme being a muramidase. It was observed however that barley chitinase reveals a resemblance to lysozyme because they have in common a central core despite lacking any obvious aminoacid sequence similarity. In barley chitinase Glu 67 appears to act as the proton donor and Glu 89 as the general base in the catalytic mechanism (Robertus and Hart, 1995; Robertus and Monzingo, 1999). The lysozyme fold appears to be extremely ancient, and recurs in the much larger chitinases; it should be kept in mind that the early works on lysozyme structure and kinetics were made with the use of the chitin hexamer as a substrate (Jollès, 1996).

Table 1. Oligomers produced by enzymatic depolymerization of chitin with chitinases (Muzzarelli *et al.* 1999)

Temperature, °C	Time, h	Oligomers
4	24-36	(GlcNAc)1-6
22	14-18	(GlcNAc)1-5
37	8-10	(GlcNAc)1-4
45	4-6	(GlcNAc)1-3

The difficulty of using lysozyme for the preparation of oligomers is partially circumvented with the chemical derivatization such as the production of glycol chitin, methylpyrrolidinone chitosan and re-acetylated chitosans, that are water-soluble at the pH values of maximum enzyme activity.

At 37°C and optimum lysozyme pH value, the K_m values for glycolchitin is 2.95 Kg/m^3 (lysozyme 0.012 g/l) (Maeda *et al.* 1996); for methylpyrrolidinone chitosan is 71 Kg/m^3 (lysozyme 0.4 g/l) (Muzzarelli *et al.* 1992) whilst Aiba (1992) indicated much lower values (in the range of 0.05 - 0.5 Kg/m^3) for chitosans and reacetylated chitosans; nevertheless he pointed out that the initial velocity data might be not indicative of the real susceptibility of a given chitosan because the hydrolysis rates change with the progress of the reaction, depending on degree of acetylation and acetyl group distribution: actually the molecular weight observed after various days of contact are in the range 95–130 kDa.

Hydrolysis of colloidal chitin by lysozyme. Method A. A colloidal chitin suspension (0.5 % w/v, 10 ml) in 0.2 M acetate buffer (pH 5.4) with 0.05 % NaN_3 is incubated with lysozyme (2 mg/ml, 1 ml) at 37°C for 3 days and then more lysozyme solution (1 ml) is added. After 3 days the suspension is centrifuged and the products in the supernatant solution are analysed by HPLC. *Method B* (large amount of lysozyme). Reacetylated chitosan (0.30-0.72, 100 mg) is dissolved as described above, to afford a 0.5 % solution. Lysozyme (20 mg) is added and the mixture is incubated at 37°C. At intervals, a portion (0.5 ml) is taken out, acetylated, and analysed by HPLC.

The highest digestibility of chitosans is due to the blocks of NAcGlc sequences. The hexamers containing 3 or more acetylated units contribute mostly to the initial degradation velocity. This information is based on the Michaelis-Menten analysis of the degradation data for various hexamers under the action of lysozyme (Nordtveit *et al.* 1994). The substrate specificities of human and hen egg white lysozymes with respect to partially N-acetylated chitosans are undistinguishable. Lysozyme does not seem to lead straightforwardly to the production of oligomers: partially N-acylated chitosans (N-acetyl-N-hexanoyl) subjected to the action of lysozyme show final average molecular weights in the range 10-100 kDa over a contact period of 24 hr (Aiba, 1994; Lee *et al.* 1995).

4.3 Unspecific Hydrolases

The efficient production of N-acetyl-D-glucosamine with the aid of cellulases derived from *Trichoderma viride* and *Acremonium cellulolyticus* was observed by HPLC analysis compared to lipase, hemicellulase, and pectinase. β-Chitin showed higher degradability than α-chitin when using cellulase from *T. viride*. The optimum pH of cellulase T was 4.0 on the hydrolysis of β-chitin. The yield of GlcNAc was enhanced by mixing of cellulases from *T. viride* and *A. cellulolyticus*. Of particular interest is the use of lipases which depolymerize chitin to the monomer with a yield as high as 62 % when used at the concentration of 20 mg/ml. Both lipase and cellulase are cheap commercial enzymes (Sashiwa *et al.* 2003).

The results with the unspecific enzymes are comparable to those obtained with crude chitinases: in fact the selective and efficient production of N-acetyl-D-glucosamine exempt from oligomers was achieved from flakes of α-chitin by using crude enzymes derived from

Aeromonas hydrophila H-2330 with 77 % yield (Sashiwa *et al.* 2002). Finely powdered chitin was completely hydrolyzed with crude chitinases from *Burkholderia cepacia* TU09 and *Bacillus licheniformis* SK-1 (EC 3.2.1.14) and acetylhexosaminidase (EC 3.2.1.52). Chitinase from *B. cepacia* produced GlcNAc in greater than 85% yield from β- and α-chitin within 1 and 7 days, respectively. *B. licheniformis* chitinase completely hydrolyzed β-chitin within 6 days, giving a final GlcNAc yield of 75%, along with 20% of chitobiose, but the yield was only 41% with α-chitin (Pichyangkura *et al.* 2002).

5. Fungal and Plant Chitinases

Accumulation of chitinous material in the edible mushroom *Agaricus bisporus* stalks was determined during postharvest storage at 4 and 25°C. The chitinous material was extracted after alkali treatment and acid reflux of alkali insoluble material and analyzed for yield, purity, degree of acetylation and crystallinity. The total glucosamine content in mushroom stalks increased from 7% dry weight at harvest to 11% and 19% after 15 days of storage at 4°C and 5 days of storage at 25°C, respectively. The yield of crude chitin isolated from stalks stored at 25°C for 5 days was 27% dry weight and consisted of 46% glucosamine and 21% neutral polysaccharides associated to chitin. The degree of acetylation of fungal chitin was from 0.76 – 0.88 (Wu *et al.* 2004).

The chitin contents of pileus and stipes of fruit bodies of common varieties of the cultivated mushrooms *A. bisporus, Pleurotus ostreatus* and *Lentinula edodes* from Hungarian and German large-scale farming activities were determined and compared. The chitin level of stipes seemed to be constant. The *A. bisporus* varieties had practically the same chitin levels, indicative that the chitin content is a permanent characteristic of the species and there are no significant differences among the varieties. The chitin levels of pileus and stipes were not significantly different in *A. bisporus*, but showed significant differences for *P. ostreatus* and *L. edodes*, the pileus having the higher and the stipe the lower chitin content. The data confirmed that saprotrophic *A. bisporus* had higher chitin level than had the wood-rotting *P. ostreatus* and *L. edodes* (Vetter, 2007). The extraction of chitin from *A. bisporus* has been found relatively difficult, because chitin is covalently associated with glucans, and therefore the final product contains at least 3 % of β-glucan (Vincendon and Desbrieres, 2002).

The competitor fungus *Trichoderma aggressivum* causes green mould disease, a potentially devastating problem of the commercial mushroom *A. bisporus*. Due to the recent appearance of this problem, very little is known about the mechanisms by which *T. aggressivum* interacts with and inhibits *A. bisporus*. The activities of chitinases produced in dual cultures of these fungi were determined over a 14-day period. Both intracellular and extracellular enzymes were studied. *A. bisporus* produced N-acetylglucosaminidases with apparent molecular masses of 111, 105, and 96 kDa. Two resistant brown strains produced greater activities of the 96 kDa N-acetylglucosaminidase than susceptible off-white and white strains. This result suggested that this enzyme might have a role in the resistance of commercial brown strains to green mould disease. *T. aggressivum* produced three N-acetylglucosaminidases with apparent molecular masses of 131, 125, and 122 kDa, a 40 kDa chitobiosidase, and a 36 kDa endochitinase. The 122 kDa N-acetylglucosaminidase showed the greatest activity and may be an important predictor of antifungal activity (Guthrie and Castle, 2006).

In recent years, genome sequencing of several fungi has been completed, for instance *Saccharomyces cerevisiae, Candida albicans, Coccidioides immitis, Neurospora crassa, Gibberella zeae, Magnaporthe grisea, Aspergillus nidulans, Aspergillus fumigatus* and *Trichoderma reesei*. By genome analysis, new and more chitinase genes have been found.

Chitinase genes of fungi are well expressed functionally active in yeasts, filamentous fungi and plants, as the following examples show. Expression of *ech42* encoding endochitinase of *Trichoderma harzianum* in *Escherichia coli* results in high chitinase activity. Transformation of *Schizosaccharomyces pombe* with *cts1* from *Saccharomyces cerevisiae* results in the appearance of about a 5-13-fold increase in chitinase activity. Introduction of *Aphanocladium album* chitinase gene into *Fusarium oxysporum* results in high chitinase levels in the host. Expression of a gene encoding endochitinase from *T. harzianum* in *T. reesei* results in a transformant that produces 20 times more chitinase than *T. reesei*. The *ThEn42* gene of *T. harzianum* endochitinase is transferred to tobacco, potato and apple, and highly expressed in different plant tissues.

Rice (*Oryza sativa* L.) has been a major component of the human diet for many thousands of years. Approximately one-third of the world's population relies on rice for a major portion of their food: rice is currently ca. 30 % of total cereal production. It is therefore non surprising that since several years transgenic rice plants are being cultivated in various countries: it is now possible to confer resistance against pathogens in cultivated rice by introducing genes encoding chitinase and thaumatin-like proteins. Diverse defense responses were studied in transgenic Pusa Basmati rice lines engineered with rice chitinase gene (*chi11*) for resistance against the sheath blight pathogen *Rhizoctonia solani*. Enhancement of phenylalanine ammonia lyase, peroxidase and polyphenoloxidase enzymes in response to the pathogen challenge under controlled conditions resulted in reduced symptom development and containment of the disease in transgenic rice lines compared to non-transgenic control plants. Loss of chlorophyll resulting from *R. solani* infection was comparatively less in transgenic plants (Sareena *et al.* 2006).

The same approach was extended to malting barley (*Hordeum vulgare* L.). Transgenes were introduced by co-bombardment with two plasmids, one carrying the rice chitinase gene *chi11* and another carrying a rice thaumatin-like protein gene *tlp*. Each gene was under the control of the maize ubiquitin (*Ubi1*) promoter. Fifty-eight primary transformants from three independent transformation events were regenerated. Progeny from one event had stable integration and expression of the rice *chi11* and *tlp*. (Tobias *et al.* 2007). In fact, there is remarkable sequence homology between the two chitinases from rice and barley: therefore the shapes of these proteins are very similar with respect to the structure of binding site and charge distribution on the surface. Both enzymes have the same shape and dimensions and have identical activity in the individual plants (Sasaki *et al.* 2001). All of these studies tend to make immediately available the plant chitinase upon attack by a pathogen, and to enable the plant to produce a more active chitinase than that usually produced by normal plants. Transgenic maize plants expressing powerful chitinases and endowed with other favorable traits such as kernel quality, nutritional content and tolerance of soil-related stresses are also widely used in agriculture.

In fungi, chitinases have autolytic, nutritional, and morphogenetic roles (Adams 2004). Most chitinolytic fungi have been found to produce more than one kind of chitinase. *Trichoderma harzianum* produces seven individual chitinases: two N-acetyl-glucosaminidases (102 and 73 kDa), four endochitinases (52, 42, 33 and 31 kDa) and one chitobiosidase (40 kDa). *Talaromyces flavus* produces at least two kinds of chitinases.

Chitinases (EC 3.2.1.14) expressed during plant-microbe interaction are involved in defense responses of host plant against pathogens. The chitinase gene from wheat has been subcloned and overexpressed in *Escherichia coli*. Molecular phylogeny analyses of wheat chitinase indicated that it belongs to an acidic form of class VII chitinase (glycosyl hydrolase family 19) and shows 77% identity with other wheat chitinases and low level identity to other plant chitinases. The three-dimensional structural model of wheat chitinase showed the presence of 10 α-helices, 3 β-strands, 21 loop turns and the presence of 6 cysteine residues that are responsible for the formation of 3 disulphide bridges. The active site units Glu94 and Glu103 may be indicated for antifungal activity (Figure 6). Expression of chitinase (33 kDa; 20 mg/l yield and 1.9 U/mg activity) was confirmed by SDS PAGE and Western hybridization analyses. Purified chitinase exerted a broad-spectrum antifungal activity against *Colletotrichum falcatum* (red rot of sugarcane) *Pestalotia theae* (leaf spot of tea), *Rhizoctonia solani* (sheath blight of rice), *Sarocladium oryzae* (sheath rot of rice) *Alternaria* sp. (grain discoloration of rice) and *Fusarium* sp. (scab of rye). Due to its innate antifungal potential, wheat chitinase can be used to enhance resistance in crop plants (Singh *et al.* 2007).

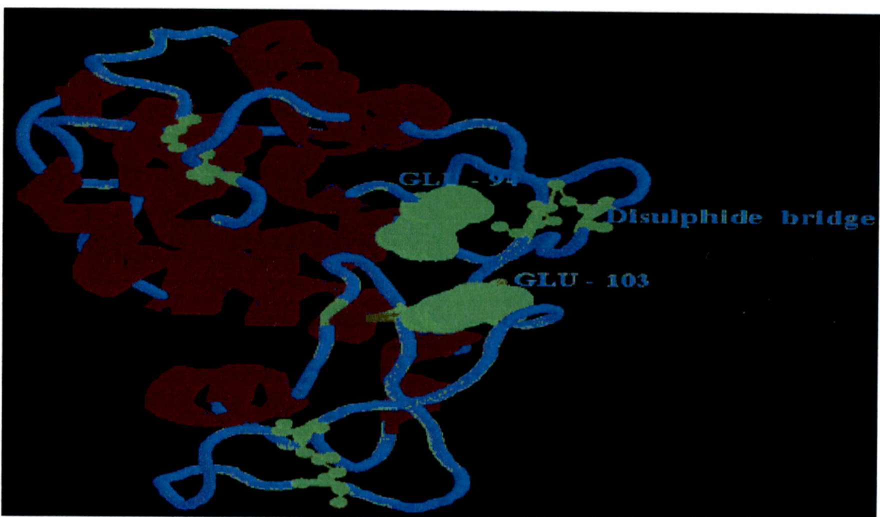

Figure 6. Predicted structure of wheat chitinase. Three-dimensional structural model of the 33-kDa wheat chitinase with 10 alpha-helices (red), 3 beta-strands (orange) and 21 loop turns. Six conserved cysteine residues are shown to form 3 disulphide bonds (yellow). The catalytic glutamic acid residues Glu94 and Glu103 (green) identified in the loop sequence connected to beta-strand may be suggested for chitinase activity. Reprinted from *Protein Expression and Purification* 56. Singh A, Kirubakaran SI, Sakthivel N. Heterologous expression of new antifungal chitinase from wheat. Pages 100-109. Copyright (2007), with permission from Elsevier.

The induction of chitinases (CHIT102, CHIT73) in *T. harzianum* takes place during parasitism on *Sclerotium rolfsii*. *T. harzianum* CHIT42 chitinase expression is strongly enhanced during interactions of *T. harzianum* with *R. solani* or with *Botrytis cinerea*. Analysis of confronting cultures of *T. harzianum* and its hosts (such as *B. cinerea* and *R. solani*) demonstrates that the transcript of *ech42* and *chit36* is observed during the pre-contact stage of the confrontation. The *ech42* expression is triggered by diffusible factors whose formation does not require contact between *T. harzianum* and *R. solani*: the diffusible is a constitutive chitinase produced by *T. harzianum,* as its action can be inhibited by allosamidin. The action of the enzyme on *R. solani* is crucial for *ech42* expression (Li, 2006).

Upon contact with the host, the *T. harzianum* mycelium coils around the host hyphae, forms hook-like structures that aid in penetrating the host cell wall, and finally absorb nutrients from host cells. Chitinases play an important role in the above mycoparasitic progress, especially in the cell wall penetration and in making available the nutrients from the confronting microorganism.

The digestibility of various chitins by the chitinase from *Bacillus sp.* PI-7S is much higher than that by lysozyme, and β-chitin is digested more smoothly than α-chitin. Chitin deacetylated under homogeneous conditions is hydrolysed by lysozyme more rapidly than that deacetylated under heterogeneous conditions. Those from shrimp shell and squid pen show the same degree of digestibility by lysozyme in spite of a difference in the crystal structure of the original chitins. The crystal structure of chitin and the degree of N-acetyl group aggregation affect the enzymatic digestibility of chitins and deacetylated chitins.

Bacterial chitinases play a role in the digestion of chitin for utilization as a carbon and energy source and recycling chitin in nature (Bhattacharya *et al.* 2007): for example, the intestinal bacteria *Vibrio scophthalmi, ichthyonenteri* and *fischeri* harboured in the Japanese flounder *Paralichtys olivaceus* can digest chitin, but *V. scophthalmi* and *ichthyonenteri* are *chiA* PCR positive while *V. fischeri* is not (Sugita and Ito, 2006). Chitin-binding proteins are common in environmental and clinical *Vibrio* strains, and they have an important general role in mediating cell interactions with chitinous surfaces including those of copepods (Vezzulli *et al.* 2007). In insects, chitinases are associated with postembryonic development and degradation of old cuticle (Merzendorfer and Zimoch, 2003). Plant chitinases are involved in defense and development. Chitinases encoded by viruses have roles in pathogenesis (Patil *et al.* 2000). On the other hand, chitinases have shown immense usefulness and applicability in agricultural and environmental sciences.

Conclusion

The structural characteristics of chitins *in vivo* permit to state that they are among the most mechanically and chemically resistant organic materials as exemplified by the squid beak, the diatom spines and the fungal hyphal tips. Nevertheless, chitins are prone to digestion under the action of specific and unspecific enzymes due to the fact that every living organism has to grow: crustaceans undergo molting and resorb most of their exoskeleton chitin to provide the substrate needed for the synthesis of a larger one; fungal hyphae have to penetrate hard materials while growing, thus the existence of every chitinous organ has to be seen as a result of concerted synthesis and resorption.

The susceptibility of chitin to hydrolytic cleavage is the basis of fundamental processes such as the recycling of nitrogen in the oceans operated by chitinolytic bacteria living in the bottom mud, and the fixation of nitrogen by nod factors in symbiosis with leguminous plants.

Mammals, which do not have chitinous tissues, recognize exogenous chitin from bacterial pathogens and nematode parasites essentially as a threat and therefore react by activating their defenses at the biochemical level (Kawada *et al.* 2007; Knight *et al.* 2007). Actually, the activation of macrophages in the presence of exogenous chitin or chitosan in the current views is a positive process (Shibata *et al.* 1997) that permits to apply to the human body manufactured chitosan items that range from targeted drug carriers to wound and ulcer dressings (Gavini *et al.* 2008; Minagawa *et al.* 2007). Human chitinases belong to the group of biochemical substances that, like lysozyme, are produced by human cells for the protection of the whole organism.

References

Adams DJ. Fungal cell wall chitinases and glucanases. *Microbiology-SGM*, 2004; 150: 2029-2035.

Aiba S. Lysozymic hydrolysis of partially N-acetylated chitosans. *International Journal of Biological Macromolecules*. 1992; 14: 225-228.

Aiba S. Preparation of N-acetylchitooligosaccharides from lysozymic hydrolysates of partially N-acetylated chitosans. *Carbohydrate Research* 1994; 261: 297-306.

Aiba S. Preparation of N-acetylchitooligosaccharides by hydrolysis of chitosan with chitinase followed by N-acetylation. *Carbohydrate Research*, 1994; 265: 323-328.

Austin PR, Brine CJ, Castle JE, Zikakis JP. Chitin: new facets of research. *Science*. 1981; 212: 749-753.

Bhattacharya D, Nagpure A, Gupta RK. Bacterial chitinases: Properties and potential. *Critical Reviews in Biotechnology*. 2007; 27: 21-28.

Bone Q, Ryan K, Pulsford AL. The structure and the composition of the teeth and grasping spines of Chaetognaths. *Journal of the Marine Biology Association* UK. 1983; 63: 929-939.

Chandumpai A, Singhpibulporn N, Faroongsarng D, Sornprasit P. Preparation and physico-chemical characterization of chitin and chitosan from the pens of the squid species, Loligo lessoniana and Loligo formosana. *Carbohydrate Polymers*. 2004; 58: 467-474.

Chretiennot-Dinet MJ, Giraud-Guille MM, Vaulot D, Putaux JL, Saito Y, Chanzy H. The chitinous nature of filaments ejected by Phaeocystis (Prymnesiophyceae). *Journal of Phycology*. 1997; 33: 666-672.

Dufresne A. Comparing the mechanical properties of high performances polymer nanocomposites from biological sources. *Journal of Nanoscience and Nanotechnology*. 2006; 6: 322-330

Fan Y, Saito T, Isogai A. Chitin nanocrystals prepared by TEMPO-mediated oxidation of alpha-chitin. *Biomacromolecules*. 2008; 9: 192-198.

Feng F, Liu Y, Hu K. Influence of alkali-freezing treatment on the solid state structure of chitin. *Carbohydrate Research*. 2004; 339: 2321–2324.

Gardner KH, Blackwell J. Sub-structures of crystalline cellulose and chitin microfibrils. *Journal of Polymer Science*. 1971; C-36: 327-340.

Gavini E, Rassu G, Muzzarelli C, Cossu M, Giunchedi P. Spray-dried microspheres based on methylpyrrolidinone chitosan as new carrier for nasal administration of metoclopramide. *European Journal of Pharmaceutics and Biopharmaceutics*. 2008; 68: 245-252.

Giraud-Guille MM, Belamie E, Mosser G. Organic and mineral networks in carapaces, bones and biomimetic materials. *Comptes Rendus Palevol*. 2004; 3: 503-513.

Goodrich JD, Wintcr WT. Alpha-Chitin nanocrystals prepared from shrimp shells and their specific surface area measurement. *Biomacromolecules*. 2007; 8: 252-257.

Guthrie JL, Castle AJ. (2006). Chitinase production during interaction of Trichoderma aggressivum and Agaricus bisporus. *Canadian Journal of Microbiology*. 2006; 52: 961-967.

Hepburn A. (Ed.). (1976). *The insect integument*. Elsevier. Amsterdam.

Heuser E. (1944). *The chemistry of cellulose*. (63–135). New York: John Wiley and Sons Inc.

Hirano S. Water-soluble glycol chitin and carboxymethylchitin. *Methods in Enzymology* .1988; 161: 408–410.

Horn SJ, Eijsink VGH. A reliable reducing end assay for chito-oligosaccharides. *Carbohydrate Polymers*. 2004; 56: 35-39.

Hu XW, Du YM, Tang YF, Wang Q, Feng T, Yang JH, Kennedy JF. Solubility and property of chitin in NaOH/urea aqueous solution. *Carbohydrate Polymers*. 2007; 70: 451-458.

Jollès P. (Ed.). (1996). *Lysozymes: Model Enzymes in Biochemistry and Biology*. Birkhauser. Basel.

Jollès P, Muzzarelli RAA. (Eds.). *Chitin and Chitinases*. Birkhauser. Basel. 1999.

Kawada M, Hachiya Y, Arihiro M, Mizoguchi E. Role of mammalian chitinases in inflammatory conditions. *Keio Journal of Medicine*. 2007; 56: 21-27.

Knight PA, Pate J, Smith WD, Miller HRP. An ovine chitinase-like molecule, chitinase-3 like-1 (YKL-40), is upregulated in the abomasum in response to challenge with thegastrointestinal nematode, Teladorsagia circumcincta. *Veterinary Immunology and Immunopathology*. 2007; 120: 55-60.

Kumar RHMN, Muzzarelli RAA, Muzzarelli C, Sashiwa H, Domb AJ. Chitosan chemistry and pharmaceutical perspectives. *Chemical Reviews*. 2004; 104: 6017-6084.

Laine RA, Jaynes JM. Ou CY. Molecular clone: chitinase gene from Vibrio parahemolyticus. *U.S. Patent*. 1994; 5,352,607.

Lavall RL, Assis OBG, Campana SP Jr. Beta-Chitin from the pens of Loligo sp.: Extraction and characterization. *Bioresource Technology*. 2007; 98: 2465-2472.

Lee K, Ha WS, Park WO. Blood compatibility and biodegradability of partially N-acylated chitosan derivatives. *Biomaterials*. 1995; 16: 1211-1216.

Li DC. Review of fungal chitinases. *Mycopathologia*. 2006; 161: 345-360.

Li J, Revol JF, Naranjo E, Marchessault RH. Effect of electrostatic interaction on phase separation behavior of chitin crystallite suspensions. *International Journal of Biological Macromolecules*. 1996; 18: 177-187.

Li J, Revol JF, Naranjo E, Marchessault RH. Effect of the degree of deacetylation of chitin on the properties of chitin crystallites. *Journal of Applied Polymers Science*. 1997; 65, 373-380.

Liu Y, Liu Z, Pan W, Wu Q. Absorption behavior and structure changes of chitin in alkali solution. *Carbohydrate Polymers*. 2008; 72: 235–239.

Maeda R, Matsumoto M, Kondo K. Kinetics of hydrolysis reactions of glycol chitin with egg white lysozyme. *Journal of Chemical Engineering Japan*. 1996; 29: 1044-1047.

Martin R, Hild S, Walther P, Ploss K, Boland W, Tomaschko KH. Granular chitin in the epidermis of nudibranch molluscs. *Biological Bulletin*. 2007; 213: 307-315.

Martin-Gil FJ, Leal JA, Gomez-Miranda B, Martin-Gil J, Prieto A, Ramos-Sanchez MC. Low temperature thermal behavior of chitins and chitin-glucans. *Thermochimica Acta*. 1992; 211: 241–254.

Mattioli-Belmonte M, Nicoli-Aldini N, DeBenedittis A, Sgarbi G, Amati S, Fini M, Biagini G, Muzzarelli RAA. Morphological study of bone regeneration in the presence of 6-oxychitin. *Carbohydrate Polymers*. 1999; 40: 23-27.

Merzendorfer H, Zimoch L. Chitin metabolism in insects: structure, function and regulation of chitin synthases and chitinases. *Journal of Experimental Biology*. 2003; 206: 4393-4412.

Minagawa T, Okamura Y, Shigemasa Y, Minami S, Okamoto Y. Effects of molecular weight and deacetylation degree of chitin/chitosan on wound healing. *Carbohydrate Polymers*. 2007; 67: 640-644.

Minke R, Blackwell J. The structure of alpha chitin. *Journal of Molecular Biology*. 1978; 120: 167-181.

Miserez A, Li YL, Waite JH, Zok F. Jumbo squid beaks: Inspiration for design of robust organic composites. *Acta Biomaterialia*. 2007; 3: 139-149.

Morin A, Dufresne A. Nanocomposites of chitin whiskers from *Riftia* tubes and poly(caprolactone). *Macromolecules*. 2002; 35: 2190-2199.

Morita I, Jinno K. (1988). Production of alkali chitin. *Japan Patent* 63092604.

Muzzarelli RAA. (Ed.). (1993). *Chitin Enzymology,* Grottammare, Italy, Atec.

Muzzarelli RAA. (Ed.). (1996). *Chitin Enzymology,* Grottammare, Italy, Atec.

Muzzarelli RAA. (Ed.). (2001). *Chitin Enzymology 2001,* Grottammare, Italy, Atec.

Muzzarelli C, Muzzarelli RAA. Reactivity of quinones towards chitosans. *Trends in Glycosciences and Glycotechnology*. 2002; 14: 229-235.

Muzzarelli RAA. (1977). *Chitin*, Oxford, Pergamon Press.

Muzzarelli RAA, Littarru GP, Muzzarelli C, Tosi G. Selective reactivity of biochemically relevant quinones towards chitosans. *Carbohydrate Polymers*. 2003; 53: 109-115.

Muzzarelli RAA, Morganti P, Morganti G, Palombo P, Palombo M, Biagini G, Mattioli-Belmonte M, Giantomassi F, Orlandi F, Muzzarelli C. Chitin nanofibrils / chitosan glycolate composites as wound medicaments. *Carbohydrate Polymers*. 2007 70: 274-284.

Muzzarelli RAA. Depolymerization of methyl pyrrolidinone chitosan by lysozyme. *Carbohydrate Polymers*. 1992; 19: 29-34.

Muzzarelli RAA, Ilari P, Xia W, Pinotti M, Tomasetti M. Tyrosinase-mediated quinone tanning of chitinous materials. *Carbohydrate Polymers*. 1994; 24: 294-300.

Muzzarelli RAA, Mattioli-Belmonte M, Miliani M, Muzzarelli C, Gabbanelli F, Biagini G. In vivo and in vitro biodegradation of oxychitin-chitosan and oxypullulan-chitosan complexes. *Carbohydrate Polymers*. 2002; 48: 15-21.

Muzzarelli RAA, Muzzarelli C, Cosani A, Terbojevich M. 6-Oxychitins, novel hyaluronan-like polysaccharides obtained by regioselective oxidation of chitins. *Carbohydrate Polymers*. 1999; 39: 361-367.

Muzzarelli RAA, Stanic V, Ramos V.(1999). Enzymatic depolymerization of chitins and chitosans. In C. Bucke, (Ed.). Methods in Biotechnology: *Carbohydrate Biotechnology Protocols*, Totowa, Humana Press.

Nair KG, Dufresne A. Crab shell chitin whisker reinforced natural rubber nanocomposites. 1. Processing and swelling behavior. *Biomacromolecules*. 2003; 4, 657-665.

Nair KG, Dufresne A. Crab shell chitin whiskers reinforced natural rubber nanocomposites. 2. Mechanical behavior. *Biomacromolecules*. 2003; 4, 666-674.

Nair KG, Dufresne A. Crab shell chitin whiskers reinforced natural rubber nanocomposites. 3. Effect of chemical modification of chitin whiskers. *Biomacromolecules*. 2003; 4, 1835-1842.

Neville AC (1975). *Biology of the arthropod cuticle*. Berlin: Springer.

Neville AC (1993). *Biology of fibrous composites. Development beyond the cell membrane*. New York. Cambridge University Press.

Noishiki Y, Nishiyama Y, Wada M, Okada S, Kuga S. Inclusion complex of beta-chitin and aliphatic amines. *Biomacromolecules*. 2003; 4: 944-949.

Noishiki Y, Takami H, Nishiyama Y, Wada M, Okada S, Kuga S. Alkali-induced conversion of beta-chitin to alpha-chitin. *Biomacromolecules*. 2003; 4: 896–899.

Nordtveit RJ, Varum KM, Smidsrod O. (1994). Degradation of partially N-acetylated chitosans with hen egg white and human lysozyme. *Carbohydrate Polymers*. 29: 163-167.

Patil RS, Ghormade V, Deshpande MV. Chitinolytic enzymes: an exploration. *Enzyme and Microbial Technology*. 2000; 26: 473-483.

Pichyangkura R, Kudan S, Kuttiyawong K, Sukwattanasinitt M, Aiba S. (2002). Quantitative production of 2-acetamido-2-deoxy-D-glucose from crystalline chitin by bacterial chitinase. *Carbohydrate Research*. 2002; 337: 557-559.

Qin Y, Hu H, Luo A, Wang Y, Huang X, Song P. Effect of carboxymethylation on the absorption and chelating properties of chitosan fibers. *Journal of Applied Polymer Science*. 2006; 99: 3110–3115.

Raabe D, Al-Sawalmih A, Yi SB, Fabritius H. Preferred crystallographic texture of alpha-chitin as a microscopic and macroscopic design principle of the exoskeleton of the lobster Homarus americanus. *Acta Biomaterialia*. 2007; 3: 882-895.

Raabe D, Romano P, Sachs C, Al-Sawalmih A, Brokmeier HG, Yi SB, Servos G, Hartwig HG. Discovery of a honeycomb structure in the twisted plywood patterns of fibrous biological nanocomposite tissue. *Journal of Crystal Growth*. 2005; 283: 1-7.

Raabe D, Romano P, Sachs C, Fabritius H, Al-Sawalmih A, Yi SB, Servos G, Hartwig HG. Microstructure and crystallographic texture of the chitin-protein network in the biological composite material of the exoskeleton of the lobster Homarus americanus. *Materials Science and Engineering*. 2006; 421: 143-153.

Richards AG. (Ed). (1951). *The integument of arthropods*. University of Minnesota Press. Minneapolis.

Rinaudo M. Chitin and chitosan: properties and applications. *Progress in Polymer Science*. 2006; 31: 603–632.

Robertus JD, Hart J. (1995). Three-dimensional structure of an endochitinase from barley. In *Enzymatic Degradation of Insoluble Carbohydrates*, Saddler, J.N. and Penner, M.H. (Eds.). 59-64. Washington DC. ACS.

Robertus JD, Monzingo AF. (1999). The structure and action of chitinases. In Jollès P and Muzzarelli RAA. (Eds.). *Chitin and Chitinases*. Birkhauser. Basel. 125-136.

Saito Y, Okano T, Chanzy H, Sugiyama J. (1995). Structural study of alpha-chitin from the grasping spines of the arrow worm Sagitta spp. *Journal of Structural Biology*. 1995; 114: 218-228.

Sareena S, Poovannan K, Kumar KK, Raja JAJ, Samiyappan R, Dudhakar D, Balasubramanian P. Biochemical responses in transgenic rice plants expressing a defense gene deployed against the sheath blight pathogen Rhizoctonia solani. *Current Science*. 2006; 91: 1529-1532.

Sasaki C, Takemura H, Kuhara S, Itoh Y, Fukamizo T. (2001). Structure and substrate binding mode of family 19 chitinase from rice, Oryza sativa L. In *Chitin Enzymology 2001*, R. A. A. Muzzarelli, (Ed.). Grottammare, Atec. (329-334).

Sashiwa H, Fujishima S, Yamano N, Kawasaki N, Nakayama A, Muraki E, Sukwattanasinitt M, Pichyangkura R, Aiba S. Enzymatic production of N-acetyl-D-glucosamine from chitin. Degradation study of N-acetylchitooligosaccharide and the effect of mixing of crude enzymes. *Carbohydrate Polymers*. 2003; 51: 391-395.

Sashiwa H, Fujishima S, Yamano N, Kawasaki N, Nakayama A, Muraki E, Hiraga K, Oda K, Aiba S. Production of N-acetyl-D-glucosamine from alpha-chitin by crude enzymes from Aeromonas hydrophila H-2330. *Carbohydrate Research*. 2002; 337: 761-763.

Shibata Y, Foster LA, Metzger WJ, Myrvik QN. Alveolar macrophage priming by intravenous administration of chitin particles, polymers of N-acetyl-D-glucosamine, in mice. *Infection and Immunity*. 1997; 65: 1734-1741.

Singh A, Kirubakaran SI, Sakthivel N. Heterologous expression of new antifungal chitinase from wheat. *Protein Expression and Purification*. 2007; 56: 100-109.

Sini TK, Santhosh S, Mathew PT. Study of the influence of processing parameters on the production of carboxymethylchitin. *Polymer*. 2005; 46: 3128–3131.

Somorin O, Nishi N, Tokura S, Noguchi J. Studies on chitin: a Preparation of benzyl and benzoylchitins. *Polymer Journal*. 1979; 11: 391–396.

Sriupayo J, Supaphol P, Blackwell J, Rujiravanit R. Preparation and characterization of alpha-chitin whisker-reinforced poly(vinyl alcohol) nanocomposite films with or without heat treatment. *Polymer*. 2005; 46: 5637-5644.

Sriupayo J, Supaphol P, Blackwell J, Rujiravanit R. Preparation and characterization of alpha-chitin whisker-reinforced chitosan nanocomposite films with or without heat treatment. *Carbohydrate Polymers*. 2005; 62: 130-136.

Sugita H, Ito Y. Identification of intestinal bacteria from Japanese flounder (Paralichthys olivaceus) and their ability to digest chitin. *Letters in Applied Microbiology*. 2006; 43: 336-342.

Sun LP, Du YM, Yang JH, Shi XW, Li J, Wang XH, Kennedy JF. Conversion of crystal structure of the chitin to facilitate preparation of a 6-carboxychitin with moisture absorption-retention abilities. *Carbohydrate Polymers*. 2006; 66: 168-175.

Tamura H, Nagahama H, Tokura S. Preparation of chitin hydrogel under mild conditions. *Cellulose*. 2006; 13: 357-364.

Thor CJB, Henderson WF. The preparation of alkali chitin. *American Dyestuff Reporter*. 1940; 29: 461–464.

Tobias DJ, Manoharan M, Pritsch C, Dahleen L. S. Co bombardment, integration and expression of rice chitinase and thaumatin-like protein genes in barley (Hordeum vulgare cv. Conlon). *Plant Cell Reports*. 2007; 26: 631-639.

Vetter J. Chitin content of cultivated mushrooms Agaricus bisporus, Pleurotus ostreatus and Lentinula edodes. *Food Chemistry*. 2007; 102: 6-9.

Vezzulli L, Pezzati E, Repetto B, Stauder M, Giusto G, Pruzzo C. A general role for surface membrane proteins in attachment to chitin particles and copepods of environmental and clinical vibrios. *Letters in Applied Microbiology*. 2007; 46: 119-125.

Vincendon M, Desbrieres J. (2002). Chitin extraction from the edible mushroom Agaricus bisporus. In R. A. A. Muzzarelli and C. Muzzarelli (Eds.), *Chitosan in Pharmacy and Chemistry*. (511-516). Grottammare, Italy: Atec.

Wan Y, Creber KAM, Peppley B, Bui VT. Ionic conductivity and tensile properties of hydroxyethyl and hydroxypropyl chitosan membranes. *Journal of Polymer Science: Part B: Polymer Physics*. 2004; 42: 1379–1397.

Wu T, Zivanovic S, Draughon FA, Sams CE. Chitin and chitosan: value-added products from mushroom waste. *Journal of Agricultural and Food Chemistry*. 2004; 52: 7905-7910.

Yi HM, Wu LQ, Bentley WE, Ghodssi R, Rubloff GW, Culver JN, Payne GF. Biofabrication with chitosan. *Biomacromolecules*. 2005; 6: 2881-2894.

Yoshimura T, Uchikoshi I, Yoshiura Y, Fujioka R. Synthesis and characterization of novel biodegradable superabsorbent hydrogels based on chitin and succinic anhydride. *Carbohydrate Polymers*. 2005; 61: 322-326.

Yui T, Taki N, Sugiyama J, Hayashi S. (2007). Exhaustive crystal structure search and crystal modeling of beta-chitin. *International Journal of Biological Macromolecules*. 2007; 40: 336-344.

Zhu BCR, Laine RA. (1997). Enzymatic depolymerization of chitin. In R. A. A. Muzzarelli (Ed.). *Chitin Handbook*. Grottammare, Italy: Atec.

In: Binomium Chitin-Chitinase: Recent Issues ISBN 978-1-60692-339-9
Editor: Salvatore Musumeci and Maurizio G. Paoletti

Chapter II

Roles of Chitinases in Nature

Neetu Dahiya[2]
Department of Biotechnology, Panjab University,
Chandigarh, India, PIN-160014

Abstract

Chitinases hydrolyze the β1-4 linkages of chitin, an unbranched polymer of β1-4 linked N-acetyl-D-glucosamine (GlcNAc). Chitin is the second most abundant polymer in nature and many organisms including prokaryotes, vertebrates, plants, fungi and insects produce chitinases. The roles of chitinases in these organisms are diverse. For example, in bacteria, chitinases play a role in nutrition and parasitism. Occurrence of multiple chitinases helps bacteria to utilize various chitinous substrates. Chitinases play a critical role in viral pathogenicity. It is suggested that viral chitinases along with cathepsin are associated with loss of integrity of host tissues permitting mature polyhedra to escape into the environment and promoting horizontal virus transmission. In fungi, the biosynthesis and hydrolysis of chitin plays an important role in formation of a functional cell wall. Chitinases are thought to have autolytic, nutritional, and morphogenetic roles as they contribute to breakage and reforming of bonds within and between polymers, leading to re-modeling of the cell wall during growth and morphogenesis. In insects, chitin functions as scaffold material so insect growth and morphogenesis are strictly dependent on the coordination of chitin synthesis and its degradation which requires strict control of the participating enzymes during development. In addition chitinases are associated with the need for partial degradation of old cuticle in crustaceans and insects. Similar to insects, crustaceans and fungi, chitinases play important role in the life cycle of several protozoan and metazoan parasites that infect humans. Some pathogens use chitinase to invade or exploit the chitin containing structures of their host to establish successful infection or transmission to another host via insect vectors. In plants, chitinases have been implicated in plant resistance against fungal pathogens. Moreover, by reducing the defense reaction of the plant, chitinases allow symbiotic interaction with nitrogen-fixing bacteria or mycorrhizal fungi. They are also involved in numerous

2 Email: ineetudahiya@yahoo.com.

physiological events. In vertebrates, chitinases are usually part of the digestive tract and recently chitinases have been found to be implicated in various human diseases such as asthma, arthritis, multiple sclerosis, Gaucher disease, Alzheimer's disease, Fabry storage diseases etc. It is also suggested that chitinases expressed in human tissues may confer protection against fungi in a manner analogous to the protection provided by lysozyme against bacteria. The complexity and functional diversity of the chitinases has made them important candidate for study. The present chapter is focused on the diverse roles of chitinases in these organisms.

1. Introduction

Chitin does not accumulate in most ecosystems, despite its abundant production, indicating that it is somehow degraded. Chitinases play an important role in the decomposition of chitin and potentially in the utilization of chitin as a renewable resource. For the complete hydrolysis of chitin to GlcNAc, the concerted action of chitinase (EC 3.2.1.14) and β-N-acetylglucosaminidase (EC 3.2.1.30) is considered to be essential (Muzzarelli, 1999; Sahai and Manocha, 1993). Chitinases are enzymes that randomly cleave glycosidic linkages of GlcNAc to produce soluble oligosaccharides, mainly chitobiose, which are further hydrolyzed to GlcNAc by β-N-acetylglucosaminidases. Based on sequence homologies chitinases fall into two groups: families 18 and 19 of glycosyl hydrolases (Henrissat, 1991). Members of family 18 employ a substrate assisted reaction mechanism (Terwisscha van Scheltinga *et al.* 1995; van Aalten *et al.* 2001), whereas those of family 19 adopt a fold and reaction mechanism similar to that of lysozyme (Monzingo *et al.* 1996), suggesting these families evolved independently to deal with chitin. During the previous decade, chitinases have received increased attention because of their wide range of applications. Multiple chitinases occur in a wide range of organisms including viruses, bacteria, fungi, insects, higher plants, and animals (Dahiya *et al.* 2005; Robertus and Monzingo, 1999; Zhu *et al.* 2001; Young *et al.* 2005; Boot *et al.* 2001). They participate in a variety of biological functions including morphogenesis, defense, nutrient digestion and pathogenesis. In vertebrates, chitinases are usually part of the digestive tract whereas in insects and crustaceans, they are associated with the need for partial degradation of old cuticle. They have been implicated in plant resistance against fungal pathogens because of their inducible nature and antifungal activities in vitro. Chitinases in fungi are thought to have autolytic, nutritional, and morphogenetic roles. In viruses, chitinases are involved in pathogenesis whereas in bacteria, chitinases play a role in nutrition and parasitism. The present chapter will focus on the diverse roles of chitinases in these organisms.

2. Bacterial Chitinases

Bacterial chitinases play a significant role in maintaining the matter cycle through making chitin usable biologically. In bacteria, chitinases play a role in nutrition and parasitism. Bacterial chitinases release N-acetylglucosamine from chitin, which makes a source of carbon but still more of nitrogen. Most of the chitinolytic organisms produce multiple isomeric forms of chitinases, which may result from posttranslational processing of single-gene product or, more often, the products of multiple genes. Existence of multiple chitinolytic enzymes have been reported in several microorganisms such as *S. marcescens* (Suzuki *et al.* 2002), *Aeromonas* sp. No. 10S-24 (Ueda *et al.* 1995), *Pseudomonas aeruginosa* K-187 (Wang and Chang, 1997), *Bacillus circulans* WL-12 (Mitsutomi *et al.* 1998), *Bacillus licheniformis* X-74 (Takayanagi *et al.* 1991), *Vibrio parahaemolyticus* (Kadokura *et al.* 2007), *Streptomyces* sp. J. 13-3 (Okazaki *et al.* 1995), and *Streptomyces griseus* HUT 6037 (Itoh *et al.* 2002). Depending on the availability of chitin source, different sets of enzymes are involved in its degradation. Although crystalline chitin is the most resistant form of chitin, certain bacteria can degrade the crystalline chitin also. However, complete degradation of chitin is supposed to be a result of the synergistic action of multiple chitinases.

A thermophilic bacterium, *B. licheniformis* X-74, possesses four chitinases, I, II, III, and IV. Chitinases II, III, and IV produced $(GlcNAc)_2$ and GlcNAc, whereas chitinase I predominantly produced $(GlcNAc)_2$. Chitinases II, III, and IV also catalyzed a transglycosylation reaction that converted $(GlcNAc)_4$ into $(GlcNAc)_6$ (Takayanagi *et al.* 1991). Suzuki *et al.* (2002) reported the synergistic action of chitinases Chi A, Chi B, and Chi C1 of *S. marcescens* 2170 on chitin degradation. They proposed that despite having similar catalytic domains, Chi A and Chi B were considered to digest chitin chains in the opposite direction. Chi A was proposed to degrade the chitin chain from the reducing end, whereas Chi B, from the nonreducing end. Addition of Chi A after treatment of powdered chitin with Chi B and vice versa generally improved chitin degradation efficiency. Kadokora *et al.* 2007 reported production of chitinase (Pa-Chi) and chitin oligosaccharide deacetylase (Pa-COD) from *Vibrio parahaemolyticus*. These studies confirmed that Pa-Chi hydrolyzes chitin to produce $(GlcNAc)_2$ and Pa-COD hydrolyzes the acetamide group of reducing end GlcNAc residue of $(GlcNAc)_2$. These findings indicate that GlcNAc-GlcN is produced from chitin by the cooperative hydrolytic reactions of both Pa-Chi and Pa-COD.

Two chitinases, Chi A and Chi B were reported from *Clostridium paraputrificum* M-21 when cultivated on ball-milled chitin and ball-milled shrimp shells (Evvyernie *et al.* 2001). A third novel chitinase gene chiC of *Clostridium paraputrificum* M-21 was characterized by Morimoto *et al.* 2007. The chi18C gene encodes 683 amino acids (signal peptide included) with a deduced molecular weight of 74,651. Chi18C is a modular enzyme composed of a family-18 catalytic module of glycoside hydrolases, two reiterated modules of unknown function, and a family-12 carbohydrate-binding module (Morimoto *et al.* 2007). Based on amino acid sequence similarity, chitinolytic enzymes are grouped into families 18, 19, and 20 of glycosyl hydrolases (Henrissat and Bairoch, 1993). Most of the bacterial chitinases belong to Family 18, except few *Streptomyces* chitinases which belong to Family 19 whereas β-N-acetylhexosaminidases from bacteria belong to family 20. The chitinases of the two families, that is, 18 and 19, do not share amino acid sequence similarity. They have completely different 3-D structures and molecular mechanisms and are therefore likely to have evolved from different ancestors (Suzuki *et al.* 1999). Bacterial chitinases are clearly separated into three major subfamilies, A, B, and C, based on the amino acid sequence of individual catalytic domains (Watanabe *et al.* 1993). Subfamily A chitinases have the presence of a third domain corresponding to the insertion of an $\alpha+\beta$ fold region between the seventh and eighth $(\alpha/\beta)_8$ barrel. On the other hand, none of the chitinases in subfamilies A and B have this insertion. Several chitinolytic bacteria that possess chitinases belonging to different subfamilies like *Serratia marcescens* (Suzuki *et al.* 1999), *Bacillus circulans* WL-12 (Tanaka and Watanabe, 1995), and *Streptomyces coelicolor* A3(2)(Saito *et al.* 1999) are known.

3. Viral Chitinases

Chitinases are present in a number of viruses. Chitinase genes have also been identified from baculovirus *Autographa californica* nucleopolyhedrovirus (AcMNPV) (Hawtin *et al.* 1995), *Bombyx mori* (Bm)NPV (Maeda, 1996), *Choristoneura fumiferana* (Cf) defective NPV (Arif and Peng, 1996), *Helicoverpa zea* (Hz)NPV (Wu and Tribe, 1996), *Orgyia pseudotsugata* (Op)NPV (Ahrens *et al.* 1997), *Cydia pomonella* granulovirus (CpGV) (Kang *et al.* 1998), Lymantria dispar (Ld)NPV (Kuzio *et al.* 1999), *Chlorella* viruses (Yamada *et al.* 1999), *Helicoverpa armigera* single-nucleocapsid nucleopolyhedrovirus (HearNPV, also called HaSNPV) (Wang *et al.* 2004) and *Epiphyas postvittana* nucleopolyhedrovirus (EppoNPV) (Young *et al.* 2005). In baculovirus, chitinases are reported to play a critical role in pathogenicity (Thomas *et al.* 2000). Evidence to date indicates that baculoviruses chitinase act together with cathepsin cause terminal host liquefaction, resulting in the complete loss of integrity of host tissues permitting mature polyhedra to escape into the environment and promoting horizontal virus transmission. A functional chitinase gene (chiA) has been identified in the genome of *Autographa californica* nucleopolyhedrovirus (AcMNPV). The chiA gene is expressed in the late stage of virus replication and its product has both endo- and exochitinase activity (Hawtin *et al.* 1995). The enzyme was found associated with viral polyhedra and is presumably released during polyhedral dissolution in the alkaline midgut of insects. This may aid in degrading the chitinous peritrophic membrane lining the insect larval midgut at an early stage of viral infection, allowing the virus more efficient access to the midgut epithelial cells (Hawtin *et al.* 1997). Chitinase is also expressed late in infection, causing dissolution of the host and assisting release of progeny virus into the environment (Hawtin *et al.* 1997). AcMNPV chitinase contains an endoplasmic reticulum (ER)-retention sequence at the C terminus (Thomas *et al.* 1998; Saville *et al.* 2002, 2004), probably involved in retaining the enzyme inside the cell until late in infection.

The chitinase gene of *Epiphyas postvittana* nucleopolyhedrovirus (EppoNPV) (Simpson and Ward, 2005) showed a high level of sequence identity to the chitinase of AcMNPV. It proved to be a canonical baculovirus chitinase with almost all properties being essentially similar to those of the well-studied AcMNPV chitinase. It contained an N-terminal secretion sequence that was cleaved upon translation and a C-terminal ER-retention sequence (RDEL) that was functional when at the C terminus of the protein. ER-retention sequences have also been identified in the chitinases of *Bombyx mori* NPV (Gomi *et al.* 1999) and *Choristoneura fumiferana* MNPV (GenBank accession no. NC_004778 [GenBank]) and are likely to be a common feature of many baculovirus chitinases. AcMNPV chitinase has been reported to have activity at high pH; in contrast, EppoNPV chitinase showed no such high-pH activity (Young *et al.* 2005). The phylogeny of viral chitinase genes has been extensively examined in comparison with chitinases derived from bacteria, fungi, nematode, actinomycetes, viruses, insects and mammals. Chitinase sequences from Granulovirus and Nucleopolyhedrovirus formed a monophyletic group that clustered with sequences from the gamma subdivision of Proteobacteria, and this pattern was supported by a highly significant branch. A chitinase from a member of the viral family Phycodnaviridae (PbCV1) fell outside the cluster of sequences from baculoviruses and the gamma division of Proteobacteria, as did other bacterial sequences and eukaryotic sequences. Thus, the phylogeny supported the hypothesis that the gene encoding chitinase was transferred from the gamma division of Proteobacteria to the common ancestor of Granulovirus and Nucleoplyhedrovirus. Surprisingly, the predicted protein sequence of the AcMNPV chiA shares extensive sequence similarity with chitinases from bacteria and, in particular, the *Serratia marcescens* chitinase A (60.5% identical residues). In the phylogenetic tree, chitinases of baculoviruses clustered with one clade of bacterial chitinases but within a larger clade that included other bacterial sequences. This topology supports the hypothesis that horizontal gene transfer occurred from bacteria to baculoviruses rather than in the opposite direction (Hughes and Friedman, 2003). *Chlorella* virus PBCV-1 encodes two putative chitinase genes, a181/182r and a260r. Phylogenetic analyses indicate that the ancestral condition of the a181/182r gene arose from the most recent common ancestor of a gene found in tobacco (Sun, *et al.* 1999).

4. Mammalian Chitinases

It was believed for a long time that chitinases had no function in humans because it was assumed that humans completely lack endogenous chitin and endogenous substrates for chitinases. Only in recent years has it become evident that chitinases also exist in humans and more has been learned about their role in human diseases. Early reports on chitinolytic activity in vertebrates (Jeuniaux, 1961) were confirmed following investigations on Gaucher disease, the most common lysosomal storage disorder in humans caused by an inherited deficiency in glucocerebrosidase. In the plasma of symptomatic patients with Gaucher disease, activity toward the artificial substrate 4-methylumbelliferyl-chitotriose is elevated several hundredfold (Hollak *et al.* 1994). The responsible enzyme, named chitotriosidase, was shown to be a true chitinase, hydrolyzing natural chitin and showing high sequence homology to chitinases from lower organisms (Hollak *et al.* 1994; Boot *et al.* 1995; Renkema *et al.* 1995). Other members of the mammalian chitinase family have been discovered since, including a second chitinase which, given its acidic pH optimum, was named acidic mammalian chitinase (AMCase) (Boot *et al.* 2001). Since chitin is an important structural component of pathogens like fungi as well as a constituent of the mammalian diet, a dual function for mammalian chitinases in innate immunity and food digestion has been envisioned (Suzuki *et al.* 2002; Boot *et al.* 2005). Several studies have tried to link a common chitotriosidase deficiency (Boot *et al.* 1998) to susceptibility for infection by chitin-containing parasites.

Mining the literature and using NCBI or ENSEMBL BLAST searches led to the identification of 44 members of the chitinase protein family from 11 different mammalian species (Bussink *et al.* 2007). Overexpression of chitinases occurs in a number of human pathologies. Chitotriosidase is the dominant chitinase in the human body that is highly expressed in specific cell types including tissue macrophages. In various disorders in which activated macrophages are implicated, elevated plasma chitotriosidase levels occur, e.g., lysosomal lipid storage disorders, sarcoidosis, visceral Leishmaniasis, extended atherosclerosis such as Tangier disease, and thalassemia (Hollak *et al.* 1994; Boot *et al.* 1999; Grosso *et al.* 2004). The high levels of chitinase reported in human granulocytes further strengthen this fact as chitin is one of the main components of fungal cell wall and granulocytes are reported to be involved in destruction of various pathogens.

The physiological function of the mammalian chitinase, AMCase, has recently attracted considerable attention due to a report linking the protein to the pathophysiology of asthma (Zhu *et al.* 2004). Acidic mammalian chitinase (AMCase) has been reported to be induced via a T helper-2 (Th2)–specific, interleukin-13 (IL-13)–mediated pathway in epithelial cells and macrophages in an aeroallergen asthma model and expressed in exaggerated quantities in human asthma. AMCase neutralization ameliorated Th2 inflammation and airway hyperresponsiveness, in part by inhibiting IL-13 pathway activation and chemokine induction. AMCase may thus be an important mediator of IL-13–induced responses in Th2-dominated disorders such as asthma (Zhu *et al.* 2004). Highly homologous plant chitinases are prominent "pathogenesis-related proteins" that are induced following attack by pathogens and take part in the defense against chitin-containing fungi (Schlumbaum *et al.* 1986; Sahai and Manocha, 1993). A similar role for chitotriosidase in the human innate immune system was suggested by van Eijk *et al.* 2005. They reported fungistatic effect of human chitotriosidase. The occurrence of deficiency in chitotriosidase is associated with susceptibility to infection with *Wuchereria bancrofti*, a filarial parasite whose microfilarial sheath contains chitin (Choi *et al.* 2001).

In addition to functional chitinases, mammals also have chi-lectins (chi-lectins are chitinases lacking enzymatic activity due to amino acid substitutions in their active site). Like the active chitinases, chi-lectins belong to family 18 of glycosyl hydrolases, consisting of a 39-kDa catalytic domain having a TIM-barrel structure, one of the most versatile folds in nature (Sun *et al.* 2001; Weirenga 2001; Fusetti *et al.* 2002; Houston *et al.* 2003). In contrast exochitinases and endochitinases, chi-lectins lack the conserved additional chitin-binding domain (Boot *et al.* 1995, 2001; Renkema *et al.* 1997). Despite the detailed knowledge regarding structure, insight into the exact physiological function of the various chi-lectins is limited. Similar to chitotriosidase and AMCase, chi-lectins are secreted locally or into the circulation and a role in inflammatory conditions is suggested. For example, human cartilage GP39 (Hcgp39/YKL-40/CHI3L1), a protein expressed by chondrocytes and phagocytes, has been implicated in arthritis, tissue remodeling, fibrosis, and cancer (Johansen 2006). Similarly, the human chi-lectin YKL-39 (CHI3L2) and the murine Ym1 (Chi3L3/ECF-L) have been associated with the pathogenesis of arthritis (Hu *et al.* 1996; Tsurugha *et al.* 2002) and allergic airway inflammation, respectively (Chang *et al.* 2001; Ward *et al.* 2001; Homer *et al.* 2006). Another chi-lectin, CHI3L1, capable of binding to chitin and chito-oligosaccharides, acts as a pathogenic mediator in acute colitis by enhancing the adhesion and invasion of intracellular bacteria to colonic epithelial cells (CECs). Other mammalian chitinases, including AMCase also possess the ability to exacerbate local inflammation by facilitating the production of chemical mediators (Kawada *et al.* 2007).

5. Fungal Chitinases

The fungal kingdom is very diverse, with species growing as unicellular yeasts to branching hyphae. Developing an outer protective layer, namely the cell wall, is critical for growth and survival of the fungal cell in the diverse environments where fungi live. The shape and integrity of the fungus is dependent upon the mechanical strength of the cell wall, which performs a wide range of essential roles during the interaction of the fungus with its environment. Furthermore, the wall is a highly dynamic structure subject to constant change, for example, during cell expansion and division in yeasts, and during spore germination, hyphal branching and septum formation in filamentous fungi. Cell wall polymer branching, cross-linking, and the maintenance of wall plasticity during morphogenesis, may depend upon the activities of a range of hydrolytic enzymes found intimately associated with the fungal cell wall (Dahiya, 2006). Chitinases are the most important enzymes involved in maintaining cell wall structure in fungi. Since chitin is hard to break due to its physicochemical properties, its degradation usually requires the action of more than one enzyme type. A number of proteins demonstrating exhibiting chitinolytic activity were identified in fungi (Mercedes *et al.* 2001; ElKatatny *et al.* 2001; Gan *et al.* 2007). These proteins present large and diverse groups of enzymes. They differ not only in spatial and temporal localization but also in their molecular structure and substrate specificity. Most of the fungal chitinases characterized to date have exochitinase, endochitinase or N-acetyl hexosaminidase activity and a number of these enzymes also exhibit transglycosylase activity. They may therefore contribute to breakage and reforming of bonds within and between polymers, leading to re-modeling of the cell wall during growth and morphogenesis.

Fungal chitinases belong to family 18 of the glycosyl hydrolase superfamily which includes chitinases from bacteria, fungi, plants, insects, mammals, and viruses. All family 18 proteins have an $(\alpha/\beta)_8$-barrel fold, where the substrate binding cleft is formed by loops positioned between the carboxyl-terminal end of the ß-strands and the amino-terminal end of the helices (Henrissat and Davies, 1997). Phylogenetic analysis of *H. jecorina* chitinases, and those from other filamentous fungi, including hypothetical proteins of annotated fungal genome databases, showed that the fungal chitinases can be divided into three groups: groups A and B (corresponding to class V and III chitinases, respectively) also contained the *Trichoderma* chitinases identified to date, whereas a novel group C comprises high molecular weight chitinases that have a domain structure similar to *Kluyveromyces lactis* killer toxins. Five chitinase genes, representing members of groups A-C, were cloned from the mycoparasitic species *H. atroviridis* (anamorph: *T. atroviride*) (Seidl *et al.* 2005). A number of chitinases have been reported from fungi. Sakurda *et al.* (1996) purified a 42kDa chitinase from *Piromyces communis* OTS1. The pH and temperature optima of enzyme were 4.0-4.5 and 40°C, respectively. It was inhibited by Ag^+, Hg^+ and allosamidin at 1mM concentration. Pinto *et al.* (1997) purified chitinase from *Metarhizium anisopliae*. The purified chitinase had a molecular weight of 30kDa and was optically active at pH 4.5-5.0 and temperature 40-45°C, respectively. A 43kDa chitinase was purified from *Trichoderma harzianum* Rifai T24. Chitinase was stable at 30°C. Its half life at 60°C was 15min (El-Katatny *et al.* 2001). Two chitinases P-1 and P-2 were purified from *Isaria japonica*. The molecular weights of enzymes P1 and P2 were 43.273kDa and 31.134kDa, respectively. The optimum pH and temperature were 3.5-4.0 and 50°C for P-1 and 4.0-4.5 and 40°C for P-2. The products from chitin hexamer obtained with P-1 were almost all dimers with only small amount of trimer whereas those with P-2 were mainly trimers with some dimer and tetramer (Kawachi *et al.* 2001). Souza *et al.* (2003) purified 43kDa endochitinase from *Colletotrichum gloeosporioides*. The pH and temperature optima were 7.0 and 50°C, respectively. Two isozymes I and II (molecular weight 67kDa) of N-acetyl β-D-glucosaminidases were purified from *Fusarium oxysporum* F3. The optimum pH of isozymes I and II were 5.0 and 6.0, respectively whereas maximum activity of both the isozymes was obtained at 40°C (Gkargkas *et al.* 2003). Nguyen *et al.* (2008) reported antifungal activity of chitinases from *Trichoderma aureoviride* DY-59 and *Rhizopus microsporus* VS-9 which inhibited microconidial germination of *Fusarium solani* effectively.

The *Saccharomyces cerevisiae* chitinase described by Correa *et al.* 1982 has been cloned and sequenced. Analysis of the derived amino acid sequence suggests that the protein contains four domains: a signal sequence, a catalytic domain, a serine/threonine-rich region, and a carboxyl-terminal domain with high binding affinity for chitin. Most of the enzyme produced by cells is secreted into the growth medium and is extensively glycosylated with a series of short O-linked mannose oligosaccharides ranging in size from Man2 to Man5. Chitinase O-mannosylation was further examined in the temperature-sensitive secretion mutants sec18, sec7, and sec6. Oligosaccharides isolated from chitinase accumulating in cells at the nonpermissive temperature revealed Man1 and Man2 associated with the sec18 mutant. sec6 and sec7 accumulated Man2-Man5 with a higher proportion of Man5 relative to the secreted protein. A significant amount of chitinase is also found associated with the cell wall through binding of COOH-terminal domain to chitin. Disruption of the gene for the enzyme leads to a defect in cell separation but does not substantially alter the level of cellular chitin (Kuranda and Robbins 1991). Indeed, chitinases are associated with the biology of insect mycopathogens. Fungal chitinases can disrupt the cuticle barrier, providing access to nutrients (Wattanalai *et al.* 2004). At a late stage of infection, internal fungal cells must emerge from the insect to produce conidiophores. At this stage the insect endocuticle is digested, suggesting that extracellular chitinases play a major role in infection. Chitinases can also inhibit the development of other microbial competitors. Lorito *et al.* 1998 showed that certain fungal endochitinases, such as the one isolated from *Trichoderma harzianum*, could act as potent anti-fungal enzymes.

6. Parasite Chitinases

Chitinases have been characterized from a number of eukaryotic pathogens including malarial parasite *Plasmodium gallinaceum*, *Plasmodium falciparum*, the nematode *Brugia malayi* and *Lesmania donovani*, where protein is believed to be involved in the transmission of those pathogens in the insect vector, presumably by degrading the chitin-bearing peritrophic membrane in the midgut. The role of chitinase in degradation of peritrophic membrane was further supported by the fact that inhibition of chitinase activity in the mosquito midgut with allosamidin, a chitinase inhibitor, blocks parasitic transmission. In *Brugia malayi* chitinase (BmCHT1) is expressed in the microfilarial stage, the first larval stage, of the organism and is thought to be important in the exsheathment process of the microfilaria. Exsheathment is required for further development of the microfilaria once ingested by the mosquito vector. The microfiliaria of *Brugia malayi* have been shown to have chitin in their sheaths. Antisera to the chitinase temporarily cleared the microfilaria from the bloodstream of infected jirds (Kabir *et al.* 2006).

7. Plant Chitinases

Chitinases are widely distributed in higher plants. Many seed plants synthesize various chitinases (Collinge *et al.* 1993; Graham and Sticklen 1994). Based on their amino acid sequences, plant chitinases are divided into five classes (Beintema, 1994): class I chitinases, consisting of an N-terminal chitin-binding domain and a catalytic domain; class II chitinases, which have only a catalytic domain homologous to that of class I chitinases; class IV chitinases, which share homology with class I chitinases but are smaller as a result of four deletions; and class III and V chitinases, which share no homology with class I, II, or IV chitinases but have distant sequence similarity to bacterial and fungal chitinases. Yamagami *et al.* (1998) have proposed an additional subclass of chitinases, designated IIIb chitinases; these constitute a subclass of class III chitinases, based on some differences in structure and function between typical class III and some bulb chitinases. According to the classification of glycosyl hydrolases by Henrissat and Bairoch (1993), enzymes of classes I, II, and IV are included in family 19, whereas class III, IIIb, and V are included in family 18. Chitinases are known as Pathogenesis Related Proteins (PR proteins) belonging to the PR-3 family (van Loon, 1999). They are strongly induced when plants respond to wounding or infection by fungal, bacterial, or viral pathogens, and there is compelling evidence that chitinases are among the important players in plant defenses against fungal infection. Plant chitinases are induced not only by pathogenesis but also by abiotic stress. Additionally plant chitinases are constitutive, developmentally regulated, and tissue- and organ-specific. One of the physiological roles of these chitinases is to protect plants against fungal pathogens by degrading chitin, a major component of the cell wall of many fungi (Schlumbaum *et al.* 1986). However, some chitinases do not show any antifungal activity (Taira *et al.* 2002), they are implicated in other physiological and developmental processes, including embryogenesis, microsporogenesis, flowering, and abscission. Chitinases have also been found to be associated with plant development. Their expression is regulated by plant hormones that can also influence germination (Rezzonico *et al.* 1998). Chitinases have been found during germination of *Pisum sativum* (Petruzzelli *et al.* 1999), *Hordeum vulgare* (Leah *et al.* 1991), *Zea mays* (Cordero *et al.* 1994), and *Triticum aestivum* (Caruso *et al.* 1999). In carrot, chitinases are involved in the generation of endogenous signals controlling early embryo development (Kragh *et al.* 1996). Also of particular interest is the observation that chitinases can cleave Nod factors (Schultze *et al.* 1994; Staehelin *et al.* 1994). Nod factors are lipo-oligosaccharides containing 3–5 GlcNAc residues which are produced by Rhizobium bacteria during symbiotic interactions and are plant morphogens. It has been suggested that host chitinases may be involved in controlling the biological activity of Nod factors by cleaving and inactivating them. Although immunochemical studies have suggested that GlcNAc-containing glycolipids are present in secondary walls of plants (Benhamou and Asselin, 1989), chitin-like substrates have never been characterized in plants. The inducibility of plant chitinase genes upon pathogen attack is well established (Collinge *et al.* 1993). Data obtained in vitro and in vivo argue for a direct role of chitinases in plant defense. Indeed various chitinases have been shown to exhibit antifungal activity in vitro, especially in combination with β-1,3 glucanase (Melchers *et al.* 1994; Ponstein *et al.* 1994).

Some plant chitinases have also been shown to be able to hydrolyse the peptidoglycan of bacterial cell wall (Bernasconi *et al.* 1987). In vivo antifungal activity of chitinases has been recently tested by constitutively expressing chitinase genes in transgenic plants. Some of these transgenics plants exhibited a higher degree of resistance to fungal pathogens when compared to controls (Broglie *et al.* 1991; Jach *et al.* 1995; Zhu *et al.* 1994) whereas, in other examples, chitinase overexpression was not correlated with an increased level of resistance (Neuhaus *et al.* 1991). In some cases, the co-expression of a chitinase gene with a glucanase gene has been demonstrated to enhance crop protection (Zhu *et al.* 1994), possibly through a synergistic effect of hydrolases on the pathogen cell wall. Recently a new type of plant chitinase, PrChi-A, have been reported from a fern. The new type of chitinase consists of two N-terminal LysM domains and a C-terminal catalytic domain of family-18 chitinases (Onaga and Taira, 2008).

8. Insect Chitinases

In insects, chitinases are mainly engaged together with β-N-acetylglucosamine in the molting process during ecdysis to degrade chitin in the cuticle, in the fore and hindgut, and peritrophic membrane of the midgut to achieve growth and development, which is hormonally regulated. The chitinases responsible for the digestion of cuticular and gut specific chitin have been biochemically and molecularly characterized from various insect species such as the tobacco hornworm, *Manduca sexta*, the stable fly, *Stomoxys calcitrans*, the silk worm, *Bombyx mori*, the common cutworm, *Spodoptera litura*, and the pupae of *Pieris rapae*, *Tenebrio molitor* and the spruce worm, *Choristoneura fumiferana* and *Helicoverpa armigera*. In mosquito *Anopheles gambiae*, a gut specific chitinase gene product has been characterized and is thought to be a regulator of plasma membrane structure and function. Similarly in *Aedes aegypti*, chitinolytic enzymes are involved in the digestion and modulation of chitin containing structures in the gut.

Insect chitinases belong to family 18 of the glycohydrolase superfamily and share a high degree of amino acid similarity. A characteristic of the family 18 chitinases is their multi-domain structure, which is consistently found in all primary structures deduced from insect genes encoding these enzymes. Substantial biochemical and kinetic data are available, and primary structures of different enzymes have been determined by nucleotide sequencing. Insect chitinases have theoretical molecular masses ranging between 40 kDa and 85 kDa and also vary with respect to their pH optima (pH 4-8) and isoelectric points (pH 5-7). The basic structure consists of three domains that include: (i) the catalytic region, (ii) a PEST-like region, enriched in the amino acids proline, glutamate, serine and threonine, and (iii) a cysteine-rich region (Kramer and Muthukrishnan, 1997). The last two domains, however, do not seem to be necessary for chitinase activity because naturally occurring chitinases that lack these regions are still enzymatically active. In agreement with these observations, C-terminus-truncated versions of the recombinant *Manduca* chitinase still exhibit catalytic activity (Wang *et al.* 1996; Zhu *et al.* 2001).

The silkworm, *Bombyx mori*, has been recently demonstrated to contain a bacterial-type chitinase gene (BmChi-h) in addition to a well-characterized endochitinase gene (BmChitinase). The deduced amino acid sequence of BmChi-h showed extensive structural similarities with chitinases from bacteria such as *Serratia marcescens* chiA and baculoviruses (v-CHIA). Comparison of BmChi-h orthologues revealed that bacterial-type chitinase genes are highly conserved among lepidopteran insects, suggesting that the utilization of a bacterial-type chitinase during the molting process may be a general feature of lepidopteran insects (Daimon *et al.* 2005).

Conclusion

Chitinases are involved in cell separation in unicellular yeast, and development and maintenance of cell wall architecture in fungi and insects. They may function in pathogen recognition leading to the activation of host defenses. In plants, there is no chitin but a number of chitinases have been reported from various plants, most of which are secreted as PR-proteins involved in plant defense system. Similarly, in mammals there is no endogenous chitin, but a number of chitinases or chitin like proteins have been reported recently which are found to be associated with a number of disease conditions. In viruses and insect parasites, chitinases are important players in the entry of parasites into the host and their dissemination. Clearly this is an important area for future research and the potential of chitinase can be utilized for development of antifungal drugs and agricultural fungicides.

References

Ahrens CH, Russell RL, Funk CJ, Evans JT, Harwood SH, Rohrmann GF. The sequence of the *Orgyia pseudotsugata* multinucleocapsid nuclear polyhedrosis virus genome. *Virol.* 1997; 229:381-99.

Arif B, Peng H. CfDEF, defective virus from *Choristoneura fumiferana* nucleopolyhedrovirus. GenBank accession no. U72030; 1996.

Beintema JJ. Structural features of plant chitinase and chitin binding proteins. *FEBS lett.* 1994; 350:159-63.

Benhamou N, Asselin A. Attempted localization of a substrate for chitinases in plant cells reveals abundant *N*-acetyl-D-glucosamine residues in secondary walls. *Biol Cell.* 1989; 67: 341-50.

Boot RG, Blommaart EF, Swart E, Ghauharali-van der Vlugt K, Bijl N. Identification of a novel acidic mammalian chitinase distinct from chitotriosidase. *J Biol Chem.* 2001; 276: 6770–6778.

Boot RG, Renkema GH, Verhoek M, Strijland A, Bliek J, *et al.* The human chitotriosidase gene. Nature of inherited enzyme deficiency. *J Biol Chem.* 1998; 273: 25680–25685.

Boot RG, Bussink AP, Verhoek M, de Boer PA, Moorman, AF. Marked differences in tissue-specific expression of chitinases in mouse and man. *J Histochem Cytochem.* 2005; 53:1283–1292.

Boot RG, Renkema GH, Strijland A, van Zonneveld AJ, Aerts JM. Cloning of a cDNA encoding chitotriosidase, a human chitinase produced by macrophages. *J Biol Chem.* 1995; 270: 26252–26256.

Broglie K, Chet I, Hollyday M, Cressman R, Biddle P, Knowlton S, Mauvais CJ, Broglie R. Transgenic plants with enhanced resistance to the fungal pathogen *Rhizoctonia solani*. *Science.*1991; 254:1194–1197.

Bussink AP, Speijer D, Aerts JMFG, Boot RG. Evolution of Mammalian Chitinase(-Like) Members of Family 18 Glycosyl Hydrolases. *Genetics*. 2007; 177: 959-70.

Chang NC, Hung SI, Hwa KY, Kato I, Chen JE *et al.* A macrophage protein, Ym1, transiently expressed during inflammation is a novel mammalian lectin. *J Biol Chem.* 2001; 276:17497–17506.

Choi EH, Zimmerman PA, Foster CB, Zhu S, Kumaraswami V, Nutman TB, Chanock SJ. Genetic polymorphisms in molecules of innate immunity and susceptibility to infection with Wuchereria bancrofti in South India. *Genes Immun*. 2001; 2:248-53.

Collinge DB, Kragh KM, Mikkelsen JD, Nielsen KK, Rasmussen U, Vad K. Plant chitinases. *Plant J.* 1993; 3:31-40.

Caruso C, Chilosi G, Caporale C, Leonardi L, Bertini L, Magro P, Buonocore V. Induction of pathogenesis related proteins in germinating wheat seeds infected with *Fusarium culmorum*. *Plant Sci*. 1999; 140:87-97.

Correa JU, Elango N, Polacheck I, Cabib E. Endochitinase, a mannan-associated enzyme from *Saccharomyces cerevisiae*. *J Biol Chem.* 1982; 257:1392-1397.

Dahiya N, Tewari R, Tiwari RP, Hoondal GS. Chitinase production in solid-state fermentation by Enterobacter sp NRG4 using statistical experimental design. *Curr Microbiol.* 2005; 51:222-28.

Dahiya N. In Mycotechnology: Current Trends and future Prospects: Fungal chitinases – An overview on regulation and cloning New Delhi India IK International Publishing house Pvt Ltd 2006; pp 106-20.

Dahiya N, Tewari R, Tiwari RP, Hoondal GS. Chitinase production in solid state fermentation by *Enterobacter* spNRG4 using statistical experimental design. *Curr Microbiol.* 2005; 51:1–9.

Daimon T, Katsuma S, Iwanaga M, Kang W, Shimada T. The BmChi-h gene a bacterial-type chitinase gene of *Bombyx mori* encodes a functional exochitinase that plays a role in the chitin degradation during the molting process. *Insect Biochem Mol Biol.* 2005; 35:1112-23.

El-Katatny MH, Gudelj M, Robra KH, Elnaghy MA, Gubitz GM. Characterizatioin of a chitinase and endo β-13 glucanase from *Trichoderma harzianum* Rifai T24 *involved in control of the phytopathogen Sclerotium rolfsii. Appl Microbiol Biotechnol.* 2001; 56:137-43.

Evvyernie D, Morimoto K, Karita S, Kimura T, Sakka K, Ohmiya K. Conversion of chitinous wastes to hydrogen gas by *Clostridium paraputrificum* M-21. *J Biosci Bioeng.* 2001; 91:339-43.

Fusetti F, von Moeller H, Houston D, Rozeboom HJ, Dijkstra BW *et al.* Structure of human chitotriosidase Implications for specific inhibitor design and function of mammalian chitinase-like lectins. *J Biol Chem* 2002; 277:25537–544.

Gkargkas K, Mamma D, Nedey G, Topakas E, Christakopoulos P, Kekos D, Macris BJ. Studies on a N-acetyl-β-D-glucosaminidase produced by *Fusarium oxysporum* F3 grown in solid-state fermentation. *Proc Biochem.* 2003; 39:1599-05.

Gomi S, Majima K, Maeda S. Sequence analysis of the genome of *Bombyx mori* nucleopolyhedrovirus. *J Gen Virol.* 1999; 80:1323-37.

Graham LS, Sticklen MB. Plant chitinases. *Can J Bot.* 1994; 72:1057-83.

Grosso S, Margollicci MA, Bargagli E, Buccoliero QR, Perrone A, Galimberti D, Morgese G et al. Serum levels of chitotriosidase as a marker of disease activity and clinical stage in sarcoidosis. *Scand J Clin Lab Invest* 2004; 64:57–62.

Hawtin RE, Arnold K, Ayres MD, Zanotto PM, Howard SC, Gooday GW, Chappell LH, Kitts PA, King LA, Possee RD. Identification and preliminary characterization of a chitinase gene in the *Autographa californica* nuclear polyhedrosis virus genome. *Virol.* 1995; 212:673-85.

Henrissat B, Bairoch A. New families in the classification of glycosyl hydrolases based on amino acid sequence similarities. *Biochem J.* 1993; 293:781-88.

Henrissat B, Davies G. Structural and sequence-based classification of glycoside hydrolases *Curr Opin Struct Biol* 1997; 7: 637-44.

Henrissat B. A classification of glycosyl hydrolases based on amino acid sequence similarities. *Biochem J.* 1991; 280:309-16.

Hollak CE, van Weely S, van Oers MH, Aerts JM. Marked elevation of plasma chitotriosidase activity: a novel hallmark of Gaucher disease. *J Clin Invest.* 1994; 93:1288-92.

Homer RJ, Zhu Z, Cohn L, Lee CG, White WI *et al.* Differential expression of chitinases identify subsets of murine airway epithelial cells in allergic inflammation. *Am J Physiol Lung Cell Mol Physiol.* 2006; 291:502-11.

Houston DR, Recklies AD Krupa JC, van Aalten DM. Structure and ligand-induced conformational change of the 39-kDa glycoprotein from human articular chondrocytes. *J Biol Chem* 2003; 278:30206–212.

Hu B, Trinh K, Figueira WF, Price PA. Isolation and sequence of a novel human chondrocyte protein related to mammalian members of the chitinase protein family. *J Biol Chem.* 1996; 271:19415-20.

Hughes AL, Friedman R. Genome-Wide Survey for Genes Horizontally Transferred from Cellular Organisms to Baculoviruses. *Mol Biol Evol* 2003; 20:979-87.

Itoh Y, Kawase T, Nikajdou N, Fukada H, Mitsutomi M, Watanabe T, Itoh Y. Functional analysis of the chitin binding domain of a family 19 chitinase from *Streptomyces griseus* HUT6037: substrate-binding affinity and cis-dominant increase of antifungal function. *Biosci Biotechnol Biochem.* 2002; 66:1084-92.

Jeuniaux C. Chitinase: an addition to the list of hydrolases in the digestive tract of vertebrates *Nature*.1961; 192:135-36.

Kabir KE, Hirowatari D, Watanabe K, Koga D. Purification and Characterization of a Novel Isozyme of Chitinase from *Bombyx mori*. *Biosci Biotechnol Biochem*. 2006; 70: 252-62.

Kadokura K, Rokutani A., Yamamoto M, Ikegami T, Sugita H, Itoi S, Hakamata W, Oku T, Nishio T. Purification and characterization of *Vibrio parahaemolyticus* extracellular chitinase and chitin oligosaccharide deacetylase involved in the production of heterodisaccharide from chitin. *Appl Microbiol Biotechnol*. 2007; 75:357-65.

Kang WK, Tristem M, Maeda S, Crook NE, O'Reilly, DR. Identification and characterization of the *Cydia pomonella* granulovirus cathepsin and chitinase genes. *J Gen Virol.* 1998; 79:2283-92.

Kawachi I, Fujieda T, Ujita M, Ishii Y, Yamagishi K, Sato H, Funaguma T, Hara A. Purification and properties of extracellular chitinases from the parasitic fungus *Isaria japonica*. *J Biosci Bioeng*. 2001; 92:544-49.

Kawada M, Hachiya Y, Arihiro A, Mizoguchi E. Role of mammalian chitinases in inflammatory conditions. *Keio J Med*. 2007; 56:21-27.

Kragh KM, De Jong AJ, Hendriks T, Bucherna N, Hojrup P, Mikkelsen JD, De Vries SC. Characterization of chitinases able to rescue somatic embryos of the temperature-sensitive carrot variant ts11. *Plant Mol Biol.* 1996; 31:631-45.

Kramer KJ, Muthukrishnan S. Insect Chitinases:Molecular biology and potential use as biopesticides. *Insect Biochem Mole Biol.* 1997; 27:887-900.

Kuranda MJ, Robbins PW. Chitinase is required for cell separation during growth of *Saccharomyces cerevisiae*. *J Biol Chem* 1991; 266:19758-767.

Lorito M, Woo SL, Garcia I, Colucci G, Harman GE, Pintor Toro JA, Filippone E, Muccifora S, Lawrence CB, Zoina A, Tuzun S, Scala F, Fernandez IG. Genes from mycoparasitic fungi as a source for improving plant resistance to fungal pathogens. *Proc Natl Acad Sci USA* 1998; 95: 7860-65.

Maeda S. *Bombyx mori* nuclear polyhedrosis virus individual isolate T3 DNA complete genome. 1996; GenBank accession no L33180.

Melchers LS, Apotheker M, van der Knaap J, Ponstein AS, Sela-Buurlage MB Bol JF, Cornelissen BJC, van den Elzen PJM, Linthorst HJM. A new class of tobacco chitinases homologous to bacterial exo-chitinases displays antifungal activity. *Plant J.* 1994; 5:469-480

Mercedes DM, Limon MC, Mejias R, Robert LM, Tahia B, Jose APT, Kubicek CP. Regulation of chitinase 33 *chit33* gene expression in *Trichoderma harzianum*. *Curr Gen.* 2001; 38:335-42.

Mitsutomi O, Isono M, Uchiyama A, Nikaidov N, Ikegami T, Watanabe T. Chitosanase activity of the enzyme previously reported as β-13-14-glucanase from *Bacillus circulans* WL 12. *Biosci Biotechnol Biochem.* 1998; 62:2107-14.

Monzingo AF, Marcotte EM, Hart PJ, Robertus JD. Chitinases chitosanases and lysozymes can be divided into prokaryotic and eukaryotic families sharing a conserved core. *Nat Struct Biol.* 1996; 3:133-40.

Morimoto K, Yoshimoto M, Karita S, Kimura T, Ohmiya K, Sakka K. Characterization of the third chitinase Chi18C of *Clostridium paraputrificum* M-21. *Appl Microbiol Biotechnol.* 2007; 73:1106-13.

Muzzarelli RAA. Analytical biochemistry and clinical significance of N-acetyl-beta-D-glucosaminidase and related enzymes In P Jolles and R A A Muzzarelli Eds Chitin and Chitinases: 1999; pp 235-247 Basel: Birkhauser Verlag.

Neuhaus JM. Plant chitinases *In* Datta SK Muthukrishnan S eds Pathogenesis-Related Proteins in Plants CRC Press Boca Raton FL. 1999; pp 77-105.

Nguyen NV, Kim YJ, Oh KT, Jung WJ, Park RD. Antifungal activity of chitinases from *Trichoderma aureoviride* DY-59 and *Rhizopus microsporus* VS-9. *Curr Microbiol.* 2008; 56:28-32.

Okazaki K, Kato F, Watanabe N, Yasuda S, Masui Y, Hayakawa S. Purification and properties of two chitinases from *Streptomyces* sp J-13-3. *Biosci Biotechnol Biochem.* 1995; 59:1586-87.

Onaga S, Taira T. *Pteris ryukyuensis*: roles of LysM domains in chitin binding and antifungal activity. *Glycobiol.* 2008; 18:414-23.

Petruzzelli L, Kunz C, Waldvogel R, Meins F Jr, Leubner-Metzger, G. Distinct ethylene- and tissue-specific regulation of β-1 3-glucanases and chitinases during pea seed germination. *Planta* 1999; 209:195-201.

Pinto S, Barreto CC, Schrank A, Ulhao CJ, Vainstein MH. Purification and characterization of an extracellular chitinase from the entomopathogen. *Metarhizium anisopliae. Can J Microbiol.* 1997; 43:322-27.

Ponstein AS, Bres-Vloemans SA, Sela-Buurlage MB, van den Elzen PJM, Melchers LS, Cornelissen BJC. A novel pathogen- and wound-inducible tobacco *Nicotiana tabacum* protein with antifungal activity. *Plant Physiol.* 1994; 104:109-18.

Renkema GH, Boot RG, Au FL, Donker-Koopman WE, Strijland A. Chitotriosidase a chitinase and the 39-kDa human cartilage glycoprotein a chitin-binding lectin are homologues of family 18 glycosyl hydrolases secreted by human macrophages. *Eur J Biochem* 1998; 251:504-09.

Rezzonico E, Flury N, Meins F Jr, Beffa R. Transcriptional down-regulation by abscisic acid of pathogenesis-related β-13-glucanase genes in tobacco cell cultures. *Plant Physiol.* 1998; 117: 585-92.

Robertus JD, Monzingo AF. The structures and actions of chitinases In P Jolles, Muzzarelli RAA. Eds Chitin and Chitinases: 1999; pp 125-136 Basel: Birkhauser Verlag.

Sahai AS, Manocha MS. Chitinases of fungi and plants: their involvement in morphogenesis and host-parasite interaction. *FEMS Microbiol Rev.* 1993; 11:317-38.

Saito A, Fujii T, Yoneyama T, Redenbach M, Ohno T, Watanabe T, Miyashita K. High-multiplicity of chitinase genes in *Streptomyces coelicolor* A32. *Biosci Biotechnol Biochem.* 1999; 63:710-18.

Sakurda M, Morgavi DP, Komatani K, Tomita Y, Onodera R. Purification and characterization of a cytosolic chitinase from *Piromyces communis* OTS1. *FEMS Microbiol Lett.* 1996; 137:75-78.

Saville GP, Patmanidi AL, Possee RD, King LA. Deletion of the *Autographa californica* nucleopolyhedrovirus chitinase KDEL motif and *in vitro* and *in vivo* analysis of the modified virus. *J Gen Virol.* 2004; 85:821-31.

Saville GP, Thomas CJ, Possee RD, King LA. Partial redistribution of the *Autographa californica* nucleopolyhedrovirus chitinase in virus-infected cells accompanies mutation of the carboxy-terminal KDEL ER-retention motif. *J Gen Virol.* 2002; 83:685-94.

Schlumbaum A, Mauch F, Vogli U, Boller T. Plant chitinases are potent inhibitors of fungal growth. *Nature* 1986; 324:365-67.

Schultze M, Staehelin C, Brunner F, Genetet I, Legrand M, Fritig B, Kondorosi E, Kondorosi A.. Plant chitinase/lysozyme isoforms show distinct substrate specificity and cleavage site preference towards lipochitooligosaccharide Nod signals. *Plant J.* 1998; 16:571-80.

Seidl V, Huemer B, Seiboth B, Kubicek CP. A complete survey of *Trichoderma* chitinases reveals three distinct subgroups of family 18 chitinases. *FEBS J.* 2005; 272:5923-39.

Souza RF, Gomes RC, Coelho RR, Alviano CS, Soares RM. Purification and characterization of an endochitinase produced by *Colletotrichum gloeosporioides. FEMS Microbiol Lett.* 2003; 222: 45-50.

Staehelin C, Schultze M, Kondorosi E, Mellor RB, Boller T, Kondorosi A. Structural modifications in *Rhizobium meliloti* Nod factors influence their stability against hydrolysis by root chitinases. *Plant J.* 1994; 5:319-30.

Sun L, Adams B, Gurnon JR, Ye Y, Van Etten JL. Characterization of two chitinase genes and one chitosanase gene encoded by Chlorella virus PBCV-1.*Virol.* 1999; 263:376-387

Sun YJ, Chang NC, Hung SI, Chang AC, Chou CC *et al.* The crystal structure of a novel mammalian lectin Ym1 suggests a saccharide binding site. *J Biol Chem* 2001; 276:17507-514.

Suzuki K, Sugawara N, Suzuki M, Uchiyama T, Katouno F, Nikaidou N, Watanabe T. Chitinases A B and C1 of *Serratia marcescens* 2170 produced by recombinant *E coli*: enzymatic properties and synergism on chitin degradation. *Biosci Biotechnol Biochem* 2002; 66:1075-83.

Suzuki K, Taiyoji M, Sugawara N, Nikaidou Henrissat B, Watanabe T. The third chitinase gene *chiC* of *Serratia marcescens* 2170 and the relationship of its product to other bacterial chitinases. *Biochem J.* 1999; 343:587-96.

Taira T, Ohnuma T, Yamagami T, Aso Y, Ishiguro M, Ishihara M. Antifungal Activity of Rye *Secale cereale* Seed Chitinases: the Different Binding Manner of Class I and Class II Chitinases to the Fungal Cell Walls. *Biosci Biotechnol Biochem.* 2002; 66:970-77.

Takayanagi T, Ajisaka K, Takiguchi Y, Shimahara K. Isolation and characterization of thermostable chitinases from *Bacillus licheniformis* X-74. *Biochim Biophys Acta.* 1991; 1078:404-10.

Tanaka H, Watanabe T. Glucanases and chitinases of *Bacillus circulans* WL-12. *J Ind Microbiol* 1995; 14:478-83.

Terwisscha van Scheltinga AC, Armand S, Kalk KH, Isogai A, Henrissat B, Dijkstra BW. Stereochemistry of chitin hydrolysis by a plant chitinase/lysozyme and X-ray structure of a complex with allosamidin: evidence for substrate assisted catalysis. *Biochem.* 1995; 34:15619-623.

Thomas CJ, Brown HL, Hawes CR, Lee BY, Min MK, King LA, Possee RD. Localization of a baculovirus-induced chitinase in the insect cell endoplasmic reticulum. *J Virol.* 1998; 72:10207-212.

Thomas CJ, Gooday GW, King LA, Possee RD. Mutagenesis of the active site coding region of the *Autographa californica* nucleopolyhedrovirus *chiA* gene. *J Gen Virol.* 2000; 81:1403-411.

Tsurugha J, Masuko-Hongo K, Kato T, Sakata M, Nakamura H *et al.* Autoimmunity against YKL-39 a human cartilage derived protein in patients with osteoarthritis. *J Rheumatol.* 2002; 29:1459-466.

Ueda M, Fujiwara A, Kawaguchi T, Arai M. Purification and some properties of six chitinases from *Aeromonas* sp No 10S- 24. *Biosci Biotechnol Biochem* 1995; 59:2162-64.

van Aalten DM, Komander D, Synstad B, Gaseidnes S, Peter MG, Eijsink VG. Structural insights into the catalytic mechanism of a family 18 exo-chitinase. *Proc Natl Acad Sci U S A.* 2001; 98:8979-84.

van Eijk M, van Roomen CPAA, Renkema GH, Bussink AP, Andrews L, Blommaart EFC, Sugar A, Verhoeven AJ, Boot RG, Johannes MFG, Aerts JMFG. Characterization of human phagocyte-derived chitotriosidase a component of innate immunity. *International Immunol.* 2005; 17:1505-512.

van Loon LC. Occurrence and properties of plant pathogenesis-related proteins. *In* SK Datta S Muthukrishnan eds Pathogenesis-Related Proteins in Plants CRC Press Boca Raton FL 1999; pp 1-19.

Wang X, Ding X, Gopalakrishnan B, Morgan TD, Johnson L, White F, Muthukrishnan S, Kramer KJ. Characterization of a 46 kDa insect chitinase from transgenic tobacco. *Insect Biochem Mol Biol* 1996; 26:1055 -64.

Wang H, Wu D, Deng F, Peng H, Chen X, Lauzon H, Arif BM, Jehle JA, Hu Z. Characterization and phylogenetic analysis of the chitinase gene from the *Helicoverpa armigera* single nucleocapsid nucleopolyhedrovirus. *Virus Research.* 2004; 100:179-89.

Wang SL, Chang WT. Purification and characterization of two bifunctional chitinases/lysozymes extracellularly produced by *Pseudomonas aeruginosa* K-187 in shrimp and crab shell powder medium. *Appl Environ Microbiol.* 1997; 63:380-86.

Ward JM, Yoon M, Anver MR, Haines DC, Kudo G *et al.* Hyalinosis and Ym1/Ym2 gene expression in the stomach and respiratory tract of 129S4/SvJae and wild-type and CYP1A2-null B6 129 mice. *Am J Pathol.* 2001; 158:323-32.

Watanabe T, Kobori K, Miyashita K, Fujii T, Sakai H, Uschida M, Tanaka H. Identification of glutamic acid 204 and aspartic acid 200 in chitinase A1 of *Bacillus circulans* WL-12 as essential residues for chitinase activity. *J Biol Chem* 1993; 268:18567-572.

Wattanalai R, Wiwat C, Boucias DG, Tartar A. Chitinase gene of the dimorphic mycopathogen *Nomuraea rileyi. J Invertebr Pathol.* 2004; 85:54-57.

Wierenga RK. The TIM-barrel fold: a versatile framework for efficient enzymes. *FEBS Lett* 2001; 492:193-98.

Wu T, Tribe D. *Helicoverpa zea* nuclear polyhedrosis virus chitinase and putative DNA-directed RNA polymerase component lef8 genes. 1996; GenBank accession no U67265.

Yamada T, Chuchird N, Kawasaki T, Nishida K, Hiramatsu S. *Chlorella* virus as a source of novel enzymes. *J Biosci Bioeng*. 1999; 88:353-61.

Yamagami T, Ishiguro M. Complete amino acid sequence of chitinase-1 and -2 from bulbs of genus Tulipa. *Biosci Biotechnol Biochem*. 1998; 62:1253-57.

Young VL, Simpson RM, Ward VK. Characterization of an exochitinase from *Epiphyas postvittana nucleopolyhedrovirus* family *Baculoviridae*. *J Gen Virol*. 2005; 86:3253-61.

Zhu X, Zhang H, Fukamizo T, Muthukrishnan S, Kramer KJ. Properties of *Manduca sexta* chitinase and its C-terminal deletions. *Insect Biochem Mol Biol*. 2001; 31:1221-30.

Zhu Q, Maher EA, Masoud S, Dixon RA, Lamb CJ. Enhanced protection against fungal attack by constitutive co-expression of chitinase and glucanase genes in transgenic tobacco. *BioTechnol*. 1994; 12: 807-12.

Zhu Z, Zheng T, Homer RJ, Kim YK, Chen NY, Cohn L, Hamid Q, Elias JA. Acidic Mammalian Chitinase in Asthmatic Th2 Inflammation and IL-13 Pathway Activation. *Science*. 2004; 304:1678-82.

In: Binomium Chitin-Chitinase: Recent Issues
Editor: Salvatore Musumeci and Maurizio G. Paoletti
ISBN 978-1-60692-339-9

Chapter III

Chitin, Chitinases and Chitinase-Like Proteins: A Hypothesis on Ancestral Relationships

Andrea Giansanti*, Fabio Mecozzi*and Salvatore Musumeci**
*Department of Physics, "La Sapienza" University of Rome, Italy
**Department of Neurosciences and Mother and Child Sciences, University of Sassari and Institute of Biomolecular Chemistry, National Research Council (CNR), Li Punti (SS), Italy

Abstract

In this contribution we reconsider the phylogeny of mammalian proteins homologous to the glycosyl hydrolase 18 family: chitinases and chitinase-like proteins. This problem has been recently dealt with in two important papers (Bussink *et al.* 2007; Funkhouser and Aronson, 2007). A clear scheme emerges from these analyses, in which chitinase-like proteins are specialized, tissue-specific, mammalian proteins that have lost the chitinolytic function and have acquired a wealth of possible new functions, mainly related to inflammatory processes. We present here preliminary results from different methods of sequence analysis based on: i) multiple alignments; ii) compression algorithms; iii) statistical over(under)-representation of short k-grams. From our preliminary exploration we formulate and discuss the hypothesis that, chitinase-like proteins are the ancestor group, present as pre-chitinase activators in an ancestral unicellular world from which active chitinases originated as a response to the emergence of chitin synthesis. Chitinase-like proteins in mammals could play a role, in inflammation and in cancer development, similar to the ancient role of activator or signalling molecules in unicellular organisms.

1. Introduction

Where do human chitinases and chitinase-like proteins come from? The answer to this phylogenetic problem has been given in Dr. Boot's chapter, in a very consistent way. Our contribution is a speculative one, it does not refer to solid published work, but on preliminary results[3]. We revisit the problem of reconstructing the phylogeny of mammalian chitinases and chitinase-like proteins, elaborating on two themes. The first one, more general, is that of chitin and chitinase co-evolution; the second one, addresses the problem of remote and fine homology recognition amongst the mammalian members of the glycosyl hydrolase (GH18) family of proteins. For the basic information on chitinases and chitinase-like proteins we refer to the above mentioned chapter by Dr. Boot and to dr. Kzhyshkowska's chapter. On the basis of a quick survey of possible phylogenetic relationships among mammalian proteins belonging to the GH18 family we suggest that inactive chitinase-like proteins are precursors of active chitinases, and that the stabilization of the chitinolytic function could have been induced by the emergence of the synthesis of chitin by living organisms, to be located, possibly, after the speciation of actynomycetes. Chitinase-like proteins, such as YKL-40, YKL-39, SI-CLP, murine YM1/2, and oviductins are able to bind chitin but ineffective as chitinolytic enzymes. These proteins are expressed in mammals and associated, as potential markers, with several pathological conditions.

A terminological distinction is worth to be made at this point: that between *remote homology* and *fine homology*. Homology in general means similarity of function, but how two proteins with a similar function can result from evolution? The more straightforward answer could be: they emerge because the genes that encode them are descendant of the same ancestor gene. From an ancestor gene, through orthologous or paralogous adaptations, new versions of the original specimen are produced by speciation and duplication events, along evolution. Mutations occur at the molecular level and selection at the level of phenotypic adaptation. In turn, phenotypic selection manifests itself, at the molecular level, as selective pressure on specific loci in genes. The recognition of homology, as similarity based on ancestry, is generally accomplished through multiple sequence alignment methods, based on the basic Smith-Waterman algorithm and derivatives (Smith and Waterman, 1981) or on the heuristic BLAST and PSI-BLAST methods (Altschul *et al.* 1997). Generally speaking, in the presence of high levels of sequence identity, these multiple alignment methods are able to robustly build phylogenetic relevant blocks or profiles of aligned amino acids, shared among groups of proteins. The search for remote homology refers to cases where, due to an ancestry very remote in time, the present-day level of sequence identity between two cognate sequences is low and it may be difficult to detect the old, weak genetic signal.

The case of fine homology refers to a different kind of evolutionary process in which genes, sharing a clear common ancestry, i.e. an high degree of sequence identity among them, originate proteins with novel functions. The shift of functions should be also recognized as a signal in present-day sequences, but requires different methods, complementary to those able to detect orthology and paralogy on the basis of multiply aligned regions. To recognize that

3 Rigorously speaking there are no such things as preliminary results, but only results. So, the rigorous reader will do better skipping this chapter, at least in the first reading of the book.

two proteins, belonging to the same (in the sense of orthology and paralogy) family, had acquired a shift of function or changed the biological context of their function is a problem of fine homology. From this point of view, the problem of understanding the possibile functional shift behind the expression, in mammals, of chitotriosidases, AMCases and chitinase-like proteins may be thought as a problem of fine homology. We do not discuss here the role of lateral or horizontal gene transfer between species, that could add further intricacies to the problem of phylogeny reconstruction; this discussion, though premature in the present context, could be interestingly developed in the next future.

It is well known that regions in a protein enzyme evolve at different rates. So, within a protein sequence belonging to the GH18 family, there are slow evolving, more conservative regions or motifs and fast evolving regions, prone to explore, over the template of a common fold, a repertoire of functional and dynamical themes. In a word, these fast evolving regions tend to enlarge the so called *designability* of a class of protein molecules, i. e. more functional sequences tend to insist on the same structural scaffold. It can be thought that mutation rates in regions of a gene can be different and can also change along the course of evolution. One could agree on the idea that, in mammals and humans in particular, an elevated, evolutionarily recent, dynamics in the fast evolving regions, combined with key substitutions at the active site, gave origin to the chitinase-like proteins. The emergence of these proteins can be interpreted as a shift of function. Chitotriosidases and AMCases, besides their ancestral ability to bind chitin and chitin-like substrates, may have acquired new functions, still to be clearly understood. Following the view that in the conservative regions is encoded the ancestral functional architecture and in the fast evolving regions the adaptive functional flexibility of a protein family, another hypothesis could be raised. Namely, that of an original non-chitinolytic function of GH18 (and possibly also of GH19) which survives in chitinase-like proteins and which specialized into a chitinase function when chitin synthesis emerged. This is the hypothesis that we want to propose and explore in this contribution. Let us note that this hypothesis, if confirmed, would shine a different light on the binomium chitin-chitinase. Chitin, in this view, is the robust ubiquitary substrate that induced the specialization of GH18 family members into chitinolytic enzymes, out of a general repertoire of functional possibilities, related to ancient defense mechanisms in unicellular organisms.

The evolutionary biology of chitinases should be strictly interwoven with the biology of chitin. It is conceivable that, in the early evolutionary stages, chitinases produced by different organisms, mainly as a basic defense against chitinous invaders or as an inner remodeling factor, should have been essentially adapted to efficiently degrade chitin by acquiring some degree of specialization, even at the molecular mechanistic level as signalled by the presence of the two broad classes of *exo- and endo-chitinases* (Horn *et al.* 2006).

This chapter has the following organization. In section 2 we discuss relevant topics for the phylogeny of GH18 family, and we make some observations on previous investigations. In section 3 we present our recent preliminary reconstructions, suggesting the new hypothesis, that is further discussed in the concluding section.

2. Phylogeny of Mammalian GH18 Proteins Revisited

Since ancestral times chitin is massively ubiquitous in the living as the structural coating of fungi, insects and of parasitic nematodes and crustaceans, but it is not present in mammals[4]. Then the molecular phylogeny of chitinases is of fundamental relevance for the reconstruction of the adaptation of their hydrolase function to different environmental and biological contexts, against the almost invariant (or at least slowly evolving) chitin substrate. Starting from active chitinases and chitinase-binding proteins also known as chi-lectins or chitolectins (Bussink *et al.* 2007; Funkhouser and Aronson, 2007) and chitinase-like proteins (Kzhyshkowska *et al.* 2007) present in the human genome and in other mammalian genomes we shall trace back their evolutionary history, using bioinformatic methods complementary to the well established multiple alignment methods. In particular, we shall use methods based on the compressibility (LZ method) and on k-gram composition (POPPs) of protein sequences; presumably more apt to unveil fine homologies.

In the case of enzymes, the residues essential for the binding of the substrate belong to the multiply aligned conservative regions. It is well known that the reconstruction of the basic phylogenetic relationships (orthology and paralogy) within a given protein family are usually based on multiple aligment methods, which rely and amplify the robust information contained in the slow evolving regions. Nevertheless, to infer functional or evolutionary clusterings that are less forced by the assumption behind the usual methods, one should get information not only from the conservative regions, but also from the highly variable patterns that are not considered by the multiple aligment methods of classic phylogenetic analysis. Let us just point out that, following such a kind of approach, we were able to propose, at the heart of malaria epidemics, the fine similarity between the chitinases of host and parasite (Giansanti *et al.* 2007).

Recent phylogenetic analyses using available sequence information of a large variety of chitinase(-like) proteins from lower eukaryotes to mammals gave new insights regarding the evolutionary history of the GH18 family (Bussink *et al.* 2007; Funkhouser and Aronson, 2007). These studies clarify a sound scheme for the evolution of mammalian GH18 chitinase and chitinase-like mammalian proteins.

All started from an early ancestor by a gene duplication event which originated the actual active chitinases: chitotriosidases and AMCases. On one branch, oviductins shifted away and, more recently, rodents' Ym's. On the other branch, the others chitinase-like proteins flank chitotriosidases. Let us summarize the present view, based on those papers (see also Dr. Boot's chapter).

4 It is worth mentioning here the website of the European Chitin Society (EUCHIS), devoted to basic and applicative research on chitin and derivatives: http://www.euchis.org/.

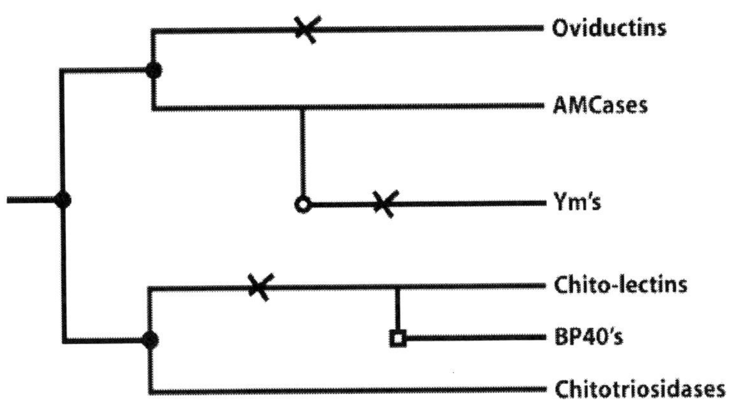

Figure 1. General phylogenetic relationship of GH18 chitinases and chito-lectins from Bussink *et al.* 2007.

Funkhouser and Aronson convincingly argued that the ancient ancestor from which the actual mammalian group of chitinase(-like) proteins originated was already present at the time of the bilaterian expansion (about 550 million years ago). The family expanded in the chitinous protostomes *C. elegans* and *D. melanogaster*, followed by a decline in early deuterostomes as chitin synthesis disappeared, only to expand again in late deuterostomes with a significant increase in gene number occuring after the avian/mammalian split. The early duplication of the active chitinases is older than the speciation leading to *X. tropicalis* which already has two active chitinases, one of which, like AMCase, is expressed in the stomach (Fujimoto *et al.* 2002). In fact, the X*enopus tropicalis* AMCase clusters with the mammalian AMCases.

All mammalian downstream orthologs to the old ancestor are now grouped either in the AMCase or in the chitotriosidase branches, see figure 1.

Chitinase-like proteins are separated in the two clades as AMCase related chito-lectins (Oviductins and Ym proteins, specific to rodents) and chitotriosidase related chito-lectins (Hcgp39 and YKL39 and the close homolog BP40, possibly specific to the artiodactyls)

The sequence of evolutionary events could be the following: The early gene duplication event induced the specialization of the two active chitinases, chitotriosidase and AMCase (acidophylic). Then, recent (after the split of mammals from avian species) duplications of chitotriosidase and AMCase genes gave rise to the respective chito-lectins that lost the original digestive and modelling enzymatic function, possibly pointing towards novel functions, still to be clearly understood.

The evolution of the various mammalian chitinase-like proteins should be a recent event, as indicated by the fact that several family 18 chitinase-like proteins are present only in certain lineages of mammals, but mutations inducing loss of chitinase activity have apparently occurred independently in many other higher genera and species besides cordata, such as plants and invertebrates.

The homologous genes encoding chitinase(-like) proteins are clustered in two distinct loci that display a high degree of *synteny*, i.e. chromosomal co-localization, among mammals. Despite the shared chromosomal location and high homology, individual genes seem to have evolved independently. Moreover, orthologs seem to be more closely related than paralogues, and calculated substitution rate ratios indicate that protein-coding sequences were subjected to purifying selection. Substantial gene specialization has occurred in time, allowing for tissue-specific expression of pH optimized chitinases and chitinase-like proteins.

Aim of this chapter is to make some remarks on the above scheme, based on the observations presented in the next section.

3. Methods and Results

In general, methods for evolutionary sequence comparison fall into two categories: alignment-based and alignment-free. The phylogenetic investigations by Bussink et al. and by Funkhouser and Aronson were based on alignments, i.e. on the CLUSTAL W algorithm (Thompson *et al.* 1994). For the sake of comparison we present here a survey using a method partly based on alignments (MUSCLE) and two alignment-free methods; of the latter, one is based on compression distances, evaluated following the well known Lempel-Ziv (LZ) algorithm (Lempel and Ziv, 1976) and the other one is based on distances inferred from word frequencies. We have considered here the same sequences considered by Boot and co-workers (Bussink *et al.* 2007) and we essentially follow their sequence identification. To have an indication regarding a possible root of the phylogenetic trees we included, as they did, the chitotriosidase sequence from *C.elegans* and chitotriosidase and AMCase from *X.tropicalis*. We have considered the two groups of the active chitinases: chitotriosidases (Chito) and acidic mammalian chitinases (AMCase), and also chitinase-like proteins: Oviductins, BP40, Hcgp39, YKL39 and murine YMs, in the appendix we collect, in a table, the abbreviations we use, and a list of gene identifications.

3.1. Phylogeny of Mammalian GH18 Proteins, Based on MUSCLE

3.1.1. Method

MUSCLE (MUltiple Sequence Comparison for Log-Expectation) (Edgar, 2004) is a computer program for creating multiple alignments of protein sequences. It is based on three steps: i) fast distance estimation using k-mer counting; ii) progressive alignment using a profile function called the log-expectation score; iii) a final refinement, using tree-dependent restricted partitioning. MUSCLE uses two distance measures for a pair of sequences: a k-mer distance, for an unaligned pair, and the Kimura distance, for an aligned pair (Kimura, 1983). A k-mer is a contiguous subsequence composed by k aminoacids. Related sequences tend to share more k-mers than expected by chance. The relative distance is derived from the fraction of k-mers in common in a compressed alphabet. Not requiring an alignment, this step results in a significant speed. Given an aligned pair of sequences, the algorithm computes the pairwise identity and converts it to an additive distance estimate, applying the Kimura correction for multiple substitutions at a single site (Kimura, 1983). Distance matrices are then clustered, using UPGMA. MUSCLE achieves high speed and accuracy, compared with other methods. The MUSCLE program and source code are freely available at: http://www.drive5.com/muscle.

3.1.2. Tree Based on MUSCLE

The alignment file obtained from running MUSCLE over the set of GH18 sequences has been subsequently processed by the PHYLIP package (Version 3.6 from: http://evolution.genetics.washington.edu/phylip.html) (Felsenstein, 1989), we have used *protdist* to generate the distance matrix and then a rooted tree was generated using the neighbor-joining algorithm (Saitou and Nei, 1987; Gascuel and Steel, 2006) and drawn using *drawgram*.

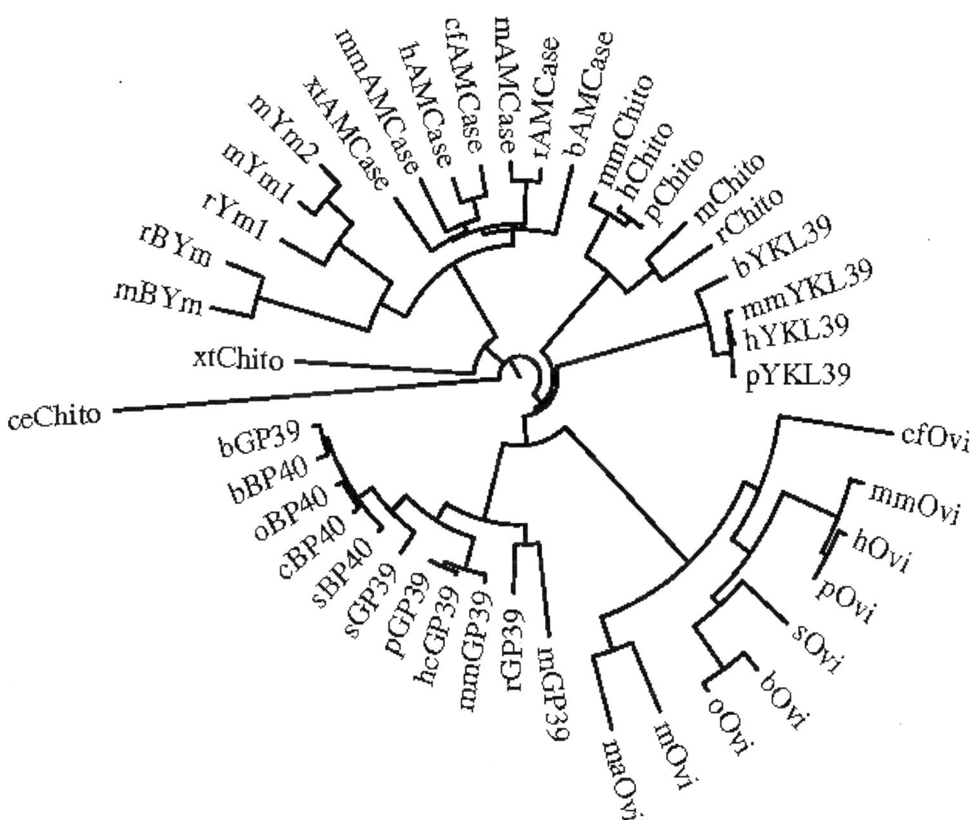

Figure 2. Phylogeney of mammalian chitinase(-like) proteins from MUSCLE.

The resulting tree is shown in figure 2, as a circular projection of the clades. It is interesting to note, in this tree, that *C.elegans* chitotriosidase can be assumed as the tree's root; note then that *X.tropicalis* chitotriosidase outlies from the group of AMCases and Yms, *X.tropicalis* AMCase belongs to this group. This observation confirms that the gene duplication, giving origin to the separation of AMCases from chitotriosidases, can be located between *C.elegans* and *X.tropicalis*, as suggested previously. Note then the cluster of chitotriosidases and YKL39s, well separated from the group of Oviductins and the GP39 and BP40 clusters. At variance with previous recontructions here oviductins seem to be well separated from AMCases. Of course, it could be argued that, since MUSCLE is not a pure sequence-based multialignment method, but partly adopts k-mer statistics, this is not a correct way of confronting previously obtained results. Nevertheless, the cladogram in figure 2 suggests that xtchito, AMCases and murine YMs form a separated cluster.

3.2. Phylogeny of Mammalian GH18 Proteins, Based on Compression Algorithms

3.2.1. Method

The well known compression algorithm by Lempel and Ziv (LZ) (Lempel and Ziv, 1976) is here reconsidered. The combination of hydropathy profiles with LZ has been recently proposed (Na Liu and Tianming Wang, 2006) as a tool for phylogenetic reconstruction, based on proteome analysis. We follow here a variant of their method, avoiding the use of hydrophobicity scales, at all. The application of LZ to a protein sequence in fasta format gives, as output, a list or dictionary of short words which appear two or more times in the sequence. We select in the dictionary only words that are not subwords, contained in longer words, and this filtered dictionary has been called the *exhaustive history* of a given sequence. Now, let c(S) be the number of words in the exhaustive history of sequence S; this number of counts is the *complexity factor* of a sequence, that is the minimum number of steps needed to re-generate S. The distance between two sequences Q and S is computed with the formula, inspired by Otu and Sayhood:

$$d(S,Q) = (c(SQ)-c(S)+c(QS)-c(Q))/(c(SQ)+c(QS)),$$

if sequence S is different from Q. The distance is set to 0 if the two sequences are equal. In this notation the "product" of two sequences QS indicates the sequence obtained by appending S to Q. In general, c(QS) is not equal to c(SQ). In our protocol, the distance matrix computed via the above formula has been elaborated with the *neighbour* program of the PHYLIP package to reconstruct a tree, which uses the neighbor-joining algorithm.

3.2.2. Tree Based on LZ Compression

The tree based on the compressibility of sequences is shown in figure 3. Here the relevant feature that emerges is that of oviductins as the originating cluster from which the other clusters emerge. A big cluster is formed by AMCases and murine YMs on one side and another big cluster is made by chitotriosidases, well separated from the group of GP39/BP40 and from the cluster of YKL39. The resolution of the method is well confirmed by the separation of each cluster and by the fine clustering inside, see for example the separation, among chitotriosidases, of those from humans and primates from the murine ones and from the ancient chitotriosidases, from *C.elegans* and *X.tropicalis*.

3.3. Phylogeny of Mammalian GH18 Proteins, Based on Popps (Protein Oligonucleotide Probability Profiles)

3.3.1. Method

The POPPs (Protein or Oligonucleotide Probability Profile) (Wise, 2002) is a collection of codes to evaluate what is statistically unusual in the composition of a protein, with reference to a large database. Then, a group of proteins can be clustered, following their peptide composition. Statistically based peptide composition provides information about a protein sequence which is complementary to that obtained from classical approaches, based on multiple alignments. We have written scripts to automatically run the first tool of the POPPs suite, called *popp_create.py*, which compares the distributions of peptides of typical length 1aa-3aa, with their distribution across a large database, i. e.

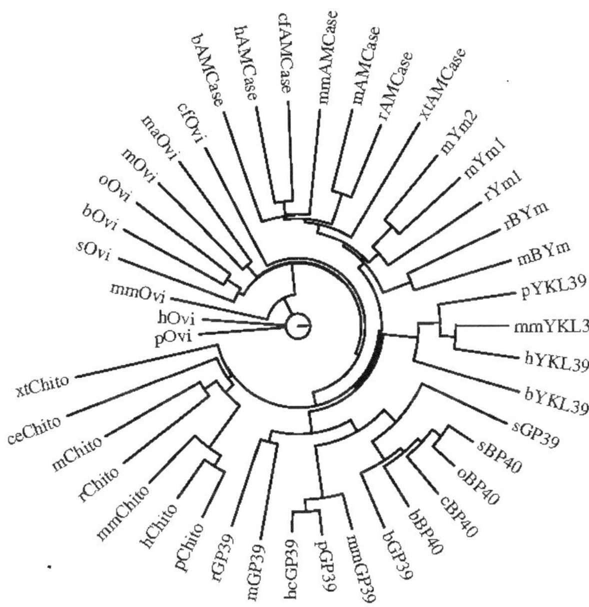

Figure 3. Phylogeny of mammalian chitinase(-like) proteins based on LZ compression algorithm.

UniProtKB/TrEMBL[5]. Non-overlapping repeats of the same peptide have been counted and, by means of a single-sided binomial distribution statistic, *popp_create.py* produced a list of peptides that are either significantly over-represented or under-represented in the test sequence. This list is a Protein or Oligonucleotide Probability Profile, or POPP. Given sequences i, j and their profiles we build a score index S_{ij} which is increased by the length of the peptide if it is present in both lists with the same sign, or decreased if the sign differs. Then, a distance between the two sequences is evaluated by the following symmetric formula:

---

5 http://www.ebi.ac.uk/trembl/

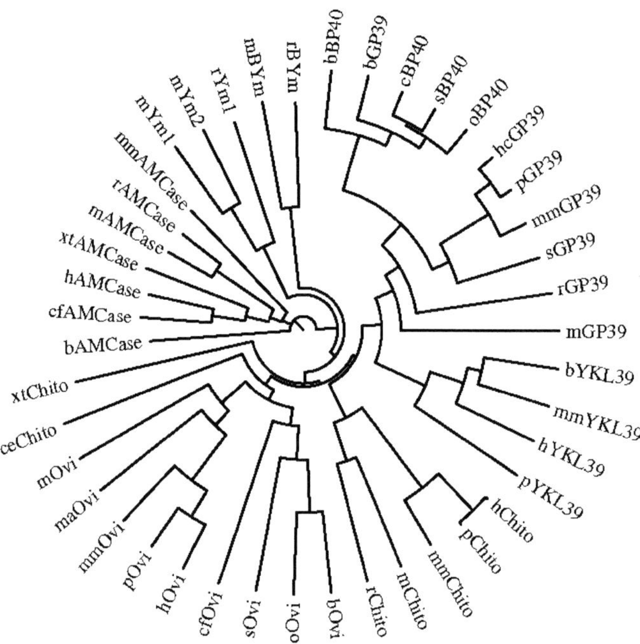

Figure 4. Phylogeny of mammalian chitinase(-like) proteins based on POPPs.

$$D_{ij} = |S_{ij} - S_{ii} + S_{ji} - S_{jj}| / (S_{ii} + S_{jj}).$$

Also in this case, to get a phylogenetic tree, once a distance matrix was obtained by POPPs, we used the PHYLIP package, as previously done.

3.3.2 Tree Based on POPPs

The unrooted tree reconstructed from the POPPs analysis is shown in figure 4. It is interesting to note that, starting from ceChito the closest neighbor is xtChito, closely related to the cluster of AMCases, to which belongs xtAMCase, the acidic active chitinase of *X.tropicalis*. The cluster closest to AMCases, is also in this reconstruction, that of murine YMs, basicophilic chitinase-like protein. It is suggestive that, in all reconstructions we show here, AMCases and YM's are close, pointing at a common mechanism of adaptation of a basic function to environments with an excess or a defect of protons. On the other side, closest to ceChito is the cluster of oviductins. The other clusters are, also in this case, well resolved: the group of mammalian chitotriosidases, with primates separated from murine species, the cluster of YKL39s and the clusters of GP39s and BP40s. It is noticeable that, within each cluster, the fine structure is substantially consistent with the previous LZ reconstruction.

The combined observations presented in this section, suggest that oviductins, GP39s and BP40s could be considered as independent clusters, on the same footing as chitotriosidases and AMCases to which the Yms seem, definitely, to be related. In the next section we discuss the potential of this point, that deserves a lot of further work.

4. Discussion

Let us now elaborate on our theme with the aim of raising questions in the hope that they could suggest useful critical comments and be of inspiration for further research. The main point is that chitinase-like could be phylogenetically prior to chitinases and not posterior. Chitinase-like proteins, in our view, would convey a very ancient phylogenetic signal, related to a group of pre-chitinase proteins, which, looking at the involvement of chitinase-like proteins in inflammation processes, should be related to the basic function of activating a cell from a normal state to an activated state, following environmental changes. So, it is natural to raise the following questions: is there, in present-time unicellular organisms like Bacteria or Archaea, a universal metabolic mechanism that shifts the cell from a normal to an activated (irritated) state? Is this mechanism invariant against extremophilic adaptation? Which is the group, if any, of pre-chitinase proteins that play a key-role in this mechanism?

The observations, made in section 3.3 above, rest on a method for sequence comparison, based on the statistics of presence of short motives. This kind of methods, based on sequence composition, are complementary to multiple aligment methods and of great potential interest in the search for fine or remote homologies, but, in general, there is a need for a clear theoretical assessment of the validity of each method and much work is still to be done.

In particular, there are specific technical problems, when reconstructing phylogenies in the presence of different substitution rates at different loci in a single gene, or in a group of related diverging genes, as mentioned in the introduction above. This theme is of particular relevance and incorporates the problem of finding out the presence of positive or negative selection pressure on specific loci in a gene. The problem has been raised and very clearly discussed in the case of GH18 chitinase (-like) proteins (Bussink *et al.* 2007), where the PAML maximum likelihood method has been used (Yang, 2007); in this approach the ratio ω of synonymous (d_s) to non-synonimous (d_n) substitutions at specific residue sites is evaluated. Just a few words on this point. Protein sequences diverge from a common ancestor sequence because mutations occur in the DNA coding them. Some mutations are fixed by selection and others, possibly, by chance and reveal themselves as substitutions at certain loci in a gene. At the codon level, synonymous vs. non-synonymous substitutions imply conservation or mutation of the corresponding aminoacid, respectively. If $\omega < 1$ then non-synonimous substitutions are deleterious at a given site, which is subjected to a purifying, negative selective pressure; if $\omega > 1$, then the site is subjected to a positive Darwinian selection. This approach has been frequently used to complement the static classification or clustering of sequences, synthesized by a phylogenetic tree, with local information on the evolutionary forces that drive the molecular adaptation or specialization of a protein sequence. Interestingly, Bussink et al. conclude, from a PAML analysis, that in mammalian GH18 chitinase(-like) proteins there is no evidence for positive selection, with the exception of some residues in oviductins. This result seems to us somehow at variance with the interpretation of the chitinase-like proteins as specialized tissue-specific chitinases that have lost their function, because of critical mutations of catalytically relevant residues. We would expect to see that residues essential for binding and catalysis are subjected to peculiar evolutionary forces. This observation would suggest a re-estimate of the different selective forces acting on different parts of GH18 chitinase(-like) proteins. By the way, it would be also interesting, and to our knowledge still not done, a complete phylogenetic reconstruction of GH19 chitinases and corresponding chitinase-like proteins, such as the GhCTL group, recently discovered in cotton (Zhang *et al.* 2004). The parallel analysis of the GH19 chitinase(-like) proteins would be relevant having in mind a broad project, to be done in the next future, devoted to reconstruct the evolutionary history of chitin synthesis (i.e., to be concrete, of chitin synthases) in living organisms. When such a phylogenetic reconstruction will be ready it would be tempting to see if there is backward convergence of chitin synthesis and of the specialization of chitinases. For sure, the gene duplication giving origin to the separation of chitotriosidases and AMCases, can be located in between the speciation of *C.elegans* and the speciation of *X.tropicalis,* as elegantly pointed out by Boot and co-workers. We expect, in a rough view, that chitin synthesis and chitinases, could have emerged after the speciation of actynomycetes, prokaryotic living forms intermediate between bacteria and fungi; it is known that they synthesize peptydoglycans but not chitin yet.

To follow our line of reasoning it is necessary to choose phylogenetic methods of great resolution. The resolving power of standard phylogeny, even based on maximul likelihood criteria, may be flawed by overlooking the problem alluded to in the introduction. Namely, that of *eterotachy*, i. e. of the different evolutionary rates and mutational propensity of different regions of genes. To cope with this problem it is essential to adopt and compare methods based both on the *gamma distribution* (Swofford *et al.* 1996) and on the *covarion* model (Wang al 2007; Zhou *et al.* 2007; Wang *et al.* 2008). We believe that these approaches will be useful in supporting or discarding the hypothesis we are presenting here, viewing the tissue specific expression of chitinase-like proteins in many inflammatory processes in mammals as the reviving of an ancient, then flexible and generic, response mechanism. Let us mention that the expression of high levels of chitotriosidase and AMCase in various inflammatory processes in mammals could be interpreted as a response against a ghost- or pre-parasite (see also Elias *et al.* 2005). This response could be the remnant of a mostly ancient mechanism of innate immunity against still undifferentiated perturbations that unicellular organisms should response to. Of course, the question still to be answered is why pre-chitinases, why this ability to bind saccharides should be relevant for defensive responses? Of particular interest is the search for signals of chinase(-like) proteins in cancer cell lines and in cancer genomes, one could think of exploring this theme within the Cancer Genome Atlas[6]. In that direction, the basic questions are: why should cancer cell express chitinases? Do cancer cells synthesize chitin? Consider also, in particular, the growing interest of YKL40 as an almost-ubiquitous biomarker and potential target in many cancer diseases (Johansen *et al.* 2006).

Connecting the diverse strands of our analyses, let us conclude. On the basis of a preliminary exploration of alternative methods for evolutionary sequence similarity, based on compression and composition properties, we suggest here a different interpretation of the chitinase-like proteins in mammals as the survivors of an ancestral uni-cellular non chitinolytic function, possibly related to inter- and intra-cellular signalling. In complex multi-cellular organisms the original signaling function, revived by chitinase-like proteins, plays a

6 http://cancergenome.nih.gov/

role in inflammatory processes that do not require the presence of a chitinous invader and that, interestingly, can be blocked by administrating chitin or of chitin-like substrates that can be recognized both by active chitinases and inactive chitinase-like proteins. The parallel study of the co-evolution of chitin synthesis enzyme networks, and of the genetics of chitinases, both in chitin-producing and chitin non-producing organisms, appears of fundamental importance for the general biology of adaptation, of immunity and, when humans are involved, also of clinical relevance. Moreover, we believe that the evolutionary genetics of chitinases and chitinase-like proteins is of relevance for a basic approach to inflammatory diseases treatment and in the prognostics of cancer diseases.

Appendix

Sequence Identification and abbreviation used

Species	Chito	AMCase	Oviductins	BP40	Hcgp39	YKL39	Ym
Homo sapiens	hChito	hAMCase	hOvi		hGP39	hYKL39	
Pan troglodytes	pChito		pOvi		pGP39	pYKL39	
Macaca mulatta	mmChito	mmAMCase	mmOvi		mmGP39	mmYKL39	
Mus musculus	mChito	mAMCase	mOvi		mGP39		mYm1-2, mBYm
Rattus norvegicus	rChito	rAMCase			rGP39		rYm1, rBYm
Bos taurus		bAMCase	bOvi	bBP40	bGP39	bYKL39	
Capra hircus				cBP40			
Ovis aries			oOvi	oBP40			
Sus scrofa			sOvi	sBP40	sGP39		
Canis familiaris		cfAMCase	cfOvi				
Mesocricetus auratus			maOvi				
Xenopus tropicalis	xtChito	xtAMCase					
Caenorhabditis.elegans	ceChito						

List of chitinase (-like) proteins investigated and gene identifications.

Active chitinases

Chitotriosidases: hChito (NP_003456), pChito (XP_514112), mmChito (XP_001103012), mChito (AAS47832), rChito (XP_001061784), xtChito (NP_001005792), ceChito (NP_508588);

AMCases: hAMCase (AAG60019), mmAMCase (XP_001104487), mAMCase (AAH34548), rAMCase (AAR28968), bAMCase (NP_777124), cfAMCase (XP_537030), xtAMCase (AAH90382).

Chitinase-like proteins

Oviductins: hOvi (AAO37816), pOvi (XP_001159353), mmOvi (NP_001036252), mOvi (NP_031722), bOvi (XP_611787), oOvi (NP_001009779), sOvi (NP_999235), cfOvi (XP_852238), maOvi (AAC53584);

BP40: bBP40 (AAP41220), cBP40 (AAL87007), oBP40 (AAQ94054), sBP40 (AAV30548);

Hcgp39: hGP39 (NP_001267), pGP39 (XP_514111), mmGP39 (XP_001103739), mGP39 (AAH03780), rGP39 (AAH91365), bGP39 (AAX46682), sGP39 (CAA87764);

YKL39: hYKL39 (NP_00102037), pYKL39 (XP_513645), mmYKL39 (XP_001093397), bYKL39 (XP_591204);

murine Ym1 and Ym2 : mYm1 (AAH61154), mYm2 (NP_660108), rBYm (XP_001069894.1.), mBYm (AAH51070), rYm1 (XP_001069857)

References

Altschul SF, Madden TL, Schaffer AA , Zhang J, Zhang Z, Miller W, Lipman DJ. Gapped BLAST and PSI-BLAST: a new generation of protein database search programs. *Nucleic Acids Res.* 1997; 25:3389-3402.

Bussink AP, Speijer D, Aerts JM, Boot RG. Evolution of mammalian chitinase(-like) members of family 18 glycosyl hydrolases. *Genetics* 2007; 177:959-70.

Edgar RC. MUSCLE: multiplesequence alignment with high accuracy and high throughput. *Nucl Acids Res*. 2004; 27:2682-2690.

Elias JA, Homer RJ, Hamid Q, Lee CG. Chitinases and chitinase-like proteins in T(H)2 inflammation and asthma. *J Allergy Clin Immunol.* 2005; 116: 497-500.

Felsenstein J. PHYLIP phylogeny inference package. *Cladistics* 1989;5:164-166.

Funkhouser JD, Aronson NN Jr. Chitinase family GH18: evolutionary insights from the genomic history of a diverse protein family. *BMC Evol Biol.* 2007; 7: 96.

Fujimoto W, Kimura K, Iwanaga T. Cellular expression of the gut chitinase in the stomach of frogs *Xenopus laevis* and Rana catesbeiana. *Biomed Res.* 2002; 23:91-99.

Gascuel O, Steel M. (2006). Neighbor Joining Revealed. Mol Biol Evol 2006; 23:1997-2000.

Giansanti A, Bocchieri M, Rosato V, Musumeci S. A fine functional homology between chitinases from host and parasite is relevant for malaria transmissibility. *Parasitol Res* 2007; 101: 639-45.

Horn SJ, Sørbotten A, Synstad B, Sikorski P, Sørlie M, Värum KM, Eijsink VG. Endo/exo mechanism and processivity of family 18 chitinases produced by *Serratia marcescens. FEBS J.* 2007; 273:491-503.

Johansen JS, Vittrup Jensen B, Roslind A, Nielsen D, Price PA. Serum YKL-40, A New Prognostic Biomarker in Cancer Patients? *Cancer Epidemiol Biomarkers Prev.* 2006; 15:194-202.

Kimura M. The Neutral Theory of Molecular Evolution. Cambridge University Press;1983.

Kzhyshkowska J, Gratchev A, Goerdt S. Human Chitinases and Chitinase-like Proteins as Indicators for Inflammation and Cancer. *Biomarker Insights* 2007; 2: 128-146.

Lempel A, Ziv J. On the complexity of finite sequences. IEEE Trans Inform Theory 1976; 22:75-81.

Na Liu, Tianming Wang. Protein-based phylogenetic analysis using hydropaty profile of amino acids. *FEBS Lett* 2006; 580:5321-5327.

Otu HH, Sayhood K. A new sequence distance measure for phylogenetic tree reconstruction. *Bioinformatics* 2003; 19:2122-2130.

Saitou N, Nei M. The neighbor-joining method: a new method for reconstruction of phylogenetic trees. *Mol Biol Evol* 1987; 4:406-425.

Smith TF, Waterman MS. Identification of common molecular subsequences *J Mol Biol* 1981; 147:195-197.

Swofford DL, Olsen GJ, Waddell PJ, Hillis DM. Phylogenetic Inference, in: Hillis, DM, Moritz C, Mable BK. Molecular Systematics: *Sinauer*;1986.

Thompson JD, Higgins D, Gibson TJ. CLUSTAL W: improving the sensitivity of progressive multiple sequence alignment through sequence weighting, position-specific gap penalties and weight matrix choice. *Nucleic Acids Res*.1994; 22: 4673-80.

Wang HC, Spencer M, Susko E, Roger AJ. Testing for covarion-like evolution in protein sequences. *Mol Biol Evol.* 2007; 24:294-305.

Wang HC, Susko E, Spencer M, Roger AJ. Topological estimation biases with covarion evolution. *J Mol Evol.* 2008; 66:50-60.

Wise MJ. The POPPs: clustering and searching using peptide probability profiles. *Bioinformatics* 2002; 18 Suppl 1:S38-45.

Yang Z. PAML 4: phylogenetic analysis by maximum likelihood. *Mol Biol Evol.* 2007; 24:1586-91.

Zhang, D., M. Hrmova, et al. (2004). Members of a new group of chitinase-like genes are expressed preferentially in cotton cells with secondary walls. *Plant Mol Biol.* 2004; 54:353-72.

Zhou Y, Rodrigue N, Lartillot N, Philippe H. Evaluation of the models handling heterotachy in phylogenetic inference. *BMC Evol Biol.* 2007; 7:206.

In: Binomium Chitin-Chitinase: Recent Issues ISBN 978-1-60692-339-9
Editor: Salvatore Musumeci and Maurizio G. Paoletti

Chapter IV

Chitinase in Fungal and Bacterial Sepsis

Helen Michelakakis[1] and Ioannis Labadaridis[2]
[1]Dept. of Enzymology and Cellular Function Institute of Child Health, Athens, Greece
[2]NICU, General Hospital of Nikea, Piraeus, Greece

Abstract

Chitinases are ubiquitous chitin fragmenting enzymes identified in several organisms. Two distinct chitinases have recently been identified in humans, chitotriosidase expressed in phagocytes and an acidic mammalian chitinase (AMCase) expressed in the gastrointestinal tract and to a lesser extent in lung.

A role for human chitotriosidase in innate immunity is suggested by several findings. *In vitro* and *in vivo* evidence link chitotriosidase overexpression by macrophages and its release from polymorphonuclear neutrophils (PMNs), via exocytosis of specific granules, to the immune response elicited in microbial infections. Initial, *in vitro* studies showing its chitinolytic activity towards the cell wall chitin of *Candida albicans* have been strengthend by later findings showing that it causes growth inhibition, hyphal tip bursting and prevention of hyphal switch in chitin containing fungi. Furthermore, administration of human recombinant chitotriosidase improved the survival of neutropenic mouse models of systemic candidiasis and aspergillosis.

Increased chitotriosidase plasma and tissue activity has been found in guinea pigs infected by *Aspergillus fumigatus*. Recently, increases in chitotriosidase activity, that run in parallel to their clinical outcome, were observed in neonates not only with systemic candidiasis and aspergillosis but also with bacterial infections. It is of interest that the highest chitotriosidase levels were observed in the neonates that succumbed to their fungal infection. Approximately 6% of the general population in Caucasians cannot synthesize an enzymatically active chitotriosidase and genetic variants in chitotriosidase were shown to be associated with gram-negative bacteremia in leukemic patients. On the other hand, no conclusive corresponding evidence exist regarding succeptibility and survival in fungal infections.

The role of AMCase in innate immunity is not well studied. Given its chitinolytic activity towards fungal cell wall chitin, it has been suggested that it might partly compensate for the above mentioned deficiency of chitotriosidase, however relevant data are missing, and the ones available link AMCase to allergic reactions rather than defence mechanisms.

Clearly more studies are required in order to fully understand the role of chitinases in fungal and bacterial infections and to assess their possible value as therapeutic agents in these infections.

1. Introduction

Chitin, a non-linear polymer of β-1,4 linked N-acetyl β-D-glycosamine is, next to cellulose, the most abundant biopolymer in nature. It is a structural component of arthropods, including crustaceans and insects, as well as mollusks, nematodes and worms. It is also found in fungal cell walls and is particularly abundant in filamentous fungi (Tharanatham and Kittur 2003).

Chitinases are ubiquitous chitin- fragmenting enzymes found in a variety of organisms which may or may not contain chitin, such as plants, fungi, bacteria and insect parasites (Flach *et al.* 1992; Sahai and Manocha 1993; Brurberg *et al.* 1996).

Several biological functions have been attributed to the chitinases of the various species. Chitinases are crucial in chitin recycling in nature and its utilization as a nutrient by various organisms (Gooday 1995, 1996; Li and Roseman 2004).

In chitin containing organisms, chitinases are key players in morphogenic processes involving their chitinous coating. In chitin containing fungi several reports support the involvement of chitinases in morphogenesis and their importance for normal growth, formation of hyphae and spore germination. Fungal cell wall morphogenesis is the result of a delicate balance between chitin synthesis and degradation, two processes which however appear to be independently regulated, at least in *Candida albicans* and *Saccharomyces cerevisiae* (Barrett-Bee and Hamilton 1984; Pedraza-Reyes and Lopez-Romero 1989; Sahai and Manocha 1993; Selvaggine *et al.* 2004).

Chitinases produced by invading microorganisms are essential in host-parasite interaction. They have been implicated, in penetration of bacterial and fungal hosts as well as in the transmission of *Plasmodium*, the malaria parasite, *Leishmania* and other parasites (Flach *et al.* 1992; Schlein *et al.* 1992; Sahai and Manocha 1993; Shahabuddin and Kaslow 1994;).

Plant chitinases are an integral component of a general disease-resistance mechanism that encompasses a number of plant pathogenesis – related proteins that are induced by pathogens and a variety of physical, chemical and environmental stresses (Kasprzewska 2003; Taira *et al.* 2005). Purified plant chitinases and β-1, 3-glucanases, another member of the pathogenesis related proteins, coinduced with chitinases, attack and digest isolated fungal cell walls that often contain both chitin and β-1, 3-glucan.

Chitin oligosaccharides are a representative elicitor, inducing defence responses in a variety of plant cells. Chitin elicitor binding protein (CEBiP) which play a key role for the perception and transduction of the chitin elicitor signal has been identified in the plasma membranes of various plant cells (Day *et al.* 2001; Okada *et al.* 2002, Kaku *et al.* 2006;). Knockdown of CEBiP gene cancelled the up-regulation by chitin elicitor of several genes including the gene encoding chitinase (Kaku *et al.* 2006). Furthermore, overexpression of a recombinant chitinase in plants resulted in their decreased susceptibility to fungal infections (Jach *et al.* 1995; Grison *et al.* 1996).

The first reports on the existence of chitinases in vertebrates did not receive the appropriate attention, since at that point the presence of chitinases was thought to be restricted to chitin containing lower life forms and the identified activities were attributed to lysozymes, which can also hydrolyse chitin (Jeuniaux 1961; Lundblad *et al.* 1974; 1979). The presence of a chitinolytic enzyme in human serum, different from lysozyme, was established a few years later (den Tandt *et al.* 1993; Overdijk *et al.* 1994).

Today, two distinct chitinases have been identified in mammals. The first is chitotriosidase, a human phagocyte-specific chitinase that was identified and characterized following the finding of its profound activity in plasma of Gaucher disease patients (Hollak *et al.* 1994; Boot *et al.* 1995; Renkema *et al.* 1995; 1997).

The second is the acidic mammalian chitinase (AMCase) which is structurally highly related to chititriosidase and is expressed in the lungs and the gastrointestinal tract (Boot *et al.* 2001; Boot *et al.* 2003). AMCase was identified in the search of possible compensatory enzyme activities, following the observation that approximately 1:20 individuals is completely deficient in chitotriosidase activity due to a 24-bp duplication in the gene encoding the enzyme protein (Boot *et al.* 1998).

The physiological role(s) of these mammalian chitinolytic enzymes is not clear. Several lines of evidence, mainly related to chitotriosidase, suggest that in analogy to their homologous plant chitinases act as pathogenesis related proteins, having a role in the human innate immune system. On the other hand, a role of AMCase in asthma has been proposed (Bussink *et al.* 2006), these data and specially those relating the chitinases to fungal and bacterial infections will be presented.

2. The Chitinolytic Activity of Human Chitinases

Mammalian chitinases, based on sequence homology and reaction mechanism, are classified as members of the family 18 of glycosyl-hydrolases. Most plant chitinases, as well as chitinases of fungi, protozoa and other species also belong to the same family (Henrissat 1991; van Aalten *et al.* 2001). The results of recent phylogenetic analysis studies suggest that both chitotriodase and AMCase have resulted from an early gene duplication event (Bussink *et al.* 2007).

Human chitotriosidase exists in two major isoforms with a molecular weight of 50kDa and 39kDa respectively. Both are active chitinases exhibiting activity towards colloidal chitin as well as artificial fluorogenic substrates which is inhibitable by allosamidin and dimethyl allosamidin in a manner similar to bacterial chitinases (Renkema *et al.* 1995).

The 50kDa form is found extracellularly originating from either macrophages or polymorphonuclear neutrophils and stored in the specific granules of the latter. It consists of a C-terminal chitin-binding domain, a hinge region and the 39kDa N-terminal domain that exhibits the chitinase activity. Thus, it is capable both of binding and degrading chitin. The 39kDa form, the N-terminal catalytic domain, is released through the intralysosomal processing of the 50kDa form and is found in the lysosomes of macrophages (Renkema *et al.* 1997).

Sequence alignement studies reveal a remarkable homology of chitotriosidase to chitinases of various species and complete conservation of the catalytic consensus sequence (Boot *et al.* 1995).

Detailed studies of the crystal structure of the native 39kDa human chitotriosidase and its complexes with a chitooligossacharide and allosamidin have been carried out. They have shown that the active site has a groove character with an elongated active site cleft compatible with the binding of long chitin polymers. Furthermore, the relatively open architecture of the active site detected, indicates that chitotriosidase acts as an endochitinase (Fusetti *et al.* 2002).

The second mammalian chitinase named acidic mammalian chitinase AMCase, like chitotriosidase is synthesized as a 50kDa protein containing a 39kDa N-terminal catalytic domain, a hinge region and a C-terminal chitin binding domain (Boot *et al.* 2001). Furthermore, like chitotriosidase it is capable of degrading artificial chitin like substrates as well as chitin showing however a distinct pH profile, being more active at acidic pH than chitotriosidase (Boot *et al.* 2001).

The *in vitro* chitinolytic activity of human chitinases has been established in several studies. Tjoelker *et al.* (2000) using recombinant chitotriosidase, including a recombinant truncate lacking the C-terminal binding sequence, showed that truncation resulted in a major decrease in hydrolytic activity against insoluble chitin in an agar diffusion assay. They suggested that the C-terminal chitin-binding domain although is not involved in the catalysis per se is likely critical *in vivo* for targeting the enzyme to its substrate which could be the chitin in the cell wall of a fungal intruder.

Boot *et al.* (2001) have shown that recombinant chitotriosidase was able to digest chitin in the cell wall of regenerating spheroplasts of *Candida albicans*. In their study Stevens *et al.* (2000), investigated the possibility of utilizing human chitinase either alone or in combination with antifungal drugs, as a non-foreign protein in the therapy of human fungal infections. They studied the fungicidal effect of human macrophage recombinant chitinase either alone or in combination with drugs such as amphotericin B, itraconazole and fluconazole, using different isolates of different fungal species. Under their experimental conditions they observed an apparent resistance to chitinase alone by all but one of the different fungal isolates studied. On the other hand, their results showed that human chitinase acted synergistically with conventional antifungal drugs for both inhibition and killing of fungi. Based on the above findings they suggested that endogenous chitinases may be a previously unrecognized factor that contributes to the outcome of every episode of clinical antifungal therapy.

The possible anti-fungal activity of a recombinant 50kDa human chitrotriosidase was further studied by van Eijk *et al.* (2005). They found that recombinant human chitotriosidase clearly inhibited the growth of *Cryptococcus neoformans*. The minimal inhibitory concentration for a 48 hours growth inhibition was $<0.25\ \mu g\ ml^{-1}$. When *Mucor rouxii* was first exposed to chitotriosidase and then placed in a hypotonic environment, formation of atypical blebs, followed by hyphal bursting was observed. It was estimated that 60% of hyphae collapsed in the hypotonic environment as a result of the exposure to chitotriosidase.

Finally, addition of the recombinant human enzyme in *Candida albicans* cultures prevented the transition from the yeast form towards the hyphal form. However, although recombinant chitotriosidase was effective against *Mucor rouxii* at a concentration of 100μg/ml, concentrations as high as 1 mg/ml were required for blocking hyphae formation in *Candida albicans*. The authors concluded that their *in vitro* results supported the antifungal effect of chitotriosidase as revealed by growth inhibition, hyphal tip bursting or prevention of hyphal switch.

However taking all relevant data into account, it appears that the connection between the chitinolytic activity of chitotriosidase and its fungicidal action is not straight forward. High concentrations of the enzyme are required for the latter, and its effectiveness in inhibiting and killing fungi seems to depend on factors such as the type of the fungus involved and also on the presence of other antifungal agents. In fact as early as 1978, David and Pope provided evidence that a mixture of glucanases greatly enchanced the chitinolytic action of chitinase on *Aspergillus fumigatus,* by degrading surface glucans and improving the access of chitinase to its substrate, chitin.

3. Chitinases and the Immune Response to Pathogens

The successful defense against pathogenic microorganisms depends on the activation and cooperation of the innate and adaptive components of the human immune system (Medzhitov 2007).

The first response to invading microorganisms is the activation of the innate immune system. It occurs within minutes of the invasion and coordinates the host defense during the initial hours and days of the infection stimulating at the same time the adaptive immune response. Although innate immunity has long been considered as non-specific and non-selective, this has been challenged by the discovery of classes of receptors on phagocytic cells that recognize specific molecular patterns that are unique to microorganisms, known as pattern recognition receptors (PRR). They have a broad specificity and can potentially bind a large number of molecules with common structural motif or pattern, referred to as pathogen-associated molecular patterns (PAMPs), although they are also present in non-pathogenic microorganisms.

PAMPs are often components of the cell wall of the invanding pathogens such as lipopolysaccharide, peptidoglycan, lipoteichoic acids, cell-wall lipoproteins in the case of bacteria and the component of fungal cell walls β-glucan in the case of fungi (Mansour and Levitz, 2002; Henmann and Rogen, 2002). Phagocytosable size chitin particles have also been shown to induce innate immunity (Shibata *et al.* 1997).

Several classes of pattern recognition receptors exist. The transmembrane toll like receptors (TRLs) are known to elicit inflammatory and antimicrobial responses to bacteria, fungi and viruses (Netea *et al.* 2004; Kullberg *et al.* 2004; Netea *et al.* 2006). Dectin-2, another transmebrane receptor, binds to β-glucan and is important to antifungal defense (Brown 2006).

Furthermore, there are intracellular (cytosolic) receptors that function in the pattern recognition of bacterial and viral pathogens. Members of this class of PRR are the proteins NOD1 and NOD2 (nucleotide-oligomerization-domains), involved in sensing bacterial peptidoglycan fragments (Fritz *et al.* 2007).

Recognition of the invading pathogens sets in motion a series of events associated with activation of the innate immune response and the concomitant stimulation of the adaptive immune response (Medzhitov 2007).

The innate-immune system consists both of cellular (neutrophils, monocytes, natural killer (NK) cells) and humoral (e.g. complement, lysozyme) components. Recent evidence suggest that chitotriosidase should be considered as a humoral component of innate immunity (van Eijk *et al.* 2005; van Eijk *et al.* 2007).

Human polymorphonuclear neutrophils (PMNs) and macrophages are specialized key components of the innate immune system. They are equipped with multiple antimicrobial mechanisms that are activated on initial contact with pathogens. They play a crucial role in defense against fungal pathogens and both intracellular and extracellular bacteria.

PMNs store in different types of granules several antimicrobial agents that are released in response to triggers such as cytokines or pathogen derived signals.

In healthy individuals, the major source of chitotriosidase are the human PMNs (Escott and Adams 1995; Boussac and Garrin 2000). More recently, van Eijk *et at.,* (2005), have studied the intracellular localization of chitotriosidase in PMNs. They showed that chitotriosidase was only released when PMNs were exposed to conditions that induced the release of the components of specific granules and using immunogold double-labelling experiments, they established the colocalization of chitotriosidase and the specific granule marker, lactoferin, in PMNs.

Furthermore, it was shown that GM-CSF both *in vitro* and when administered to healthy individuals, induced the release of chitotriosidase from PMNs, which occurred in parallel to the release of lactoferin. The promotion by GM-CSF of the release of chitotriosidase by PMNs is interesting in connection to the reported beneficial effects of GM-CSF administration in patients with fungal infections (Antachopoulos and Roilides, 2005).

Recently, trigerring of chitotriosidase release from PMNs in relation to TLR-stimulation was studied. With the exception of TLR_3, human neutrophils express all other TLRs (Fumikata *et al.* 2003) and in their study van Eijk *et al.* (2007) investigated the effect on chitotriosidase release of different ligands stimulating different TLRs. The only ligand amongst those studied that induced a potent dose-dependent chitotriosidase release by human neutrophils, was peptidoglycan (pGN), a PAMP of bacteria. Investigating the signaling cascades involved in the process, a time-dependent induction of several kinases was found. They included protein kinase B (PKB-Akt), p38 mitogen-activated protein kinase (p38 MAPK) and extracellular signal-regulated kinase – 1/2 (ERK 1/2). Inhibition of either PI3 kinase or p38-MAP kinase resulted in 50% inhibition of chitotriosidase release, the release being completely blocked when the kinases were inhibited simultaneously. The authors also provided evidence that the release of chitotriosidase was operated through TLR-2 and not NOD2 activation. On the other hand, only minor increases in the release of chitotriosidase were observed using either the TLR-4 ligand LPS or the TLR2 ligands PAM_3CSK_4 (TLR 2/1 ligand) or zymozan (TLR-2/6 ligand).

Monocytes do not express chitotriosidase, induction of mRNA and protein is, however, observed *in vitro* upon their differentiation to macrophages. Mature human macrophages express chitotriosidase that is being excreted as a 50kDa protein and stored intralysossomaly in the processed 39kDa form (Renkema *et al.* 1997). It appears that lysosomal stress is an important inducer of chitotriosidase synthesis by macrophages (Hollak *et al.* 1994; Guo *et al.* 1995; Michelakakis *et al.* 2004).

Macrophages are long-lived cells that remain resident in tissues for extended time periods and are involved in inflammation and immune responses. They possess the machinery for antigen presentation, however their main contribution to immune response may be their central role in the inhibition and killing of pathogens (Mansour and Levitz 2002; De Leo 2004; Medzhitov 2007). Activation of macrophages can occur following the activation of TLRs by their microbial ligands as well as by a variety of cytokines and most notably IFN-γ.

A number of studies have investigated the production of chitotriosidase following activation of monocytes and macrophages by a variety of activating molecules.

In order to determine the factors involved in chitotrioside expression in maturing monocytes, van Eijk *et al.* (2005), cultured human monocytes in culture medium alone or supplemented with either M-CSF, IL-4, IFN-γ or GM-CSF. The expression of chitotriosidase was investigated by western blot analysis, assaying of enzyme activity and mRNA quantization. It was shown that in maturing monocytes M-CSF had little effect on the production of chitotriosidase whereas both IFN-γ, associated with classical activation of macrophages, and IL-4, associated with alternative activation of macrophages, prevented induction of chitotriosidase. On the other hand the presence of GM-CSF during the maturation process of monocytes to macrophages superinduced the production and release of chitotriosidase. These findings in agreement with an earlier report by Hashimoto *et al.* (1999), that in a serial analysis of gene expression, showed increased chitotriosidase transcripts in human macrophages matured in the presence of GM-CSF. The increase was higher than that observed in the presence of M-CSF.

In matured macrophages, IFN-γ, TNF-a and prolactin exposure resulted in increased expression and activity of chitotriosidase which was transient and time dependent (Malaguarnera *et al.* 2004; Di Rosa *et al.* 2005). However, prolonged stimulation of chitotriosidase expressing macrophages (24 and 48 hrs) with IFN-γ inhibited expression, IL-4 having a similar effect (van Eijk *et al.* 2005). Significant suppression of chitotrioside mRNA synthesis, upon treatment of macrophages with IL-10, as compared to IFN-γ, TNF-a or LPS treated cells has also been reported (Di Rosa *et al.* 2005).

In vitro prevention of the induction of chitotriosidase in maturing monocytes under conditions mimicking pathogen infection has been observed. Thus, stimulation of human monocytes with TLR ligands e.g. LPS for TLR4, TLR or NOD2- activation by MDP inhibited enzyme induction (van Eijk *et al.* 2007).

Furthermore, in the same study it was shown that TLRs stimulation by ligands such as LPS, pIC or pGN in mature chitotriosidase expressing mactophages reduced chitotriosidase activity in cell lysates by approximately 25%. On the other hand, in their study Di Rosa *et al.* (2005), showed a 300 – fold increase in chitotriosidase activity and a concomitant chitotriosidase mRNA increase upon incubation of human macrophages with LPS, that gradually decreased over a period of 24 hrs. The discrepancy between the above studies could result from differences in the experimental procedures and/or the different source of LPS used in each of them.

In contrast to the effect of TLR stimulation, it has been observed that stimulation of chitotriosidase – expressing macrophages with the NOD2 ligand MDP results in a significant increase in chitotriosidase activity in a dose and time dependent manner. It was significant at MDP concentration of 20 μg/ml following 5 hrs of exposure and it gradually diminished to control levels after 48 hrs of stimulation by MDP (van Eijk *et al.* 2007).

Thus *in vitro* results provide evidence that expression and release of chitotriosidase can be induced under conditions mimicking infections by pathogens.

Cytokines as well as PRRs are involved in this process, which depends on the type of phagocyte, its overall state of activation and the time of its exposure to the elicitors involved. It is of interest that both release from PMNs and expression and release from macrophages can be induced by bacteria PAMPs.

3. *In Vivo* Evidence

The first report of the behaviour of chitinases in fungal infections *in vivo* came from the work of Overdijk *et al.* (1996). They showed that chitinase activity increased in the blood of guinea pigs after their intravenous infection with *Aspergillus fumigatus*. The enzyme was of guinea pig origin and the observed increase appeared to depend on the time after injection and the size of the infecting fungal inoculum. Furthermore, they showed that treatment with amphotericin B and itracozanole but not flucanazole, resulted in reduction of chitinase activity. However, due to the small number of aminals studied, it was not possible to evaluate any correlation of the reduction of chitinase activity to the survival time of the individual animals. In a subsequent study increased chitinase activity was also observed in different tissues following infection of guinea pigs with *Aspergillus fumigatus*. (Overdijk *et al.* 1999).

Similar to these observations were the findings in neonates with proven fungal infection. It was shown that in neonates with proven *Candida albicans* and *Aspergillus niger* infections chitotriosidase activity was increased in plasma and/or urine on diagnosis of the fungal infection. Furthermore, serial estimations of chitotriosidase activity showed that it changed in parallel to the clinical outcome of the neonates. In neonates that recovered, a gradual decrease and eventual normalization in chitotriosidase levels occured. On the other hand, in neonates that succumbed to their infection, a continuous increase in chitotriosidase activity was observed, reaching the highest values found in the study, which were 16x – 150x the upper normal limit (Labadaridis *et al.* 1998; Labadaridis *et al.* 2005). Thus based on the above observations it appears that increase in chitotriosidase activity was not sufficient by itself for the control of fungal infections. This should not be surprising, since defense mechanisms involve a combination of effectors and the final outcome depends on their successful interaction (Mansour and Levitz 2002). On the other hand inability to synthesize a catalytically active enzyme did not appear to have any adverse effect. The only neonate that was incapable of synthesizing catalytically active chitotriosidase, survived and fully recovered from its fungal infection. In a preliminary report, a statistically significant positive correlation was found between IL-10 levels and chitotriosidase activity in neonatal infections

(Labadaridis *et al.* 2005). IL-10, an anti-inflammatory cytokine associated with Th2 response and alternative activation of macrophages, is an indicator of poor prognosis in fungal infections, and in this preliminary report the highest IL-10 levels were found in neonates that died (Gordon 2003; Kullberg *et al.* 2004; Labadaridis *el al.*, 2005).

As it has already been referred to, an estimated 6% of Caucasians are deficient in chitotriosidase activity due to a 24-base pair duplication in exon 10 that distrupts protein production (Boot *et. al.*, 1998). Homozygote individuals are healthy and the consequences of chitotriosidase deficiency are not clear (Lee *et al.* 2007). Furthermore, recently polymorphisms associated with reduced chitotriosidase activity have been described (Lee *et al.* 2007). In their study, Masoud *et al.* (2002), hypothesizing that chitotriosidase deficient individuals may be more vulnerable to fungal infections, investigated the prevalence of homozygosity for the mutation in a cohort of survivors of *Candida sepsis*. They found that the prevalence of homozygosity amongst survivors is similar to that of the general population. However, since only survivors were included in the study, the results cannot be considered conclusive. Nontheless, they are in agreement with the findings in other studies that showed lack of association of chitotriosidase deficiency with candidal infections in children with acute myeloid leukemia and adults with leukemia (Lehrnbercher *et al.* 2005; Choi *et al.* 2005). Furthermore, chitotriosidase deficiency did not appear to affect the recovery of a neonate with disseminated *Candida albicans* infection (Labadaridis *et al.* 2005). On the other hand, in their study of Lehrnbecher *et al.* (2005) observed an association between the functional variant of chitotriosidase and bacterial infection. In particular they found that the risk of gram-negative bacteremia is significantly higher in patients carrying the 24-bp duplication either in hetero- or homozygosity, suggesting that chitotriosidase deficiency renders children with acute myeloid leukemia more succeptible to gram- negative infections. In that context it is of interest that increased chitotriosidase activity, the magnitude of which did not relate to the outcome of the infection, was detected on diagnosis of gram-negative and Gram-positive bacteria infection in 12/15 neonates included in the study of Labadaridis *et al.* (2005). This is in agreement with the *in vitro* observations that bacterial TLR ligands induce the release of chitotriosidase from PMNs and that NOD_2 stimulation by MDP induced a 2-fold increase of chitotriosidase activity in macrophages. The mechanism(s) eliciting the *in vivo* chitotriosidase response and the origin of the activity measured in fungal and bacterial infections are not known. However, the *in vivo* results show that increase in chitotriosidase activity is not a specific response to fungal infection but rather the outcome of the overall activation of the immune system. Furthermore, the results suggest that chitotriosidase may have a more pleiotropic effect in innate immune response than previously assumed.

The *in vivo* efficacy of recombinant human chitotriosidase in controlling fungal infections was studied in immunocompetent, immunosuppressed and neutropenic mice infected with either *Candida albicans* or *Aspergillus fumigatus*. The enzyme was administered i.p. and promoted survival of candidiasis mice in a dose-dependent manner in all groups of mice tested. Doses as high as 100 mg/kg^{-1} were required for an 80% increase in survival. The effect on aspergillosis mice was similar, although not so pronounced. Combination therapy with amphotericin B resulted in a significantly better promotion of survival in both types of infection, suggesting a possible synergistic action *in vivo,* in agreement with the results of relevant *in vitro* studies, (Vaddi *et al.* 1998; Stevens *et al.* 2000; van Eijk *et al.* 2007).

Conclusions

Establishing the biological role(s) of chitinases in humans is a challenging task. The available data clearly establish the chitinolytic properties of the enzymes as well as their synthesis by cells of the innate immune system, and have led to the suggestion, that similar to their plant counterparts, human chitinases are involved in the defense against chitin containing pathogens e.g. fungi. *In vitro* data, although contradictory at times, show that chitotriosidase can be overexpressed and released by phagocytes in response to ligands activating the immune response not only towards fungi but also bacteria. In agreement with this, increased levels of chitotriosidase activity was seen in neonates with fungal and bacterial infections. But what is the biological concequences of such an increase? Assuming the fungal cell wall chitin is its target in fungal infections, which is its target in bacterial infections? Is it involved in inhibiting the growth and eventually killing of the pathogen intruders or its increase is just a concequence of the overall activation of the immune cells?

In vitro and *in vivo* data suggest a limited fungicidal capacity for chitotriosidase, which is increased in the presence of other antifungal agents. The high frequency of the 24-bp duplication associated with absence of chitotriosidase activiy and the indentification of yet other polymorphisms that reduce its enzymic activity poses further questions with respect to the biological significance of human chitotriosidase. Is too much or too little that is important? It is of interest that maximal increase of chitotriosidase activity was observed in those patients that succumbed to their fungal infections. Furthermore, the inability to synthesize an active enzyme does not seem to be associated with increased risk and/or diminished survival in the case of fungal infections and only the data from one study suggest increased succeptibility to bacterial infections.

It has been proposed that AMCase activity could compensate for the absence of chitotriosidase in the defense against pathogens. However, no evidence supporting such a role is present in the literature.

It is clear therefore, that further studies are needed before the role(s) of chitotriosidase and AMCase in the fungal and bacterial sepsis is illucidated.

References

Antachopoulos C and Roilides E. Cytokines and fungal infections. *Brit J Hematol.* 2005; 129: 583-96.

Barrett-Bee K and Hamilton M. The detection and analysis of chitinase activity from the yeast form of *Candida albicans*. *J Gen Microbiol.* 1984; 130: 1857-61.

Boot RG, Blommaart EFC, Swart E, Ghauharali-Van der Vlugt K, Bijl N, Moe C, Place A, Aerts JMFG. Identification of a novel acidic mammalian chitinase distinct from chitotriosidase. *J Biol Chem.* 2001; 276: 6770-8.

Boot RG, Bussink AP, Verhoek M, de Boer PAJ, Moorman AFM, Aerts JMFG. Marked differences in tissue-specific expression of chitinases in mouse and man. *J Histochem Cytochem.* 2005; 53: 1283-92.

Boot RG, Renkema GH, Strijland A, van Zonneveld AJ, Aerts JMFG. Cloning of cDNA enconding chitotriosidase, a human chitinase produced by macrophages. *J Biol Chem.* 1995; 270: 2652-6.

Boot RG, Renkema GH, Verhoek M, Strijland A, Bliek J, de Meulemeester TM, Mannens MM, Aerts JM. The human chitotriosidase gene: Nature of inherited enzyme deficiency. *J Bio Chem.* 1998; 273: 25680-5.

Boussac M and Garin J. Calcium dependent secretion in human neutrophils. A proteomic approach. *Electrophoresis* 2000; 21: 665-72.

Brown GD. Dectin-1: a signalling non-TLR pattern-recognition receptor. *Nature Rev Immunol.* 2006; 6: 33-43.

Brurberg MB, Nes IF, Eijsink VGH. Comparative studies of chitinases A and B from *Serratia marcescens*. *Microbiology* 1996; 142: 1581-9.

Bussink AP, Van Eijk M, Renkema H, Aerts JM, Boot RG. The biology of the Gaucher cell: The cradle of human chitinases. *Intern Rev Cytol* .2006; 252: 71-128.

Bussink AP, Speijer D, Aerts JMFG, Boot RG. Evolution of mammalian chitinase (like) members of family 18 glycosylhydrolases. *Genetics* 2007; 177: 959-70.

Choi EH, Taylor JG, Foster CB, Walsh TJ, Anttila VJ, Ruutu T, Palotie A, Chanock SJ. Common polymorphisms in critical genes of innate immunity do not contribute to the risk of chronic disseminated candidiasis in adult leukemia patients. *Med Mycol.* 2005; 43: 349-53.

Day RB, Okada M, Ito Y, Tsukada K, Zaghouani H, Shibuya N, Stacey G. Binding site for chitin oligosaccharides in the soybean plasma membrane. *Plant Physiol.* 2001; 126: 1162-73.

Davies DAL and Pope AMS. Mycolase, a new kind of systemic antimycotic. *Nature* 1978; 273: 235-6.

De Leo FR. Modulation of phagocyte apoptosis by bacterial pathogens. *Apoptosis* 2002; 2004: 399-413.

den Tandt WR, Scharpe S, Overdijk B. Evaluation on the hydrolysis of methylumbelliferyl-tetra-N-acetylchitotetraoside by various glycosidases: A comparative study. *Int J Biochem.* 1993; 25: 113-9.

Di Rosa M, Musumeci M, Scuto A, Musumeci S, Malaguarnera L. Effect of interferon-γ, interleukin-10, lipopolysaccharide and tumor necrosis factor-a on chitotriosidase synthesis in human macrophages. *Clin Chem Lab Med.* 2005; 43: 499-502.

Escott GM and Adams DJ. Chitinase activity in human serum and leukocytes. *Infect Immun* 1995; 63: 4770-3.

Flach J, Pilet PE, Jolles P. What's new in chitinase research. *Experientia* 1992; 48: 701-716.

Fritz JH, Ferroro RL, Philpott DG, Giraldin SE. NOD-like proteins in immunity, inflammation and disease. *Nature Immunol.* 2007; 7: 1250-7.

Fumikata H, Means TK. Luster AD. Toll-like receptors stimulate human neutrophil function. *Blood* 2003; 102: 2660-9.

Fusetti F, von Moeller H., Houston D, Rozeboom HJ, Dijkstra BW, Boot RG, Aerts JMFG, van Aalten DMF. Structure of human chitotriosidase. Implications for specific inhibitor design and function of mammalian chitinase-like lectins. *J Biol Chem.* 2002; 277: 25537-44.

Gordon S. Alternative activation of macrophages. Nat. Rev. *Immunol.* 2003; 3: 23-35.

Grison R, Grezesbesset B, Schneider M, Lucante N, Olsen L, Leguay JJ, Toppan A. Field tolerance to fungal pathogens of *Brassica napus* constituvely expressing a chimeric chitinase gene. *Nat Biotechnol.* 1996; 14: 643-6.

Guo Y, He W, Boer AM, Wevers RA, de Bruijn AM, Groener JE, Hollak CE, Aerts JM., Galjaard H, van Diggelen OP. Elevated plasma chitotriosidase activity in various lysosomal storage disorders. *J Inher Metab Dis.* 1995; 18: 717-22.

Hashimoto S, Suzuki T, Dong HY, Yamazaki N, Matsushima K. Serial analysis of gene expression in human monocytes and macrophages. *Blood* 1999; 94: 837-44.

Henrissat B. A classification of glycosyl hydrolases based on amino acid sequence similarities. *Biochem J.* 1991; 280: 309-16.

Heumann D, Rogier T. Initial responses to endotoxins and Gram-negative bacteria. *Clin Chim Acta.* 2002; 323: 59-72.

Hollak CEM, Van Weely S, Van Oers MHJ, Aerts JMFG. Marked elevation of plasma chitotriosidase activity. A novel halmark of Gaucher disease. *J Clin Invest.* 1994; 93: 1288-92.

Jach G, Gornhardt B, Mundy J, Logemann J, Pinsdorf E, Leah R, Schell J, Maas C. Enchanced quantitative resistance against fungal disease by combinatorial expression of different barley antifungal proteins in transogenic tobacco. *Plant J.* 1995; 8: 97-109.

Jeuniaux C. Chitinases an addition to the list of hydrolases in the digestive tract of vertebrates. *Nature.* 1961; 192: 135-6.

Kaku H, Nishizawa Y, Ishii-Minami N, Akimoto-Tomiyama C, Dohmae N, Takio K, Minami E, Shibuya N. Plant cells recognize chitin fragments for defense signaling through a plasma membrane receptor. *Proc Nat Acad Sci.* 2006; 103: 11086-91.

Kasprzewska A. Plant chitinases: Regulation and function . *Cell Mol Biol Lett.* 2003; 8: 809-24.

Kullberg BJ, Oude Lashof AML, Netea MG. Design of efficacy trials of cytokines in combination with antifungal drugs. *Clin Infect Dis.* 2004; 39: S218-23.

Labadaridis J, Dimitriou E, Costalos C, Aerts J, van Weely S, Donker-Koopman WG, Michelakakis H. Serial chitotriosidase activity estimations in neonatal systemic candidiasis. *Acta Paediatr.* 1998; 87: 605.

Labadaridis J, Dimitriou E, Theodorakis M, Kafalidis G, Velegraki A, Michelakakis H. Chitotriosidase in neonates with fungal and bacterial infections. *Arch Dis Child Fetal Neonatal Ed. 2005; 90: F531-2.*

Labadaridis J, Theodoraki M, Dimitriou E, Spanou K, Sarafidou J, Triantaphylidis G, Michelakakis H. Chitotriosidase activity and IL-10 levels in neonatal infections. *Ped Res.*2005; 58: A213.

Lee SC and Hwang BK. Induction of some defense-related genes and oxidative burst is required for the establishement of systemic acquired resistance in *Capsicum annuum. Planta* 2005; 221: 790-800.

Lee P, Waalen J, Crain K, Smargon A, Beutler E. Human chitotriosidase polymorphisms 2007; 39: 353-60.

Lehrnbecher T, Bernig T, Hanisch M, Koehl U, Behl M, Reinhardt D, Creutzig U, Klingebiel T, Chanock SJ. Common genetic variants in the interleukin-6 and chitotriosidase genes are associated with the risk for serious infection in children undergoing therapy for acute myeloid leukemia. *Leukemia* 2005; 19: 1745-50.

Lundblad G, Elander M, Lind J, Sletlengrem K. Bovine serum chitinase. *Eur J Biochem.* 1979; 15: 455-60.

Lundblad, G, Hedertedt B, Lind J, Steby B. Chitinase in goat serum: Preliminary purification and characterization. *Eur J Biochem.* 1974; 46: 367-76.

Malaguarnera L, Musumeci M, Licata F, Di Rosa M, Messina A, Musumeci S. Prolactin induces chitotriosidase gene expression in human monocyte-derived mactophages. *Immunol Lett.* 2004; 94: 57-63.

Mansour MK and Levitz SM. Interactions of fungi with phagocytes. *Curr Opin Microbiol.* 2002; 5: 359-65.

Masoud M, Rudensky B, Elstein D, Zimran .A. Chitotriosidase deficiency in survivors of Candida sepsis. *Blood Cells Molec Dis.* 2002; 29: 116-8.

Michelakakis H, Dimitriou E, Labadaridis I. The expanding spectrum of disorders with elevated plasma chitotriosidase activity. An Update. *J Inher Metab Dis.* 2004; 27:705 706.

Medzhitov R. Recognition of microorganisms and activation of the immune response. *Nature.* 2007; 449: 819-29.

Netea MG, van der Graaf C, van der Meer JWM, Kullberg BJ. Toll like receptors and the host defense against microbial pathogens: bringing specifity to the innate immune system. *J Leuk Bio.* 2004; 75: 749-55.

Netea MG, van der Meer JWM, Kullberg BJ. Role of the dual interaction of fungal pathogens with pattern recognition receptors in the activation and modulation of host defence. *Clin Microbiol Infect.* 2006; 12: 404-9.

Okada M, Matsumura M, Ito Y, Shibuya N. High-affinity binding proteins for N-acetylchitooligosaccharide elicitor in the plasma membranes from wheat, barley and carrot cells: conserved presence and correlation with the responsiveness to the elicitor. *Plant Cell Physiol.* 2002; 505-12.

Overdijk KB, van Steijn GN, den Tandt WR. Partial purification and further characterization of the novel endoglucosaminidase from human serum that hydrolyses 4methylumbelliferyl -N-acetyl-β-D-chitotetraoside (MU-TACT hydrolase). *Int J Biochem.* 1994; 26: 1369-75.

Overdijk KB, van Steijn GJ, Odds FC. Chitinase levels in guinea pig blood are increased after systemic infection with *Aspergillus fumigatus*. *Glycobiol.* 1994; 6: 627-34.

Overdijk B, van Steijn GJ, Odds FC. Distribution of chitinase in guinea pig tissues and increases in levels of this enzyme after systemic infection with *Aspegillus fumigatus*. *Microbiol.* 1999; 145: 259-69.

Pedraza-Reyes M and Lopez-Romero E. Purification and some properties of two forms of chitinase from mycelian cells of *Mucor rouxii*. *J Gen Microbiol.* 1989; 135: 211-8.

Pirttila AM, Laukkanen H, Hohtola A. Chitinase production in pine callus (*Pinus sylvestris L.*): A defense reaction against endophytes? *Planta.* 2002; 214: 848-52.

Renkema GH, Boot RG, Muijsers AO, Donker-Koopman WE, Aerts JMFG. Purification and characterization of human chitotriosidase, a novel member of the chitinase family of proteins. *J Biol Chem.* 1995; 270: 2198-202.

Renkema GH, Boot RG, Strijland A, Donker-Koopman WE, van den Berg M, Muijsers AO, Aerts JM. Synthesis, sorting and processing into distinct isoforms of human macrophage Chitotriosidase. *Eur J Biochem.* 1997; 244: 279-85.

Sahai AS and Manocha MS. Chitinases of fungi and plants: their involvement in morphogenesis and host-parasite interaction *FEMS Microbio. Rev.* 1993; 11: 317-38.

Schlein Y, Yacobson RL, Messer G. Leishmania infections damage the feeding mechanism of the sandfly vector and implement parasite transmission by bite. *Proc Natl Acad Sci USA.* 1992; 89: 9944-8.

Schlumbaum A, Mauch F, Vogeli U, Boller T. Plant chitinases are potent inhibitors of fungal growth. *Nature.* 1986; 324: 365-7.

Selvaggini S, Munro CA, Paschoud S, Sanglard D, Gow NA. Independent regulation of chitin synthase and chitinase activity in *Candida albicans* and *Saccharomyces cerevisiae. Microbiology.* 2004; 150: 922-8.

Shahabuddin M and Kaslow DC. *Plasmodium*: Parasite chitinase and its role in malaria transmission. *Exp Parasitol.* 1994; 79: 85-8.

Shibata Y, Metzger WJ, Murvik Q.N. Chitin particle-induced cell-mediated immunity is inhibited by soluble mannan: mannose receptor-mediated phagocytosis initiates IL-12 production. *J Immunol.* 1997; 159: 2462-7.

Stevens DA, Brammer HA, Meyer DW, Steiner BH. Recombinant human chitinase. *Curr Opin Anti-Infective Investig. Drugs* 2000; 2: 399-404.

Taira T, Toma N, Ichi M, Takeuchi M, Ishihara M. Tissue distribution, synthesis stage, and ethylene induction of pineapple (*Ananas conosus*) chitinase. *Biosci Biotechnol Biochem.* 2005; 69: 852-4.

Tharanathan RN, Kittur FS. Chitin the undisputed biomolecule of great potential. *Crit Rev Food Sci Nutr.* 2003; 43: 61-87.

Tjoelker LW, Gosting L, Frey S, Hunter CL, Trong HL, Steiner B, Brammer H, Gray PW. Structural and functional definition of the human chitinase chitin binding domain. *J Biol Chem.* 2000; 275: 514-20.

van Aalten DM, Kommander D, Synstad B, Gaseidnes S, Peter MG, Eijsink VG. Structural insights into the catalytic mechanism of a family 18 exo-chitinase. *Proc Natl Acad Sci. USA* 2001; 98: 8979-84.

van Eijk M, van Roomen CPAA, Renkema GH, Bussink AP, Andrews L, Blommaart EFC, Sugar A, Verhoeven AJ, Boot RG, Aerts JMFG. Characterization of human phagocyte-derived chitotriosidase, a component of innate immunity. *Int Immunol.* 2005; 17: 1505-12.

van Eijk M, Scheij SS, van Roomen CPAA, Speijer D, Boot RG, Aerts JMFG. TLR- and NOD_2 dependent regulation of human phagocyte – specific chitotriosidase. *FEBS Letters* 2007; 581: 5389-95.

Vaddi K, Preston C, Sugar AM, Aerts JMFG, Friedman B. Antifungal efficacy of recombinant human chitinase in murine candidiasis and aspergillosis. Preceedings of the 38th Interscience Conference on Antimicrobial Agents and Chemotherapy. 1998; 38: Abs. J-53.

In: Binomium Chitin-Chitinase: Recent Issues
Editor: Salvatore Musumeci and Maurizio G. Paoletti
ISBN 978-1-60692-339-9

Chapter V

The Origin and Genetics of Human Chitinases

R.G. Boot, D. Speijer, A.P. Bussink and J.M.F.G. Aerts
Department of Medical Biochemistry,
Academic Medical Center, University of Amsterdam, The Netherlands

Abstract

Chitin, the polymer of β-1,4 linked β-N-Acetyl-glucosamine (GlcNAc), is the most abundant biopolymer in marine environments and the second most abundant in nature, after cellulose. The degradation of chitin is mediated by chitinolytic hydrolases such as chitinases (EC.3.2.1.14) and β-hexosaminidases (EC.3.2.1.52). Based on sequence homologies chitinases fall into two groups: families 18 and 19 of glycosyl hydrolases.

Though mammals lack endogenous chitin, chitinases and chi-lectins, highly homologous proteins lacking enzymatic activity due to catalytic amino acid substitutions, are found in a wide variety of mammalian species. All belong to the family 18 of glycosyl hydrolases (GH18).

Despite the wealth of structural information that is available regarding the mammalian chitinase protein family, insight into their exact physiological role(s) remains limited.

This review gives an overview of all mammalian family members with special emphasis on their occurrence and expression in humans.

Recent molecular phylogenetic analyses suggest that both active mammalian chitinases (chitotriosidase and AMCase) result from an early gene duplication. Further gene duplication events, followed by loss of function mutations, allowed the evolution of the chi-lectins. The homologous genes coding family 18 glycosyl hydrolases are clustered in two distinct chromosomal loci.

The phylogenetic analyses suggest that the evolution of this gene family is in accordance with a form of multigene family evolution referred to as "birth-and-death evolution under strong purifying selection". Finally, several chitinase family members are present only in certain lineages of mammals and their tissue specific expression patterns differ profoundly between species, exemplifying recent evolutionary adaptations in the chitinase protein family.

1. Introduction

Polysaccharides are present in all organisms, mostly for structural purposes. Chitin, the linear polymer of β-1,4 linked β-N-Acetyl-glucosamine (GlcNAc), is the most abundant biopolymer in marine ecosystems and, after cellulose, the second most copious polysaccharide in nature. Quantitatively, most chitin is found in arthropods that utilize chitin as the main constituent of their exoskeleton, protecting them from environmental influences and providing structural support. Chitin is considered to be a relatively minor, yet structurally important, component of the fungal cell wall (Bownan and Free, 2006). Chitin is also present in algae and protists (reviewed in Mulisch, 1993), in the septum between mother and daughter yeast cells (Cid *et al.* 1995) and in the microfilarial sheaths of parasitic nematodes (Fuhrman and Piessens,1985; Araujo *et al.* 1993).

Chitin is an insoluble linear polymer of considerable mechanical and chemical strength, mainly due to hydrogen bonding between the >NH and the >C=O groups of the N-acetyl groups of aligning chitin polymers, that can occur in parallel (α), anti-parallel (β) or a mixture (γ) of aligned chains. These properties of chitin are ideal in serving as a protective structural coating in a variety of organisms, in which chitin is often covalently bound to other glycopolymers, such as α-glucan or β-glucan in the cell wall of many fungi (Debono and Gordee, 1994; Cabib *et al.* 2007). Importantly, to our knowledge, there are no reports describing the existence of chitin in higher plants and vertebrates, even though chitin synthase-like proteins are present in vertebrates and plants (Weigel and DeAngelis, 2007). A more detailed description of chitin chemistry and biology can be found in the chapter of Muzzarelli in this Book.

Chitin degradation is mediated by groups of distinct enzymes with different specificities. First, most organisms contain β-hexosaminidases that are capable of releasing monomeric GlcNAc residues from the non-reducing end of the chitin polymer. Second, some eukaryotes contain a di-N-acetylchitobiase, a lysosomal reducing-end exohexosaminidase, involved in degradation of asparagine-linked oligoscaccharides on glycoproteins that is also capable of cleaving chitin (Aronson *et al.* 1989; Aronson and Kuranda, 1989). Finally, there are the chitinases that are able to cleave within the chitin polymer. Endochitinases are defined as enzymes hydrolyzing the glycosidic bonds randomly within the chitin polymer, releasing mainly soluble, low-molecular weight chito-oligomers (Sahai and Manocha, 1993). The less common exochitinases catalyze the successive removal of chitobiose units from the non-reducing end of chitin polymers (Robbins *et al.* 1988).

Based on sequence homologies chitinases fall into two groups: families 18 and 19 of glycosyl hydrolases (Henrissat, 1991). Besides the true chitinases also the chitobiases are members of family 18 of glycosyl hydrolases, but the overall homology to the other members of the chitinase family is relatively low (Funkhouser and Aronson, 2007).

Members of family 18 employ a substrate assisted reaction mechanism (Terwisscha van Scheltinga *et al.* 1995; van Aalten *et al.* 2001), whereas those of family 19 adopt a fold and reaction mechanism similar to that of lysozyme (Monzingo *et al.* 1996). All chitinases with the exception of several plant chitinases are grouped together in family 18 of the glycoside superfamily, all of which contain the theoretical catalytic domain consensus sequence DXXDXDXE.

Chitinases have been detected in chitin-containing organisms and in species that do not contain chitin, for example in a variety of bacteria, plants, vertebrates and even viruses (Sahai and Manocha, 1993; Flach. *et al.* 1992, Gooday, 1995). Various biological functions have been attributed to the chitinases in the different species such as food processing, defense mechanisms and morphogenesis (see Bussink *et al.* 2006 for a recent overview). In this review we will focus on the human chitinases, and their evolutionary and cellular origin. Their genetic organization will also be discussed.

2. Human Chitinases

2.I. Chitotriosidase

Since chitin seems not to be made in mammals, it was initially assumed that chitinases would also be absent, being restricted to species that do contain the polymer. Even after the pioneering work of Lundblad *et al.* in the 1970's who detected chitinase activity in bovine and goat serum, a similar activity in human serum was still attributed to activity of lysozymes that are also able to hydrolyse chitin, albeit at slow rates (Lundblad *et al.* 1974; Lundblad, 1979). Later, den Tandt, Overdijk and co-workers, using the artificial fluorescent substrate 4MU-chitotetraoside noticed that human serum contains an enzyme able to hydrolyse this substrate distinct from other hydrolases, including lysozyme (den Tandt *et al.* 1988; den Tandt *et al,.* 1993). This so called MU-TACT hydrolase was subsequently shown to have hydrolytic activity towards chitin as well (Overdijk and van Steijn, 1994). Attempts to purify the enzyme responsible from human plasma have been unsuccessful.

A breakthrough came with the discovery of a markedly increased chitotriosidase activity in Gaucher patients (Hollak *et al.* 1994). This allowed the first successful purification and molecular characterization of a human chitinase. Two major active isoforms with molecular masses of 50 and 39 kDa have been purified from the spleen of a Gaucher patient (Renkema *et al.* 1995). Subsequently its cDNA and gene were cloned using a macrophage cDNA library (Boot *et al.* 1995). The gene coding for chitotriosidase is about 20 kb long, located on chromosome 1q32 and consisting of 13 exons.

Macrophages secrete a 50 kDa active enzyme, which consists of a catalytic domain and a C-terminal chitin-binding domain (Chitin binding domain type 2, PROSITE signature PS50940) separated by a hinge region (Renkema *et al.* 1997; Tjoelker *et al.* 2000). A minor fraction of the 50 kDa enzyme is intra-cellularly processed into a 39 kDa isoform that lacks the chitin binding domain and accumulates in the lysosome (Renkema *et al.* 1997). Alternative splicing is capable of generating a mRNA coding for the 50 kDa enzyme (exons 1-12 without exon 11) as well as two distinct mRNA's coding directly for 39 kDa catalytic active isoforms (exons 1-12 and exons 1-13 without exons 11 and 12) that differ only at the extreme carboxy-terminal amino acids (Boot *et al.* 1998; Bussink *et al.* 2006). So far the major source of this enzyme in man seems to be the professional phagocytes such as macrophages and neutrophilic granulocytes (van Eijk *et al.* 2005). The latter cells only synthesize chitotriosidase as bone-marrow derived precursor cells, since these cells contain the chitotriosidase mRNA, and store the protein within their specific granules (Boussac and

Garin, 2000; van Eijk *et al.* 2005). So far we have been unable to demonstrate chitotriosidase mRNA in circulating mature neutrophilic granulocytes. Only appropriate triggering of these cells leads to a release of the specific granules and hence of chitotriosidase (van Eijk *et al.* 2005; van Eijk *et al.* 2007).

The crystal structures of the native 39 kDa human chitotriosidase and of complexes with a chitooligosaccharide and allosamidin have been studied in detail (Fusetti *et al.* 2002; Rao *et al.* 2003). The structure consists of two domains. The core domain has a $(\beta/\alpha)_8$ (TIM) barrel as observed in the other family 18 chitinase structures for hevamine, chitinases A (ChiA) and B (ChiB) from *Seratia marcescens*, and CTS1 from *Coccidioides immitis*, although helix $\alpha 1$ is missing (Fusetti *et al.* 2002). An additional α/β domain, composed of six antiparallel β-strands and one α-helix, is inserted in the loop between strand β7 and helix α7, which gives the active site a groove character. Like all other family 18 chitinases, the chitotriosidase has the DXDXE motif at the end of strand β4 with Glu-140 being the catalytic acid. Two disulfide bridges were observed between residues 26-51 and 307-370. The crystal structures reveal an elongated active site cleft, compatible with the binding of long chitin polymers. Given the relatively open active site architecture, chitotriosidase appears to function as an endochitinase rather than an exochitinase. The complex with NAG_2 followed by modelling of a longer chitooligosaccharide indeed revealed that the active site would be able to accommodate longer chitin polymers, which agrees with its ability to degrade various forms of polymeric chitin.

2.2. Chitotriosidase Transglycosylation Activity

Chitotriosidase shows an unexpected inhibition of catalytic activity at high 4MU-chito-oligosaccharide substrate concentrations. This remarkable phenomenon was explained by the demonstration that chitotriosidase is not only capable of catalyzing hydrolysis of the chitooligosaccharide substrate, but can also transglycosylate it. Although chitinases are not able to hydrolyse 4MU-N-acetyl-glucosaminide, it was demonstrated that fluorescent 4MU was formed by recombinant chitotriosidase in the presence of PNP-chitobioside or chitoologiosaccharide, an observation that could only be explained by the occurrence of transglycosylation (Aguilera *et al.* 2003; Chou *et al.* 2006). This observation eventually led to the design of a novel substrate that can not be transglycosylated and is equally well hydrolysed: 4MU–(4-deoxy)-chitobioside. With this novel substrate a convenient and more accurate assay of enzymatic activity of chitinase became feasible (Aguilera *et al.* 2003).

Despite the occurrence of transglycosylation, chitotriosidase can release surprising amounts of chito-triose from 4MU-chitotriose. Based on the specific activity towards this substrate (Renkema *et al.* 1995), each molecule can catalyze hydrolysis over 4000 times each second under standard assay conditions. However, presumably due to a decrease in substrate availability, activity towards chitin is thought to be considerably less.

Stereospecific transglycosylation as demonstrated for chitotriosidase is a common feature to glycoside hydrolases (Holtje, 1996). The fact that mammalian chitinases also show transglycosylation raises the question whether the phenomenon has any physiological importance or is a mere catalytic imperfection.

2.3. Chitotriosidase Deficiency

Screening of patients with Gaucher disease revealed that a recessively inherited deficiency is commonly encountered, due to a 24 bp duplication in exon 10 of the chitotriosidase gene that results in abnormal splicing (Boot *et al.* 1998). The 24 bp duplication activates a 3' cryptic splice site that, although the authentic splice site is still intact, leads to a spliced mRNA with an in-frame deletion of 87 nucleotides from exon 10. This mRNA codes for a protein that lacks an internal stretch of 29 amino acids which is enzymatically inactive as determined in transfected COS cells. Analysis of mRNA from macrophages of individuals that are homozygous for this 24 bp duplication also clearly demonstrated reduced amounts of chitotriosidase mRNA in these cells (Boot *et al.* 1998). The crystal structures shed light on the inactivation of the enzyme through this inherited genetic deficiency. The common mutation results in a completely inactive enzyme in which residues Val-344 to Gln-372 missing. The residues 344-372 correspond to the C-terminal half of helix $\alpha 7$, the entire strand $\beta 8$, and almost the entire $\beta 8$-$\alpha 8$ loop. Deletion of these important secondary structure elements possibly would lead to misfolded, and therefore inactive, protein (Fusetti *et al.* 2002; Rao *et al.* 2003).

In the Dutch population, the observed carrier frequency of approximately 35% results in 6% of the individuals being completely deficient in chitotriosidase activity (Boot *et al.* 1998). A recent survey of this polymorphism in different ethnic groups demonstrated a large variability. Comparing various ethnic groups, the duplication was rare in individuals of African ancestry but more common in individuals of Asian descent (Malaguarnera *et al.* 2003; Lee *et al.* 2007).

2.4. Human AMCase

The high incidence of this chitotriosidase deficiency in man stimulated us to examine rodent tissues for chitinolytic activity in search for possible alternative chitinases. We indeed identified and characterized a second mammalian chitinase christened acidic mammalian chitinase (AMCase) that is also present in man (Boot *et al.* 2001). This 50 kDa enzyme is structurally highly related to chitotriosidase, however, unlike chitotriosidase this enzyme has an acidic pH optimum and is very acid stable (Boot *et al.* 2001; Bussink *et al.* 2008). Although the extreme acidic pH optimum (pH 2) of AMCase seems more specific for rodents, the enzyme occurs in the gastrointestinal tract and the lungs of both man and mice (Boot *et al.* 2001). The enzyme appears to be adapted to function in the extreme environment of the stomach where it may fulfill a role in defense and/or digestion of chitin-containing organisms.

The crystal structure of AMCase (mice or human) has not yet been resolved. With the aid of the published coordinates of human chitotriosidase Bussink *et al.* created models of human and mouse AMCase using the online SWISSPROT server (Bussink *et al.* 2008). The quality of the models based on a variety of stereochemical parameters was determined by PROCHECK and no major deviations from optimal geometry were detected. The overall structure shows high similarity with the experimentally determined crystal structure of chitotriosidase. The majority of substitutions in AMCase compared to chitotriosidase reside mainly on the protein surface. This suggests they are changes conferring stability in altered milieus rather than adaptations directly influencing the reaction mechanism. Indeed, calculation of the surface potential of both mouse and human AMCase shows that its substitutions result in a predominantly negative charge at the surface, as compared to chitotriosidase. This most likely restricts intramolecular repulsion as a consequence of excessive protonation at low pH. Moreover, AMCase contains two additional cysteines compared to chitotriosidase, a third disulfide bond most likely enhances the fold integrity in an acidic environment. Indirect evidence for this is provided by the observation that AMCase shows a different electrophoretic behaviour in the absence of a reducing agent (as compared to chitotriosidase), suggesting a difference in disulphide bonding between the two mammalian chitinases (Boot *et al.* 2001: Bussink *et al.* 2008).

Recently, AMCase attracted considerable attention due to reports linking the enzyme to the pathogenesis of asthma and other allergic inflammatory conditions of the lung (Zhu *et al.* 2004; Xu *et al.* 2003; Sandler *et al.* 2003; Zimmermann *et al.* 2004).

Intravenous injection of Schistosoma mansoni eggs was found to cause massive expression of AMCase in the lung of wildtype mice and animals with an exaggerated Th2 response, which is dominated by the cytokines IL4 and IL13. This induction did not occur in mice with a bias towards a Th1 response or IL13-knockout mice (Sandler *et al.* 2003). Zimmermann *et al.* reported highly induced AMCase mRNA levels in mouse models of experimental asthma either induced by ovalbumin (OVA) or by Aspergillus fumigatus antigen (Zimmermann *et al.* 2004). This induction was mediated by the STAT6 signalling pathway, again suggesting a role for IL4 or IL13 (Zimmermann *et al.* 2004). Very recently it was shown for an aeroallergen asthma mouse model that AMCase is induced in the lung via a Th2-specific, IL13-mediated pathway (Zhu *et al.* 2004). Interestingly, AMCase activity appeared instrumental in the pathogenesis of asthma. Inhibition of AMCase, either by a specific antibody or the specific chitinase inhibitor allosamidin, alleviated the Th2 mediated inflammatory damage that occurs in asthma (Zhu *et al.* 2004). It has been suggested that inhibition of chitinase activity may render an attractive new therapy of asthma (Zhu *et al.* 2004; Couzin, 2004).

A more recent report, however, suggests that chitin itself induces the accumulation in tissue of IL-4-expressing innate immune cells, including eosinophils and basophils, when given to mice. Moreover, this response was absent if the chitin was pre-treated with AMCase, or when chitin was given to mice overexpressing AMCase (Reese *et al.* 2007). The data presented in this latter report seem to conflict with the observations of Zhu *et al.* Whether AMCase is the good or the bad guy in asthma pathophysiology will need more investigation.

3. Polymorphisms in Human Chitinase Genes and Association with Diseases

In order to establish the physiological function of the mammalian chitinases, several groups searched for polymorphisms and other genetic defects in the chitinase genes or for increased levels of the enzyme in plasma in groups of patients with particular diseases. Due to the presence of chitin in coatings of several pathogens, such as fungi and nematodes, it has been suggested that chitotriosidase serves as component of innate immune responses (Renkema *et al.* 1995). The evidence for a role of chitotriosidase in such responses remains limited. Besides the highly increased levels in plasma of Gaucher patients, chitotriosidase activity is elevated in plasma of children suffering from acute infection with *Plasmodium falciparum* malaria (Barone *et al.* 2003). Since chitotriosidase is also elevated in other disorders involving red blood cells such as beta-thalassemia it is not clear whether increased red blood cell turnover might be responsible for the observed elevations in these diseases (Barone *et al.* 1999; Michelakakis *et al.* 2004).

Increases in plasma chitotriosidase have also been found in sera of individuals suffering from visceral *Leishmaniasis* (Hollak *et al.* 1994). Susceptibility to *Wuchereria Bancrofti*, which causes lymphatic filariasis, is associated with the common genetic deficiency in chitotriosidase in South India (Choi *et al.* 2001). However this correlation seems to be absent in Papua New Guinea (Hise *et al.* 2003; Hall *et al.* 2007). Virtually no heterozygotes or homozygotes for the common chitotriosidase gene defect seem to occur in areas endemic for certain parasites such as the Sub-Sahara, suggesting its activity is important under these conditions (Malaguarnera *et al.* 2003). Increased plasma chitotriosidase levels have also been reported during neonatal herpes virus infection (Michelakakis *et al,.* 2004). Recently, it has been been found that genetic variants in chitotriosidase are associated with Gram-negative bacteremia in children undergoing therapy for AML and that neonates with a bacterial infection show increases in chitotriosidase activity (Lehrnbecher *et al.* 2005, Michelakakis *et al.* 2004). These observed associations suggest that chitotriosidase has more pleotropic effects in innate immunity than previously appreciated. Deficiency of chitotriosidase was reported to be unrelated to the incidence of Candida sepsis, however, as the authors state, only survivors were included in this study. Therefore the evidence must be considered inconclusive (Masoud *et al.* 2002). Supporting evidence for a role in anti-fungal activity of chitotriosidase exists. First, chitotriosidase activity was found to be elevated in plasma of neonates upon systemic Candidiasis and Aspergilosis (Labadaridis *et al.* 1998; Michelakakis *et al.* 2004). Second, chitotriosidase was found to inhibit growth of *Cryptococcus neoformans*, to cause hyphal tip lysis in *Mucor rouxii* and to prevent the occurrence of hyphal switch in *Candida albicans* (Van Eijk *et al.* 2005). These data supported the earlier observed chitinolytic activity towards cell wall chitin of *C. albicans* (Boot *et al.* 2001). Moreover, it has been demonstrated that recombinant human chitotriosidase showed synergy with existing anti-fungal drugs such as the polyene amphotericin B, the azoles itraconazole and flucanozole and cell wall synthesis inhibitors LY-303366 and nikkomycin Z (Stevens *et al.* 2000). Recombinant human chitotriosidase clearly improved survival in neutropenic mouse models of systemic Candidiasis and systemic Aspergillosis (van Eijk *et al.* 2005). In addition, AMCase also shows chitinolytic activity towards fungal cell wall chitin (Boot *et al.* 2001). The discovery of this chitinase in man has opened up the possibility that a deficiency in chitotriosidase might be (partly) compensated by activity of the latter enzyme, possibly explaining some of the conflicting observations discussed.

Besides the common chitotriosidase deficiency due to the 24 bp insertion in exon 10, other chitotriosidase gene mutations associated with altered enzymatic function have been reported. The group of Desnick reported three novel mutations in the chitotriosidase gene, the E74K (cDNA 220G>A mutation), the G102S (cDNA 304G>A) and a complex mutation in exon 10 leading to one amino acid substitution and missplicing (cDNA 1060G>A (G354R), 1155G>A (L385L) combined with a deletion of an intronic 'gtaa' near the 5' splice site) (Grace *et al.* 2007). The mutant proteins have been found to have reduced activity as assessed with transient transfection experiments. The complex mutation leads to production of unstable protein without detectable enzyme activity. The E74K enzyme and the G102S mutations have 2 fold and 4 fold reduced activity respectively (Grace *et al.* 2007). Only the G102S mutation is relatively common (~30% of alleles), both other mutations were found to be less frequent (0-6% of the alleles depending on the ethnic background) (Grace *et al.* 2007). In our hands, the catalytic efficiency of recombinant Ser102 CHIT1 was ~70% of that of wild-type Gly102 CHIT1 when measured with 4MU-chitotrioside at non-saturating concentration. However, the activity was normal with 4MU-deoxychitobioside as substrate at saturating concentration (Bussink *et al.* submitted for publication). These results are in line with other reports that describe that the G102S and A442G mutations were not significantly associated with a reduction of enzyme activity and should be considered as polymorphisms (Lee *et al.* 2007). In addition, this research group found that the G354R and A442V mutations, occurring predominantly in subjects of African ancestry, were associated with reduced chitotriosidase activity (Lee *et al.* 2007). Furthermore, they also found an association of chitotriosidase deficiency and susceptibility to tuberculosis in individuals from Asian ancestry but not in subjects of European or African ancestry (Lee *et al.* 2007). It was also demonstrated that patients of Asian ancestry with atopy (having 3 or more of the following medical indications: allergic rhinitis, contact dermatitis, drug or food allergies or asthma) have an increased prevalence of chitotriosidase deficiency. The same trend was observed in subjects of European ancestry although the data were not statistically significant (Lee *et al.* 2007). A German study, however, failed to detect an association of chitotriosidase polymorphisms and asthma (Bierbaum *et al.* 2006).

By sequencing of the human AMCase gene 12 high-frequency polymorphisms were identified, that included three known and two novel amino acid variants. Analysis of these polymorphisms in children and adults with or without asthma, demonstrated an association of the K17R variant and the nearby non-coding polymorphism rs3818822 with asthma (Bierbaum *et al.* 2005). The nomenclature in this instance is somewhat confusing since the authors used the cDNA sequence of a splice variant and start counting at the first methionine in exon 5. Using the correct full-length cDNA sequence the substitution should be referred to as K125R (counting from the start methionine and including the signal peptide). A more comprehensive overview of AMCase polymorphisms in relation to asthma is given in the chapter of. Seibold and Burchard in this Book.

4. Chi-Lectins

Besides the active chitinases, highly homologous mammalian proteins lacking enzymatic activity due to substitution of active site catalytic residues have been identified. Despite their lack of enzymatic activity, these proteins have retained active site carbohydrate binding, and hence have been named chi-lectins (Renkema *et al.* 1998; Houston *et al.* 2003; Bussink *et al.* 2006).

Like the active chitinases, chi-lectins belong to family 18 of glycosyl hydrolases, consisting of a 39 kDa catalytic domain with a TIM-barrel structure (Sun *et al.* 2001; Fusetti *et al.* 2002; Fusetti *et al.* 2003; Houston *et al.* 2003). In contrast to both chitinases, chi-lectins lack the conserved additional chitin-binding domain (Boot *et al.* 1995; Renkema *et al.* 1997; Boot *et al.* 2001).

Despite the detailed knowledge regarding structure, insight in the exact physiological function of the various chi-lectins is limited (reviewed by Bussink *et al.* 2006). Like chitotriosidase and AMCase, chi-lectins are secreted locally or into the circulation and thus a role in inflammatory processes has been suggested. For example, human cartilage GP39 (Hcgp39/ YKL-40/ CHI3L1), a chi-lectin expressed by chondrocytes and phagocytes has been implicated in arthritis, tissue-remodelling, fibrosis, cancer and asthma (Johansen *et al.* 1993; Hakala *et al.* 1993; Verheijden *et al.* 1997; Recklies *et al.* 2002; Chupp *et al.* 2007 ; reviewed by Johansen, 2006). Similarly, the human chi-lectin YKL-39 (CHI3L2) and the murine chi-lectins Ym1 (Chi3L3/ ECF-L) and Ym2 (Chi3L4) have been associated with the pathogenesis of arthritis (Hu *et al.* 1996; Tsuruha *et al.* 2002) and allergic airway inflammation, respectively (Chang *et al.* 2001; Ward *et al.* 2001; Homer *et al.* 2006; Webb *et al.* 2001).

The high molecular weight oviductin glycoproteins (OGPs), consist of the amino-terminal 39 kDa catalytic domain followed by a heavily glycosylated Ser/Thr rich domain. They are secreted by nonciliated oviductal epithelial cells and have been shown to affect sperm maturation (capacitation), motility and viability. They also seem to play a role in fertilization itself and early embryo development through interactions with the oocyte (reviewed by Buhi, 2002).

5. Human Family 18 Chitinase Genes

The human genome contains 9 genes coding for family 18 of glycosyl hydrolases, of which only two code for true active chitinases: chitotriosidase and AMCase. Both the chitotriosidase gene (CHIT1) and the AMCase gene (CHIA) are present on human chromosome 1, at locus 1q32 and 1p13 respectively (Bussink *et al.* 2007). At the chitotriosidase locus next to the chitotriosidase gene itself the CHI3L1 gene is present, coding for the chi-lectin HCgp39. The AMCase locus contains the gene OVGP1 coding for the chi-lectin oviductin, and the gene CHI3L2 coding for the chi-lectin YKL39. The gene organization of the human chitinases is shown in figure 1, which also contains a depiction of the organization in mouse for comparison.

Next to the AMCase gene, two other genes (LOC728204 and LOC149620), are present at this locus and due to the high homology to AMCase we designated them AMCase-like pseudogenes, since essential exons are missing compared to the AMCase gene (Bussink *et al.* 2007).

The two distantly related family 18 glycosyl hydrolase genes are located at different loci.

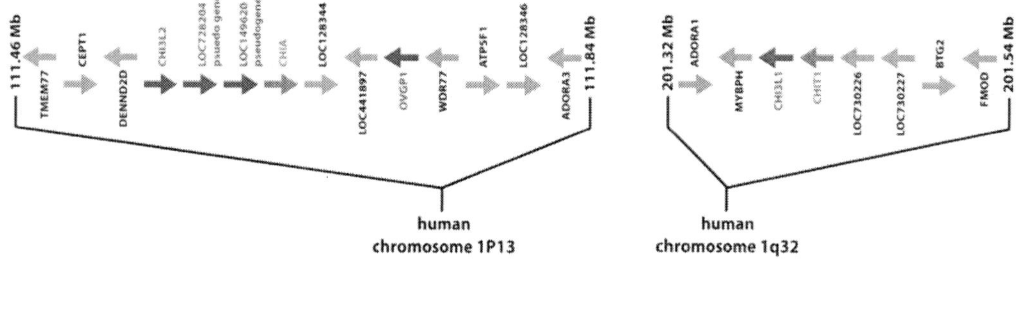

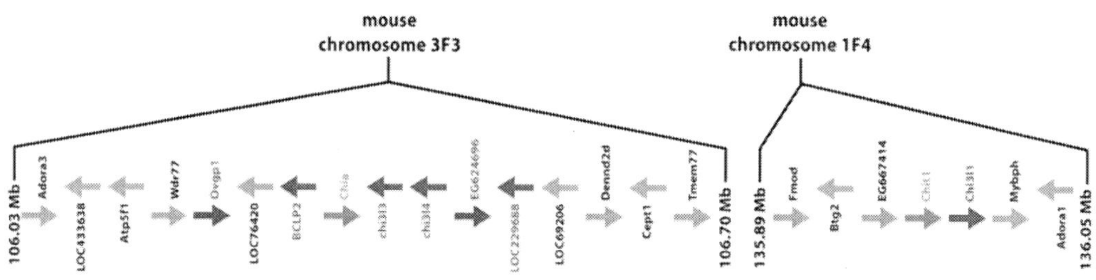

Figure 1. Synteny of loci encoding chitinase(-like) proteins between mice and men. Schematic overview of the synteny of mouse locus 1F4 with human 1q32 and mouse locus 3F3 with human 1p13. The orientation and position of the genes are indicated with arrows. The genes of members of the chitinase protein family are depicted by coloured arrows (true chitinases green, oviductin genes blue, the chi-lectin Gp39 genes red, other chi-lectins purple), whereas all other genes in the loci are depicted in grey.

The gene CTBS coding for chitobiase is located at chromosome 1p22, and the gene CHID1 coding for the stabilin-1 interacting chitinase-like protein is located at chromosome 11p15. The overall homology between these genes and that of the chitinases and chi-lectins is very low (Aronson *et al.* 1989; Kzhyshkowska *et al.* 2005; Funkhouser and Aronson, 2007).

The chromosomal location of the chitinase(-like) genes have also been mapped in other species such as mice. There is genomic synteny between mice and man regarding the chitinase(-like) genes (Bussink *et al.* 2007; Funkhouser and Aronson, 2007; see also figure 1).

The chitotriosidase locus (1q32 in humans and 1F4 in mice), containing the genes CHIT1 and CHI3L1, is flanked by the genes coding for adenosine A1 receptor (ADORA1) and fibromodulin (FMOD), and is syntenic (Bussink *et al.* 2007).

The AMCase loci, corresponding to locus 1p13 in humans and 3F3 in mice are also syntenic. Again the human and murine regions are flanked by the same genes, in this case coding for the adenosine A3 receptor (ADORA3) and the transmembrane protein 77 (TMEM77). The presence of an ADORA paralogue on chitotriosidase and AMCase loci suggests that the genes encoding the two active chitinases result from a large-scale duplication.

The AMCase locus reveals major differences between mouse and man. Firstly, additional open reading frames exist in the mouse genome encoding Ym1, Ym2, Bclp2 and BYm (encoded by Chi3l3, Chi3l4, BCLP2 and LOC229688, respectively), whereas these genes are absent from human chromosome 1p13 (Bussink *et al.* 2007; Funkhouser and Aronson, 2007).

The opposite holds true for CHI3L2 encoding YKL39. Secondly, additional AMCase-like pseudogenes, LOC728204 and LOC149620, are only present on human chromosome 1 and not in the mouse genome. Finally, the orientation of many genes in the human and mouse AMCase loci differ (see figure 1), suggesting the occurrence of multiple and diverse recombination events.

A remarkable observation is that there are differences in iso-electric points of the mammalian members of the chitinase protein family. The calculated iso-electric points are also conserved between the individual members of this family. For example all AMCase have an acidic pI (ranging from 4.68 in the dog to 5.42 in man), making them (near) neutral in an acidic environment. This is in complete concurrence with the observed expression in the mammalian stomach as observed for mouse and human AMCase (Boot *et al.* 2001). The same is the case for the members of the rodent Ym family with pI's around 5. It is of interest that at least one of the Ym members is expressed in the rodent stomach (Nio *et al.* 2004). The oviductins, which are expressed in the slightly basic oviduct, have a very basic pI's (ranging from 8.49 in mice to 9.44 in sheep). The gp39 and the YKL39 members have also a more basic pI but less pronounced as compared to the oviductins (ranging from 8.19 in *Macaca mulatta* to 9.14 in pig for the gp39 members and from 7.24 in man to 8.89 in cow for the YKL39 members). A much larger variation in pI exists among the chitotriosidases from different species (ranging from 5.94 in mice to 8.49 in the rat). The chitinases not only mostly retain their specific pI's when comparing different species, they also conserve cysteines, involved in structural disulfide bond formation. Taken together, the overall homology, conservation of cysteines and isoelectric point indicate interspecies retention of protein structure among orthologs.

6. Chitinase(-Like) Protein Family Tissue Expression

The expression of the various chitinase(-like) family members differs between species. In man chitotriosidase expression seems to be restricted to professional phagocytes. In healthy tissue samples chititriosidase expression is predominantly found in tissues rich in phagocytes such as lymphnodes, bone marrow and lung (Boot *et al.* 2005). As stated above, the expression can be highly induced. This is the case for example in Gaucher disease, in which lipid laden alternatively activated macrophages produce and secrete massive amounts of the enzyme (Boven *et al.* 2004). The expression pattern observed in man is in sharp contrast to that of mice in which chitotriosidase could not be detected in macrophages. High expression in this species is observed in the tongue, stomach and epidermis, with slightly lower amounts also in the trachea (Boot *et al.* 2005; Su *et al.* 2002). Using RT-PCR Zheng *et al.* could also demonstrate chitotriosidase mRNA in tissue samples from testis, bone marrow and brain (Zheng *et al.* 2005). With the aid of non-radioactive in situ hybridization (ISH) chitotriosidase expression was demonstrated in the stratified squamous epithelial cells of the tongue and non-glandular fore-stomach of mice. Expression of chitotriosidase in the mouse intestine was low as assessed with Northern blot analysis, however with ISH specific expression could be detected in the Paneth cells in the crypts Lieberkühn (Boot *et al.* 2005).

In both man and mice chitotriosidase activity could be detected in the plasma, however the exact source of mouse plasma chitotriosidase remains unknown.

AMCase was found to be expressed *largely* in the human stomach and to a lesser extend in the normal lung (Boot *et al.* 2001; Su *et al.* 2002). Using ISH, expression of AMCase was not readily apparent in human lung samples derived from control patients with nonpulmonary disease, but expression was readily detected in epithelial cells and macrophages in lung tissue samples taken from patients with asthma (Zhu *et al.* 2004). The exact cellular source of AMCase in the human stomach is unknown so far. AMCase in the mouse has a somewhat similar expression pattern when compared to that observed in man. High expression is detected in the gastric chief cells at the bottom of the gastric glands and in serous-type secretory cells of the parotid gland and von Ebner's gland in the salivary glands of the mouth and tongue respectively (Suzuki *et al.* 2002; Goto *et al.* 2003; Boot *et al.* 2005).

In the normal mouse lung AMCase expression could be detected, and its expression can be highly induced in response to (allergic and parasitic) inflammation and/or cytokines such as IL13 (Xu *et al.* 2003; Sandler *et al.* 2003; Zhu *et al.* 2004; Zimmermann *et al.* 2004; Boot *et al.* 2005; Homer *et al.* 2006). It is expressed in the distal airway epithelium and macrophages of these lungs (Homer *et al.* 2006). Intriguingly, in the cow AMCase is produced and secreted by the liver and present in large amounts in serum (Suzuki *et al.* 2001). It seems likely that expression is driven by yet other changes in the promoter region of this gene, adding to the rapidly growing complexity of this family.

In humans, HC-gp39 is amongst others expressed by articular chondrocytes and synovial cells, and as for human chitotriosidase, HC-gp39 expression is also observed in matured macrophages and specific granules of human neutrophils (Hakala *et al.* 1993; Renkema *et al.* 1998; Volck *et al.* 1998). In normal tissues expression is high in whole blood, bone marrow, lung, salivary glands, and the uterus corpus (Su *et al.* 2002). Less is known about expression of this gene in mice, though it has been demonstrated in the involuting mammary gland and the normal lung and bone (reviewed by Rathcke and Vestergaard, 2006; Su *et al.* 2002).

Human YKL-39 expression is expressed in human chondrocytes and synoviocytes, and its expression is induced in osteoarthritic cartilage (Hu *et al.* 1996; Steck *et al.* 2002). Tissue expression was found to be high in adipocytes, lymphoblastic leukemia cells and thymus (Su *et al.* 2002). This gene is not present in the mouse genome.

The genes coding for the highly homologous Ym1 and Ym2 are lacking from the human genome, and expression of Ym1 in normal mouse tissues was found to be high in bone, bone marrow, stomach and lung. While Ym2 expression was also found in these tissues and to a lesser extend in the salivary glands and tongue (Ward *et al.* 2001; Su *et al.* 2002). As is the case for the AMCase, YM1 is also expressed by alternatively activated tissue macrophages in allergic inflammation mostly driven by IL13 (Homer *et al.* 2006; Webb *et al.* 2001; Ward *et al.* 2001; Nio *et al.* 2004). In the allergically inflamed lung, expression was observed in the more proximal airway epithelial cells in contrast to AMCase that is expressed more distally (Homer *et al.* 2006). It has, like human chitotriosidase and human HCgp39, also been found in the specific granules of neutrophilic granulocytes in mice (Harbord *et al.* 2002). High expression of YM1 and/or Ym2 in different tissues could lead to eosinophilic crystal formation that could contribute to the observed pathology seen in situations with high Ym1/2 expression (Guo *et al.* 2000; Harbord *et al.* 2002). ISH analysis of normal mouse tissue showed expression of Ym1 in macrophages of spleen, lung and bone marrow and immature neutrophilic granulocytes, while Ym2 expression was restricted to stratified squamous epithelial cells and especially in the junctional region between fore-stomach and the glandular stomach (Nio *et al.* 2004). This leads to the continuous expression of chitinase-like proteins along the lining of the mouse stomach, beginning with chitotriosidase (expressed in the non-glandular fore-stomach in stratified squamous epithelium), followed by Ym2 (expressed in the stratified epithelium of the junction) and finally AMCase expression (by the glandular chief cells at the bottom of the glands).

The oviductins are expressed in the oviduct secretory epithelium in most mammals and appear to be regulated by oestrogen. A universal characteristic of the oviductins is their association with the zona pellucida and perivitelline space of oocytes and embryos (reviewed by Buhi, 2002). The expression pattern of this gene is the most invariable of all chitinase(-like) genes in all mammals tested.

7. Evolutionary Aspects of the Chitinase(-Like) Protein Family

Recent phylogenetic analyses using available sequence information of a large variety of chitinase(-like) proteins from lower eukaryotes to mammals gave new insights regarding the evolutionary history of the family 18 of glycosyl hydrolases (Funkhouser and Aronson, 2007; Bussink *et al.* 2007).

Funkhouser and Aronson showed that the current human family of chitinase(-like) protein genes originated from early ancestors already present at the time of bilatarian expansion (about 550 million years ago). The family expanded in the chitinous protostomes *Caenorhabditis elegans* and *Drosophila melanogaster*, followed by a decline in early deuterostomes as chitin synthesis disappeared, only to expand again in late deuterostomes with a significant increase in gene number occuring after the avian/mammalian split (Funkhouser and Aronson, 2007).

Both phylogenetic studies mentioned, revealed clustering of all mammalian orthologs that are grouped in either the AMCase or the chitotriosidase clade. Besides the active chitinases, both clades contain their 'own' chi-lectins allowing discrimination between so-called AMCase based chi-lectins and chitotriosidase based chi-lectins (Funkhouser and Aronson, 2007; Bussink *et al.* 2007). For example, the chitotriosidase clade contains the HCgp39 and YKL39 homologs. It also includes a second gp39 like gene that is highly homologous (96% nucleotide identity) named BP40 that seems specific for the artiodactyls (even-toed ungulates or hoofed mammals). Sequence analysis also revealed that the putative N-glycosylation site near the amino-terminus in the HCgp39 homologs is completely conserved, and suggests that BP40 results from an artiodactyl specific duplication of the HCgp39 gene (Bussink *et al.* 2007). The AMCase clade contains all known mammalian oviductins as well as the rodent specific Ym proteins (figure 2).

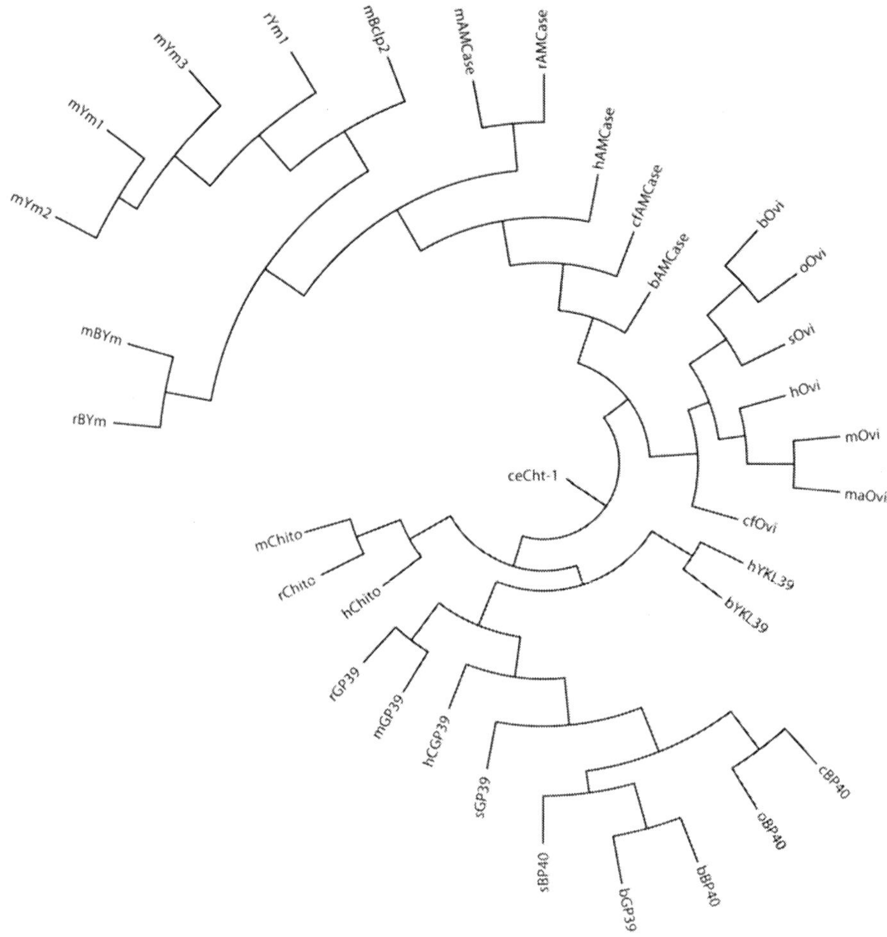

Figure 2. Phylogenetic relationship of mammalian chitinase(-like) genes. The tree was rooted with ceCht-1 (Caenorhabditis elegans chitinase-1). Abbreviations: Chito, Chitotriosidase; AMCase, Acidic Mammalian Chitinase; hcGP39, human cartilage glycoprotein 39; Ovi, Oviductin; BYm, Basic Ym; Bclp2, Brain chitinase-like protein-2; h, Homo sapiens; m, Mus musculus; r, Rattus norvegicus; b, Bos taurus; c, Capra hircus; o, Ovis aries; s, Sus scrofa; cf, Canis familiaris; ma, Mesocricetus auratus.

The phylogenetic analyses combined with the genomic synteny data point to a common evolutionary model for the appearance of the different mammalian family 18 chitinase proteins. Firstly, a gene duplication event allowed the specialization of the two active chitinases, chitotriosidase and AMCase. Duplications of both chitotriosidase and AMCase genes, followed by loss of enzymatic function mutations, led to the subsequent evolution of their respective chi-lectins.

The duplication of the active chitinase most likely has an ancient origin as *Xenopus* already has two active chitinases, one of which, like AMCase, is expressed in the stomach (Fujimoto *et al.* 2002). Indeed, the *Xenopus* AMCase homologue clusters with the mammalian AMCases (Bussink *et al.* 2007; Funkhouser and Aronson, 2007). This suggests that the gene duplication allowing evolution of chitotriosidases and AMCases occurred very early in tetrapod evolution in the wake of the development of the acidic stomach. The evolution of the various mammalian chi-lectins is most likely a more recent event. In this context, the occurrence in several plant species of chi-lectins homologous to chitinases that has been recently reported (van Damme *et al.* 2007) is also of note. Mutations resulting in loss of chitinolytic activity have apparently occurred independently in a large variety of lineages.

Theoretically, it is conceivable that the various types of chi-lectins in mammalian species have interdependently evolved by concerted evolution, a process driven by unequal crossover and gene conversion (reviewed in Nci and Rooncy, 2005). Thc lack of conscrvation of gene orientation within the AMCase locus (see figure 1) indeed shows that instances of recombination must have occurred. However, phylogenetic analyses show that orthologues are far more closely related than paralogues, which does not substantiate concerted evolution as an important contributing mechanism underlying the diversification of chi-lectins.

The phylogenetic analyses provide information on the evolutionary relationships of the chitinase(-like) protein coding genes. Insight in the selective forces shaping these relationships can, in principle, be derived from analysis of their substitution rate ratios. For example, as the AMCases have evolved to function in the acid environment of the mammalian stomach, it might be envisioned that there have been episodes of positive selection after the initial gene duplication. This would allow rapid adaptation to this environment as has been demonstrated for the mammalian stomach lysozymes (Messier and Stewart, 1997; Yang, 1998). Moreover, the repeated occurrence of the chi-lectin mutation (i.e. the mutation leading to loss of enzymatic function while retaining binding capacity) could suggest site-specific positive selection. Calculation of the substitution rate ratios, however, showed that, in contrast to the mammalian lysozyme family, there is no direct evidence for the occurrence of positive selection (Bussink *et al.* 2007). Only in the case of the mammalian oviductins there is evidence that some of the amino acids positions are subjected to positive selection but this could not be demonstrated for the other members of this protein family (Swanson *et al.* 2001; Bussink *et al.* 2007).

Based on the observed relationships and selective pressures, the evolution of this gene family is in accordance with a form of multigene family evolution now referred to as "birth-and-death evolution under strong purifying selection" (reviewed in Nei and Rooney, 2005).

The study by Bussink *et al.* also revealed the existence of a previously unidentified chi-lectin in mice, referred to as hypothetical protein (BYm). Despite the fact that it belongs to the group of AMCase- (or acidic-) lectins, it displays a basic iso-electric point, suggesting a substantial specialization. The absence of the gene in animals other than rodents points to a, relatively, recent gene duplication, with rapid adaptation to a new physiological role, possibly with bouts of positive selection. This nicely illustrates the remarkable ongoing evolution of chitinase(-like) proteins in mammals (Bussink *et al.* 2007).

8. The Chitinase(-Like) Protein Family: The Interplay between Structure and Evolution

The marked conservation of structural features of the catalytic domain of the family 18 chitinases may be the result of severe restrictions in changes compatible with preservation of catalytic function. Mammalian chitinases appear to be strongly subjected to negative (purifying) selection, substantially more so than the functionally similar lysozymes. This is nicely illustrated by the fact that comparison of substitution rate ratios between lysozyme from Rhesus macaque and that of hominoids point strongly towards positive selection whereas such comparisons for chitotriosidase and AMCase did not (see above and Bussink *et al.* 2007). A large number of constrained sites in both chitinases should mask any signal of positive selection when the omega values are averaged over all sites and longer time periods. Site-specific models for ω calculation did not give values above 1, though positive selection in the diversification of the two chitinases and their respective chi-lectins clearly occurred. To detect a very strong indicator of positive selection such as $\omega > 1$, it is necessary not only to look at a specific site but also at specific (relatively brief) time intervals (Sharp, 1997; Zhang *et al.* 2005).

Conclusion

In conclusion, our investigation of family 18 glycosyl hydrolases has revealed that active chitinases and chi-lectins are widespread and conserved in the mammalian kingdom. An ancient gene duplication first allowed the specialization of two active chitinases, chitotriosidase and AMCase, and subsequent gene duplications followed by loss of enzymatic function mutations, have led to the evolution of a broad spectrum of chi-lectins in mammals, involved in very diverse functional roles.

This diversification is nicely illustrated by the fact that the chitinase(-like) proteins fulfill quite different physiological roles in mice and man, as has been shown in numerous studies. This process is driven not only by alterations in the coding sequences themselves, but to a large extend by changes in the regulatory sequences, leading to extreme differences with regards to tissue distribution and amounts of expression.

The common theme in the role of most chitinase(-like) proteins seems to be inflammation. Most are highly regulated by 'outside' stimuli, nicely in accordance with their expression in epithelium and immune cells.

Though our insights with regard to the chitinase(-like) protein family have been profoundly altered over the last years, much remains to be discovered.

References

van Aalten DM, Komander D, Synstad B, Gaseidnes S, Peter MG, Eijsink VG. Structural insights into the catalytic mechanism of a family 18 exo-chitinase. *Proc Natl Acad Sci USA.* 2001; 98: 8979-8984.

Aguilera B, Ghauharali-van der Vlugt K, Helmond MT, Out JM, Donker-Koopman WE, Groener JE, Boot RG, Renkema GH, van der Marel GA, van Boom JH, Overkleeft HS, Aerts JM. Transglycosidase activity of chitotriosidase: improved enzymatic assay for the human macrophage chitinase. *J Biol Chem.* 2003; 278: 40911-40916.

Araujo AC, Souto-Padrón T, De Souza W. Cytochemical localization of carbohydrate residues in microfiliae of Wuchereria bancrofti and Brugia malayi. *J Histochem Cytochem*. 1993; 41: 571-578.

Aronson NN, Kuranda MJ. Lysosomal degradation of Asn-linked glycoproteins. *FASEB J*. 1989; 3: 2615-2622.

Aronson NN, Backes M, Kuranda MJ. Rat liver chitobiase: purification, properties, and role in the lysosomal degradation of Asn-linked glycoproteins. *Arch Biochem Biophys*. 1989; 272: 290-300.

Barone R, Di Gregorio F, Romeo MA, Schilirò G, Pavone L. Plasma chitotriosidase activity in patients with beta-thalassemia. *Blood Cells Mol Dis*. 1999; 25: 1-8.

Barone R, Simpore J, Malaguarnera L, Pignatelli S, Musumeci S. Plasma chitotriosidase activity in acute Plasmodium falciparum malaria. *Clin Chim Acta*. 2003; 331: 79-85.

Bierbaum S, Nickel R, Koch A, Lau S, Deichmann KA, Wahn U, Superti-Furga A, Heinzmann A. Polymorphisms and haplotypes of acid mammalian chitinase are associated with bronchial asthma. *Am J Respir Crit Care Med*. 2005; 172: 1505-1509.

Bierbaum S, Superti-Furga A, Heinzmann A. Genetic polymorphisms of chitotriosidase in Caucasian children with bronchial asthma. *Int J Immunogenet*. 2006; 33: 201-204.

Boot RG, Renkema GH, Strijland A, van Zonneveld AJ, Aerts JM. Cloning of a cDNA encoding chitotriosidase, a human chitinase produced by macrophages. *J Biol Chem*. 1995; 270: 26252-26256.

Boot RG, Renkema GH, Verhoek M, Strijland A, Bliek J, de Meulemeester TM, Mannens MM, Aerts JM. The human chitotriosidase gene. Nature of inherited enzyme deficiency. *J Biol Chem*. 1998; 273: 25680-25685.

Boot RG, Blommaart EF, Swart E, Ghauharali-van der Vlugt K, Bijl N, Moe C, Place A, Aerts JM. Identification of a novel acidic mammalian chitinase distinct from chitotriosidase. *J Biol Chem*. 2001; 276: 6770-6778.

Boot RG, Bussink AP, Verhoek M, de Boer PA, Moorman AF, Aerts JM. Marked differences in tissue-specific expression of chitinases in mouse and man. *J Histochem Cytochem*. 2005; 53: 1283-1292.

Boussac M, Garin J. Calcium-dependent secretion in human neutrophils: a proteomic approach. *Electrophoresis*. 2000; 21: 665-672.

Boven LA, van Meurs M, Boot RG, Mehta A, Boon L, Aerts JM, Laman JD. Gaucher cells demonstrate a distinct macrophage phenotype and resemble alternatively activated macrophages. *Am J Clin Pathol*. 2004; 122: 359-369.

Bowman SM, Free SJ. The structure and synthesis of the fungal cell wall. *Bioessays*. 2006; 28: 799-808.

Buhi WC. Characterization and biological roles of oviduct-specific, oestrogen-dependent glycoprotein. *Reproduction* 2002; 123: 355-362.

Bussink AP, van Eijk M, Renkema GH, Aerts JM, Boot RG. The biology of the Gaucher cell: the cradle of human chitinases. *Int Rev Cytol*. 2006; 252: 71-128.

Bussink AP, Speijer D, Aerts JM, Boot RG. Evolution of mammalian chitinase(-like) members of family 18 glycosyl hydrolases. *Genetics*. 2007; 177: 959-970.

Bussink AP, Vreede J, Aerts JM, Boot RG. A single histidine residue modulates enzymatic activity in acidic mammalian chitinase. *FEBS Lett*. 2008; 582: 931-935.

Cabib E, Blanco N, Grau C, Rodríguez-Peña JM, Arroyo J. Crh1p and Crh2p are required for the cross-linking of chitin to beta(1-6)glucan in the Saccharomyces cerevisiae cell wall. *Mol Microbiol.* 2007; 63: 921-935.

Chang NC, Hung SI, Hwa KY, Kato I, Chen JE, Liu CH, Chang AC. Macrophage protein, Ym1, transiently expressed during inflammation is a novel mammalian lectin. *J Biol Chem.* 2001; 276: 17497-17506.

Choi EH, Zimmerman PA, Foster CB, Zhu S, Kumaraswami V, Nutman TB, Chanock SJ. Genetic polymorphisms in molecules of innate immunity and susceptibility to infection with Wuchereria bancrofti in South India. *Genes Immun.* 2001; 2: 248-253.

Chou YT, Yao S, Czerwinski R, Fleming M, Krykbaev R, Xuan D, Zhou H, Brooks J, Fitz L, Strand J, Presman E, Lin L, Aulabaugh A, Huang X. Kinetic characterization of recombinant human acidic mammalian chitinase. *Biochemistry.* 2006; 45: 4444-4454.

Chupp GL, Lee CG, Jarjour N, Shim YM, Holm CT, He S, Dziura JD, Reed J, Coyle AJ, Kiener P, Cullen M, Grandsaigne M, Dombret MC. Aubier M, Pretolani M, Elias JA. A chitinase-like protein in the lung and circulation of patients with severe asthma. *N Engl J Med.* 2007; 357: 2016-2027.

Cid VJ, Duran A, del Rey F, Snyder MP, Nombela C, Sanchez M. Molecular basis of cell integrity and morphogenesis in Saccharomyces cerevisiae. *Microbiological Reviews.* 1995; 59: 345-386.

Couzin J. Unexpectedly, ancient molecule tied to asthma. *Science.* 2004; 304: 1577.

Van Damme EJ, Culerrier R, Barre A, Alvarez R, Rougé P, Peumans WJ. A novel family of lectins evolutionarily related to class V chitinases: an example of neofunctionalization in legumes. *Plant Physiol.* 2007 ; 144: 662-672.

Debono M, Gordee RS. Antibiotics that inhibit fungal cell wall development. *Annual Review of Microbiology.* 1994; 48: 471-497.

van Eijk M, van Roomen CP, Renkema GH, Bussink AP, Andrews L, Blommaart EF, Sugar A, Verhoeven AJ, Boot RG, Aerts JM. Characterization of human phagocyte-derived chitotriosidase, a component of innate immunity. *Int Immunol.* 2005; 17: 1505-1512.

van Eijk M, Scheij SS, van Roomen CP, Speijer D, Boot RG, Aerts JM. TLR- and NOD2-dependent regulation of human phagocyte-specific chitotriosidase. *FEBS Lett.* 2007; 581:5389-5395.

Flach J, Pilet PE, Jolles P. What's new in chitinase research? Experientia. 1992; 48: 701-716.

Fuhrman JA, Piessens WF. Chitin synthesis and sheath morphogenesis in Brugia malayi microfilariae. *Molecular and Biochemical Parasitology.* 1985; 17: 93-104.

Fujimoto W, Kimura K, Iwanaga T. Cellular expression of the gut chitinase in the stomach of frogs Xenopus laevis and Rana catesbeiana. *Biomed Res.* 2002; 23: 91-99.

Funkhouser JD, Aronson NN. Chitinase family GH18: evolutionary insights from the genomic history of a diverse protein family. *BMC Evol Biol.* 2007; 7; 96

Fusetti F, von Moeller H, Houston D, Rozeboom HJ, Dijkstra BW, Boot RG, Aerts JM, van Aalten DM. Structure of human chitotriosidase. Implications for specific inhibitor design and function of mammalian chitinase-like lectins. *J Biol Chem.* 2002; 277: 25537-25544.

Fusetti F, Pijning T, Kalk KH, Bos E, Dijkstra BW. Crystal structure and carbohydrate-binding properties of the human cartilage glycoprotein-39. *J Biol Chem.* 2003; 278: 37753-37760.

Gooday GW. Diversity of roles for chitinases in nature. In M. B. Zakaria Editor, W. M. W. Muda Editor and M. P. Editor Chitin and chitosan - the versatile environmentally friendly modern materials. 1995. (pp. 191-202). Pangi, Malaysia: Kebangsaan Malaysia Penerbit Universiti

Goto M, Fujimoto W, Nio J, Iwanaga T, Kawasaki T. Immunohistochemical demonstration of acidic mammalian chitinase in the mouse salivary gland and gastric mucosa. *Arch Oral Biol.* 2003; 48: 701-707.

Grace ME, Balwani M, Nazarenko I, Prakash-Cheng A, Desnick RJ. Type 1 Gaucher disease: null and hypomorphic novel chitotriosidase mutations-implications for diagnosis and therapeutic monitoring. *Hum Mutat.* 2007; 28: 866-873.

Guo L, Johnson RS, Schuh JC. Biochemical characterization of endogenously formed eosinophilic crystals in the lungs of mice. *J Biol Chem.* 2000; 275: 8032-8037.

Hakala BE, White C, Recklies AD. Human cartilage gp-39, a major secretory product of articular chondrocytes and synovial cells, is a mammalian member of a chitinase protein family. *J Biol Chem.* 1993; 268: 25803-25810.

Hall AJ, Quinnell RJ, Raiko A, Lagog M, Siba P, Morroll S, Falcone FH. Chitotriosidase deficiency is not associated with human hookworm infection in a Papua New Guinean population. *Infect Genet Evol.* 2007; 7: 743-747.

Harbord M, Novelli M, Canas B, Power D, Davis C, Godovac-Zimmermann J, Roes J, Segal AW. Ym1 is a neutrophil granule protein that crystallizes in p47phox-deficient mice. *J Biol Chem.* 2002; 277:5468-5475.

Henrissat BA. Classification of glycosyl hydrolases based on amino acid sequence similarities. *Biochem J.* 1991; 280: 309-316.

Hise AG, Hazlett FE, Bockarie MJ, Zimmerman PA, Tisch DJ, Kazura JW. Polymorphisms of innate immunity genes and susceptibility to lymphatic filariasis. *Genes Immun.* 2003; 4: 524-527.

Hollak CE, van Weely S, van Oers MH, Aerts JM. Marked elevation of plasma chitotriosidase activity. A novel hallmark of Gaucher disease. *J Clin Invest.* 1994; 93: 1288-1292.

Holtje JVLytic transglycosylases. *EXS.* 1996; 75: 425-429.

Homer RJ, Zhu Z, Cohn L, Lee CG, White WI, Chen S, Elias JA. Differential expression of chitinases identify subsets of murine airway epithelial cells in allergic inflammation. *Am. J Physiol Lung Cell Mol Physiol.* 2006; 291: L502-511.

Houston DR, Recklies AD, Krupa JC, van Aalten DM. Structure and ligand-induced conformational change of the 39-kDa glycoprotein from human articular chondrocytes. *J Biol Chem.* 2003; 278: 30206-30212.

Hu B, Trinh K, Figueira WF, Price PA. Isolation and sequence of a novel human chondrocyte protein related to mammalian members of the chitinase protein family. *J Biol Chem.* 1996; 271: 19415-19420.

Johansen JS, Jensen HS, Price PA. A new biochemical marker for joint injury. Analysis of YKL-40 in serum and synovial fluid. *Br J Rheumatol.* 1993; 32: 949-955.

Johansen JS. Studies on serum YKL-40 as a biomarker in diseases with inflammation, tissue remodelling, fibroses and cancer. *Dan Med Bull.* 2006; 53: 172-209.

Kzhyshkowska J, Mamidi S, Gratchev A, Kremmer E, Schmuttermaier C, Krusell L, Haus G, Utikal J, Schledzewski K, Scholtze J, Goerdt S. Novel stabilin-1 interacting chitinase-like protein (SI-CLP) is up-regulated in alternatively activated macrophages and secreted via lysosomal pathway. *Blood.* 2005; 107: 3221-3228.

Labadaridis J, Dimitriou E, Costalos C, Aerts J, van Weely S, Donker-Koopman WE, Michelakakis H. Serial chitotriosidase activity estimations in neonatal systemic candidiasis. *Acta Paediatr.* 1998; 87: 605.

Lee P, Waalen J, Crain K, Smargon A, Beutler E. Human chitotriosidase polymorphisms G354R and A442V associated with reduced enzyme activity. *Blood Cells Mol Dis.* 2007; 39:353-360.

Lehrnbecher T, Bernig T, Hanisch M, Koehl U, Behl M, Reinhardt D, Creutzig U, Klingebiel T, Chanock SJ. Common genetic variants in the interleukin-6 and chitotriosidase genes are associated with the risk for serious infection in children undergoing therapy for acute myeloid leukemia. *Leukemia.* 2005; 19: 1745-1750.

Lundblad G, Hedertedt B, Lind J, Steby M. Chitinase in goat serum. Preliminary purification and characterization. *Eur J Biochem.* 1974; 46: 367-376.

Lundblad G, Elander M, Lind J, Slettengren K. Bovine serum chitinase. *Eur J Biochem.* 1979; 100: 455-460.

Malaguarnera L, Simpore J, Prodi DA, Angius A, Sassu A, Persico I, Barone R, Musumeci S. A 24-bp duplication in exon 10 of human chitotriosidase gene from the sub-Saharan to the Mediterranean area; role of parasitic diseases and environmental conditions. *Genes Immun.* 2003; 4: 570-574.

Masoud M, Rudensky B, Elstein D, Zimran A. Chitotriosidase deficiency in survivors of Candida sepsis. *Blood Cells Mol Dis.* 2002; 29: 116-118.

Messier W, Stewart CB. Episodic adaptive evolution of primate lysozymes. *Nature.* 1997; 385: 151-154.

Michelakakis H, Dimitriou E, Labadaridis I. The expanding spectrum of disorders with elevated plasma chitotriosidase activity: an update. *J Inherit Metab Dis.* 2004; 27: 705-706.

Monzingo AF, Marcotte EM, Hart PJ, Robertus JD. Chitinases, chitosanases, and lysozymes can be divided into procaryotic and eucaryotic families sharing a conserved core. *Nature Structural Biology.* 1996; 3: 133-140.

Mulisch M. Chitin in protistan organisms. *Eur. J. Protistol.* 1993; 29: 1-18.

Nei M, Rooney AP. Concerted and birth-and-death evolution of multigene families. *Annu Rev Genet.* 2005; 39: 121-152.

Nio J, Fujimoto W, Konno A, Kon Y, Owhashim M, Iwanaga T. Cellular expression of murine Ym1 and Ym2, chitinase family proteins, as revealed by in situ hybridization and immunohistochemistry. *Histochem Cell Biol.* 2004; 121: 473-482.

Overdijk B, van Steijn GJ. Human serum contains a chitinase: identification of an enzyme, formerly described as 4-methylumbelliferyl-tetra-N-acetylchitotetraoside hydrolase (MU-TACT) hydrolase. *Glycobiology.* 1994; 4: 797-803.

Rao FV, Houston DR, Boot RG, Aerts JM, Sakuda S, van Aalten DM. Crystal structures of allosamidin derivatives in complex with human macrophage chitinase. *J Biol Chem.* 2003; 278: 20110-20116.

Rathcke CN, Vestergaard H. YKL-40, a new inflammatory marker with relation to insulin resistance and with a role in endothelial dysfunction and atherosclerosis. *Inflamm Res.* 2006; 55: 221-227.

Recklies AD, White C, Ling H. The chitinase 3-like protein human cartilage glycoprotein 39 (HC-gp39) stimulates proliferation of human connective-tissue cells and activates both extracellular signal-regulated kinase- and protein kinase B-mediated signalling pathways. *Biochem J.* 2002; 365: 119-126.

Reese TA, Liang HE, Tager AM, Luster AD, Van Rooijen N, Voehringer D, Locksley RM. Chitin induces accumulation in tissue of innate immune cells associated with allergy. *Nature*. 2007; 447: 92-96.

Renkema GH, Boot RG, Muijsers AO, Donker-Koopman WE, Aerts JM. Purification and characterization of human chitotriosidase, a novel member of the chitinase family of proteins. *J Biol Chem.* 1995; 270: 2198-2202.

Renkema GH, Boot RG, Strijland A, Donker-Koopman WE, van den Berg M, Muijsers AO, Aerts JM. Synthesis, sorting, and processing into distinct isoforms of human macrophage chitotriosidase. *Eur J Biochem.* 1997; 244: 279-285.

Renkema GH, Boot RG, Au FL, Donker-Koopman WE, Strijland A, Muijsers AO, Hrebicek M, Aerts JM. Chitotriosidase, a chitinase, and the 39-kDa human cartilage glycoprotein, a chitin-binding lectin, are homologues of family 18 glycosyl hydrolases secreted by human macrophages. *Eur J Biochem.* 1998; 251: 504-509.

Robbins PW, Albright C, Benfield B. Cloning and expression of Streptomyces plicatus chitinase (chitinase-63) in Escherichia coli. *J Biol Chem.* 1988; 263: 443-447.

Sahai AS, Manocha MS. Chitinases of fungi and plants: their involvement in morphogenesis and host-parasite interaction. *FEMS Microbiology Reviews.* 1993; 11: 317-338.

Sandler NG, Mentink-Kane MM, Cheever AW, Wynn TA. Global gene expression profiles during acute pathogen-induced pulmonary inflammation reveal divergent roles for Th1 and Th2 responses in tissue repair. *J Immunol.* 2003; 171: 3655-3667.

Sharp PM. In search of molecular Darwinism. *Nature.* 1997; 385: 111-112.

Steck E, Breit S, Breusch SJ, Axt M, Richter W. Enhanced expression of the human chitinase 3-like 2 gene (YKL-39) but not chitinase 3-like 1 gene (YKL-40) in osteoarthritic cartilage. *Biochem Biophys Res Commun.* 2002; 299: 109-115.

Stevens DA, Brammer HA, Meyer DW, Steiner BH. Recombinant human chitinase. *Curr Opin Anti-Infective Investig Drugs.* 2000; 2: 399–404.

Su AI, Cooke MP, Ching KA, Hakak Y, Walker JR, Wiltshire T, Orth AP, Vega RG, Sapinoso LM, Moqrich A, Patapoutian A, Hampton GM, Schultz PG, Hogenesch JB. Large-scale analysis of the human and mouse transcriptomes. *Proc Natl Acad Sci USA.* 2002; 99: 4465-4470.

Sun YJ, Chang NC, Hung SI, Chang AC, Chou CC, Hsiao CD. The crystal structure of a novel mammalian lectin, Ym1, suggests a saccharide binding site. *J Biol Chem. 2001*; 276: 17507-17514

Suzuki M, Morimatsu M, Yamashita T, Iwanaga T, Syuto B. A novel serum chitinase that is expressed in bovine liver. *FEBS Lett.* 2001; 506: 127-130.

Suzuki M, Fujimoto W, Goto M, Morimatsu M, Syuto B, Iwanaga T. Cellular expression of gut chitinase mRNA in the gastrointestinal tract of mice and chickens. *J Histochem Cytochem.* 2002; 50: 1081-1089.

Swanson WJ, Yang Z, Wolfner MF, Aquadro CF. Positive Darwinian selection drives the evolution of several female reproductive proteins in mammals. *Proc Natl Acad Sci USA.* 2001; 98: 2509-2514.

den Tandt WR, Inaba T, Verhamme I, Overdijk B, Brouwer J, Prieur D. Non-identity of human plasma lysozyme and 4-methylumbelliferyl-tetra-N-acetyl-β-D-chitotetraoside hydrolase. *Int J Biochem.* 1988; 20: 713-719.

den Tandt WR, Scharpe S, Overdijk B. Evaluation on the hydrolysis of methylumbelliferyl-tetra-N-acetylchitotetraoside by various glucosidases. A comparative study. *Int J Biochem.* 1993; 25: 113-119.

Terwisscha van Scheltinga AC, Armand S, Kalk KH, Isogai A, Henrissat B, Dijkstra BW. Stereochemistry of chitin hydrolysis by a plant chitinase/lysozyme and X-ray structure of a complex with allosamidin: evidence for substrate assisted catalysis. *Biochemistry.* 1995; 34: 15619-15623.

Tjoelker LW, Gosting L, Frey S, Hunter CL, Trong HL, Steiner B, Brammer H, Gray PW. Structural and functional definition of the human chitinase chitin-binding domain. *J Biol Chem.* 2000; 275: 514-520.

Tsuruha J, Masuko-Hongo K, Kato T, Sakata M, Nakamura H, Sekine T, Takigawa M, Nishioka K. Autoimmunity against YKL-39, a human cartilage derived protein, in patients with osteoarthritis. *J Rheumatol.* 2002; 29: 1459-1466.

Verheijden GF, Rijnders AW, Bos E, Coenen-de Roo CJ, van Staveren CJ, Miltenburg AM, Meijerink JH, Elewaut D, de Keyser F, Veys E, Boots AM. Human cartilage glycoprotein-39 as a candidate autoantigen in rheumatoid arthritis. *Arthritis Rheum.* 1997; 40: 1115-1125.

Volck B, Price PA, Johansen JS, Sorensen O, Benfield TL, Nielsen HJ, Calafat J, Borregaard N. YKL-40, a mammalian member of the chitinase family, is a matrix protein of specific granules in human neutrophils. *Proc Assoc Am Physicians.* 1998; 110: 351-360.

Ward JM, Yoon M, Anverm MR, Haines DC, Kudo G, Gonzalez FJ, Kimura S. Hyalinosis and Ym1/Ym2 gene expression in the stomach and respiratory tract of 129S4/SvJae and wild-type and CYP1A2-null B6, 129 mice. *Am J Pathol.* 2001; 158: 323-332.

Webb DC, McKenzie AN, Foster PS. Expression of the Ym2 lectin-binding protein is dependent on interleukin (IL)-4 and IL-13 signal transduction: identification of a novel allergy-associated protein. *J Biol Chem.* 2001; 276: 41969-41976.

Weigel PH, DeAngelis PL. Hyaluronan synthases: a decade-plus of novel glycosyltransferases. *J Biol Chem.* 2007; 282: 36777-36781.

Xu Y, Clark JC, Aronow BJ, Dey CR, Liu C, Wooldridge JL, Whitsett JA. Transcriptional adaptation to cystic fibrosis transmembrane conductance regulator deficiency. *J Biol Chem.* 2003; 278: 7674-7682.

Yang Z. Likelihood ratio tests for detecting positive selection and application to primate lysozyme evolution. *Mol Biol Evol.* 1998; 15: 568-573.

Zhang J, Nielsen R, Yang Z. Evaluation of an improved branch-site likelihood method for detecting positive selection at the molecular level. *Mol Biol Evol.* 2005; 22: 2472-2479.

Zheng T, Rabach M, Chen NY, Rabach L, Hu X, Elias JA, Zhu Z. Molecular cloning and functional characterization of mouse chitotriosidase. *Gene.* 2005; 357: 37-46.

Zhu Z, Zheng T, Homer RJ, Kim YK, Chen NY, Cohn L, Hamid Q, Elias JA. Acidic mammalian chitinase in asthmatic Th2 inflammation and IL-13 pathway activation. *Science.* 2004; 304: 1678-1681.

Zimmermann N, Mishra A, King NE, Fulkerson PC, Doepker MP, Nikolaidis NM, Kindinger LE, Moulton EA, Aronow BJ, Rothenberg ME. Transcript signatures in experimental asthma: identification of STAT6-dependent and -independent pathways. *J Immunol.* 2004; 172: 1815-1824.

In: Binomium Chitin-Chitinase: Recent Issues
Editor: Salvatore Musumeci and Maurizio G. Paoletti
ISBN 978-1-60692-339-9

Chapter VI

Polymorphism of Chitotriosidase in Human Populations

Andrea Angius* *
Institute of Population Genetics, National Research Council, (CNR), Alghero, Italy

Abstract

CHIT1 has been the first human gene encoding a chitinolytic enzyme to be discovered. CHIT1 gene product, designated as chitotriosidase (Chit) is a member of the chitinase family and it synthesized by activated macrophages. Sequence homology studies indicate that CHIT1 gene is conserved across the evolutionary scale and consequently has an important biological role. Recently, a genetic polymorphism (a 24 bp duplication in exon 10) was found to be responsible for the common deficiency in Chit activity, frequently encountered in different populations. The presence of the duplication in individuals from various ethnic groups suggests that this mutation is relatively old.

Here we discuss the analysis of the CHIT1 gene in some ethnic groups from the Mediterranean, African to Asian areas, to evaluate whether the CHIT1 gene polymorphism H correlates with the changes in environmental features. From a population point of view, the understanding of the variability of the CHIT1 variants improve the knowledge on origin and diffusion of the gene from an original population to other people living in different world areas.

We can also use the study of CHIT1 variants to perform a correlation between the mean Chit enzyme activity with a particular genotype and the origin ancestry of population. The median enzyme activity in wild-type subjects was significantly higher in subjects of European ancestry, than subjects of African and Asian ancestry. Moreover, genomic analysis of individuals heterozygous or wild type for the H polymorphism with little or absence of enzyme activity allows to identified several polymorphisms related to

* Address correspondence to: Andrea Angius, Ph.D., Institute of Population Genetics, National Research Council, (CNR), S.P. 55, Km. 8,400, Loc. Tramariglio, 07041 Alghero, Italy, Tel +39 079 946706-08, Fax +39 079 946714, E-mail angius@igp.cnr.it

the Chit activity. The presence of mutations e/o polymorphisms, as the G354R and the A442V, occurring predominantly in subjects of African ancestry directly influence the Chit activity.

Abbreviations

Chit	Chitinase;Chitotriosidase I; Methylumbelliferyl-tetra-N-acetylchitotetraoside hydrolase
CHIT1	Chitinase 1 gene
AMCase	Acidic mammalian chitinase;
T	24 bp duplication wild type allele
H	24 bp duplication mutant allele

1.Introduction

Chitinases are enzymes that hydrolyze chitin: a glycopolymer of β-(1,4)-linked N-acetylglucosamine present as a structural component in the coating of the cell wall of fungi Kuranda and Robbins, 1991), the sheath of nematodes (Wu *et al.* 2001), and protozoan parasites (Vinetz *et al.* 1999), and in the gut lining of many insects (Cohen, 1993). They are expressed in different parasites as nematodes, fungi and insects. It has been thought for a long time that chitinases have no important physiological function in humans. During the last decade, a substantial number of studies attempted to clarify its cellular functions, which have not been fully defined yet. Nevertheless, at least two distinguished types of human chitinases have been identified: chitotriosidase (Chit) (Boot *et al.* 1995) and acidic mammalian chitinase (AMCase) (Boot *et al.* 2001). Both enzymes cleaves chitin contained in many different human parasites. So far, only little is known about their function in man and mainly in their implication in various human diseases associated with chitinases activity alteration, such as asthma, arthritis, multiple sclerosis, Gaucher disease and Alzheimer's disease. The expression pattern of both chitinases varies completely, Chit is exclusively produced by phagocytes, whereas AMCase is expressed in alveolar macrophages and in the gastrointestinal tract (Boot *et al.* 2005).

Because of its phagocyte-specific expression, CHIT1 is supposed to play a role in innate immunity. This hypothesis might be confirmed by the observation that recombinant human Chit targets chitin-containing fungi (van Eijk *et al.* 2005). Furthermore, genetic association was found between a 24 bp duplication of CHIT1 and susceptibility to several parasite infections (Choi *et al.* 2001) and Gram-negative bacteraemia in children with acute myeloid leukaemia (Lehrnbecher *et al.* 2005). This 24 bp duplication completely demolish Chit activity by activating a 3-prime splice site that leads to an in-frame deletion of 87 nucleotides (Boot *et al.* 1998). On the other hand, AMCase has been shown to be very important in the pathogenesis of bronchial asthma in mice models and also to be highly expressed in human asthmatics lungs (Zhu *et al.* 2004).

Interestingly, the 24 bp duplication and other biallelic variants, both in CHIT1 and AMCase, showed very different allelic distribution in Caucasian, African and Asian populations. Maintenance of certain allelic variants in the population over time might reflect selective pressures in previous generations. The study of the different frequencies of the allelic variants highlights how the understanding of these, and the population movements, can be useful in tracing the dispersal of disease-causing mutant alleles, and how these data could be applied to predicting the segregation of mutant alleles within populations. Herein is a discussion of how genetic variants might explain, at least in part, the origin and the diffusion of some chitinase polymorphisms and their implication and/or association to the role of these genes in the different function meanings in various diseases.

2. Chitinases Genomic Structure and Gene Function

Chit has been the first human analogue of chitinases to be discovered. In about 6% of Caucasians, the enzyme shows very low activity without apparent symptoms (pseudodeficiency) and a very marked increase in Chit activity in the plasma of type I Gaucher disease patients: the median activity of Chit in Gaucher disease was more than 600 times the median value in plasma of healthy volunteers (Hollak *et al.* 1994).

Renkema *et al.* (1995) purified and characterized the Chit protein from the spleen of a Gaucher patient. Two major isoforms with isoelectric points of 7.2 and 8.0 and molecular masses of 50 and 39 kDa, respectively, were found to have identical N-terminal amino acid sequences. An antiserum raised against the purified 39-kD Chit precipitated all isozymes. This finding suggested that a single gene may encode the different isoforms of Chit (Boot *et al.* 1998)

The Chit cDNA was cloned utilizing a macrophage library (Boot *et al.* 1995) showing a nucleotide sequence that predict a protein identical to the previously purified Chit. By linkage studies, the CHIT1 locus have been mapped to 1q31-qter between flanking markers D1S191 and D1S245. Fluorescence in situ hybridization (FISH) method refined the CHIT1 locus to 1q31-q32. The CHIT1 gene consists of 12 exons and spans about 20 kb of genomic DNA (Boot *et al.* 1995). Several CHIT1 gene regions have high homology to sequences present in chitinases from different species belonging to family 18 of glycosylhydrolases (Henrissat and Bairoch, 1993, Boot *et al.* 1995). Other human members of the chitinase protein family were identified: oviductin (human oviduct-specific glycoprotein) (Arias *et al.* 1994), human cartilage glycoprotein 39 (HCgp-39/ YKL40) (Hakala *et al.* 1993), YKL39 (Hu *et al.* 1996), and TSA 1902 (Saito *et al.* 1999). Even though a significant homology was shared between the four human chitinases, the glycosyl hydrolase activity was present only in CHIT1 product. The biological function of this family has not been fully elucidated. It has been suggested that they might have a role in a tissue-remodeling processes (Hakala *et al.* 1993), or chemotaxis (Owhashi *et al.* 2000, Malinda *et al.* 1999).

The CHIT1 gene is present and functional in rodents and primates. Comparison of human CHIT1 to mouse CHIT revealed 79% identity at the nucleotide level and 72% identity (80% similarity) at the protein level. Additionally, human and rat proteins show 82% similarity, the same which was found between mouse and rat proteins. Sequence homology found among human, mouse and rat suggests a functional conservation supported by the presence of the

CHIT1 gene in our closest living relative (chimpanzee, Pan troglodytes, etc) (Gianfrancesco and Musumeci, 2004, Funkhouser and Aronson, 2007).

Recently, a second chitinase has been identified in humans (Boot *et al.* 2001): it is characterized by an acidic pI and extreme stability at acid pH 2, (Boot *et al.* 2001, Chou *et al.* 2006) and for this reason, it was called acidic mammalian chitinase (AMCase). This protein is relatively abundant in the gastrointestinal tract and lung, supporting a possible role as a food processor in stomach and its involvement in lung inflammation (Boot *et al.* 2001, Chou *et al.* 2006). This protein has 52% sequence identity with the human macrophage chitinase and also contains the additional α/β folds (Boot *et al.* 2001). Given the different expression patterns and the fact that this additional mammalian chitinase has a pH optimum of around 2, it is likely that it plays a different role compared to its human analogue CHIT1 (see details in the chapter of Dr. Seibold).

3. Allelic Variants of CHIT1 Gene

Now the presence of the complete DNA sequence of humans in international genomic database based on the results of the Human Genome Project, allows us to simply obtain the complete sequence and all the information regarding the polymorphic variants of the CHIT1 gene. The schematic representation of the CHIT1 gene in figure 1 showed that about 90 biallelic variants were found and confirmed in various subjects. Obviously, the large part of them lies in the intronic part of the gene and only 34 polymorphic sites showed a heterozygosity > 0.1. Five missense variation were described and located in exon 3,4 6, 10 and 11, but only the rs2297950 and rs1065761 had an heterozigosity of 0.408 and 0.167, respectively, while the others were rare variants. Additionally, we have 1 synonymous change in exon 9.

The first discovered and more studied allelic variant of the CHIT1 gene was the 24 bp duplication in exon 10 [7] resulting in activation of a cryptic 3-prime splice site that causes the deletion of amino acids 344–372. The internal deletion in the mutant CHIT1 will prevent the formation of a proper barrel conformation, with resulting loss of chitinolytic activity. As a result, the enzyme is totally inactive in homozygous subjects for the duplication (Boot *et al.* 1998; Canudas *et al.* 2001). Chit deficiency has no clear clinical evidences but it was proposed that the 24 bp duplication can be associated with susceptibility to particular infection of parasitic diseases (Malaguarnera *et al.* 2003, Choi *et al.* 2001).

In order to better explain the Chit gene function in higher organisms and the role of the known polymorphisms the presence of the 24 duplication was evaluated in primates amplifying the corresponding region from seven species of primates: chimpanzee, gorilla, orangutan, gibbon, baboon, a common marmoset and black macaque (Gianfrancesco and Musumeci, 2004).

---

7 Reference deletion/insertion polymorphism (ref DIP: rs3831317): Deleted Nucleotide AGGGACTG GGCGGGGCCATGGTCT; Contig position 53677328 bp; Map to Genome Build: 36.2 dbSNP database homepage: http://www.ncbi.nlm.nih.gov/SNP/index.html.

All the primates studied displayed identical PCR products corresponding to the wild-type allele. No 24-base pair duplication, homozygous or heterozygous, was found suggesting that this polymorphism was probably created during human evolution.

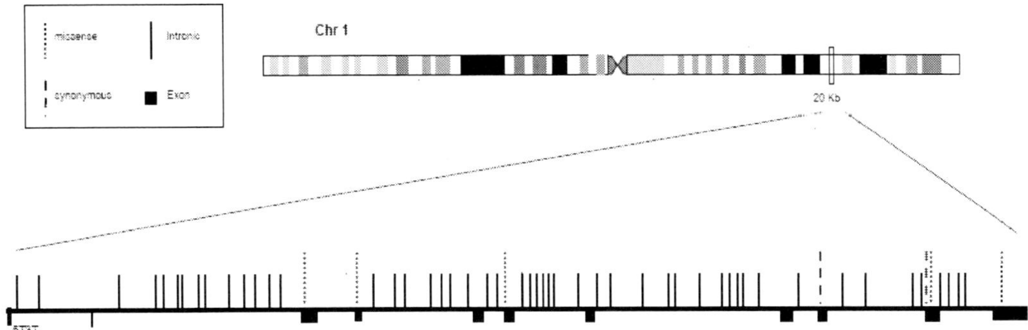

Figure 1. Schematic representation of CHIT 1 gene genomic structure. The size of the exons ranges from 30 to 461 bp. The polymorphic sites described in SNP database on NCBI Reference Assembly, were indicated by vertical lines as described in legend.

4. Variability of the 24 Bp Duplication and its Geographical Distribution in Different Populations

The study of the variants of the CHIT1 gene from a population point of view can help to better understand the origin and the diffusion from an original population to other people living in different geographical areas of the world. The aim of this kind of studies was to increase our knowledge of the distribution of the 24 bp duplication CHIT1 alleles on different continents and to interpret its variability. This kind of studies could be also a useful instrument to better explain if the distribution of the polymorphisms of the CHIT1 can be considered neutral with respect to natural selection or related to a protective factor in populations living under environmental conditions favourable to parasitic diseases.

Different groups extensively analyzed the 24 bp insertion polymorphism in subjects of European, Asian and African ancestry and frequencies of this variant were significantly different between the groups, being highest among Asians and lowest among African subjects (Boot *et al.* 1998, Chien *et al.* 2005, Choi *et al.* 2001, Choi *et al.* 2005, Hise *et al.* 2003, Lee *et al.* 2007, Malaguarnera *et al.* 2003, Piras *et al.* 2007, Rodrigues *et al.* 2004). In Europe the average frequency of the wild type allele is about 80-90% while in Asia the same allele showed values about 40-50%. The highest allele frequency was found among South China (0.64) with lower frequencies among those with European (Basque: 0.12) and African ancestry (Benin: 0.00) (Table 1). The number of subjects analyzed by various groups is probably enough to have a valuable frequency in the population studied: for example the observed allele frequency of subjects of European ancestry described by Lee *et al.* (2007) was comparable to previously published allele frequencies in European subjects (Piras *et al.* 2007) that ranged from 12 to 27%.

Particularly interesting is the CHIT1 gene distribution in Mediterranean and European areas: the frequencies of the H allele ranged from 10.5% (Morocco) to 25.0% (France). There was an absence of H/H homozygotes only in Basque and Morocco samples, while in the other populations, the homozygotes frequencies ranged from 1% (Corsica) to 9.3% (continental France). Usually, in all the observed distributions of European genotypes fell within Hardy–Weinberg predictions. These populations possess significant heterogeneity among them but the variability of the 24 bp duplication allele frequencies does not seem to be related to geography.

Table 1. Allele frequencies of the CHIT1 24 bp insertion in exon 10 in different populations and ethnic groups

Population	Subjects	Allele frequency		Reference
		wt*	H**	
Holland	171	0.77	0.23	Boot *et al.* (1998)
Ashkenazi Jews	68	0.77	0.23	Boot *et al.* (1998)
South India	67	0.60	0.40	Choi *et al.* (2001)
Papua New Guinea	906	0.88	0.12	Hise *et al.* (2003)
Benin	100	1.00	0.00	Malaguarnera *et al.* (2003)
Burkina Faso	100	0.98	0.02	Malaguarnera *et al.* (2003)
Sicily	100	0.73	0.27	Malaguarnera *et al.* (2003)
Sardinia	107	0.79	0.21	Malaguarnera *et al.* (2003)
Portugal	295	0.78	0.22	Rodrigues *et al.* (2004)
Finland	50	0.80	0.20	Choi *et al.* (2005)
Taiwan	82	0.42	0.58	Chien *et al.* (2005)
Continental France	128	0.75	0.25	Piras *et al.* (2007)
Spain	103	0.77	0.23	Piras *et al.* (2007)
Basque country	60	0.88	0.12	Piras *et al.* (2007)
Morocco	90	0.89	0.11	Piras *et al.* (2007)
Turkey	95	0.81	0.19	Piras *et al.* (2007)
Corsica	194	0.87	0.13	Piras *et al.* (2007)
Continental Italy	99	0.81	0.19	Piras *et al.* (2007)
European ancestry	984	0.83	0.17	Lee *et al.* (2007)
African ancestry	536	0.93	0.07	Lee *et al.* (2007)
Asian ancestry	2054	0.44	0.56	Lee *et al.* (2007)
Korea	80	0.42	0.58	Lee *et al.* (2007)
Japan	326	0.46	0.54	Lee *et al.* (2007)
North China	36	0.50	0.50	Lee *et al.* (2007)
South China	272	0.36	0.64	Lee *et al.* (2007)
Southeast Asia	1020	0.43	0.57	Lee *et al.* (2007)
Middle East	26	0.65	0.35	Lee *et al.* (2007)

* wt: wild-type (intact) allele; ** H: mutant allele (24 bp duplication).

A specific analysis was performed in Sardinia (Piras *et al.* 2007), where the average H allele frequency of the entire cohort of subjects was 17.5%, whereas the genotype frequencies were 3.5, 27.9, and 68.6 for H/H, wt/H, and wt/wt, respectively. Also in this case the comparison with previous data (Malaguarnera *et al.* 2003) showed no significant difference. The subsequent division of the sample according to the municipality of origin, pointed out the presence of frequency differences of H allele that decreased with altitude from 24.6% to 11.4% showing important variation at micro-geographic level within the island of Sardinia.

The spatial distribution of II allele frequency suggests the absence of natural selection on the CHIT1 gene both at micro or macro-geographic level. To quantify geographic variation in allele frequencies, Hall *et al.* (2007) applied the calculation of the Fixation index (Fst) values [8] to five wide geographical regions, using the weighted mean allele frequency for each area (Europe/Mediterranean, Africa, South Asia, East Asia and Papua New Guinea). The value of Fst calculated in all populations for CHIT1, was very similar to the median value (0.11 vs 0.12) for about 200 polymorphisms uniformly distributed over the genome (Soranzo *et al.* 2005) typed in populations from similar ancestry selected from the Human Genome Diversity Panel (http://www.stanford.edu/group/morrinst/hgdp.html). The worldwide variation in allele frequencies suggests that there has not been directional selection at this locus and that the worldwide variation in allele frequencies is consistent with genetic drift (Akey *et al.* 2002).

The analyses of the published results seems to validate the hypothesis that the H allele originated from East Asia, where the highest frequencies are present, and then migrate to the West. The lowest frequency of this allele in Africa might confirm this theory. The current frequency in smaller groups, suchs as Basques, might than also be explained by specific founder effects. An alternative possibility is represented by a diffusion of the mutant allele out of Africa, where the largest genetic variation can be found, and an effect of the migration waves and genetic drift that increased the frequency in Europe and Asia. In order to verify the hypothesis validity, we absolutely need to investigate various aspects regarding this variant. It is necessary to better explain the H allele function and the selective pressures acting on chitinases and flanking genes possibly due to the effect of the linkage disequilibrium and, additionally, we require a larger number of populations sampled that are helpful for spatial autocorrelation analysis on the H allele worldwide distribution.

5. Serum Chitotriosidase Activity and Polymorphisms

Usually, the mean Chit enzyme activity is normal in the wild type (TT) subjects, half value in heterozygotes (HT) for the 24 duplication and practically absent in homozygous mutant (HH) individuals. Recent results pointed out that the enzyme activity is variable related to the origin ancestry of population samples. Particularly, the median enzyme activity

8 Fixation index (Fst) is a measure of population differentiation based on genetic polymorphism data (such as Single Nucleotide Polymorphism). It is a concept developed in the 1920s by Sewall Wright. This statistic compares the genetic variability within and between population and is frequently used in the field of population genetics. compares the genetic variability within and between population and is frequently used in the field of population genetics.

in wild-type subjects was significantly higher in subjects of European ancestry, than subjects of African and Asian ancestry (Lee *et al.* 2007). Furthermore, values of enzyme activity could be less or absent in homozygous wild-type or heterozygous individuals, mostly in Asian and African samples. A possible explanation of these phenomena could be that the presence of other mutations or polymorphisms might exist that would account for the enzyme deficiency. The direct sequence of subjects heterozygous or wild type for the H polymorphism with little or absence of enzyme activity identified several polymorphisms, previously documented in genomic databases (see figure 1): A442G or V (exon 12), G354R (exon 10), G102S (exon 4) and a promoter polymorphism −1432A>G (Lee *et al.* 2007).

Sequence comparison between various species evidences that the amino acid at position 354 is conserved between human, mice and frog indicating a presumable pressure selection to a glycine in this position, while 442 alanine change in a glycine in mouse and frog.

The four polymorphisms showed differences in allele frequencies between the analyzed populations except for G102S (see Table 2). In addition, the G354R allele was never observed with the H allele and the low frequency could not permit further analyses. The A442G allele was found predominantly but not exclusively with the wild-type T allele in subjects of all ancestries. On the other hand, the A442V allele was never found with the H allele and showed the highest frequency in Africa.

Further analyses focused to explain the association with enzyme activity in presence of the 24 bp duplication, excluded the association with a reduce enzyme activity to the G102S and A442G variants, while the G354R and the A442V polymorphisms were significantly related to the reduction of serum Chit activity.

Recently, Grace *et al.* (2007) screened Type 1 Gaucher Disease (GD) patients for CHIT1 genotype and plasma enzyme levels. The complete sequencing of CHIT1 genes in four patients who possess very low plasma Chit activities and the wild-type allele revealed two novel mutations in the CHIT1 gene, the E74K (220G>A) and a complex exon 10 lesion, leading to one amino acid substitution and missplicing (1060G>A; 1155G>A) combined with a deletion of 4 intronic nucleotides close to the 5' splice site (1156+5_1156+8delGTAA). Besides, they confirm the G102S and validate the reduced activity previously described. The E74K mutation was rare and present only in 3 GD Ashkenazi Jewish patients.

The complex exon 10 mutation occurred in 2 GD Caribbean patients and was present in 0 to 6% of alleles among normal controls from different populations.

Table 2. Allele frequencies of CHIT1 G102S, G354R, A442G and A442V polymorphisms in different ethnic groups (Lee *et al.* 2007)

CHIT1	nt 304 G>A (G102S)	nt 1072 G>A (G354R)	nt 1325 C>G (A442G)	nt 1325 C>T (A442V)
	Allele frequency	Allele frequency	Allele frequency	Allele frequency
European	49/180 (0.27)	1/426 (0.002)	65/598 (0.11)	2/598 (0.003)
African	39/150 (0.26)	39/432 (0.09)	58/522 (0.11)	44/522 (0.08)
Asian	213/904 (0.24)	0/362 (0.0)	16/382 (0.04)	0/382 (0.0)

In vitro expression demonstrated that the E74K had approximately 51% of wild-type Chit activity. RNA studies indicated that the complex exon 10 lesion allele also caused missplicing.

In summary, the recent findings suggested a variable association of the HH genotype to the mean enzyme activity, related to the population origin. Europeans possess higher serum enzyme activity compared to Africans and lower than Asiatic people. These results were apparently in contrast with Barone *et al.* (2003) who observed that people from Benin and Burkina Faso exhibited higher Chit enzyme activity than Caucasians from Sicily and Sardinia. We cannot exclude the presence of variable association of serum levels to the HH genotype between African people of different ethnic origin and also a high variability between the European populations as suggested by the variable frequencies in the samples present in Table 1.

Furthermore, the other significant associations of enzyme activity and E74K, G354R and A442V polymorphisms in man demonstrates that other variants in CHIT1 could influence the level of serum Chit activity. Based on the published data, we can suppose that there may be further polymorphisms, probably in the promoter region, or other splice sites that could be causative of the reduction of enzyme activity. Noticeably, additional data on the complete sequence of the CHIT1 gene region in HH samples from various continents, especially on Asia and China, can better explain the relation between genomic variants and serum levels.

Conclusions

In the past few years, a number of groups have clearly delineated the genomic characterization of the CHIT1 in humans and mammals and defined a number of variants of the CHIT 1 gene. The 24 bp insertion polymorphism was extensive evaluated in Europe, Asia and Africa and the pattern of frequency distribution suggests an Asiatic origin of the H allele as most plausible.

On the other hand, the analyses of serum Chit activity versus genetic polymorphisms evidenced that there is a direct correlation between several variants and the reduced enzyme activity that is relievable in ethnic differences and evidence association with specific diseases.

Clearly additional studies are required to answer many open questions on Chit but the increased activity in plasma could represents one ancestral response of innate immunity that we need to clarify for clinical and biochemical applications. The next goal could be represented by the use of enzyme activity, genetic variation in serum as a biomarker into the molecular pathogenesis of various diseases whether ChT activation might have different functional meanings.

Acknowledgments

Thanks to all collaborating colleagues and friends of the laboratory (Persico I, Sassu A, Prodi DA, Simpore J) that help me in the collection of the samples and directly contribute to the genetic analysis. Thanks so much to Salvatore Musumeci for give me the opportunity to start the study of chitinase and for the continuous support, encouragement and precious suggestions.

References

Akey JM, Zhang G, Zhang K, Jin L, Shriver MD. Interrogating a high-density SNP map for signatures of natural selection. *Genome Res.* 2002; 12:1805–1814.

Arias EB, Verhage HG, Jaffe RC. Complementary deoxyribonucleic acid cloning and molecular characterization of an estrogen-dependent human oviductal glycoprotein. *Biol Reprod.* 1994; 51:685–694.

Barone R, Simporè J, Malaguarnera L, Pignatelli S, Musumeci S. Plasma chitotriosidase activity in acute Plasmodium falciparum malaria. *Clin Chim Acta* 2003; 331:79–85.

Boot RG, Blommaart EF, Swart E, Ghauharali-van der Vlugt K, Bijl N, Moe C, Place A, Aerts JM. Identification of a novel acidic mammalian chitinase distinct from chitotriosidase. *J Biol Chem.* 2001; 276:6770–6778.

Boot RG, Bussink AP, Verhoek M, de Boer PA, Moorman AF, Aerts JM. Marked differences in tissuespecific expression of chitinases in mouse and man. *J Histochem Cytochem.* 2005; 53:1283–1292.

Boot RG, Renkema GH, Strijland A, van Zonneveld AJ, Aerts JM. Cloning of a cDNA encoding chitotriosidase, a human chitinase produced by macrophages. *J Biol Chem.* 1995; 270: 26252–26256.

Boot RG, Renkema GH, Verhoek M, Strijland A, Bliek J, de Meulemeester TM, Mannens MM, Aerts JM. The human chitotriosidase gene. Nature of inherited enzyme deficiency. *J Biol Chem.* 1998; 273: 25680–25685.

Canudas J, Cenarro A, Civeira F, Garci-Otin AL, Aristegui R, Diaz C, Masramon X, Sol JM, Hernandez G, Pocovi M. Chitotriosidase genotype and serum activity in subjects with combined hyperlipidemia: effect of the lipid-lowering agents, atorvastatin and bezafibrate. *Metabolism.* 2001; 50: 447–450.

Chien YH, Chen JH, Hwu WL. Plasma chitotriosidase activity and malaria. *Clin Chim Acta.* 2005; 353(1-2):215-217.

Choi EH, Taylor JG, Foster CB, Walsh TJ, Anttila VJ, Ruutu T, Palotie A, Chanock SJ. Common polymorphisms in critical genes of innate immunity do not contribute to the risk for chronic disseminated candidiasis in adult leukemia patients. *Med Mycol.* 2005; 43(4):349-53.

Choi EH, Zimmerman PA, Foster CB, Zhu S, Kumaraswami V, Nutman TB, Chanock SJ. Genetic polymorphisms in molecules of innate immunity and susceptibility to infection with Wuchereria bancrofti in South India. *Genes Immun.* 2001; 2: 248–253.

Chou YT, Yao S, Czerwinski R, Fleming M, Krykbaev R, Xuan D, Zhou H, Brooks J, Fitz L, Strand J, Presman E, Lin L, Aulabaugh A, Huang X. Kinetic characterization of recombinant human acidic mammalian chitinase. *Biochemistry.* 2006; 45: 4444–4454.

Cohen E. Chitin synthesis and degradation as targets for pesticide action. *Arch Insect Biochem Physiol.* 1993; 22: 245–261.

Funkhouser JD, Aronson NN Jr. Chitinase family GH18: evolutionary insights from the genomic history of a diverse protein family. *BMC Evol Biol.* 2007; 7:96.

Gianfrancesco F, Musumeci S. The evolutionary conservation of the human chitotriosidase gene in rodents and primates. *Cytogenet Genome Res.* 2004;105(1):54-6.

Grace ME, Balwani M, Nazarenko I, Prakash-Cheng A, Desnick RJ. Type 1 Gaucher disease: null and hypomorphic novel chitotriosidase mutations-implications for diagnosis and therapeutic monitoring. *Hum Mutat.* 2007; 9:866-73.

Hakala BE, White C, Recklies AD. Human cartilage gp-39, a major secretory product of articular chondrocytes and synovial cells, is a mammalian member of a chitinase protein family. *J Biol Chem.* 1993; 268: 25803–25810.

Hall AJ, Quinnell RJ, Raiko A, Lagog M, Siba P, Morroll S, Falcone FH. Chitotriosidase deficiency is not associated with human hookworm infection in a Papua New Guinean population. *Infect Genet Evol.* 2007; 6:743-7.

Henrissat B and Bairoch A. New families in the classification of glycosyl hydrolases based on amino acid sequence similarities. *Biochem. J.* 1993; 293: 781–788.

Hise AG, Hazlett FE, Bockarie MJ, Zimmerman PA, Tisch DJ, Kazura JW. Polymorphisms of innate immunity genes and susceptibility to lymphatic filariasis. *Genes Immun.* 2003; 7:524-7.

Hollak CE, van Weely S, van Oers MH, Aerts JM. Marked elevation of plasma chitotriosidase activity. A novel hallmark of Gaucher disease. *J Clin Invest.* 1994; 93: 1288–1292.

Hu B, Trinh K, Figueira WF, Price PA. Isolation and sequence of a novel human chondrocyte protein related to mammalian members of the chitinase protein family, *J Biol Chem.* 1996; 271: 19415–19420.

Kuranda MJ and Robbins PW. Chitinase is required for cell separation during growth of Saccharomyces cerevisiae. *J Biol Chem.* 1991; 266: 19758–19767.

Kwiatkowski D. The molecular genetic approach to malarial pathogenesis and immunity. *Parassitologia* 1999; 41: 233–240.

Lee P, Waalen J, Crain K, Smargon A, Beutler E. Human chitotriosidase polymorphisms G354R and A442V associated with reduced enzyme activity. *Blood Cells Mol Dis.* 2007;39(3):353-60.

Lehrnbecher T, Bernig T, Hanisch M, Koehl U, Behl M, Reinhardt D, Creutzig U, Klingebiel T, Chanock SJ, Schwabe D. Common genetic variants in the interleukin- 6 and chitotriosidase genes are associated with the risk for serious infection in children undergoing therapy for acute myeloid leukemia. *Leukemia* 2005; 19: 1745–1750.

Malaguarnera L, Simpore J, Prodi DA, Angius A, Sassu A, Persico I, Barone R, Musumeci S. A 24-bp duplication in exon 10 of human chitotriosidase gene from the sub-Saharan to the Mediterranean area: role of parasitic diseases and environmental conditions. *Genes Immun.* 2003; 4: 570–574.

Malinda KM, Ponce L, Kleinman HK, Shackelton LM, Millis AJ. Gp38k, a protein synthesized by vascular smooth muscle cells, stimulates directional migration of human umbilical vein endothelial cells. *Exp Cell Res.* 1999; 250: 168–173.

Owhashi M, Arita H, Hayai N. Identification of a novel eosinophil chemotactic cytokine (ECF-L) as a chitinase family protein. *J Biol Chem.* 2000; 275: 1279–1286.

Piras I, Melis A, Ghiani ME, Falchi A, Luiselli D, Moral P, Varesi L, Calò CM, Vona G. Human CHIT1 gene distribution: new data from Mediterranean and European populations. *J Hum Genet.* 2007; 52(2):110-6.

Renkema GH, Boot R G, Muijsers AO, Donker-Koopman WE, Aerts JM. Purification and characterization of human chitotriosidase, a novel member of the chitinase family of proteins. *J Biol Chem.* 1995; 270: 2198–21202.

Rodrigues MR, Sa Miranda MC, Amaral O. Allelic frequency determination of the 24-bp chitotriosidase duplication in the Portuguese population by real-time PCR. *Blood Cells Mol Dis.* 2004; 33: 362–364.

Saito A, Ozaki K, Fujiwara T, Nakamura Y, Tanigami A. Isolation and mapping of a human lung-specific gene, TSA1902, encoding a novel chitinase family member. *Gene* 1999; 239: 325–331.

Soranzo N, Bufe B, Sabeti PC, Wilson JF, Weale ME, Marguerie R, Meyerhof W, Goldstein DB. Positive selection on a high-sensitivity allele of the human bitter-taste receptor TAS2R16. *Curr Biol.* 2005; 15(14):1257-65.

van Eijk M, van Roomen CPl, Renkema GH, Bussink AP, Andrews L, Blommaart EF, Sugar A, Verhoeven AJ, Boot RG, Aerts JM. Characterization of human phagocyte-derived chitotriosidase, a component of innate immunity. *Int Immunol.* 2005; 17: 1505–1512.

Vinetz JM, Dave SK, Specht CA, Brameld KA, Xu B, Hayward R, Fidock DA. The chitinase PfCHT1 from the human malaria parasite Plasmodium falciparum lacks proenzyme and chitin-binding domains and displays unique substrate preferences. *Proc Natl Acad Sci USA* 1999; 96: 14061–14066.

Wu Y, Egerton G, Underwood AP, Sakuda S, Bianco AE. Expression and secretion of a larval-specific chitinase (family 18 glycosylhydrolase) by the infective stages of the parasitic nematode, Onchocerca volvulus. *J Biol Chem.* 2001; 276: 42557–42564.

Zhu Z, Zheng T, Homer RJ, Kim YK, Chen NY, Cohn L, Hamid Q, Elias JA. Acidic mammalian chitinase in asthmatic Th2 inflammation and IL-13 pathway activation. *Science* 2004; 304: 1678–1682.

In: Binomium Chitin-Chitinase: Recent Issues
Editor: Salvatore Musumeci and Maurizio G. Paoletti
ISBN 978-1-60692-339-9

Chapter VII

The Gaucher Cell and Chitotriosidase, the Phagocyte Chitinase

J.M. Aerts and R.G. Boot
Department of Medical Biochemistry,
Academic Medical Center, University of Amsterdam, The Netherlands

Abstract

Gaucher disease (GD) has been the cradle of the human phagocyte chitinase, also known as chitotriosidase (CHIT1). GD is caused by deficiency of glucocerebrosidase, the enzyme responsible for the lysosomal breakdown of the lipid glucosylceramide. The disease is characterized by the accumulation in various tissues of pathological, lipid laden macrophages, so-called Gaucher cells. The search for suitable markers of Gaucher cells resulted in the identification of a thousand-fold increased chitinase activity in plasma from symptomatic Gaucher patients. Biochemical investigations identified a single responsible enzyme, named chitotriosidase based on its ability to hydrolyze 4-methylumbelliferyl-chitotrioside. Next, the properties of the chitotriosidase protein and gene were characterized. In the wake of the identification of chitotriosidase, the existence in mammals of another chitinase (AMCase) was discovered. This review focuses on the current knowledge on the features of the chitotriosidase protein and gene, the potential function of the enzyme in innate immunity and its value as disease marker in conditions involving macrophages. Attention is also paid to the biology of the Gaucher cell, the lipid-laden macrophage that so massively overproduces chitotriosidase.

1. Gaucher Disease and Gaucher Cells

1.1. Inherited Lysosomal Storage Disorders: Gaucher Disease

The eukaryotic cell contains membrane-enclosed compartments for the degradation of cellular macromolecules, so-called lysosomes (DeDuve, 2005). The physiological importance of lysosomes is illustrated by a group of inherited diseases in which deficiencies in one or more lysosomal pathways exist (reviewed in Meikle *et al.* 2004; Vellodi, 2005). The most frequently encountered inherited lysosomal storage disorder in man is glucosylceramidosis, better known as Gaucher disease. The clinical features of the disease were first described in detail by Philippe C. E. Gaucher more than a century ago (Gaucher, 1882). The identification of glucosylceramide (glucocerebroside) as the primary storage material in Gaucher disease was accomplished early last century (Aghion, 1934). Glucosylceramide is the common intermediate in the degradation of gangliosides and globosides which takes place intralysosomally by the stepwise action of exo-glycosidases. In 1965 Patrick and Brady *et al.* showed independently that the primary defect in Gaucher disease is a marked deficiency in activity of the lysosomal enzyme glucocerebrosidase (Brady *et al.* 1965; Patrick, 1965). This hydrolase (also known as acid beta-glucosidase or glucosylceramidase (GBA1), EC 3.2.1.45) catabolizes glucosylceramide to ceramide and glucose.

The clinical presentation of Gaucher disease is remarkably heterogeneous with respect to age of onset, nature and progression of the symptoms. Based on clinical features generally three variants are distinguished (Sidransky, 2004; Cox and Shofield, 1997). The non-neuronopathic form of GD, referred to as type 1 GD, is by far the most common. The incidence world-wide is about 1 in 50.000-200.000 births (Gieselmann, 1995). A markedly increased incidence exists in Ashkenazi Jewish populations (Beutler and Grabowski, 1995). Type 1 GD may become manifest within the first years, but hardly symptomatic individuals above the age of 70 have also been described. The major symptoms in type 1 GD result from lipid-laden macrophages in specific tissues, causing gross enlargement of spleen and liver (hepatosplenomegaly), displacement of normal bone marrow cells (pancytopenia) and damage to the bones (Beutler and Grabowski, 2001). The acute neuronopathic manifestation of Gaucher disease is called type 2. This variant is rare, and without ethnic predisposition. The average age of onset of severe hepatosplenomegaly is about 3 months, which is rapidly accompanied by progressive neurological complications, being usually lethal within the first two years of life (Barranger and Ginns, 1989; Beutler and Grabowski, 1995). Type 3, a subacute neuronopathic form of GD, is also relatively rare and occurs panethnically. The neurological symptoms of this type are similar to those observed in type 2 GD, but with a later onset and lesser severity. More recently it has become clear that a complete lack of glucocerebrosidase activity results in the so-called collodion baby phenotype characterized by ichtyotic skin (Sidransky, 2004).

During the last decades the mutations that underlie Gaucher disease have been identified by analysis of the glucocerebrosidase gene (Beutler and Grabowski, 2001). Numerous distinct mutations in the glucocerebrosidase gene have been identified (Beutler and Gelbart, 1997; Horowitz and Zimran, 1994). Six mutant alleles account for more than 95% of the defective glucocerebrosidase alleles in the Ashkenazi Jewish Gaucher patient population and about 70% of the mutant alleles in the various non-Jewish Caucasian Gaucher patient populations (Horowitz and Zimran, 1994; Boot *et al.* 1997). The most prevalent mutation in Jewish as well as non-Jewish Caucasian populations is the N370S mutation, the result of an adenine to guanine substitution at cDNA position 1226 (Tsuji *et al.* 1988). This mutation leads to the synthesis of normal amounts of enzyme that is largely correctly routed to lysosomes (Ohashi *et al.* 1991). However, the N370S enzyme is abnormal in catalytic features, showing under most conditions a markedly reduced specific activity. At sufficiently acidic pH and in the presence of activator protein the N370S mutant glucocerebrosidase shows a considerable residual activity (Van Weely *et al.* 1993). Homozygosity or heterozygosity for this allele is always associated with the type 1 form of the disease (Tsuji *et al.* 1988; Beutler and Grabowski, 2001). It has been shown that many N370S homozygotes have such a mild form of the disease that they do not seek medical advice and remain therefore undiagnosed, so-called asymptomatic patients (Aerts *et al.* 1993). The second most frequent mutation is the L444P mutation, often resulting in neurological symptoms in homozygotes. Unlike the N370S protein, this mutation appears to result in impaired trafficking and degradation in the ER (Ohashi *et al.* 1991). Although some relation exists between particular genotypes and phenotypes, clinical manifestations can differ markedly within the same genotype. Several phenotypically discordant identical twins with Gaucher disease have been documented (Cox and Shofield, 1997; Lachmann *et al.* 2005). Clearly, epigenetic factors also play a key role in Gaucher disease manifestation (Aerts *et al.* 1993).

1.2. Gaucher Cells and Pathophysiology

Although glucocerebrosidase activity is comparably reduced in all cell types of Gaucher patients, the lysosomal storage of glucosylceramide is restricted to cells of the monocyte/macrophage lineage, at least in the type 1 variant. The predominant lipid accumulation in macrophages can be ascribed to the role of these cells in degradation of senescent red and white blood cells that are rich in glycosphingolipids (Parkin and Brunning, 1982; Naito *et al.* 1988). The glucosylceramide-laden cells show a characteristic morphology with an eccentric nucleus and a "wrinkled tissue paper" like appearance due to the massive presence of lipid in tubular deposits. These storage cells are called Gaucher cells and are present in various locations, predominantly the bone marrow, spleen, liver and parenchyma of lymph nodes. The massive accumulation of storage cells in the bone marrow causes displacement of the normal haematopoietic cells (Figure 1).

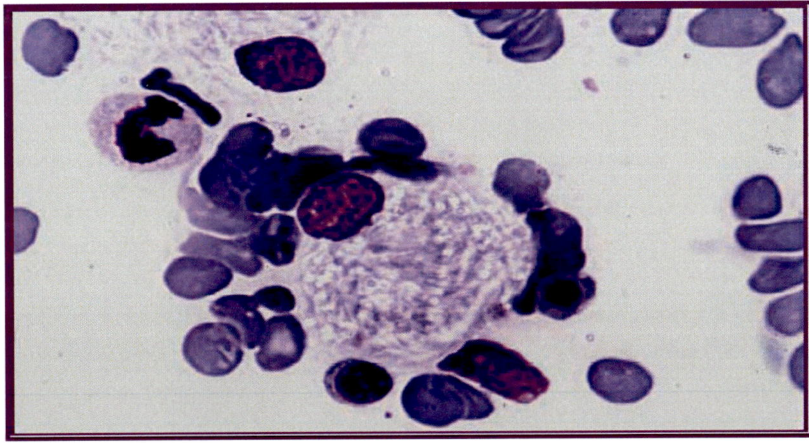

Figure 1.

In other tissues, infiltration of Gaucher cells may lead to fibrosis, infarction, necrosis and scarring. The sheer presence of storage cells does not fully explain the entire pathology of Gaucher disease. Gaucher cells are not inert storage containers, but metabolically active cells that produce and secrete proteins that drive pathophysiological processes. It is now generally believed that the complex mixture of factors, like cytokines, chemokines and hydrolases, originating from storage cells themselves or from surrounding macrophages contributes to the characteristic pathophysiology of Gaucher disease.

Gaucher cells resemble alternatively activated macrophages. The cells very strongly express IL-1Ra and CCL18, which are typical markers of alternatively activated macrophages (Boven *et al.* 2004). Gaucher cells show high levels of lysosomal acid phosphatase, HLA class II, CD68, the scavenger/lipid receptor CD36 and signal-regulatory protein (SIRP) alpha. Differential gene expression techniques applied to Gaucher spleen samples have identified increased levels of cathepsins S, C and K originating from the Gaucher cells (Moran *et al.* 2000). Typical pro-inflammatory molecules such as interleukin IL-1beta, IL-1 alpha, IL-12p40, tumor necrosis factor (TNF) alpha, interferon (IFN) gamma and MCP-1 are not expressed by Gaucher cells (Boven *et al.* 2004). Importantly, histochemistry of Gaucher spleen sections has revealed that in storage lesions the core of mature, alternatively activated, Gaucher cells is surrounded by recruited pro-inflammatory macrophages. The marked production of CCL18 by Gaucher cells is thought to play a role in the ongoing recruitment of monocytes to storage lesions.

The blend of Gaucher cells and their surrounding monocytes/macrophages in tissue lesions explains the generation of a variety of cytokines (Boven *et al.* 2004). Michelakakis and coworkers were the first to report on elevated levels of TNF-alpha in plasma of type 2 and 3 Gaucher patients, and to a lesser extent in samples from type 1 Gaucher patients (Michelakakis *et al.* 1996). Allen *et al.* could not confirm the finding of elevated plasma TNF-alpha in type 1 Gaucher disease, but did observe increases in IL-6 and IL-10 (Allen *et al.* 1977). In another study, Hollak and coworkers established that IL-8 and macrophage colony stimulating factor (M-CSF) can be markedly increased in plasma of type 1 Gaucher patients (Hollak *et al.* 1997a). In addition, plasma of many Gaucher patients contains up to sevenfold increased concentrations of the monocyte/macrophage activation marker soluble CD14 (sCD14). This finding supports the idea that activation of monocytes/macrophages occurs in symptomatic Gaucher patients. Another marker for macrophage activation is soluble CD163 (sCD163). The sCD163 plasma levels in type 1 Gaucher patients were found to be far above the levels in normal subjects (Moller *et al.* 2002; Moller *et al.* 2004). A low-grade inflammatory profile has been reported for Gaucher disease. Patients show significant elevations in fibrinogen, accelerated erythrocyte sedimentation rate and C-reactive protein (Rogowski *et al.* 2005). In addition evidence has been reported for low grade activation of coagulation and the complement cascade in Gaucher patients (Hollak *et al.* 1997b; Vissers *et al.* 2007). Very recently marked elevated concentrations of Macrophage Inflammatory Proteins (MIPs) in serum of Gaucher patients were documented (van Breemen *et al.* 2007). The elevation in MIP 1beta is of particular interest. The protein appears to be not produced by the mature storage cells but by surrounding phagocytes. A correlation of elevated MIP 1beta levels with ongoing skeletal disease in Gaucher patients has been observed. It is of interest to note that also in multiple myeloma MIP 1alpha and MIP 1beta have been implicated in affecting the delicate balance between bone degradation by osteoclasts and bone synthesis by osteoblasts (van Breemen *et al.* 2007).

In conclusion, it is now generally believed that the complex mixture of factors, like cytokines, chemokines and hydrolases, originating from storage cells themselves or from surrounding classically activated macrophages contributes to the characteristic pathophysiology of Gaucher disease. Next to the putative role of MIP proteins in bone homeostasis, high levels of cathepsin K may underlie the skeletal complications observed in Gaucher patients (Hollak and Aerts, 2007).

1.3. Therapy of Gaucher Disease: Correction and/or Removal of Gaucher Cells

Following his discovery of lysosomes, DeDuve proposed already in 1964 that treatment of lysosomal storage disorders by supplementation with the missing enzyme might be envisioned, (De Duve, 2005). Inspired by this concept, Brady, Barranger and their co-workers at the National Institutes of Health elegantly exploited the presence of the so-called mannose receptor on the cell surface of macrophages to improve targeting of therapeutic enzyme to lysosomes of lipid-laden macrophages in Gaucher patients. For this purpose, the oligosaccharide chains of glucocerebrosidase isolated from placenta were modified by enzymatically exposing the covered mannose residues (Furbish *et al.* 1981). The concept of a mannose-terminated glucocerebrosidase resulted in the development of the registered therapeutic enzyme alglucerase (Ceredase, Genzyme Corporation, MA). Intravenous administration of alglucerase was found to result in major clinical improvements (Barton *et al.* 1990; Barton *et al.* 1991). In the mid-nineties, Ceredase was replaced by treatment with recombinant produced glucocerebrosidase from CHO cells (Cerezyme, Genzyme Corporation, MA), with similar therapeutic results (Grabowski *et al.* 1995). Due to the inability of Cerezyme to pass the blood-brain barrier, unfortunately enzyme replacement therapy (ERT) does not prevent the lethal neuropathology in type 2 GD patients (Erikson *et al.* 1993). Some arrest of neurological deterioration and even in some cases signs of neurological improvements have been reported for type 3 GD patients (Campbell *et al.* 2004).

An alternative approach for therapeutic intervention of type 1 Gaucher and other glycosphingolipid storage disorders is substrate reduction therapy (SRT; also termed substrate deprivation therapy; see for a review Radin, 1996; Aerts *et al.* 2006). The approach aims to reduce the rate of glycosphingolipid biosynthesis to levels which match the impaired catabolism. It is conceived that patients who have a significant residual lysosomal enzyme activity could gradually clear lysosomal storage material and therefore should profit most from reduction of substrate biosynthesis. Two classes of inhibitors of glycosphingolipid biosynthesis have presently been described, both of which inhibit the ceramide-specific glucosyltransferase, (also termed glucosylceramide synthase; GlcT-1; UDP-glucose: N-acylsphingosine D-glucosyl-transferase, EC 2.4.1.80). The enzyme catalyses the transfer of glucose to ceramide, the first step in the biosynthesis of glucosphingolipids. The first class of inhibitors is formed by analogues of ceramide. The prototype inhibitor is PDMP (D, L-threo-1-phenyl-2-decanoylamino-3-morpholino-1-propanol). More specific and potent analogues have been subsequently developed based on substituting the morpholino group for a pyrrolodino function and by substitutions at the phenyl group: 4-hydroxy-1-phenyl-2-palmitoylamino-3-pyrrolidono-1-propanol (p-OH-P4) and ethylenedioxy-1-phenyl-palmitoylamino-3-pyrrolidino-1-propanol (EtDo-P4) (Shayman *et al.* 2004; McEachern *et al.* 2007). The second class of inhibitors of glucosylceramide synthase is formed by N-alkylated iminosugars. Such type of compounds were already in common use as inhibitors of N-glycan processing enzymes and the potential application of N-butyldeoxynojirimycin as HIV inhibitor had been studied in AIDS patients. Platt and Butters at the Glycobiology Institute in Oxford were the first to recognize the ability of N-butyldeoxynojirimycin to inhibit glycosylceramide synthesis at low micromolar concentrations (Platt *et al.* 1994). The same researchers demonstrated in knock out mouse models of Tay–Sachs disease significant reductions in glycosphingolipid storage in the brain (Platt *et al.* 1997). Preclinical studies in animals and the previous clinical trial in AIDS patients have indicated (transient) adverse effects in the gastrointestinal tract, probably related to the ability of NB-DNJ to inhibit disaccharidases on the intestinal brush border. Overkleeft and coworkers in their search for inhibitors of glucosidases have serendipitously developed a more potent inhibitor of glucosylceramide synthase. Adamantane-pentyl-deoxynojirimycin (AMP-DNM) was found to inhibit glycosphingolipid biosynthesis at low nanomolar concentrations (Overkleeft *et al.* 1998) and able to prevent globotriaosylceramide accumulation in a Fabry knock out mouse model without overt side effects (Aerts *et al.* 2003).

The first clinical study of N-butyldeoxynojirimycin (NB-DNJ) was an open-label phase I/II trial with 28 adult type 1 Gaucher patients (Cox *et al.* 2000). Improvements in hepatomegaly and hematological abnormalities as well as corrections in plasma levels of glucosylceramide and biomarkers of Gaucher disease activity were reported, although the extent of the response was less spectacular than generally observed with high dose enzyme replacement therapy (see for a review Aerts *et al.* 2006). As expected, a dose-response relationship is demonstrable for NB-DNJ in type 1 Gaucher patients. Administration of three times daily 50 mg NB-DNJ is far less effective than 100 mg daily doses (Heitner *et al.* 2002). NB-DNJ (Zavesca, Actelion) is now registered in Europe and the U.S.A. for treatment of mild to moderately affected type 1 Gaucher patients that are unsuitable to receive enzyme replacement therapy (Cox *et al.* 2003). The sustained effects of prolonged substrate reduction therapy have recently been reported (Elstein *et al.* 2004; Pastores *et al.* 2005; Elstein *et al.* 2007). Provided that iminosugars or other inhibitors of glucosylceramide synthase prove to be safe in the long term, they should have a role to play in the management of glycosphingolipid storage disorders, including Gaucher disease.

1.4. Search for Plasma Markers of Gaucher Cells: Discovery of Chitotriosidase

Following the use of ERT and SRT, an urgent need developed for surrogate markers of Gaucher cells. Such biomarkers would allow accurate monitoring of the progress of the disease and efficacy of therapy. The ideal biomarker is detectable in plasma and directly reflects the presence of storage cells. Although abnormalities in levels of tartrate resistant acid phosphatase (TRAP), angiotensin-converting enzymes, hexosaminidase and lysozyme have all been reported, none of these enzymes appear to meet this criterion (reviewed in Aerts and Hollak, 1997). Overlap between levels of these enzymes in patients versus controls further restricts their use as biomarkers in Gaucher disease. In an attempt to identify novel secondary biochemical abnormalities, a thorough screening of plasma enzyme activities in plasma of symptomatic individuals versus a variety of substrates was conducted. This led to the discovery in plasma of Gaucher patients of a thousand-fold increased capacity to hydrolyse the fluorogenic substrate 4-methylumbelliferyl–chitotrioside (Hollak *et al.*, 1994). The responsible enzyme was named chitotriosidase. Further studies revealed that plasma chitotriosidase originated from the lipid-laden macrophages of Gaucher patients. As a result of this, chitotriosidase activity levels do not reflect one particular clinical symptom, but rather reflect the total body burden on Gaucher cells. Although chitotriosidase activity can be rapidly and sensitively measured using 4-methylumbelliferyl-chitotrioside as substrate, the ability of the enzyme to transglycosylate complicates the enzyme assay. The use of a slightly modified substrate provides a much more convenient method for measuring activity of chitotriosidase (Aguilera *et al.* 2003; Schoonhoven *et al.* 2007).

Plasma chitotriosidase activities are greatly increased in symptomatic Gaucher patients, but not in asymptomatic glucocerebrosidase-deficient individuals. Chitotriosidase values drop sharply upon ERT, coinciding with clinical improvements (Hollak *et al.* 1994). To assess the utility of chitotriosidase activity measurements as a biomarker for treatment efficacy, the relationship between plasma chitotriosidase activity and clinical parameters has been studied (Hollak *et al.* 2001). On the basis of this investigation, it has been proposed that in patients in whom initiation of treatment is questionable, based solely on clinical parameters, a chitotriosidase activity above 15 000 nmol/ml/hour may serve as an indicator of a high Gaucher cell burden and an indication for the initiation of treatment. A reduction of less than 15% after one year of treatment should be a reason to consider a dose increase. Furthermore, a sustained increase in chitotriosidase at any point during treatment should alert the physician to the possibility of clinical deterioration and the need for dose adjustment, and hence are of great potential in both diagnosis and monitoring of the disease. The regular monitoring of plasma chitotriosidase levels in Gaucher patients is presently used world-wide to assist in clinical management of these patients (Deegan and Cox, 2005; Deegan *et al.* 2005; Vellodi *et al.* 2005; Cabrera-Salazar *et al.* 2004). A pitfall regarding the use of chitotriosidase as Gaucher cell biomarker results from the complete absence of the chitotriosidase activity in about 6 % of all individuals, including Gaucher patients (see also below). This results from homozygosity for a null allele of the chitotriosidase gene (Boot *et al.* 1998). Plasma chitotriosidase levels in heterozygotes for this mutation (about 35 % of all individuals) underestimate the actual presence of Gaucher cells in patients. Determination of chitotriosidase genotype in Gaucher patients is therefore required.

Although plasma chitotriosidase activity is now the most used biomarker in GD, there is still need for other biomarkers foremost because of the high incidence of deficiency. Recently a marked elevation in plasma of GD patients has also been described for the chemokine CCL18 (Boot *et al.* 2004; Deegan *et al.* 2005). Both chitotriosidase and CCL18 are secreted by Gaucher cells and the plasma levels of both markers change comparably during therapy.

Monitoring of plasma CCL18 can therefore be a useful alternative to monitor response to therapy in Gaucher patients deficient in chitotriosidase (Deegan *et al.* 2005).

2. Chitotriosidase

2.1. Molecular Features of the Chitinolytic Enzyme

The discovery of the markedly increased chitotriosidase activity in Gaucher patients allowed the purification and molecular characterization of the responsible protein. Two major isoforms with molecular masses of 50 and 39 kDa have been purified from the spleen of a Gaucher patient (Renkema *et al.* 1995). Both purified isoforms were shown to be completely functional chitinases, exhibiting activity towards colloidal chitin as well as artificial fluorogenic substrates, activity that could be inhibited by allosamidin and demethyl allosamidin, in a manner similar to bacterial chitinases (Renkema *et al.* 1995). Using degenerate primers based on conserved regions in chitinases from several species, the gene was cloned from a macrophage cDNA library constructed from mRNA isolated form long-term cultured peripheral blood monocytes that spontaneously differentiate into activated macrophages that produce large quantities of chitotriosidase (Boot *et al.* 1995). Sequence alignments showed that chitotriosidase is remarkably homologous to chitinases of various species, in particular the catalytic region consensus sequence (D-x-x-D-x-D-x-E) is completely conserved. Alignment of chitotriosidase with other chitinases also showed that the enzyme consists of a 39-kDa catalytic domain connected with a C-terminal chitin binding domain through a short linker region, again in a manner similar to other chitinases. It was found that the enzyme is synthesized as a 50-kDa protein that is either secreted into the medium or, alternatively, processed into the 39-kDa enzyme in the lysosome where it accumulates. To a quantitatively minor extent, a 39-kDa isoform containing only one extra C-terminal residue can also be synthesized as a result from alternative splicing (Renkema *et al.* 1997; Boot *et al.* 1995).

The locus of the chitotriosidase gene was assigned to 1q31-32 by fluorescent *in situ* hybridization using the genomic clone as a probe (Boot *et al.* 1998). Next it was established that the commonly encountered recessively inherited deficiency in chitotriosidase is the result of a 24- base pair duplication causing aberrant splicing (Boot *et al.* 1998). The observed carrier frequency for the duplication of about 35% is consistent with the finding that about 6% of individuals are deficient in chitotriosidase activity.

The high incidence of the chitotriosidase deficiency in man prompted questions concerning redundancy of chitotriosidase. Further investigations led to the discovery of a second mammalian chitinase named Acidic Mammalian Chitinase (AMCase) (Boot *et al.* 2001). Similar to chitotriosidase, AMCase shows chitinolytic activity towards chitin, releasing mainly soluble chitobiose fragments and is sensitive to inhibition by allosamidin. Like chitotriosidase, AMCase is synthesized as a 50 kDa protein that contains a 39 kDa catalytic domain, separated from a C-terminal chitin binding domain by a hinge region. Although the sequence similarity between the human chitinases is high, AMCase exhibits a distinct pH activity profile, being most active at acidic pH (Boot *et al.* 2001).

2.2. Phagocyte Specific Expression of Chitotriosidase

Human chitotriosidase is exclusively expressed by human phagocytes, namely in macrophages and neutrophils (Hollak *et al.* 1994; Escott and Adams, 1995). Monocytes do not express chitotriosidase, but in vitro cell culture results in the induction of message and protein after 4 to 10 days, depending on the donor. Furthermore, tissue macrophages express chitotriosidase as has for instance been demonstrated in Gaucher disease and atherosclerotic plaques (Renkema *et al.* 1995; Boot *et al.* 1999). Chitotriosidase is stored in human neutrophils as protein within their specific granules and only bone-marrow derived precursors show message for the protein (van Eijk *et al.* 2005; unpublished observation). A recent study revealed that Toll-like receptor (TLR), but not NOD2 activation, regulates chitotriosidase release by neutrophils. Furthermore, both TLR and NOD2 activation results in diminished induction by monocytes. Lastly, NOD2 activation, but not TLR stimulation, induces chitinase expression in macrophages (van Eijk *et al.* 2007).

The expression of chitotriosidase and AMCase differs among mammals. In contrast to the situation in man, mouse chitotriosidase is not present in phagocytes, but predominantly in the lining cells of the tongue and stomach (Boot *et al.* 2005; Zheng *et al.* 2005). Human AMCase is mainly expressed in the stomach and to a lesser extent in the lung, whereas mouse AMCase is expressed in tongue, stomach and alveolar macrophages (Boot *et al.* 2001; Suzuki *et al.* 2002; Boot *et al.* 2005).

Lysosomal stress is an important inducer of chitotriosidase in macrophages. By far the highest levels of chitotriosidase are found in Gaucher disease, but other diseases characterized by lysosomal accumulation of glycosphingolipids or other lipid species show increase levels as well. Examples of this are Niemann-Pick A/B, Niemann-Pick C, Krabbe, GM1 gangliosidosis, Cholesteryl ester storage disease, Wolman disease, Morquio B, and Tangier disease (Guo *et al.* 1995; Aerts *et al.* 2005). Elevated levels have also been found in fucosidosis, galactosialidosis, glycogen storage disease IV and Alagille syndrome (Michelakakis *et al.* 2004). In atherosclerosis, a pathological process in blood vessel walls, accumulation of cholesterol, which occurs in foam cells, induces chitotriosidase and the chi-lectin HC-gp39 (Boot *et al.* 1999). Individuals suffering from thalassemia, which is a result from a defect of beta-globin chain synthesis a component of hemoglobin, are treated with blood transfusions. Due to the uptake of these transfusion cells it has been suggested that accumulation of lipid or iron occurs in lysosomes, giving rise to induction of the chitinase (Barone *et al.* 1999). Furthermore, increased levels are detected in multiple sclerosis, an auto-immune disease with accumulation of myelin in macrophages (Czartoryska *et al.* 2001). In the systemic granulomatous disorder sarcoidosis increases have been reported as well (Hollak *et al.* 1994; Grosso *et al.* 2004).

2.3. Human Chitinase Activity in Innate Immunity and Allergic Responses

Innate immunity. It has been suggested that chitotriosidase serves as component of innate immune responses (Renkema *et al.* 1995). There is indeed evidence in favour of an anti-fungal activity of chitotriosidase. First, chitotriosidase activity has been found to be raised in plasma of neonates upon systemic Candidiasis and Aspergillosis (Labadaridis *et al.* 1998; Labadaridis *et al.* 2005). Second, chitotriosidase was found to inhibit growth of *C. neoformans*, to cause hyphal tip lysis in *M. rouxii* and to prevent the occurrence of hyphal switch in *C. albicans* (Van Eijk *et al.* 2005). These data strengthen the earlier observed chitinolytic activity towards cell wall chitin of *C. albicans* (Boot *et al.* 2001). In addition, it has been found that recombinant human chitotriosidase showed synergy with existing anti-fungal drugs such as the polyene amphotericin B, the azoles itraconazole and flucanozole and cell wall synthesis inhibitors LY-303366 and nikkomycin Z (Stevens *et al.* 2000). Further proof of an important anti-fungal action has been found in neutropenic mouse models of systemic Candidiasis and systemic Aspergillosis, the main causes of mortality in immuno-compromised individuals. Recombinant human chitotriosidase clearly improved survival in these mouse models (van Eijk *et al.* 2005). The observations made with chitotriosidase are not entirely surprising given the well documented anti-fungal role of chitinases in plants (Schlumbaum *et al.* 1986). Possibly recombinant chitotriosidase may be attractive from a clinical perspective to treat life-threatening fungal infections. Especially, since Gaucher patients seem to tolerate well thousand-fold elevated serum levels. In addition, AMCase also shows chitinolytic activity towards fungal cell wall chitin (Boot *et al.* 2001). Deficiency in chitotriosidase might be partly compensated for by the presence of the latter enzyme. The incidence of Candida sepsis has indeed been reported not to be related to deficiency in chitotriosidase (Masoud *et al.* 2002). The potential role of chitotriosidase in asthma and allergic responses, and its activity against pathogens like bacteria, nematodes and *Plasmodium falciparum* is covered in other chapters of this book.

Conclusion

Fundamental investigations on Gaucher disease have led to the serendipitous discovery of chitotriosidase, the human phagocyte chitinase, and in its wake the discovery of AMCase. Both chitinases seem to play important roles in susceptibility to specific infections as well as allergic responses. Further studies are warranted to clarify the exact roles of chitinases in these processes. Meanwhile, chitotriosidase can be exploited as biomarker for monitoring certain disease conditions involving macrophages, such as lysosomal storage disorders, sarcoidosis, multiple sclerosis, thalassemia and infectious diseases. Attention should also be focussed to the therapeutic potential of recombinant chitotriosidase as agent in life threatening fungal infections.

References

Aerts JM, Van Weely S, Boot R, Hollak CE, Tager JM. Pathogenesis of lysosomal storage disorders as illustrated by Gaucher disease. *J Inherit Metab Dis*. 1993; 16: 288-291.

Aerts JM, Hollak CE. Plasma and metabolic abnormalities in Gaucher's disease. *Baillieres Clin Haematol.* 1997; 10: 691-709.

Aerts JM, Hollak C, Boot R, Groener A. Biochemistry of glycosphingolipid storage disorders: implications for therapeutic intervention. *Philos Trans R Soc Lond B Biol Sci.* 2003; 358: 905-914.

Aerts JM, Hollak CE, van Breemen M, Maas M, Groener JE, Boot RG. Identification and use of biomarkers in Gaucher disease and other lysosomal storage diseases. *Acta Paediatr* Suppl. 2005; 94: 43-46.

Aerts JM, Hollak CE, Boot RG, Groener JE, Maas M. Substrate reduction therapy of glycosphingolipid storage disorders. *J Inherit Metab Dis.* 2006; 29: 449-456.

Aghion A. La maladie de Gaucher dans l'enfance. 1934 Thèse, Paris.

Aguilera B, Ghauharali-van der Vlugt K, Helmond MT, Out JM, Donker-Koopman WE, Groener JE, Boot RG, Renkema GH, van der Marel GA, van Boom JH, Overkleeft HS, Aerts JM. Transglycosidase activity of chitotriosidase: improved enzymatic assay for the human macrophage chitinase. *J Biol Chem.* 2003; 278: 40911-40916.

Allen MJ, Myer BJ, Khokher AM, Rushton N, Cox TM. Pro-inflammatory cytokines and the pathogenesis of Gaucher's disease: increased release of interleukin-6 and interleukin-10. *QJM* 1997; 90: 19-25.

Barranger JA, Ginns EI. Glucosylceramide lipidosis: Gaucher disease. In: The Metabolic Basis of Inherited Disease, 6th edition, 1989; pp. 1677–1698 (Eds. Scriver, Beaudet, Sly, Valle), McGraw–Hill.

Barton NW, Furbish FS, Murray GJ, Garfield M, Brady RO. Therapeutic response to intravenous infusions of glucocerebrosidase in a patient with Gaucher disease. *Proc Natl Acad Sci USA.* 1990; 87: 1913-1916.

Barton NW, Brady RO, Dambrosia JM, Di Bisceglie AM, Doppelt SH, Hill SC, Mankin HJ, Murray GJ, Parker RI, Argoff CE, Grewal RP, Yu KT Replacement therapy for inherited enzyme deficiency macrophage-targeted glucocerebrosidase for Gaucher's disease. *N Engl J Med.* 1991; 324: 1464-1470.

Beutler E. Grabowski GA. Gaucher disease. In: The metabolic and molecular bases of inherited disease, 7th edition, 1995; pp. 2641-2670 (Eds. Scriver, Beadet, Sly, Valle), McGraw-Hill.

Beutler E, Gelbart T. Hematologically important mutations: Gaucher disease. *Blood Cells Mol Dis.* 1997; 23: 2-7.

Beutler E, Grabowski GA. Gaucher disease. In: The Metabolic and Molecular Bases of Inherited Disease, 8th edition, 2001; pp. 3635-3668 (Eds: Scriver, Beadet, Sly, Valle) New York: McGraw-Hill.

Boot RG, Renkema GH, Strijland A, van Zonneveld AJ, Aerts JM. Cloning of a cDNA encoding chitotriosidase, a human chitinase produced by macrophages. *J Biol Chem.* 1995; 270: 26252-26256.

Boot RG, Hollak CE, Verhoek M, Sloof P, Poorthuis BJ, Kleijer WJ, Wevers RA, van

Oers MH, Mannens MM, Aerts JM, van Weely S. Glucocerebrosidase genotype of Gaucher patients in The Netherlands: limitations in prognostic value. *Hum Mutat.* 1997; 10: 348-358.

Boot RG, Renkema GH, Verhoek M, Strijland A, Bliek J, de Meulemeester TM, Mannens MM, Aerts JM. The human chitotriosidase gene. Nature of inherited enzyme deficiency. *J Biol Chem.* 1998; 273: 25680-25685.

Boot RG, van Achterberg TA, van Aken BE, Renkema GH, Jacobs MJ, Aerts JM, de Vries CJ. Strong induction of members of the chitinase family of proteins in atherosclerosis: chitotriosidase and human cartilage gp-39 expressed in lesion macrophages. *Arterioscler Thromb Vasc Biol.* 1999; 19: 687-694.

Boot RG, Blommaart EF, Swart E, Ghauharali-van der Vlugt K, Bijl N, Moe C, Place A, Aerts JM. Identification of a novel acidic mammalian chitinase distinct from chitotriosidase. *J Biol Chem.* 2001; 276: 6770-6778.

Boot RG, Verhoek M, De Fost M, Hollak CE, Maas M, Bleijlevens B, van Breemen MJ, van Meurs M, Boven LA, Laman JD, Moran MT, Cox TM, Aerts JM. Marked elevation of the chemokine CCL18 / PARC in Gaucher disease: a novel surrogate marker for assessing therapeutic intervention. *Blood* 2004; 103: 33-39.

Boot RG, Bussink AP, Verhoek M, de Boer PA, Moorman AF, Aerts JM. Marked differences in tissue-specific expression of chitinases in mouse and man. *J Histochem Cytochem.* 2005; 53: 1283-1292.

Boven LA, van Meurs M, Boot RG, Mehta A, Boon L, Aerts JM, Laman JD. Gaucher cells demonstrate a distinct macrophage phenotype and resemble alternatively activated macrophages. *Am J Clin Pathol.* 2004; 122: 359-369.

Brady RO, Kanfer JN, Shapiro D. Metabolism of glucocerebrosides. II. Evidence of an enzymatic deficiency in Gaucher's disease. Biochem Res Commun. 1965; 18: 221-225.

Bussink AP, van Eijk M, Renkema GH, Aerts JM, Boot RG. The biology of the Gaucher cell: the cradle of human chitinases. *Int Rev Cytol.* 2006; 252: 71-128.

Cabrera-Salazar MA, O'Rourke E, Henderson N, Wessel H, Barranger JA. Correlation of surrogate markers of Gaucher disease. Implications for long-term follow up of enzyme replacement therapy. *Clin Chim Acta.* 2004; 344: 101-107.

Campbell PE, Harris CM, Vellodi A. Deterioration of the auditory brainstem response in children with type 3 Gaucher disease. *Neurology.* 2004; 63: 385-387.

Cox TM, Schofield JP. Gaucher's disease: clinical features and natural history. *Baillieres Clin Haematol.* 1997; 10: 657-689.

Cox T, Lachmann R, Hollak C, Aerts J, van Weely S, Hrebicek M, Platt F, Butters T, Dwek R, Moyses C, Gow I, Elstein D, Zimran A. Novel oral treatment of Gaucher's disease with N-butyldeoxynojirimycin (OGT918) to decrease substrate biosynthesis. *Lancet.* 2000; 355: 1481-1485.

Cox TM, Aerts JM, Andria G, Beck M, Belmatoug N, Bembi B, Chertkoff R, Vom Dahl S, Elstein D, Erikson A, Giralt M, Heitner R, Hollak C, Hrebicek M, Lewis S, Mehta A, Pastores GM, Rolfs A, Miranda MC, Zimran A. Advisory Council to theEuropean Working Group on Gaucher Disease. The role of the iminosugar N-butyldeoxynojirimycin (miglustat) in the management of type I (non-neuronopathic) Gaucher disease: a position statement. *J Inherit Metab Dis.* 2003; 26: 513-526.

Czartoryska B, Fiszer U, Lugowska A. Chitotriosidase activity in cerebrospinal fluid as a marker of inflammatory processes in neurological disease. *J Lab Med.* 2001; 25: 77-81.

de Duve C. The lysosome turns fifty. *Nat Cell Biol. 2005*; 7: 847-849.

Deegan PB, Moran MT, McFarlane I, Schofield JP, Boot RG, Aerts JM, Cox TM. Clinical evaluation of chemokine and enzymatic biomarkers of Gaucher disease. *Blood Cells Mol Dis.* 2005; 35: 259-267.

Deegan PB, Cox TM. Clinical evaluation of biomarkers in Gaucher disease. *Acta Paediatr* Suppl. 2005; 94: 47-50.

Elstein D, Hollak C, Aerts JM, van Weely S, Maas M, Cox TM, Lachmann RH, Hrebicek M, Platt FM, Butters TD, Dwek RA, Zimran A. Sustained therapeutic effects of oral miglustat (Zavesca, N-butyldeoxynojirimycin, OGT 918) in type I Gaucher disease. *J Inherit Metab Dis.* 2004; 27: 757-766.

Elstein D, Dweck A, Attias D, Hadas-Halpern I, Zevin S, Altarescu G, Aerts JF, van Weely S, Zimran A. Oral maintenance clinical trial with miglustat for type I Gaucher disease: switch from or combination with intravenous enzyme replacement. *Blood.* 2007; 110: 2296-2301.

Erikson A, Johansson K, Mansson JE, Svennerholm L. Enzyme replacement therapy of infantile Gaucher disease. *Neuroped.* 1993; 24: 237-238.

Escott GM, Adams DJ. Chitinase activity in human serum and leukocytes. *Infect Immun.* 1995; 63: 4770-4773.

Furbish FS, Steer CJ, Krett NL, Barranger JA. Uptake and distribution of placental glucocerebrosidase in rat hepatic cells and effects of sequential deglycosylation. *Biochim Biophys Acta.* 1981; 673: 425-434.

Gaucher PCE. De l'épithélioma primitif de la rate. Hypertrophie idiopathique de la rate sans leucémie. 1882 ; Thèse, Paris.

Gieselmann V. Lysosomal storage diseases. *Biochim Biophys Acta.* 1995 ; 1270: 103-136.

Grabowski GA, Barton NW, Pastores G, Dambrosia JM, Banerjee TK, McKee MA, Parker C, Schiffmann R, Hill SC, Brady RO. Enzyme therapy in type 1 Gaucher disease: comparative efficacy of mannose-terminated glucocerebrosidase from natural and recombinant sources. *Ann Intern Med.* 1995 ; 122: 33-39.

Grosso S, Margollicci MA, Bargagli E, Buccoliero QR, Perrone A, Galimberti D, Morgese G, Balestri P, Rottoli P. Serum levels of chitotriosidase as a marker of disease activity and clinical stage in sarcoidosis. *Scand J Clin Lab Invest.* 2004; 64: 57-62.

Guo Y, He W, Boer AM, Wevers RA, de Bruijn AM, Groener JE, Hollak CE, Aerts JM, Galjaard H, van Diggelen OP. Elevated plasma chitotriosidase activity in various lysosomal storage disorders. *J Inherit Metab Dis.* 1995; 18: 717-722.

Heitner R, Elstein D, Aerts J, Weely S, Zimran A. Low-dose N-butyldeoxynojirimycin (OGT 918) for type I Gaucher disease. *Blood Cells Mol Dis.* 2002; 28: 127-133.

Hollak CE, van Weely S, van Oers MH, Aerts JM. Marked elevation of plasma chitotriosidase activity. A novel hallmark of Gaucher disease. *J Clin Invest.* 1994; 93: 1288-1292.

Hollak CE, Evers L, Aerts J, van Oers MH. Elevated levels of M-CSF, sCD14 and IL8 in type 1 Gaucher disease. *Blood Cells Mol Dis.* 1997a; 23: 201-212.

Hollak CE, Levi M, Berends F, Aerts JM, van Oers MH. Coagulation abnormalities in type 1 Gaucher disease are due to low-grade activation and can be partly restored by enzyme supplementation therapy. *Br J Haematol.* 1997b; 96: 470-476.

Hollak CE, Aerts JM. Clinically relevant therapeutic endpoints in type 1 Gaucher disease. *J Inherit Met Dis.* 2001; 24: 97-105.

Hollak CE, Aerts JM. In: Gaucher Disease: Diagnosis and Laboratory Features. 2007; pp.249-289. (Eds. Futerman, Zimran) Boca Raton: CRC Press.

Horowitz M, Zimran A. Mutations causing Gaucher disease. Hum Mutat. 1994; 3 : 1-11.

Lachmann RH, Grant IR, Halsall D, Cox TM. Twin pairs showing discordance of phenotype in adult Gaucher's disease. *QJM.* 2004; 97: 199-204.

Labadaridis J, Dimitriou E, Costalos C, Aerts J, van Weely S, Donker-Koopman WE, Michelakakis H. Serial chitotriosidase activity estimations in neonatal systemic candidiasis. *Acta Paediatr*. 1998; 87: 605

Labadaridis I, Dimitriou E, Theodorakis M, Kafalidis G, Velegraki A, Michelakakis H. Chitotriosidase in neonates with fungal and bacterial infections. *Arch Dis Child Fetal Neonatal Ed.* 2005; 90: F531-F532.

Masoud M, Rudensky B, Elstein D, Zimran A. Chitotriosidase deficiency in survivors of Candida sepsis. *Blood Cells Mol Dis.* 2002; 29: 116-118.

McEachern KA, Fung J, Komarnitsky S, Siegel CS, Chuang WL, Hutto E, Shayman JA, Grabowski GA, Aerts JM, Cheng SH, Copeland DP, Marshall J. A specific and potent inhibitor of glucosylceramide synthase for substrate inhibition therapy of Gaucher disease. *Mol Genet Metab*. 2007; 91: 259-267.

Michelakakis H, Spanou C, Kondyli A, Dimitriou E, van Weely S, Hollak CE, van Oers MH, Aerts JM. Plasma tumor necrosis factor-a (TNF-a) levels in Gaucher disease. *Biochim Biophys Acta.* 1996; 1317: 219-222.

Michelakakis H, Dimitriou E, Labadaridis I. The expanding spectrum of disorders with elevated plasma chitotriosidase activity: an update. *J Inherit Metab Dis.* 2004; 27: 705-706.

Meikle PJ, Fietz MJ, Hopwood JJ. Diagnosis of lysosomal storage disorders: current techniques and future directions. *Expert Rev Mol Diagn.* 2004; 4: 677-691.

Moller HJ, Aerts H, Gronbaek H, Peterslund NA, Hyltoft Petersen P, Hornung N, Rejnmark L, Jabbarpour E, Moestrup SK. Soluble CD163: a marker molecule for monocyte/macrophage activity in disease. *Scand J Clin Lab Invest* Suppl. 2002; 237: 29-33.

Moller HJ, de Fost M, Aerts H, Hollak C, Moestrup SK. Plasma level of the macrophage-derived soluble CD163 is increased and positively correlates with severity in Gaucher's disease. *Eur J Haematol.* 2004; 72: 135-139.

Moran MT, Schofield JP, Hayman AR, Shi GP, Young E, Cox TM. Pathologic gene expression in Gaucher disease: up-regulation of cysteine proteinases including osteoclastic cathepsin K. *Blood.* 2000; 96: 1969-1978.

Naito M, Takahashi K, Hojo H. An ultrastructural and experimental study on the development of tubular structures in the lysosomes of Gaucher cells. *Lab Invest.* 1988; 58: 590-598.

Ohashi T, Hong CM, Weiler S, Tomich JM, Aerts JM, Tager JM, Barranger JA. Characterization of human glucocerebrosidase from different mutant alleles. *J Biol Chem.* 1991; 266: 3661-3667.

Overkleeft HS, Renkema GH, Neele J, Vianello P, Hung IO, Strijland A, van den Burg A, Koomen GJ, Pandit UK, Aerts J. Generation of specific deoxynijirimycin-type inhibitors of the non-lysosomal glucosylceramidase. *J Biol Chem.* 1998; 273: 26522-26527.

Parkin JL, Brunning RD. Pathology of the Gaucher cell. In: A century of delineation and research. 1982; pp. 151-175. (Eds. Desnick, Gatt, Grabowski) Liss, New York.

Pastores GM, Barnett NL, Kolodny EH. An open-label, noncomparative study of miglustat in type I Gaucher disease: Efficacy and tolerability over 24 months of treatment. *Clin Ther.* 2005; 27: 1215-27.

Patrick AD. A deficiency of glucocerebrosidase in Gaucher's disease. *Biochem J.* 1965; 97: 17c-18c.

Platt FM, Neises GR, Dwek RA, Butters TD. N-butyldeoxynojirimycin is a novel inhibitor of glycolipid biosynthesis. *J Biol Chem.* 1994; 269: 8362-8365.

Platt FM, Neises GR, Reinkensmeier G, Townsend MJ, Perry VH, Proia RL, Winchester B, Dwek RA, Butters TD. Prevention of Lysosomal storage in Tay-Sachs mice treated with N-butyldeoxynojirimycin. *Science.* 1997; 276: 428-431.

Radin NS. Treatment of Gaucher disease with an enzyme inhibitor. *Glycoconj J.* 1996; 13, 153-157.

Renkema GH, Boot RG, Muijsers AO, Donker-Koopman WE, Aerts JM. Purification and characterization of human chitotriosidase, a novel member of the chitinase family of proteins. *J Biol Chem.* 1995; 270: 2198-2202.

Renkema GH, Boot RG, Strijland A, Donker-Koopman WE, van den Berg M, Muijser, AO, Aerts JM. Synthesis, sorting, and processing into distinct isoforms of human macrophage chitotriosidase. *Eur J Biochem.* 1997; 244: 279-285.

Rogowski O, Shapira I, Zimran A, Zeltser D, Elstein D, Attias D, Bashkin A, Berliner S. Automated system to detect low-grade underlying inflammatory profile: Gaucher disease as a model. *Blood Cells Mol Dis*. 2005; 34: 26-29.

Schlumbaum A, Mauch F, Vogeli U, Boller T. Plant chitinases are potent inhibitors of fungal growth. Nature 1986; 324: 365–367.

Schoonhoven A, Rudensky B, Elstein D, Zimran A, Hollak CE, Groener JE, Aerts JM. Monitoring of Gaucher patients with a novel chitotriosidase assay. *Clin Chim Acta.* 2007; 381: 136-139.

Shayman JA, Abe A, Hiraoka M. A turn in the road: How studies on the pharmacology of glucosylceramide synthase inhibitors led to the identification of a lysosomal phospholipase A2 with ceramide transacylase activity. *Glycoconj J.* 2004; 20: 25-32.

Sidransky E. Gaucher disease: complexity in a "simple" disorder. *Mol Genet Metab.* 2004; 83: 6-15.

Stevens DA, Brammer HA, Meyer DW, Steiner BH. Recombinant human chitinase. *Curr Opin Anti-Infective Investig Drugs.* 2000; 2: 399–404.

Suzuki M, Fujimoto W, Goto M, Morimatsu M, Syuto B, Iwanaga TJ. Cellular expression of gut chitinase mRNA in the gastrointestinal tract of mice and chickens. *Histochem Cytochem.* 2002; 50: 1081-1089.

Tsuji S, Martin BM, Barranger JA, Stubblefield BK, LaMarca ME, Ginns EI. Genetic heterogeneity in type 1 Gaucher disease: Multiple genotypes in Ashkenazic and non-Ashkenazic individuals. *Proc Natl Acad Sci USA*. 1988; 85: 2349-2352.

van Breemen MJ, de Fost M, Voerman JS, Laman JD, Boot RG, Maas M, Hollak CE, Aerts, JM, Rezaee F. Increased plasma macrophage inflammatory protein (MIP)-1alpha and MIP-1beta levels in type 1 Gaucher disease. *Biochim Biophys Acta.* 2007; 1772: 788-796.

van Eijk M, van Roomen CP, Renkema GH, Bussink AP, Andrews L, Blommaart EF, Sugar A, Verhoeven AJ, Boot RG, Aerts JM. Characterization of human phagocyte-derived chitotriosidase, a component of innate immunity. *Int Immunol.* 2005; 17: 1505-1512.

van Eijk M, Scheij SS, van Roomen CP, Speijer D, Boot RG, Aerts JM. TLR- and NOD2-dependent regulation of human phagocyte-specific chitotriosidase. *FEBS Lett.* 2007; 581: 5389-5395.

van Weely S, van den Berg M, Barranger JA, Sa Miranda MC, Tager JM, Aerts JM. Role of pH in determining the cell-type specific residual activity of glucocerebrosidase in type 1 Gaucher disease. *J Clin Invest.* 1993; 91: 1167-1175.

Vellodi A. Lysosomal storage disorders. Br J Haematol. 2005; 128 : 413-431.

Vellodi A, Foo Y, Cole TJ. Evaluation of three biochemical markers in the monitoring of Gaucher disease. *J Inherit Metab Dis.* 2005; 28: 585-592.

Vissers JP, Langridge JI, Aerts JM. Analysis and quantification of diagnostic serum markers and protein signatures for Gaucher disease. *Mol Cell Proteomics.* 2007; 6: 755-766.

Zheng T, Rabach M, Chen NY, Rabach, L, Hu X, Elias JA, Zhu Z. Molecular cloning and functional characterization of mouse chitotriosidase. *Gene.* 2005; 357: 37-46.

In: Binomium Chitin-Chitinase: Recent Issues ISBN 978-1-60692-339-9
Editor: Salvatore Musumeci and Maurizio G. Paoletti

Chapter VIII

"Chitinase Activity in Atherosclerosis Disease"

Ana Cenarro and Fernando Civeira[9]
Laboratorio de Investigación Molecular
Hospital Universitario Miguel Servet, I+CS
Zaragoza, Spain

Abstract

Atherosclerosis is an inflammatory disease in which macrophages play a very important role in its pathogenesis. Chitotriosidase is one of the proteins highly secreted by activated macrophages. Moreover, chitotriosidase is highly expressed by macrophages within the vascular atherosclerosis plaques, suggesting that this enzyme could be involved in the inflammatory process associated with modified LDL particles in the arteries. Several groups have recently demonstrated that serum chitotriosidase activity is related to the extension of atherosclerosis, and predicts the risk of new cardiovascular events with a predictive value similar to CRP, and, when combined with CRP, the risk prediction of new cardiovascular events and the identification of a lower risk group seem to improve. The mechanism of these associations is not fully understood but could be related, as occurs with other chitinases, through the contribution of chitotriosidase to a T helper 2 immune response to oxidized LDL.

1. Introduction

Atherosclerosis is a complex disease in which a large number of mechanisms are involved in its pathogenesis and complications. However, the research efforts have clearly demonstrated that two major phenomena are the main actors in the pathobiology of the atherosclerosis disease: 1) a sustained high blood low density lipoprotein (LDL) cholesterol

9 E-mail: civeira@unizar.es.

concentration, and 2) a complex inflammatory reaction that takes places within the arterial wall when the LDL particles deposit and modify in the subendothelial space. So, at present time atherosclerosis is considered an inflammatory vascular disease mainly driven by LDL.

Many prospective epidemiological studies, human interventions studies using different cholesterol-lowering regimens, human diseases associated with hypercholesterolemia and animal models with hyperlipemia have demonstrated that cholesterol, mainly the subfraction transported in the LDL particles, is a key factor in the development of atherosclerosis. A highly demonstrative example of the close relation between cholesterol and atherosclerosis is familial hypercholesterolemia. Patients with homozygous familial hypercholesterolemia with very high LDL cholesterol concentrations develop severe and very premature atherosclerosis even in absence of any other risk factor indicating that at certain LDL cholesterol concentration, cholesterol is enough to trigger the disease (Civeira *et al.* 2004). However, in the general population, even with high LDL cholesterol as occurs in heterozygous patients with familial hypercholesterolemia, the progression of atherosclerosis vary considerably among subjects with similar LDL cholesterol levels. The mechanisms involved in the variation in the progression of atherosclerosis among subjects is not fully understood but is highly related to the inflammatory response that a certain level of cholesterol induces in each subject. Probably the other risk factors, other than cholesterol, are behind this heterogeneity and participate in the internalization, modification of LDL and inflammatory response. The subendothelial formation of pro-inflammatory oxidized products from LDL particles plays very important roles (Ross R, 1999). Oxidation of LDL is a complex phenomenon promoted, mainly, by macrophages and endothelial cells within the subendothelial extracellular matrix. Lipoxygenases, myeloperoxidases, inducible nitric oxide synthase and NADPH oxidases have been proposed as candidate enzymes for LDL oxidation "in vivo". Oxidized LDL (oxLDL) contains a large variety of lipid peroxidation products and reactive aldehydes which generate multiple immunogenic oxidation-specific epitopes and are recognized by macrophages, the primary cell of the innate immunity involved in the recognition of foreign molecular patterns by pattern recognition receptors (PRRs) as a rapid first line of defense. Hence, macrophages are major players in the vascular inflammatory response to LDL cholesterol, and they are involved in all phases of atherosclerosis, from the initiation, progression and complications of the disease. The uptake of modified lipoproteins by macrophages leads to the formation of foam cells, the most characteristic cell of the atherosclerotic lesion.

Hypercholesterolemia induces a modification in the endothelium which stimulates the production of monocyte chemoattractant molecules leading to integrin-dependent adhesion of monocytes to the endothelium vascular, mainly by cell-adhesion molecule 1, and diapedesis into the subendothelium (Linton *et al.* 2003). Inside the arterial wall monocytes differentiate into macrophages and internalize the excess of LDL throughout their PRRs. Macrophage PRRs include: Toll-like receptors, which behave as signaling receptors triggering the expression of pro-inflammatory cytokines, and scavenger receptors, endocytic cell surface receptors that bind and internalize pathogen-associated molecular patterns, such as oxLDL, bacterial components or apoptotic cells. OxLDL is rapidly recognized and internalized by macrophage scavenger receptors, which in contrast with the LDL receptor, do not down-regulate in response to increased cellular cholesterol content, leading to foam-cell formation

(Steinberg *et al.* 2002). Scavenger receptors are a group of transmembrane proteins with large sequence variation, which are classified in eight classes (A-H). However, one class A (SR-A), one class B (CD36) and the class E scavenger receptor (lectin-like oxidized LDL receptor-1, LOX-1) account for over 90% of oxLDL uptake (Miller *et al.* 2003). Circulating monocytes express scavenger receptors at very low levels, but they are highly up-regulated by the oxLDL located in the subendothelial space, leading to macrophage activation, foam-cell formation and secretion of pro-inflammatory cytokines and activation of adaptive immunity.

2. Chitotriosidase in Atherosclerosis Lesions

Rolf G. Boot and cols. in 1999 were the first group to show a very potent induction of chitotriosidase in human atherosclerosis vessel wall. They hypothesized that, in a similar fashion as in Gaucher disease, lipid accumulation in atherosclerosis would be associated with an increase in chitotriosidase expression within the arterial wall. They collected tissue samples from coronary arteries and aortas from heart transplantation recipients and donors with and without atherosclerosis and studied the chitotriosidase mRNA expression by radioactive "in situ" hybridization. No expression was observed in the normal specimens; however, in the atherosclerosis lesions the macrophage infiltrates were associated with high chitotriosidase expressions although not all macrophages showed the same pattern, with a tendency to higher expression in advanced lesions (Boot *et al.* 1999).

Moreover, these authors studied the chitotriosidase activity in different tissue extracts from arteries with and without atherosclerosis. The chitotriosidase activity was increased up to 55-fold in atherosclerosis vessel compared to normal arteries and again with high heterogeneity among the different samples, but with good correlation between chitotriosidase mRNA expression and chitotriosidase activity in the extracts of vascular tissue (Boot *et al.* 1999).

3. Chitotriosidase Activity Correlates with Atherosclerosis Disease in Humans

Considering that chitotriosidase is produced mostly in humans by activated macrophages, and that chitotriosidase activity is elevated in atherosclerotic tissue, we hypothesized that serum chitotriosidase activity could be related to the amount of lipid-loaded macrophages in the atherosclerotic arterial wall. Then we proposed to investigate if serum chitotriosidase activity was increased in subjects suffering an ischemic stroke of atherothrombotic etiology and in subjects with ischemic heart disease and, in that way, if chitotriosidase activity could be related to the extension of atherosclerosis. To evaluate our hypothesis, we analyzed the serum chitotriosidase activity and the CHIT1 gene polymorphism in a group of subjects with ischemic stroke of atherothrombotic origin, in a group of subjects with ischemic heart disease, and in a group of control subjects (Artieda *et al.* 2003). The atherothrombotic stroke group consisted of 153 nonrelated Spanish subjects younger than 71 years of age with an ischemic stroke attributable to occlusion or stenosis of atheromatous etiology in an intracranial or extracranial artery. The ischemic heart disease group consisted of 124 nonrelated Spanish subjects with stabilized unstable angina pectoris. A group of 148 nonrelated Spanish subjects with normal lipid profile and without symptomatic atherosclerosis disease was also included in the study and considered the control group. There were no differences in the allelic and genotype distributions between the studied groups, so we could rule out the possibility that chitotriosidase activity differences observed between groups were attributable to distinct genotype distributions.

There were statistical differences between atherothrombotic stroke group versus control groups and between IHD versus control groups independently of the CHIT1 genotype. Subjects in the atherosclerosis groups showed higher chitotriosidase activities than controls after adjustment for age and CHIT1 genotype (figure 1).

To assess the relationship between chitotriosidase activity and the arterial intima-media thickness, we analyzed the carotid stenosis data of the ATS subjects proved by duplex sonography. Chitotriosidase activities for 3 subgroups with different stenosis grade were as follows: 66.9±9.5 nmol/mL·h for stenosis ≤30%, 88.7±8.3 nmol/mL·h for stenosis 31% to 60%, and 107.7±11.8 nmol/mL·h for stenosis >60%. Statistical differences were found between subjects with stenosis of 31% to 60% and subjects with stenosis >60% compared with subjects with stenosis ≤30%, as shown in figure 2 (Artieda *et al.* 2003).

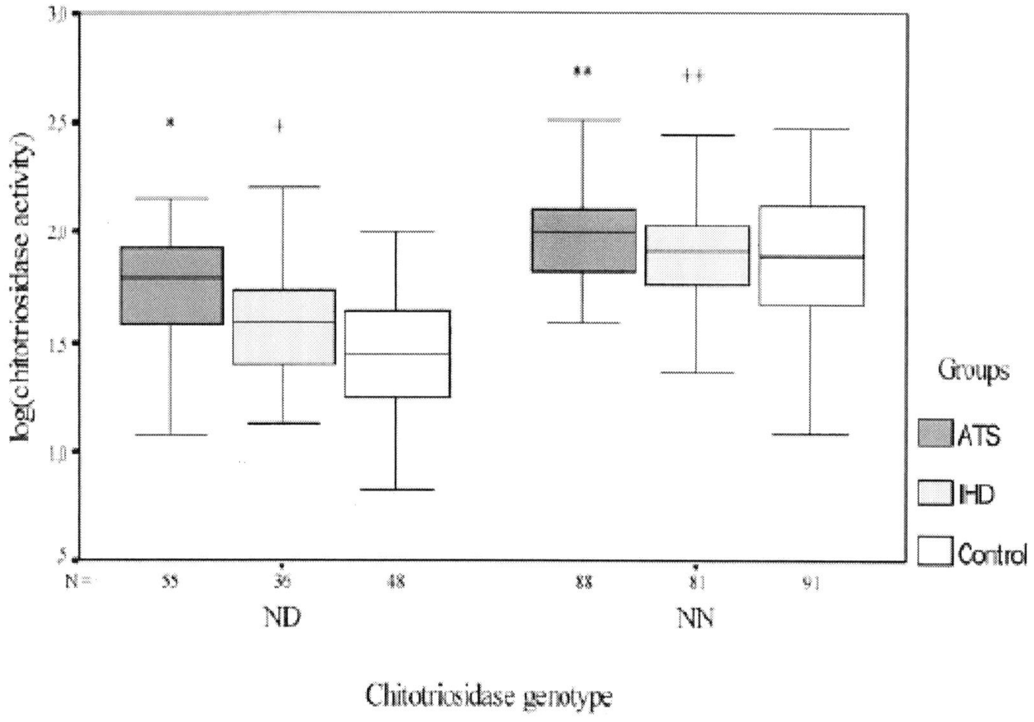

Figure 1. Chitotriosidase activity in subjects with atherosclerosis (Adapted from Artieda *et al.* 2003).

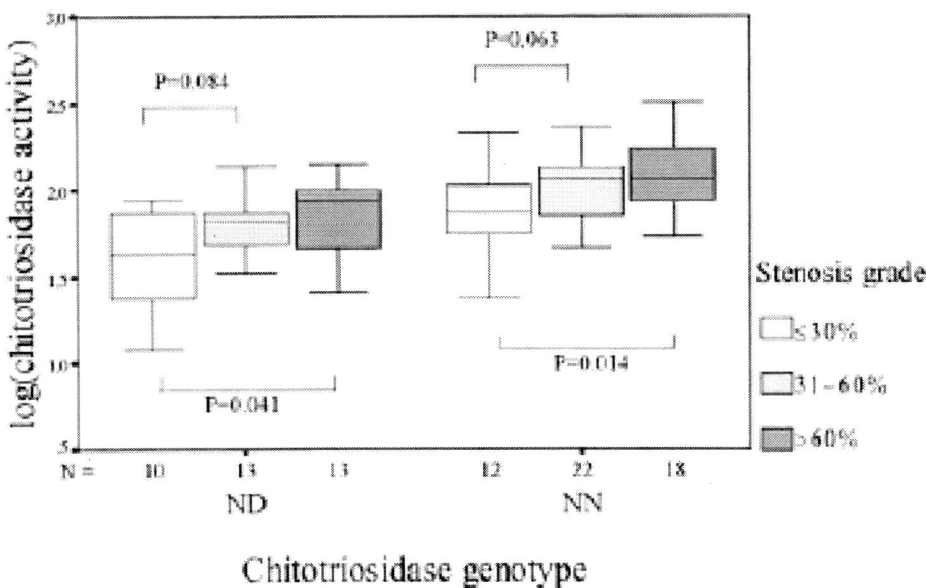

Figure 2. Chitotriosidase activity according to carotid stenosis in subjects with ischemic stroke (Adapted from Artieda *et al.* 2003).

These results have been recently confirmed by others in a different and independent population with coronary artery disease (Karadag *et al.* 2008). In this latter work, the chitotriosidase activity in serum was studied 200 subjects undergoing coronary angiography and was significantly higher in patients with coronary disease than in controls. As in our previous study with carotid arteries, in this work the extent of coronary stenosis was significantly associated with the serum chitotriosidase activity.

We found a positive correlation between chitotriosidase activity and age. This age-dependent increase in serum chitotriosidase activity has been previously described in subjects with different lysosomal disorders and in the general population and could be explained by the ongoing accumulation of lipid-laden macrophages during the gradual progression of atherosclerosis in relation to age.

Chitotriosidase activity measurement in clinical practice has several important limitations. First, it is necessary to adjust serum activity with chitotriosidase genotype because of the high frequency of the defective allele in the general population. Second, there is a large range of serum chitotriosidase activity values in control subjects, even sharing the same genotype, and an overlap exists between serum chitotriosidase activity values from control and affected subjects. Third, little is known about physiological role of chitotriosidase and the mechanisms that could modify its activity in humans, and they could be relevant in clinical practice. For these reasons, chitotriosidase activity measurements in the diagnosis of atherosclerosis disease have relatively weak predictive value as a marker of atherosclerotic lesion.

In summary, serum chitotriosidase activity is related to the severity of the atherosclerotic lesions, suggesting a possible role as a marker of atherosclerotic extension, and this increase of serum chitotriosidase activity demonstrates that it is feasible to measure functional aspects of macrophages from blood samples of patients with atherosclerosis (Artieda *et al.* 2003).

4. Correlation between Chitotriosidase Activity with C-Reactive Protein and other Atherosclerosis Risk Factors

Chitotriosidase can be considered an inflammatory protein because it is secreted by activated macrophages, but its production happens after at least 1 week of cell culture and increases with time; therefore, it does not behave as an acute reactive protein but rather as a chronic inflammatory marker. We have also investigated the relation of chitotriosidase activity with C-reactive protein (CRP), a well-known marker of overall systemic inflammation, in control subjects, in patients with Gaucher's disease, and in subjects with coronary disease. CRP has been shown in multiple prospective studies to predict risk of coronary disease, stroke, peripheral vascular disease and sudden cardiac death (Ridker *et al.* 2001). Chitotriosidase activity was 170-fold increased in Gaucher patients compared with control subjects. There were statistical differences in chitotriosidase activity between Gaucher versus control subjects but no statistical differences in CRP values were found (Cenarro *et al.* 1999). In patients with Gaucher's disease, only chitotriosidase activity was highly increased, CRP showed the same levels as control subjects, and there was no correlation between chitotriosidase activity and CRP in both groups. Therefore, these inflammatory markers seem to be regulated by different mechanisms.

We also investigated the correlation between chitotriosidase activity in coronary patients. As occurred in Gaucher patients and in controls, chitotriosidase activity did not correlate with CRP after adjustment for CHIT1 genotype and age, furthermore we could not find any association between serum chitotriosidase activity and major cardiovascular risk factors, including, LDL cholesterol, HDL cholesterol, triglycerides, lipoprotein(a), hypertension, smoking or diabetes mellitus (Artieda *et al.* 2007).

5. Predictive Value of Serum Chitotriosidase Activity of New Cardiovascular Events

Because chitotriosidase is elevated in human atherosclerotic plaques and this increase is related to the severity of atherosclerosis independently of CRP, we studied the predictive value of serum chitotriosidase activity additive to high sensitivity CRP in a group of 133 coronary patients followed up for a new cardiovascular event for a mean of 4 years (Artieda *et al.* 2007).

Serum chitotriosidase activity and CRP were significantly increased in coronary disease patients at baseline. Besides, chitotriosidase activity was increased in subjects with a major coronary event (nonfatal myocardial infarction, nonfatal ischemic stroke, coronary revascularization procedure or death from cardiovascular cause). The event-free survival curves showed that baseline chitotriosidase activity was predictive of new cardiovascular events occurring after 2 years of follow-up than of new events in the first two years (figure 3).

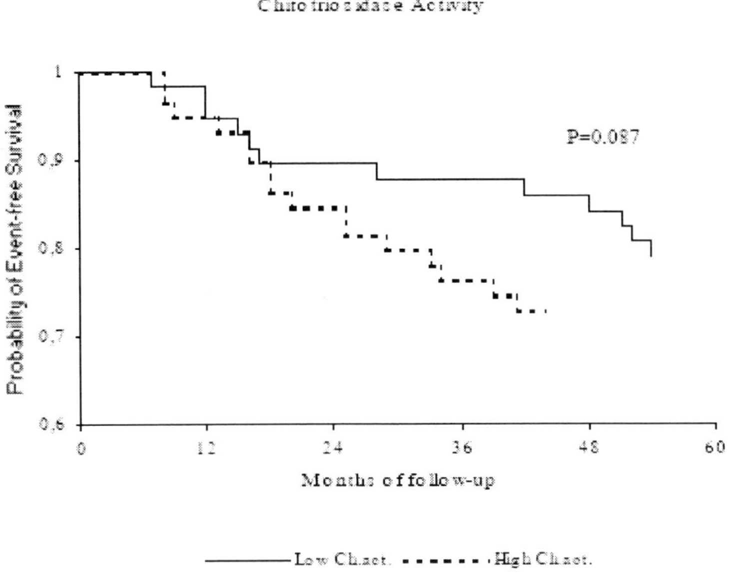

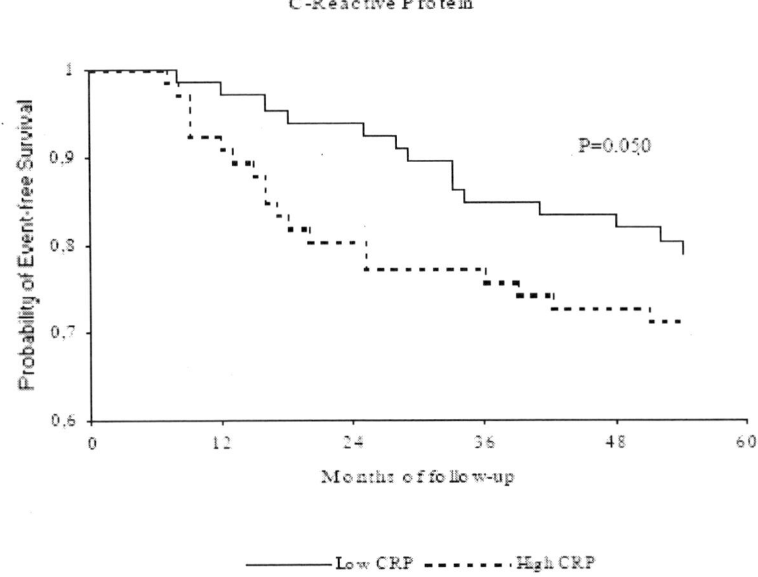

Figure 3. Kaplan-Meier curves for event-free survival according to the baseline of chitotriosidase activity in two defined groups (low and high chitotriosidase activity) (a), and C-reactive protein (CRP) in two defined groups (low and high CRP) (b). (Adapted from Artieda *et al.* 2007).

The magnitude of the prediction was comparable to the obtained for CRP. Interestingly, both markers showed additive prediction values. The relative risk of suffering a new event in the low chitotriosidase and low CRP group was 0.57 compared to the high chitotriosidase and high CRP group (Artieda *et al.* 2007).

The results of this study would suggest the important role of activated macrophages in various complications of atherosclerosis and also that chitotriosidase activity could reflect the state of activation of macrophages, possibly within atherosclerotic lesions, and it could be used as a clinical marker of such activation.

Plasma chitotriosidase activity and CRP are two inflammatory proteins, but related to different aspects of inflammation. CRP is an acute-phase response protein, of hepatic origin, that activates the classical complement pathway after aggregation or binding to several ligands (ref). The plasma cytokine interleukin-6 (IL-6) is the main inductor of hepatic synthesis of CRP and is produced by different cells, including smooth muscle cells, macrophages, lymphocytes and adipocytes (refs). In contrast, chitotriosidase is not an acute-phase response protein, is expressed in chronically activated tissue macrophages, such as the lipid-laden storage cells that accumulate in Gaucher´s disease or atherosclerotic plaques.

Considering that CRP is an excellent cardiovascular risk predictor, its lack of association with chitotriosidase activity probably reflects that increased levels of CRP in high risk vascular patients are not related with lipid laden chronically activated vascular macrophages, but with other IL-6 producing cells including adipocytes. In fact, CRP correlates with triglycerides, obesity, insulin sensitivity and fasting glucose, and also with the risk to developing type 2 diabetes, indicating a close relationship with the metabolic syndrome (Pradhan *et al.* 2001).

We can conclude that serum chitotriosidase activity predicts the risk of new cardiovascular events with a predictive value similar to CRP. This new cardiovascular risk marker is independent of CRP and, when combined, the risk prediction of new cardiovascular events and the identification of a lower risk group seem to improve.

6. Chitotriosidase and Interindividial Variation Response to Oxidized LDL

Many evidences support that the inflammatory response to modified LDL particles within the artery wall is initiated by macrophages and the atherosclerosis can be dramatically diminish when the inflammatory response is blunted. Moreover, the type of inflammatory response to modified LDL can vary among subjects and macrophages could be responsible of such variation. There is a great human inter-individual variation in the vascular and systemic inflammation associated with oxLDL. Moreover, a large variation in human foam-cell formation, intracellular cholesterol content and cytokine gene expression in individuals exists when induced by oxLDL "in vitro", suggesting that this differential response is due to individual characteristics (Artieda *et al.* 2005). Thus, the magnitude and type of macrophage responses to oxLDL could determine part of the observed inter-individual inflammatory variability associated with hypercholesterolemia and oxLDL, and could be helpful to improve individual cardiovascular risk, which is closely related with the levels of systemic inflammation. In a recent work, we have studied the inter-individual differences in macrophage scavenger receptor gene expression and the inflammatory variability in response to oxLDL (Martín-Fuentes *et al.* 2007). We quantified the gene expression by quantitative RT-PCR of scavenger receptors and inflammatory molecules from macrophages isolated from 18 volunteer subjects and incubated with oxLDL for different times. The individual gene expression profile of the studied scavenger receptors at each incubation time was highly variable, showing a wide fold-change range. We also identified subjects as high and low-responders for each scavenger receptor gene expression, showing a different inflammation response pattern. CD36 and LOX-1 gene expression correlated positively with IL-1β an inflammatory molecule, and SR-A correlated negatively with IL-8 (inflammatory) and positively with PPARγ and NF-κBIA (anti-inflammatory molecules). These data would suggest that the type of scavenger receptor could determine the macrophage activation: more pro-inflammatory when associated to CD36 and LOX-1 than when associated with SR-A (Martín-Fuentes *et al.* 2007). Therefore, we demonstrated that inter-individual variability of macrophage scavenger receptor gene expression in response to oxLDL and the subsequent induced intracellular signals could determine the observed inflammatory variability in response to oxLDL. According with this study, and as chitotriosidase is a macrophage activation marker, we hypothesized that chitotriosidase gene (CHIT1) genotype could be involved in the immune response against oxLDL. A common CHIT1 polymorphism leads to a null allele and therefore a defective protein without chitinase activity. The nature of this common deficiency in chitotriosidase activity is a 24-base pair (bp) duplication in exon 10 of the CHIT1 gene that results in activation of a cryptic 3' splice site, generating a mRNA with an in-frame deletion of 87 nucleotides, encoding a protein that lacks an internal stretch of 29 amino acids. This genetic polymorphism is responsible for the recessive inherited deficiency of chitotriosidase activity, which is found in individuals from various ethnic origins. In white populations, 30% to 40% of individuals are carriers of this abnormal CHIT1 allele and, approximately, 6% are homozygotes (refs ?).

To carry out this investigation, we selected 26 subjects and we determined the CHIT1 genotype. Also, we isolated and cultured their monocytes/macrophages with oxLDL, and we analyzed the gene expression of several scavenger receptors and inflammatory molecules. Macrophages from homozygous subjects for normal allele of CHIT1 presented more gene expression of CD36, IL-1beta and TNF-alfa, and a tendency to overexpress IL-8 gene. At protein level, IL-8 concentration in macrophages from homozygous subjects for the normal allele of chitotriosidase was significantly higher than that of macrophages from subjects carrying a defective allele of chitotriosidase. Our results confirm the important function of chitotriosidase in the inflammatory process, as subjects homozygous for normal chitotriosidase allele, that is, with full activity of the enzyme, present a more inflammatory phenotype in response to a proatherogenic stimulus, as oxidized LDL, overexpressing the scavenger receptor CD36 and the inflammatory genes, IL-1beta and TNF-alfa, and producing more IL-8 protein than subjects with minor or null enzymatic activity. These results would suggest that presence of a chitotriosidase defective allele, as production of chitotriosidase enzyme is diminished in macrophages, would modulate the immune response type against oxidized LDL. Taking into account the important role that inflammation plays in atherosclerotic process, CHIT1 genotype could determine the inflammatory response in an individual subject, and could explain, at least in part, the observed interindividual variability in response to oxidized LDL, and therefore, it could be a factor contributing to susceptibility to develop atherosclerosis.

7. Influence of Chitotriosidase Gene Polymorphism in the T Helper 2 Inflammatory Response against LDL

Chitin is a main component of the protective coats found in fungal cell walls, in the exoskeletons of crustaceans and insects, and in the microfilarial sheaths of nematodes. Chitinase production is a common mechanism in the immune response against chitin-containing pathogens in most species. The existence of two endogenous chitinases, chitotriosidase and acidic mammalian chitinase (AMCase), has been demonstrated in humans (Boot et al 2005). AMCase is expressed in alveolar macrophages and in the gastrointestinal tract, and chitotriosidase is one of the most quantitative proteins secreted by activated tissue macrophages, such as occurs in Gaucher disease and in various lysosomal storage disorders (Guo et al 1995). The physiological role of chitotriosidase is not fully understood, but its phagocyte-specific expression supports a role in innate immunity (Boot *et al.* 2005).

Recently, Zhu and colleges have demonstrated, in a murine asthma model and in patients with asthma, that AMCase expression is increased and involved in the pathogenesis of the disease by contributing to a T helper 2 (Th2) immune response (Zhu *et al.* 2004).

Considering the importance of the Th2 response in LDL-cholesterol induced atherosclerosis and the involvement of chitotriosidase activity in the progression of atherosclerotic vascular lesions, we have investigated if CHIT1 genotype and chitotriosidase activity could also be involved in the immune response to oxLDL. To carry out this study, we determined the CHIT1 genotype, plasma chitotriosidase activity and several inflammatory markers in a group of subjects with genetically defined heterozygous familial hypercholesterolemia, characterized by elevated levels of LDL-cholesterol in plasma. The elevation in serum LDL-cholesterol levels, secondary to the LDLR defect, results in formation of oxidized LDL (oxLDL) and other modified LDL species, and in their uptake by the macrophage scavenger receptors, leading to massive lipid accumulation and foam cell formation.

Our results suggest that, in subjects with very high LDL-cholesterol levels of genetic origin, such as those with heterozygous FH, the CHIT1 24-bp duplication genotype modulates the type of immune response to oxLDL, probably through a decrease in macrophage chitotriosidase production (in preparation).

In our study, we have found that heterozygous FH subjects homozygous for the normal allele (NN) of the CHIT1 gene, and therefore, full macrophage chitotriosidase activity, showed a higher prevalence of positive oxidized LDL antibodies (OLAB), more tendon xanthomas and lower serum IgE levels than those carrying defective CHIT1 alleles (in preparation). The oxLDL antibodies status in FH was assessed in two prior studies. Hulthe and colleagues measured the antibody titers to oxLDL in 51 subjects with heFH and 45 controls (Hulthe *et al.* 1998), and Paiker and colleagues did the same in 26 homozygous FH and 44 heterozyous FH (Paiker *et al.* 2000). In agreement with our results, these studies found large inter-individual differences in the antibody titers, which were unrelated to LDL-cholesterol levels, suggesting a complex relationship between LDL-cholesterol and autoimmune responses to oxLDL in FH. Interestingly, oxLDL antibody titers appeared to be lower in patients with a history of myocardial infarction in one study (Hulthe *et al.* 1998), suggesting that antibodies against oxLDL could have a protective effect. Our finding of reduced inflammation (lower IL-6 levels) in subjects with higher oxLDL antibodies titers agrees with this observation (in preparation).

Our results suggest that Chitotriosidase is involved in T helper cell immune-type responses, and that Chitotriosidase activity seems to favor Th2 responses, similarly to AMCase in the mouse asthma model (Zhu *et al.* 2004). Although the Th1–Th2 system is very adaptable in humans, the general pattern is that the type 1 helper T (Th1) response activates macrophages, initiates an inflammatory response similar to delayed hypersensitivity, and characteristically functions in the defense against chitin containing pathogens. In contrast, Th2 responses elicit the activation of B lymphocytes, antibody production and allergic inflammation.

These data are in agreement with the alleged implication of an exaggerated Th2-mediated response in airway inflammation in human and murine asthma. It is believed that the Th2 response was originally evolved to deal with parasites rich in chitin, thus it is conceivable that AMCase is involved in the pathogenesis of the Th2 immune response. This hypothesis was partially confirmed by the finding of increased chitinase activity in the lungs of antigen-sensitized mice, the potent stimulation of AMCase by Th2 cells, and markedly decreased Th2 inflammation by administration of anti-AMCase sera in the asthmatic mice model (Zhu *et al.* 2004). Moreover, chitinase activity was not detected in normal human lung samples, but was readily detected in lung samples from patients with asthma. The role of chitinases in the Th2 mediated inflammation was also highlighted by Sandler et al. in a study of acute pathogen-induced pulmonary inflammation in a Th1 or Th2 environmental and genetically polarized animal model (Sandler *et al.* 2003). In both cases, several members of the chitinase-like family were markedly induced in the lung after *Schistosoma mansoni* sensitization in the Th2-polarized mice. These authors and others have suggested the involvement of the chitinase family in wound healing because chitinase peak expression correlates with development, tissue remodeling and fibrosis. We speculate that the differences in the presence of tendon xanthomas according to the CHIT1 genotype found in our study could be partially explained by the concomitant different inflammatory response. Tendon xanthomas (TX) are pathognomonic deposits of lipid and connective tissue commonly found in patients with severe hyperlipidemia, as occurs in FH. Monocyte-derived foam cells due to intracellular accumulation of oxLDL, extracellular unesterified and esterified cholesterol and connective tissue are the main components of TX. The highly fibrotic component of TX suggest that subjects with TX develop a type of inflammation with less proclivity to tissue damage and more to wound healing, as occurs in the Th2 immune response. In fact, different macrophage inflammatory responses to oxLDL in heFH subjects with and without TX have been previously described by our group (Artieda *et al.* 2005).

The grade of inflammation associated with increased LDL-cholesterol or LDL-cholesterol reduction obtained with lipid lowering drugs varies considerably among individuals, suggesting that other genetic or environmental factors are associated with the type of the inflammatory response to LDL. Therefore, the CHIT1 genotype and ensuing chitotriosidase activity are associated with the type of immune response to oxLDL in FH, and could explain, at least in part, the variable inflammatory response to oxLDL among individuals.

8. Effect of High Blood Cholesterol Treatment on Chitotriosidase Activity

Because serum chitotriosidase activity could be related to the number of activated lipid-laden macrophages, and lipid-lowering drugs can reduce macrophage number and macrophage lipid content inside the plaques then we studied if lipid lowering treatment could modify this chitotriosidase activity in serum of hyperlipidemic patients. With this objective, we studied the serum chitotriosidase activity and the common chitotriosidase gene polymorphism, before and after lipid-lowering treatment in a group of subjects enrolled in the Atozvastatin Versus Bezafibrate in Mixed Hyperlipidemia (ATOMIX) study, a 1-year, double-blind, comparative, and randomized study comparing the efficacy of atorvastatin and bezafibrate in mixed hyperlipidemia (Gomez-Gerique *et al.* 2002). However, Chitotriosidase activity remained fairly constant throughout the study. There was no difference in serum chitotriosidase activity between treatment groups after 6 months of treatment. Besides, No relationship was found between total cholesterol, triglycerides, LDL cholesterol and HDL cholesterol concentrations, and the serum chitotriosidase activity at baseline in either treatment group. Our results did not support the idea that chitotriosidase activity could be used as a biologic marker of atherosclerotic plaque modification related to hypolipidemic treatment, at least after only several months of treatment (Canudas *et al.* 2001).

Conclusions

Chitotriosidase is highly expressed in activated macrophages within the atherosclerosis vascular lesions in humans. Serum chitotriosidase activity is a good inflammatory marker of clinical atherosclerosis and is related the disease extent. In a prospective study, chitotriosidase activity predicted new cardiovascular events independently of hs-CRP, and when combined the risk prediction improved for both markers probably because chitotriosidase is not an acute-phase response protein and is expressed in chronically activated macrophages. Finally, CHIT1 genotype could be responsible, at least in part, of the interindividual variation in the inflammatory of macrophages in response to oxLDL.

References

Artieda M, Cenarro A, Gañán A, Jericó I, Gonzalvo C, Casado JM, Vitoria I, Puzo J, Pocoví M, Civeira F. Serum chitotriosidase activity is increased in subjects with atherosclerosis disease. *Arterioscler Thromb Vasc Biol.* 2003;23:1645-52.

Artieda M, Cenarro A, Junquera C, Lasierra P, Martínez-Lorenzo MJ, Pocoví M, Civeira F. Tendon xanthomas in familial hypercholesterolemia are associated with a differential inflammatory response of macrophages to oxidized LDL. *FEBS Lett.* 2005;579:4503-12.

Artieda M, Cenarro A, Gañán A, Lukic A, Moreno E, Puzo J, Pocoví M, Civeira F. Serum chitotriosidase activity, a marker of activated macrophages, predicts new cardiovascular events independently of C-reactive protein. *Cardiology.* 2007;108:297-306.

Boot RG, van Achterberg TA, van Aken BE, Renkema GH, Jacobs MJ, Aerts JM, Vries CJ. Strong induction of members of the chitinase family of proteins in atherosclerosis: chitotriosidase and human cartilage gp-39 expressed in lesion macrophages. *Arterioscler Thromb Vasc Biol.* 1999;19:687–94.

Boot RG, Bussink AP, Verhoek M, de Boer PA, Moorman AF, Aerts JM. Marked differences in tissue-specific expression of chitinases in mouse and man. *J Histochem Cytochem.* 2005;53:1283-92.

Canudas J, Cenarro A, Civeira F, García-Otín AL, Arístegui R, Díaz C, Masramon X, Sol JM, Hernández G, Pocoví M. Chitotriosidase genotype and serum activity in subjects with combined hyperlipidemia: effect of the lipid-lowering agents, atorvastatin and bezafibrate. *Metabolism.* 2001;50:447–50.

Cenarro A, Pocoví M, Giraldo P, García-Otín AL, Ordovás JM. Plasma lipoprotein responses to enzyme-replacement in Gaucher's disease. *Lancet.* 1999;353:642–3.

Civeira F; International Panel on Management of Familial Hypercholesterolemia. 2004 Guidelines for the diagnosis and management of heterozygous familial hypercholesterolemia. *Atherosclerosis.* 2004;173:55-68.

Gómez-Gerique JA, Ros E, Oliván J, Mostaza JM, Vilardell M, Pintó X, Civeira F, Hernández A, da Silva PM, Rodríguez-Botaro A, Zambón D, Lima J, Díaz C, Arístegui R, Sol JM, Chaves J, Hernández G, for the ATOMIX Investigators. Effect of atorvastatin and benzafibrate on plasma levels of C-reactive protein in combined (mixed) hyperlipemia. *Atherosclerosis.* 2002;162:245–51.

Guo Y, He W, Boer AM, Wevers RA, de Bruijn AM, Groener JE, Hollak CE, Aerts JM, Galjaard H, van Diggelen OP. Elevated plasma chitotriosidase activity in various lysosomal storage disorders. *J Inher Metab Dis.* 1995;18:717–22.

Hulthe J, Wikstrand J, Lidell A, Wendelhag I, Hansson GK, Wiklund O. Antibody titers against oxidized LDL are not elevated in patients with familial hypercholesterolemia. *Arterioscler Thromb Vasc Biol.* 1998;18:1203-11.

Karadag B, Kucur M, Isman FK, Hacibekiroglu M, Vural VA. Serum chitotriosidase activity in patients with coronary artery disease. *Circ J.* 2008;72:71-5.

Linton MF, Fazio S. Macrophages, inflammation, and atherosclerosis. *Int J Obes.* 2003;27:535-40.

Martin-Fuentes P, Civeira F, Recalde D, Garcia-Otin AL, Jarauta E, Marzo I, Cenarro A. Individual variation of scavenger receptor gene expression in human macrophages with oxidized LDL is associated with a differential inflammatory response. *J Immunol.* 2007;179:3242-8.

Miller YI, Chang MK, Binder CJ, Shaw PX, Witztum JL. Oxidized low density lipoprotein and innate immune receptors. *Curr Opin Lipidol.* 2003;14:437-45.

Paiker JE, Raal FJ, von Arb M. Auto-antibodies against oxidized LDL as a marker of coronary artery disease in patients with familial hypercholesterolaemia. *Ann Clin Biochem.* 2000;37:174-8.

Pradhan AD, Manson JE, Rifai N, et al. C-reactive protein, interleukin 6, and risk of developing type 2 diabetes mellitus. *JAMA* 2001;286:327-34.

Ridker PM, Stampfer MJ, Rifai N.Novel risk factors for systemic atherosclerosis: a comparison of C-reactive protein, fibrinogen, homocysteine, lipoprotein(a), and standard cholesterol screening as predictors of peripheral arterial disease. JAMA 2001; 285:2481-5.

Ross R. Atherosclerosis: an inflammatory disease. *N Engl J Med.* 1999;340: 115-26.

Sandler NG, Mentink-Kane MM, Cheever AW, Wynn TA. Global gene expression profiles during acute pathogen-induced pulmonary inflammation reveal divergent roles for Th1 and Th2 responses in tissue repair. *J Immunol.* 2003;171:3655-3667.

Steinberg, D. 2002. Atherogenesis in perspective: hypercholesterolemia and inflammation as partners in crime. *Nat Med.* 2003;8:1211-17.

Zhu Z, Zheng T, Homer RJ, Kim YK, Chen NY, Cohn L, Hamid Q, Elias JA. Acidic mammalian chitinase in asthmatic Th2 inflammation and IL-13 pathway activation. *Science.* 2004;304:1678-82.

In: Binomium Chitin-Chitinase: Recent Issues
Editor: Salvatore Musumeci and Maurizio G. Paoletti
ISBN 978-1-60692-339-9

Chapter IX

Chitotriosidase Activity in Juvenile Idiopathic Arthritis and Juvenile Sarcoidosis

Juergen Brunner[10]
Pediatric Rheumatology, Department of Pediatrics, Medical University Innsbruck, Anichstrasse 35, A-6020 Innsbruck, Austria

Abstract

Juvenile idiopathic arthritis (JIA) is an inflammatory joint disease of unknown aetiology. The pathogenesis is driven by T and B-cells. The role of macrophages remains unclear. Sarcoidosis is a chronic granulomatous inflammation. The clinical spectrum in childhood is heterogeneous. Angiotensin converting enzyme (ACE) activity is used as a marker for disease activity. An unknown agent activates resident T-cells and macrophages, which subsequently release cytokines and chemokines which prime and activate neighbouring cells and are chemotactic for mononuclear cells. Human chitotriosidase is produced in macrophages. Chitotriosidase belongs to the chitinase protein family and is secreted by activated macrophages. The chitinases are able to catalyze the hydrolysis of chitin or chitin-like substrates such as 4-methylumbelliferyl chitotrioside. Serum chitotriosidase levels could represent the activity of macrophages in the synovial fluid in JIA. Serum chitotriosidase concentrations may be a useful marker for monitoring disease activity in sarcoidosis.

10 Email: juergen.brunner@uki.at.

1. Introduction

1.1. Juvenile Idiopathic Arthritis (JIA)

Juvenile idiopathic arthritis (JIA) is a chronic inflammatory rheumatic disorder of unknown cause. Macrophages are playing a key role in synovitis (Gogarty and Fitzgerald, 2007). The enzyme chitotriosidase can be assayed to quantify the degree of macrophage activation. Chitotriosidase (Chit) belongs to the chitinase protein family and is secreted by activated macrophages. The chitinases are able to catalyze the hydrolysis of chitin or chitin-like substrates such as 4-methylumbelliferyl chitotrioside. Human chitotriosidase seem to be involved in the degradation of chitin-containing pathogens such as fungi, nematodes, and insects (Guo *et al.* 1995). Increased levels of Chit have been observed in a number of lysosomal storage diseases (Guo *et al.* 1995; Czartoryska *et al.* 1998, 2000; Den Tandt and van Hoof, 1996; Michelakakis *et al.* 2004; Renkema *et al.* 1995). Chit is also an activity marker for juvenile sarcoidosis (Brunner *et al.* 2006).

1.2. Juvenile Sarcoidosis

Sarcoidosis is an inflammatory disorder of unknown aetiology identifiable by the formation of confluent noncaseating granulomas (Shetty and Gedalia, 1998). It is characterized by lymphocyte and macrophage activation and migration into involved organs. Sarcoidosis in children is rare and spectrum of juvenile sarcoidosis is heterogeneous, ranging from asymptomatic patients to disastrous organ involvement (Shetty and Gedalia 1998; Ramana *et al.* 2002). In early childhood sarcoidosis (EOS), the triad of rash, arthritis and uveitis is typical.In adolescence the diagnostic findings are resembling to that in adults with primary pulmonary manifestation. The amounts of angiotensin converting enzyme (ACE) activity, increase in the serum of patients with sarcoidosis.

2. Chitotriosidase Activity in JIA

Chit activity was determined in 84 sera of 47 patients with oligo- and polyarticular JIA. Chit activity was determined using the substrate 4-methylumbelliferyl β-DNN'N''-triacetylchitotrioside (4-MU-TCT, SIGMA Chemical Co). The substrate was incubated with the serum in a citrate/phosphate buffer. The reaction was stopped by adding a Na_2CO_3-buffer (74 ml 0,5 molar Na_2CO_3 plus 4 ml 0,5 molar $NaHCO_3$; pH 10,3 at 37°C). The fluorescence of 4-methylumbelliferone was evaluated by fluorimeter at excitation 360 nm and emission 450 nm. Chit activities are ranging from 6 to 157 nmol/h/ml serum. The Chit activity in normal healthy donors was < 500 nmol/h/ml. The Chit activity in blood is in normal range in JIA. The cut off was 200 nmol/h/ml. The Chit activity does not correlate neither with the clinical subgroup of JIA nor with the clinical disease activity or inflammation markers (Brunner *et al.* 2008).

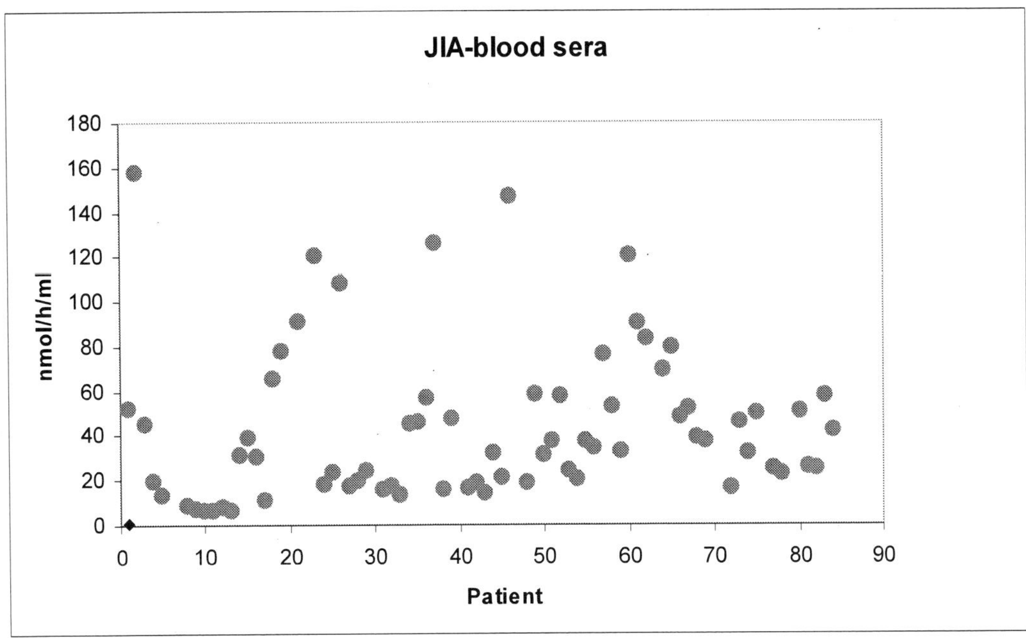

Figure 1. Chit activities are ranging from 4 to 195 nmol/h/ml in the plasma of patients with JIA (Brunner *et al.* 2008).

The Chit activity is higher in SF than blood. The activity was elevated up to 965 nmol/h/ml (Brunner *et al.* 2008).

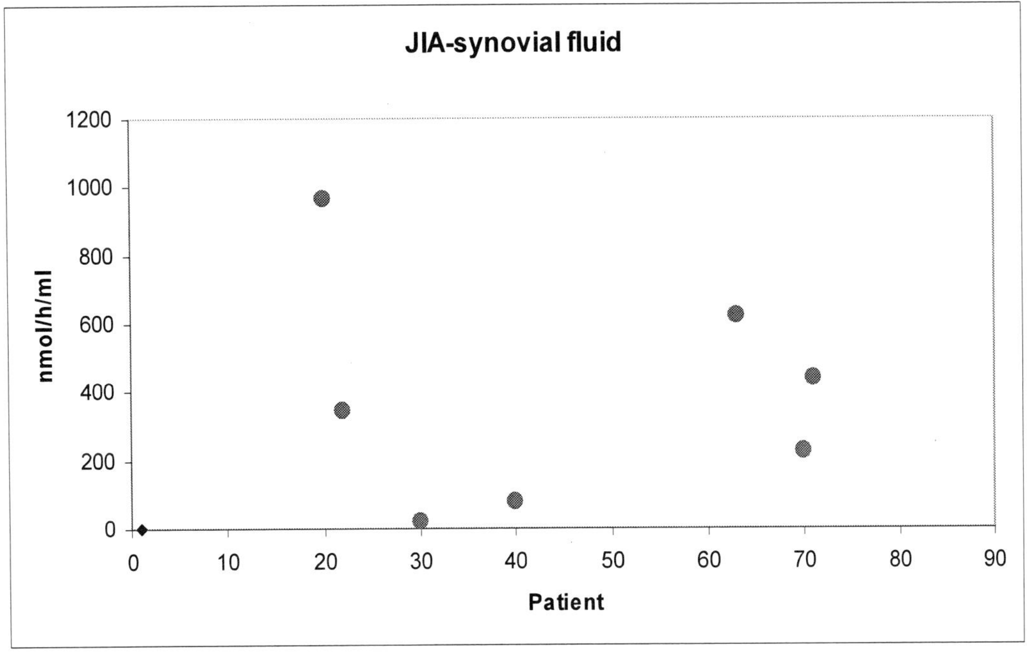

Figure 2. The Chit activity is elevated in SF up to > 1000 nmol/h/ml (Brunner *et al.* 2008).

The enzyme Chit is of interest for clinical reasons, because it is selectively secreted by activated macrophages. Chit was elevated in the SF of some patients with JIA. These patients had a complicated clinical course of the disease and elevated CrP and ESR. Chit represents the role of macrophages in the synovia in JIA and Chit might be a predictor for the disease course.

3. Chitotriosidase Activity in Juvenile Sarcoidosis

Biochemical markers in sarcoidosis are related to the activity of inflammatory effector cells at sites of granuloma formation. The best-known marker is ACE activity (Allen, 1991). In this study serum chitotriosidase levels were significantly higher in active sarcoidosis than in inactive disease and in normal controls. Serum chitotriosidase levels could be a marker for disease activity in sarcoidosis. The pathogenetic reason for increased chitotriosidase serum levels in sarcoidosis might belong to activated macrophages. This remains to be established. In conclusion, although the data presented in this observation (Brunner *et al.* 2007) need to be validated by further investigation, the results indicate that serum chitotriosidase concentrations may be a useful marker for monitoring disease activity in sarcoidosis.

References

Allen RK. A review of angiotensin converting enzyme in health and disease. *Sarcoidosis* 1991;8:95-100.

Brunner J, Sergi C, Muller T, Gassner I, Prufer F, Zimmerhackl LB. Juvenile sarcoidosis presenting as Crohn's Disease. *Eur J Pediatr.* 2006; 165:398-401

Brunner J, Scholl-Burgi S, Prelog M, Zimmerhackl LB. Chitotriosidase as a marker of disease activity in sarcoidosis. *Rheumatol Int.* 2007; 27:1185-1186.

Brunner JK, Scholl-Burgi S, Hossinger D, Wondrak P, Prelog M, Zimmerhackl LB. Chitotriosidase activity in juvenile idiopathic arthritis. *Rheumatol Int.* 2008;28:949-950.

Czartoryska B, Tylki-Szymanska A, Gorska D. Serum chitotriosidase activity in Gaucher patients on enzyme replacement therapy (ERT). *Clin Biochem.* 1998; 31:417-420.

Czartoryska B, Tylki-Szymanska A, Lugowska A. Changes in serum chitotriosidase activity with cessation of replacement enzyme (cerebrosidase) administration in Gaucher disease. *Clin Biochem.* 2000;33:147-149.

den Tandt WR, van Hoof F. Marked increase of methylumbelliferyl-tetra-N-acetylchitotetraoside hydrolase activity in plasma from Gaucher disease patients. *J Inherit Metab Dis.* 1996;19:344-350.

Gogarty M, Fitzgerald O. Immunohistochemistry of the inflamed synovium. *Methods Mol Med.* 2007;135:47-63.

Guo Y, He W, Boer AM, Wevers RA, de Bruijn AM, Groener JE, Hollak CE, Aerts JM, Galjaard H, van Diggelen OP. Elevated plasma chitotriosidase activity in various lysosomal storage disorders. *J Inherit Metab Dis.* 1995;18:717-722.

Michelakakis H, Dimitriou E, Labadaridis I. The expanding spectrum of disorders with elevated plasma chitotriosidase activity: an update. *J Inherit Metab Dis*. 2004;27:705-706.

Renkema GH, Boot RG, Muijsers AO, Donker-Koopman WE, Aerts JM. Purification and characterization of human chitotriosidase, a novel member of the chitinase family of proteins. *J Biol Chem*. 1995;270:2198-2202.

Ramanan AV, Thimmarayappa AD, Baildam EM. Muscle involvement in childhood sarcoidosis and need for muscle biopsy. *Ann Rheum. Dis*. 2002;61:376-377.

Shetty AK, Gedalia A. Sarcoidosis: a pediatric perspective. *Clin Pediatr*. 1998;37:707-17.

In: Binomium Chitin-Chitinase: Recent Issues
Editor: Salvatore Musumeci and Maurizio G. Paoletti ISBN 978-1-60692-339-9

Chapter X

Chitinases in Neurological Diseases

Stefano Sotgiu[11]
Associate Professor of Neurology, Department of Neurosciences and Mother and Child Sciences, University of Sassari, Viale San Pietro, 10; I-07100, Sassari, Italy

Abstract

Chitotriosidase (Chit) is a member of mammalian chitinase family with structural homology to chitinases from other species. Chit has yet unexplored roles in the immune network occurring in ischemic, inflammatory and degenerative neurological diseases, in which the macrophage-microglia activation is known to be pathogenic. Its prominent archaic hydrolytic function on chitin may only be a windscreen beyond which new functions can be discovered to support its clinical importance.

Chit is synthesized and secreted by activated macrophages and immature neutrophils and its natural substrate, chitin, is a N-acetylglucosamine polymer of fungi cell wall and several human parasites. In principle, as Chit plays a major role in defence mechanisms against chitin-containing pathogens, the clinical monitoring of its activity may be relevant in human infectious diseases. Contrary to this theoretical assumption, plasma Chit activity has been shown to have a positive correlation with normal ageing and to have application, as a lipid-laden macrophage marker, in the monitoring of non pathogen-mediated diseases such as Gaucher and Fabry storage diseases. Our study group have recently suggested that Chit elevation represents an useful marker of other, non-infectious, neurological diseases such as stroke, Alzheimer's disease (AD) and multiple sclerosis (MS). Peripheral and intrathecal Chit activity in MS have been also found to strongly correlate with MS severity. These findings are reviewed along with new unpublished data.

11 E-mail: *stefanos@uniss.it* or *stesot@hotmail.com*.

1. Introduction

Initial biochemical investigations on Gaucher disease (GD), a lysosomal storage disease, unexpectedly led to the discover that plasma samples of patients had a several hundred-fold elevated ability to hydrolyze chitin, a natural polymer of beta 1,4-linked N-acetylglucosamine naturally found in fungi cell wall and coatings many human parasites (Hollack *et al.* 1994). Later, it was observed that lipid-laden activated macrophages accumulating in GD tissues were able of secreting extraordinarily high levels of an enzyme able to cleave chitin and artificial chitotrioside substrates, and therefore named chitotriosidase (Renkema *et al.* 1997 and 1998). The measurement of plasma Chit activity has now recently found application also for the clinical monitoring of patients with Fabry disease (Vedder *et al.* 2006).

Chitotriosidase (Chit), is a member of mammalian chitinase family, synthesized by activated macrophages and neutrophil progenitors (van Eijk *et al.* 2005) as a 50 kDa protein, proteolitically cleaved and predominantly secreted. An important question arises when considering that the natural substrate chitin is absent in humans. Nevertheless, as the Chit gene is present and evolutionary conserved in primates and rodents (Gianfrancesco and Musumeci, 2004), this argues in favour of an important biological role of this enzyme though the significance in humans still remains enigmatic. In fact, physiological condition such as aging are found to be correlated with a progressive increase of plasmatic Chit activity (Bouzas *et al.* 2003). This might suggest that the innate immune system is involved in protecting the healthy human organism from the cell damage that occur during aging, perhaps linked to oxidative processes (Droge, 2002). In pathological conditions, and also thanks to our contribution, Chit has now achieved a growing important role as marker of some neurological diseases, perhaps by virtue of the peculiar immune condition which characterises the brain.

2. Chitotriosidase within the Central Nervous System (CNS)

Given its peculiar immune "priviledge" the healthy CNS status is tightly regulated and maintained at a low functional level in order to prevent immune-mediated damage to occur. Among brain cells, microglia and a few other cell types such as astrocytes represent components of the innate immune system which promptly activate a complex immune cascade soon after specific or unspecific stimuli, either ischemic-oxidative or inflammatory, face the brain. Despite its natural substrate, chitin, absent in human brain, Chit is found elevated in some brain diseases as described later. Recently, however, chitin-like substances are increasingly found to accumulate within the brain, in given circumstances.

Glucosamine, the basic unit of chitin, is synthesized by virtually all cells and has various physiological properties. It has been used to treat human osteoarthritis for its beneficial effect in the reconstruction of joint cartilage and its immunoregulatory ability. Glucosamine inhibits pro-inflammatory cytokines from antigen presenting cells (APC), suppresses T cell response by interfering with functions of APC and shows a direct inhibitory effect on antigen-independent CD3-induced T cell proliferation (Ma *et al.* 2002; Zhang *et al.* 2005).

Glucosamine administration in the animal model of MS known as experimental autoimmune encephalomyelitis (EAE), significantly reduces macrophage infiltration within the inflamed brain, reduces microglia activation, nitric oxide and inflammatory cytokines production such as IFN-γ, IL-17 and TNF-α, resulting in resistance to acute EAE (Zhang *et al.* 2005). However, glucosamine, which is formed from glucose, has been seen to form chitin-like polymers. Likewise, the markedly augmented glucose metabolism occurring during inflammation within the brain could induce an increased chitin-like substances formation.

2.1. Stroke

It is now very well established that after an acute brain ischemia, an early activation of glial and endothelial cells and their transcription of TNF-α and IL-6 are able to induce a cascade of inflammatory pathways which transforms local endothelia into a pro-thrombotic state and allows peripheral mono- and poly-morphonuclear cytotoxic chemoattraction into the lesion site (Rothwell *et al.* 1997; Castillo *et al.* 2004). This early event correlates with the worsening of the cerebral damage as a clear relationship between the extent of the brain damage and the early increase of TNF-α plasma level is generally reported in stroke (Sotgiu *et al.* 2006a). Besides the increase of innate immune cytokines, some authors also found an increased Chit activity in stroke. However, perhaps due to an unselective patients recruitment, Chit activity in stroke patients was closely related to concomitant infections. Infection itself is able to worse the outcome of the stroke (Palasik *et al.* 2005). Nevertheless, these initial results suggest the relevant role that infiltrated macrophages, or resident microglia activated by the ischemic event, may have inside the brain in the absence of infectious diseases. Therefore, we have conducted another study in patients with acute ischemic stroke. Differently from previous findings, our study, performed on more than 40 consecutive patients without concomitant infections (Sotgiu *et al.* 2005), confirmed that plasma Chit activity significantly correlated with disability scores ($p<0.01$) and with the infarct size ($p< 0.05$) as measured by cranial computerised tomography and suggested Chit production to be of central origin being patients with signs of infection ruled out (Figure 1).

With the exception of IL-6, our results indicate that Chit, like TNF-α and other pro-inflammatory cytokines, is a marker of microglia activation occuring during a stroke, which is independent of pre-existing inflammatory or infectious conditions. Many questions remain open, i.e. why chemo attracted or resident macrophages such as microglia do activate their gene to produce Chit, and, does Chit have a direct influence on the tissue damage or simply reflects an epiphenomenon? We cannot argue for any such hypotheses yet. The increased activity of Chit may represent an epiphenomenon of one ancestral response of the innate immunity whose hydrolytic function on chitin is only an archaic demonstration of its old functions. However, even in this case, Chit can hold clinical importance in ischemic stroke.

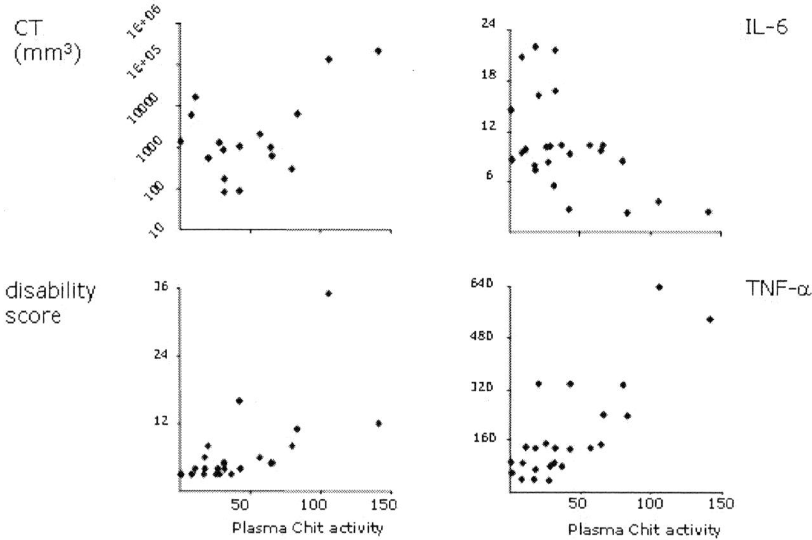

Figure 1. Chit activity in stroke patients directly correlates with infarct size as measured by Computerized Tomography (CT), with the disability score ($p<0.01$ for both), and with TNF-α level, oarticularly in the subgroup with worse relative to that with better stroke outcome ($p<0.001$). An inverse correlation is found with the IL-6 level. Adapted from Sotgiu *et al.* 2005.

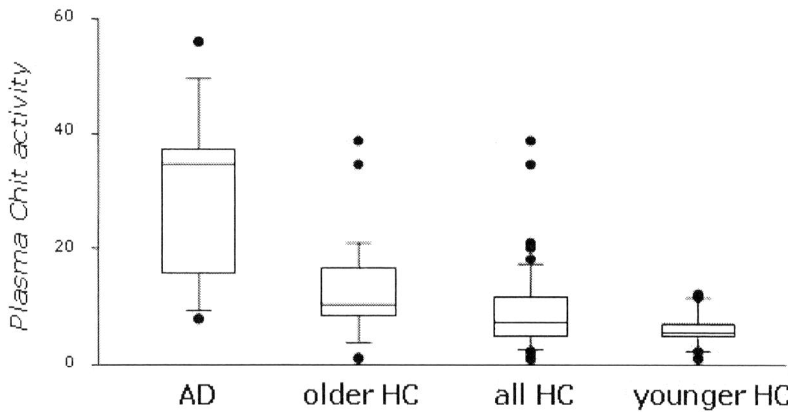

Figure 2. Chit activity in plasma of 30 AD patients and 69 healthy controls (HC): HC are subdivided in two groups of 40 younger (mean age 35.6 years) and 29 older individuals (mean age 60); plasma Chit level in AD is significantly higher as compared to that of the whole HC group ($p<0.001$) and that of the older control subgroup ($p<0.01$). Moreover, plasma Chit activity directly correlated with individual's age in the whole control group ($r=0.72$, $p<0.05$), but not in AD patients ($r=0.3$). Adapted from Sotgiu *et al* 2007.

2.2. Alzheimer's Disease (AD)

Amyloid plaques, senile plaques and amyloid angiopathy of AD brains have been described to co-localise with chitin-like glucosamine polymers. Such polymers are suggested to induce the formation of pathogenic amyloid fibrils (Castellani *et al.* 2005). Intriguingly, Calcofluor was used on brain tissue samples at autopsy from AD brains to characterise this relationship, as it excites upon ultraviolet light exposure and has a high affinity for chitin and chitin-like substanccs *in vivo*.

Based on this original finding, we firstly conducted an association study of Chit in plasma of people with AD (Sotgiu *et al.* 2007) analyzing Chit activity in 30 AD patients and about 70 healthy individuals. Results were unexpectedly interesting and indicate that plasma Chit values in the AD group were not only significantly higher than those obtained for the whole control group, but also as compared to the older individuals of the control subgroup ($p<0.01$). Moreover, plasma Chit activity level correlates with the individual's age in the whole control group ($p<0.05$), but not in the AD group, confirming an age-related, physiological increase (Figure 2).

β-amyloid brain deposition is involved in the AD neuro-degenerative cascade through either a direct neurotoxicity or an immune network (Lue *et al.* 2006; Minagar *et al.* 2002). The high plasma Chit level of AD, as found in our study, may well reflect the peripheral response of a strong brain macrophage/microglia activation. In this light, the high Chit expression as demonstrated also by Di Rosa et al (2006) may have a dual role: either it represents a mere epiphenomenon of a strong macrophage activation due to β-amyloid deposition or it reflects a scavenging protective immune activity against potentially pathogenic chitin-like glucosamine polymers which have been demonstrated histochemically in AD brains (Castellani *et al.* 2005).

Thus, encouraged by these findings, we performed a collaborative immunocytochemistry study, by means of Calcofluor as previously described by Castellani and colleagues, so as to explore the presence of chitin-like substances within AD, multiple sclerosis (MS) and healthy control (HC) brains. On one hand we could confirm the presence of abundant chitin-like substances in brains of AD, which may well relate with the detection of peripheral Chit activity (Sotgiu *et al.* submitted). Conversely, we failed to demonstrate the deposition of chitin-like substances both in MS and in normal brains (Figure 3).

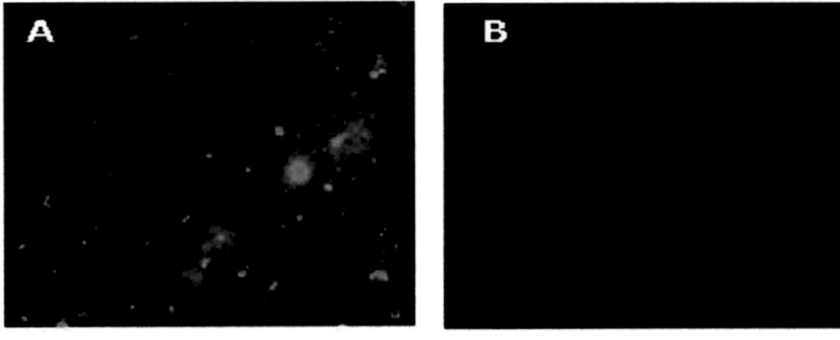

Figure 3. Calcoflour hystochemistry in AD and MS brain sections. Calcoflour stainings gives intense signals with a predominant plague pattern in AD samples (A). In MS tissue sections no labelling is detectable (B). Sotgiu *et al.* unpublished data.

One can argue that Chit activity in MS is induced by mechanisms different than those operating in AD. The different histochemical content of chitin-like substances between MS and AD brain samples may be a reflection of these mechanisms. The intrathecal, microglia-derived Chit activity found in MS patients (Sotgiu *et al.* 2006b) could counterbalance the naturally occurring glucosamine aggregation as well as its transformation into detrimental chitin. The deposition of chitin-like substances in AD brain could be result from a reduced immune response which characterizes the most severe clinical expression of disease (Motta *et al.* 2007), or impaired chitin cleavage in relation also to the age of the AD patients. In the light of this consideration, Chit production may have a protective rather than detrimental role in the CNS.

2.3. Multiple Sclerosis (MS)

MS is a chronic inflammatory/degenerative disease of the CNS. Most studies in humans, principally based on the animal EAE model, claim that at its patho-physiological basis stand the adaptive T- and B-cell antigen-specific responses (Corcione *et al.* 2004). Other studies, however, indicate that the innate immune response predominates in most MS lesion subtypes, which is played by both resident (microglia) and peripheral infiltrating macrophages (Lucchinetti *et al.* 2000; Kornek and Lassmann, 2003). TNF-α, IL-6, nitric oxide (NO), reactive oxygen species (ROS) and other macrophage-derived products unfortunately show only a modest correlation with the MS clinical activity, and are, as yet, of a very limited usefulness in clinical routine (Hendriks *et al.* 2005).

Through a case-control study, we demonstrated that Chit activity is increased in blood and, particularly, in cerebrospinal fluid (CSF) of MS patients (Sotgiu *et al.* 2006b) and that it correlates with the extent of CNS damage as scored by the Extended Disability Status Scale (EDSS; Kurtzke, 1987) (Figure 4).

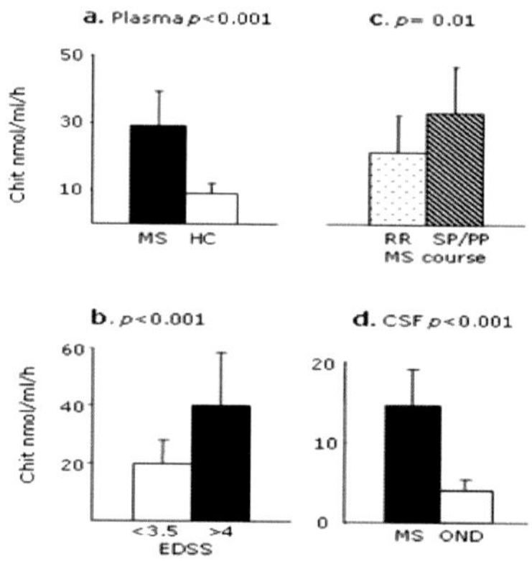

Figure 4. Plasma and CSF Chit activity in MS, HC and other neurological diseases: a) Chit level is significantly higher in plasma of MS patients compared to matched HC; b) plasma Chit activityt is significantly higher in the MS group with higher EDSS; c) progressive MS patients (SP and PP) have significantly higher Chit plasma activity as compared to RR patients; d) Chit level is significantly higher in CSF of MS patients as compared to patients with other neurological diseases. Adapted from Sotgiu *et al.* 2006b.

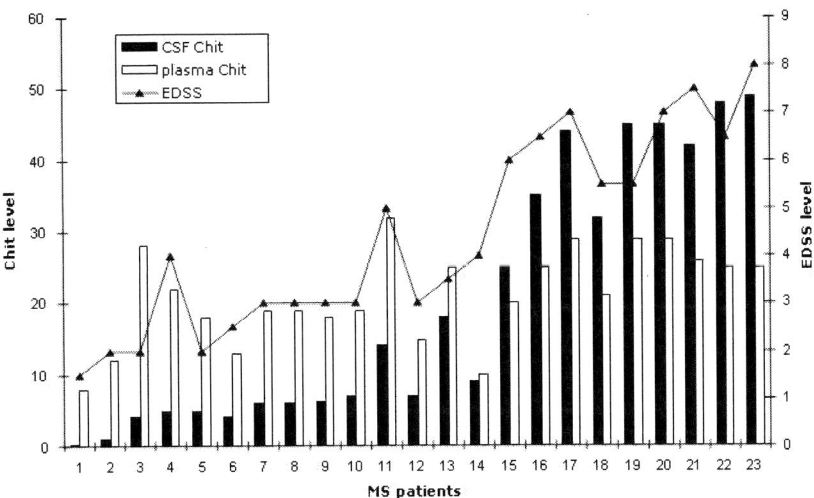

Figure 5. Correlation between Chit level in CSF and plasma of MS patients and EDSS (triangles) on right Y axis in 23 patients with MS (X axis), Columns indicate the plasma (blank bars) and CSF (black bars) Chit activity (nmol/ml/h, left Y axis). Adapted from Sotgiu *et al.* 2006b.

Also, by calculating a Chit Quotient (CSF Chit/plasma Chit) and comparing the study-entry and study-end EDSS score of the patients with the Chit Quotient at the time of CSF withdrawal, we demonstrated that CSF Chit activity is of intrathecal and not of peripheral origin (Figure 5).

This study not only confirms the important role of macrophages in MS phenomenology, but also allows to propose Chit determination in CSF and plasma as a diagnostic and monitoring method in a MS laboratory. As a result of our analysis, plasma and, to a larger extent, CSF Chit levels better correlate with the extent of CNS damage as compared to the previously proposed macrophage-derived markers such as TNF-α, IL-6, IL-1, NO, and ROS.

As its CSF production is unrelated to its plasma level and to the albumin quotient, we argued that Chit activity is compartmentalised in MS. Despite this enzymatic activity derives from infiltrating macrophages, we think that resident microglia might be induced to produce Chit as these cells can gradually transform phenotypically into lipid-laden, activated macrophage (Trapp *et al.* 1999). This intrathecal, microglia-derived Chit activity found in MS patients could counterbalance the naturally occurring glucosamine aggregation and its transformation into detrimental chitin (see Figure 3). Therefore, Chit production in MS may have a protective rather than detrimental role in the CNS. However, even if Chit activity only represents the hallmark of an innate macrophage response without detectable effects, it could represent a biomarker to be considered in the future follow-up of MS patients.

3. Gene Polymorphism of Chitotriosidase in MS

Human Chit gene, located on chromosome 1q31-32 has been cloned and characterized (Boot e al, 1998). Chit contains several regions with high homology to those present in chitinases from different species belonging to family 18 of glycosyl hydrolases (Boot *et al.* 2005). One specific 24 bp duplication in exon 10, known as CHIT1 and which generate a mRNA with an in-frame deletion of 87 nucleotides, abolishes Chit activity and occurs in various ethnic groups with 30% heterozygote and 4% homozygote frequency in Caucasian population (Boot *et al.* 1998; Malaguarnera *et al.* 2003). Recently Canudas and colleagues (2001) studying Chit polymorphism in the Spanish population found an inverse correlation between the mean serum Chit activity and the CHIT1 allele.

These findings were validated in a recent study demonstrating a heterozygote frequency for the 24 base pairs duplication of 44% in Sicily and 32.7% in Sardinia, whilst the homozygous mutants were 5.45 % and 3.73 %, respectively. Interestingly, a low incidence of Chit mutation was found in Burkina Faso (heterozygous 0% and 2% respectively) and no subject was homozygous for Chit deficiency (Malaguarnera *et al.* 2003). These results suggest that Chit might play a major role in defence mechanisms against chitin containing pathogens, particularly in endemic area for malaria and parasitic diseases (Choi *et al.* 2001).

Based on such evidences we investigate on the possible phenotypic effect (i.e. on the clinical expression of disease) of the known CHIT1 polymorphism (Boot *et al.* 1998). On one hand our results demonstrated that this high-protector polymorphism (wild type) does not preferentially segregate in the normal controls as relative to MS patients, which might imply that CHIT1 polymorphism, but not other yet unstudied CHIT alleles, is indifferent in the complex genetic predisposition to MS. However, in the wild type homozygous carriers results of Chit measurement confirmed that Chit activity is significantly elevated in MS as relative to HC and that it correlates with the MS clinical course (higher in the more severe secondary-progressive-SP as compared to the more benign relapsing-remitting-RR course of the disease). Moreover, it is worth noting that the average Chit level in the 7 patients homozygous for the mutant CHIT1 allele was significantly lower than that of heterozygous and that they all show a RR course and a lower disability score as measured by the EDSS (Figure 6).

	W/W	W/w	w/w	All
RR	114	64	7	185
SP	29	23		52
RP	2	3		5
PP	2			2

Figure 6. Clinical course in 244 MS patients stratified for different CHIT1 genotyhpes (W= wild type CHIT1; w= polymorphic CHT1). Sotgiu et al. unpublished data.

Conclusions

The increase of Chit activity in plasma and CSF of neurological patients likely depends upon macrophage activation at both peripheral and intrathecal levels. To the best of our knowledge, the exact role of Chit in the pathogenesis of MS remains at least not clear yet. Histopathological studies allow us to argue in favour of a protective role of Chit in the CNS. On the contrary, the study of the CHIT1 gene mutation, as well as previous case-control and observational studies demonstrated, lead us to the conclusion that Chit secretion is associated to a worse outcome of MS. Anyway, much as still to be done in this intriguing area and, even if it is unclear whether Chit is a protagonist or a bystander, a friend or foe of the brain, it could well represent a simple marker of monitoring disease progression.

Acknowledgments

Thanks to the patients who generously contributed to the studies here described and all collaborating colleagues and friends (Giannina Arru, Maria Laura Fois, Rita Barone, Andrea Angius and, last but not least, Bruno Bonetti). Particularly warm thanks are addressed to my friend Salvatore Musumeci, who allowed me to enter the fascinating world of chitinases, stimulated the growth of cultural weapons to face the intriguing association between chitinases and neurological diseases and who continuously encouraged me in the study.

References

Boot RG, Renkema GH, Verhoek M, Strijland A, Bliek J, de Meulemeester TM, Mannens MM, Aerts JM.The human chitotriosidase gene. Nature of inherited enzyme deficiency. *J Biol Chem*. 1998;273:25680-5.

Bouzas L, Carlos Guinarte J, Carlos Tutor J. Chitotriosidase activity in plasma and mononuclear and polymorphonuclear leukocyte populations. *J Clin Lab Anal.* 2003;17:271-5.

Canudas J, Cenarro A, Civeira F, Garci-Otin AL, Aristegui R, Diaz C, Masramon X, Sol JM, Hernandez G, Pocovi M. Chitotriosidase genotype and serum activity in subjects with combined hyperlipidemia: effect of the lipid-lowering agents, atorvastatin and bezafibrate. *Metabolism.* 2001;50:447-50.

Castellani RJ, Siedlak SL, Fortino AE, Perry G, Ghetti B, Smith MA.Chitin-like polysaccharides in Alzheimer's disease brains. Curr Alzheimer Res. 2005;2(4):419-23.

Castillo J, Rodriguez I: Biochemical changes and inflammatory response as markers for brain ischaemia: molecular markers of diagnostic utility and prognosis in human clinical practice. *Cerebrovasc Dis.* 2004;17(Suppl. 1):7-18.

Choi EH, Zimmerman PA, Foster CB, Zhu S, Kumaraswami V, Nutman TB, Chanock SJ. Genetic polymorphisms in molecules of innate immunity and susceptibility to infection with Wuchereria bancrofti in South India. *Genes Immun*. 2001; 2:248-53.

Corcione A, Casazza S, Ferretti E, Giunti D, Zappia E, Pistorio A, Gambini C, Mancardi GL, Uccelli A, Pistoia V. Recapitulation of B cell differentiation in the central nervous system of patients with multiple sclerosis. *Proc Natl Acad Sci USA.* 2004;101:11064-9.

Di Rosa M, Dell'Ombra N, Zambito AM, Malaguarnera M, Nicoletti F, Malaguarnera L. Chitotriosidase and inflammatory mediator levels in Alzheimer's disease and cerebrovascular dementia. *Eur J Neurosci.* 2006;23:2648-56.

Droge W. The plasma redox state and ageing. Ageing Res Rev. 2002; 1: 257-78

Gianfrancesco F, Musumeci S. The evolutionary conservation of the human chitotriosidase gene in rodents and primates. *Cytogenet Genome Res.* 2004;105:54-6.

Hendriks JA, Teunissen CE, de Vries HE, Dijkstra CD. Macrophages and neurodegeneration. *Brain Research Reviews* 2005;48:185-95.

Hollak CE, Van Weely S, Van Oers MHJ, Aerts JMFG. Marked elevation of plasma chitotriosidase activity. A novel hallmark of Gaucher disease. *J Clin Invest.* 1994; 93: 1288-92.

Kornek B, Lassmann H. Neuropathology of multiple sclerosis: new concepts. *Brain Research Bulletin.* 2003;61:321-26.

Kurtzke JF. Rating neurologic impairment in multiple sclerosis: An expanded disability status scale (EDSS). *Neurology.* 1983;33:1444-52.

Lucchinetti C, Bruck W, Parisi J, Scheithauer B, Rodriguez M, Lassmann H. Heterogeneity of MS lesions: implications for the pathogenesis of demyelination. *Ann Neurol.* 2000; 47: 707-17.

Lue LF, Kuo YM, Roher AE, Brachova L, Shen Y, Sue L, Beach T, Kurth JH, Rydel RE, Rogers J. Soluble amyloid β peptide concentration as a predictor of synaptic change in Alzheimer's disease. *Am J Pathol.* 2006; 155:853–62.

Ma L, Rudert W A, Harnaha J, Wright M, Machen J, Lakomy R, Qian S, Lu L, Robbins PD, Trucco M, Giannoukakis N. Immunosuppressive effects of glucosamine. *J Biol Chem.* 2002:277:39343-39349.

Malaguarnera L, Simpore J, Prodi DA, Angius A, Sassu A, Persico I, Barone R, Musumeci S. A 24-bp duplication in exon 10 of human chitotriosidase gene from the sub-Saharan to the Mediterranean area: role of parasitic diseases and environmental conditions. *Genes* Immun. 2003 ; 4:570-4.

Minagar A, Shapshak P, Fujimura R, Ownby R, Heyes M, Eisdorfer C. The role of macrophage/microglia and astrocytes in the pathogenesis of three neurologic disorders: HIV-associated dementia, Alzheimer disease, and multiple sclerosis. *J Neurol Sci.* 2002;202:13-23

Motta M, Imbesi R, Di Rosa M, Stivala F, Malaguarnera L. Altered plasma cytokine levels in Alzheimer's disease: correlation with the disease progression. *Immunol Lett.* 2007; 30;114:46-51

Palasik W, Fiszer U, Lechowicz W, Czartoryska B, Krzesiewicz M, Lugowska A: Assessment of relations between clinical outcome of ischemic stroke and activity of inflammatory processes in the acute phase based on examination of selected parameters. *Eur Neurol.* 2005;53:188-193.

Renkema GH, Boot RG, Au FL, Donker-Koopman WE, Strijland A, Muijsers AO, Hrebicek M, Aerts JM. Chitotriosidase, a chitinase and the 39-kDa human cartilage glycoprotein, a chitin-binding lectin, are homologues of family 18 glycosyl hydrolases secreted by human macrophages. *Eur J Biochem*. 1998; 251: 504-9.

Renkema GH, Boot RG, Strijland A , Donker-Koopman WE, van den Berg M, Muijsers AO, Aerts JM. Synthesis, sorting and processing into distinct isoforms of human macrophage chitotriosidase. *Eur J Biochem.* 1997; 244: 279-85.

Rothwell NJ, Loddick SA, Stroemer P.Interleukins and cerebral ischaemia. *Int Rev Neurobiol.* 1997;40:281-98.

Sotgiu S, Barone R, Zanda B, Arru G, Fois ML, Arru A, Rosati G, Marchetti B, Musumeci S. Chitotriosidase in patients with acute ischemic stroke. *Eur Neurol.* 2005;54:149-53.

Sotgiu S, Zanda B, Marchetti B, Fois ML, Arru G, Pes GM, Salaris FS, Arru A, Pirisi A, Rosati G. Inflammatory biomarkers in blood of patients with acute brain ischemia. *Eur J Neurol.* 2006a;13:505-13.

Sotgiu S, Barone R, Arru G, Fois ML, Pugliatti M, Sanna A, Rosati G, Musumeci S. Intrathecal chitotriosidase and the outcome of multiple sclerosis. *Mult Scler* 2006b;12:551-557.

Sotgiu S, Piras MR, Barone R, Arru G, Fois ML, Rosati G, Musumeci S. Chitotriosidase and Alzheimer's disease. Curr Alzheimer Res 2007;4:295-6.

Trapp BD, Bö L, Mörk S, Chang A. Pathogenesis of tissue injury in MS lesions. *J Neuroimmunol.* 1999;98:49-56.

van Eijk M, van Roomen CP, Renkema GH, Bussink AP, Andrews L, Blommaart EF, Sugar A, Verhoeven AJ, Boot RG, Aerts JM. Characterization of human phagocyte-derived chitotriosidase, a component of innate immunity. *Int Immunol.* 2005;17:1505-12.

Vedder AC, Co x-Brinkman J, Hollak CE, Linthorst GE, Groener JE, Helmond MT, Scheij S, Aerts JM. Plasma chitotriosidase in male Fabry patients: A marker for monitoring lipid-laden macrophages and their correction by enzyme replacement therapy. *Mol Genet Metab.* 2006;89:239-44;

Zhang G-X, Yu S, Gran B, Rostami A. Glucosamine Abrogates the Acute Phase of experimental Autoimmune Encephalomyelitis by Induction of Th2 Response. *J Immunology* 2005;175:7202-7208.

In: Binomium Chitin-Chitinase: Recent Issues ISBN 978-1-60692-339-9
Editor: Salvatore Musumeci and Maurizio G. Paoletti

Chapter XI

Chitin in Alzheimer's Disease

Luis F. Gonzalez-Cuyar and Rudy J. Castellani[12]
Department of Pathology, Division of Neuropathology, University of Maryland,
Baltimore, Maryland USA 21201

Abstract

Alzheimer's disease is the most common cause of dementia, affecting over four million patients in the Unites States and fifteen million worldwide. As the average life expectancy increases in the United States and worldwide, the number of patients will proportionally increase, affecting over thirteen million individuals in the United States by 2050. The diagnosis of Alzheimer's disease carries significance, because the life span of these patients is halved when compared with healthy population controls. The therapeutic efforts are directed towards eradicating brain lesions, without the accurate knowledge of whether these are actually pathogenic. Therefore, currently, our understanding of this neurodegenerative disease precludes us from attaining a more elusive cure. Several drugs are currently utilized in the treatment of Alzheimer's disease, and when started early, the progression might be momentarily halted. No significant information has emerged from clinical trials involving immune therapy. Authors in our group have demonstrated that Amyloid β deposition, a histopathological landmark in Alzheimer's disease brains, may confer a protective effect against oxidative damage induced by reactive oxygen species. Chitin confers antioxidant properties of comparable strength to super oxide dismutase.

For several decades the hypothesis of Amyloid β mediated pathogenesis has perhaps diverted us from the real issues in this disease's etiology, hence, devoting millions of dollars and research hours into the Amyloid cascade hypothesis. Some investigators have shifted their efforts from the Amyloid β dogma into other possible pathophysiological processes such as oxidative stress. As a matter of fact, oxidative stress has taken a significant role in the study of several neurodegenerative diseases such as Creutzfeldt Jakob, Pick's disease, diffuse Lewy body dementia and Cerebrotendinous Xanthomatosis. Decreased glucose metabolism, deficiency in antioxidant metals such as zinc, and mitochondrial abnormalities in the electron transport chain mediate the generation of toxic reactive oxygen species that coupled with redox active metals leads to

12 Corresponding Author: Rudy J. Castellani, MD, Department of Pathology, 22 South Greene Street, Baltimore, Maryland 21201, Tel: 410-328-5555, Fax: 410-328-5508, Email: rcastellani@som.umaryland.edu.

free radical damage. Antioxidant vitamin supplementation with Vitamin E and C has been included in the treatment of Alzheimer's patients. Furthermore, some investigators have demonstrated that in fact oxidative stress is an early process in neurodegeneration and that it precedes Amyloid β deposition.

Deranged glucose utilization and subsequent hyperglycemia in Alzheimer's disease patients is mediated by diminished numbers of cellular glucose transporters and down regulation of genes involved in the oxidative phosphorylation of glucose. This in turn leads to the activation of the hexosamine pathway and thus increases the synthesis of glucosamine polymers as described for other diseases such as Diabetes Mellitus. In this pathway there is synthesis of O linked glycoproteins from glucose by means of fructose and fructose-6 phosphate. These glucosamine polymers are the building blocks of chitin.

The neuropathological examination of Alzheimer's disease brains involves the quantification and location of histopathological landmarks such as Amyloid plaques, neurofibrillary tangles, and Amyloid angiopathy. Studies in familial and sporadic Alzheimer's disease patients have localized chitin and chitin-like polysaccharides in both the Amyloid plaque as well as in the neurofibrillary tangles by utilization of Calcofluor; a fluorochrome that is notable for identifying chitin *In vivo* by interacting with the β 1-4 bonds that make up the chitin polymer. Amyloid is a highly insoluble molecule which stains with Congo Red and displays apple green birefringence when exposed to polarized light.

These properties, which were initially attributed to the Amyloid protein conformation, are shared by commercial chitin and could perhaps represent that chitin imparts these biochemical features to the cerebral Amyloid deposition. The role of chitin in the Amyloid plaque is unknown. However, chitin might function as a primer of Amyloid deposition as inferred by the staining profiles with Calcofluor and with Amyloid β immunohistochemistry. Therefore chitin might be a protective accumulation against oxidative stress.

Keywords: Chitin, Alzheimer's disease, presenilin, Amyloid β, Calcofluor, neurodegenerative, oxidative stress, neurofibrillary tangle

1. Introduction

1. Historical Background

1.1. Alzheimer's Disease

Alois Alzheimer (1864-1915) was born in Marketbreit-am-Main, Bavaria in present day Germany (Berrios 1990; Goedert *et al.* 2007) . Alzheimer was twenty-two years old when he was appointed to the Municipal Asylum for mentally ill and epileptic patients in Frankfurt-am-Main in 1888. He began working under Franz Nissl (known for the Nissl stain for visualization of neuronal cell bodies) who had just been named the assistant medical director of the Frankfurt asylum (Wilkins *et al.* 1969; Berrios 1990; Goedert *et al.* 2007). This proved not only to be a productive professional relationship, but also a life long friendship. Known as clinicians by day and histopathologists by night, they had set out to link histology with clinical presentations(Goedert *et al.* 2007). They believed that the treatment spun from the histopathological characterization would in turn lead to a better understanding of the disease process.

It is documented that Alzheimer had an initial interest in progressive spinal muscular atrophy and cerebral atherosclerosis. Although through the years he had several academic appointments, in 1903 he moved to Heidelberg and reunited with Nissl under German psychiatrist Emil Kraepelin (co-discoverer of Alzheimer's disease). While in Munich he assembled an impressive research group including Creuzfeldt, Jakob, and Lewy.

Alzheimer described for the first time, what later would be known as his disease, at the 37th conference of Sounth-West German Psychiatrists in Tübingen and was also published in 1907 (Berrios 1990; Maurer *et al.* 1997; Moller *et al.* 1998; Small *et al.* 2006; Zilka *et al.* 2006; Goedert *et al.* 2007). The first patient examined by Alois Alzheimer is refered to as August D, who was a 51 year old female patient who was first treated in 1901 (Moller *et al.* 1998). The patient had an impairment to formulate new memories as well as the inability to recognize individuals that through her life had been close to her. Additionally, progressive cognitive impairment, delusions and hallucinations rendered her socially incompetent (Goedert *et al.* 2007).

Alzheimer's patient, August D, expired in 1906 and after a lengthy postmortem neuro-histopathological examination, Alzheimer, with help of the Bielschowsky method, described plaques and neurofibrillary tangles (Moller *et al.* 1998; Small *et al.* 2006; Zilka *et al.* 2006; Goedert *et al.* 2007). Increased intra neuronal fibrillary silver staining and subsequent "ghost tangles" were also identified (Garcia-Marin *et al.* 2007; Goedert *et al.* 2007). It is also reported that Solomon Carter Fuller, an American psychiatrist, was the first to describe the neurofibrillary tangle (Berrios 1990; Kaplan *et al.* 2000). Plaques had previously been described by Bloq, Marinesco and Redlich in patients with dementia and epilepsy (Wilkins *et al.* 1969; Kaplan *et al.* 2000). In the eighth edition of his textbook of psychiatry, Kraepelin separated Alzheimer's disease from senile dementias and named it after Alois Alzheimer (Schorer 1985; Berrios 1990; Goedert *et al.* 2007).

In 1911 the case of August D was reported with a second case involving the patient known as Johann F (Moller *et al.* 1998; Nunomura *et al.* 2006; Small *et al.* 2006; Zilka *et al.* 2006; Goedert *et al.* 2007). Johann F was a 56 year old widowed laborer with clumsiness, difficulty of expression, paragraphia, and difficulty carrying out activities of daily living (Moller *et al.* 1998; Tanzi 2005; Zilka *et al.* 2006). Post mortem neuropathological examination of the Johann F case revealed gyral atrophy of the temporal lobe with enlargement of the sulci and no evidence of cerebral atherosclerosis (Moller *et al.* 1998). Histological examination interestingly showed that the amount of plaques recognized varied proportionally with the degree of cerebral atrophy (Moller *et al.* 1998). Hence, the temporal lobe contained the largest amount of plaques with the smallest number being observed in the occipital lobe. Additionally, plaques were documented in other structures such as the lentiform nucleus, thalamus, and striatum. It is important to note that Alzheimer described that neurofibrillary degeneration was not apparent in the Johann F case, and therefore stated that the relationship of neurofibrillary tangles to plaques in the disease could not be established (Moller *et al.* 1998).

2. Evolution and Transformation of the B Amyloid Hypothesis

From the initial documentation of the disease, a large body of evidence has accumulated around the Amyloid β cascade hypothesis (Joseph *et al.* 2001; Tanzi 2005; Nunomura *et al.* 2006; Walsh *et al.* 2007). In particular, research has pointed towards germline mutations and biochemical processing which are linked to the familial or autosomal dominant form of Alzheimer's disease (Priller *et al.* 2006; Glenner *et al.* 1984; Vassar *et al.* 1999; De Strooper *et al.* 1998). Mutations associated with the deposition of Amyloid β affect the Amyloid β precursor protein (APP), as well as the transmembrane proteases presenilin 1 (PSEN1) and presenilin 2 (PSEN2). APP is a neuronal membrane protein which can be expressed in other tissues (in a variety of tissues). APP's main function appears to be in synapse formation and maintenance of its functionality (Priller *et al.* 2006). The gene for the Amyloid precursor protein is harbored in chromosome 21q21 (Glenner *et al.* 1984). The post translational processing of the protein is directed by ξ secretases and β secretases. β Amyloid synthesis is commenced by extracellular domain cleavage of the APP by β secretase which yields a fragment named APP C terminal fragment C99 (Vassar *et al.* 1999). Presenilin 1 has been implicated in ξ secretase proteolytic cleavage of the C terminal fragments of APP (De Strooper *et al.* 1998). Presenilins are transmembrane proteases that are additionally involved in intracellular calcium signaling (Cowburn *et al.* 2007). Consequently, trisomy of chromosome 21, as seen in Down's syndrome patients, carries an inherent increase in the synthesis of Amyloid β, thus predisposing these patients to Alzheimer's disease changes at a very young age (Castellani *et al.*).

Elderly non-demented individuals have been demonstrated to posses a higher T-cell mediated response against Amyloid β (Solomon 2008; Janus *et al.* 2000). This has lead some researchers to attempt to eradicate the Amyloid β utilizing immunological approaches (Solomon 2008). Murine models have been subjected to active immunization in an effort to clear human Amyloid β from the brain, with some of these animal regaining lost functional deficits (Janus *et al.* 2000; Morgan *et al.* 2000). Human trials were suspended after 5% of the patients developed meningoencephalitis (Nicoll *et al.* 2003).

Given that the Amyloid β peptide is the focus of therapeutic intervention, it is important to note that several factors have lead to the modification of the Amyloid β cascade hypothesis. First, plaque formation does not correlate well with neurological deficits, in addition to the fact that plaque formation does not correlate well with dementia. The ensuing modification to the Amyloid β hypothesis states that Amyloid β is required for the assemblage of pathogenic, previously unidentified, soluble oligomers, with an apparent paramount role in synaptic degradation (Mucke *et al.* 2000).

3. Current Perspective on Alzheimer's Disease

Worldwide, Alzheimer's disease is the most prevalent neurodegenerative disease, affecting fifteen million patients (Castellani *et al.* ; Hebert *et al.* 2003; Tschape *et al.* 2006; Webber *et al.* 2007). In the United States an approximated four and a half million Americans are affected by the disease (Castellani *et al.*). It is important to note that as the average life expectancy gradually increases there is also a directly proportional concomitant increase in the incidence of the disease. Hebert et al (Hebert *et al.* 2003) estimated that by 2050 over eight million people in the United States would be over 85 years of age (currently four million), and over thirteen million individuals will be affected by the Alzheimer's disease, provided that preventive methods remain unavailable. Recent estimates in the United States show that the combined direct and indirect cost of Alzheimer's disease exceeds sixty billion dollars yearly. Taken in parallel with the estimated increased incidence of the disease some authors have proposed that the total cost of the disease to the United States for present and future generations could approximate 1.75 trillion dollars (Ernst *et al.* 1994). The Canadian Journal of Neurological Sciences additionally demonstrates the devastating effect on the country's economy (Tator *et al.* 2007). The diagnosis of Alzheimer's disease cuts approximately half the life expectancy of patients in some studies (Larson *et al.* 2004).

Despite these compelling statistics and millions of research monies spent in investigating Alzheimer's disease, it is essentially unresolved, and potential therapeutic methods remain elusive. Some authors suggest that the basis of the current misunderstanding of the pathogenesis of Alzheimer's disease rests on incorrect assumptions gravitating around the Amyloid cascade hypothesis (Castellani *et al.* ; Joseph *et al.* 2001). Several drugs are utilized individually or in combination to symptomatologically treat Alzheimer's disease. Metanalyses of several clinical trials demonstrated that cholinesterase inhibitors neither delay the onset nor slow the progression of Alzheimer's disease (Raschetti *et al.* 2007).

3.1. Clinical Presentation Outcomes and Treatment of Alzheimer's Disease

Alzheimer's disease is a neurodegenerative disease and is currently the number one cause of dementia (Desai *et al.* 2005; Qiu *et al.* 2007). DSM-IV lists symptoms of Alzheimer's disease which are characterized by progressive cognitive impairment such as apraxia, aphasia, and agnosia. Additionally, patients have decreased ability to form and/or recall new memories or information. Patients also have difficulties carrying out daily activities.

Current FDA approved treatments for Alzheimer's disease include cholinesterase inhibitors (ChEI), such as donezepil and rivastigmine, as well as memantine, a N methyl D aspartate (NDMA) receptor antagonist (Farlow *et al.* 2007). It is believed that early commencement of therapy with ChEI permits patients to maintain a high level of cognition as well as to reduce the rate of decline.

4. Neuropathology of Alzheimer's Disease

From the time of the first documented case by Alois Alzheimer more than a century ago, light microscopy has been an essential tool in the postmortem evaluation of demented patients. However, the past several decades of investigation of the role of lesions such as Amyloid β, have perhaps given an inaccurate vision of the pathogenesis of Alzheimer's disease (Castellani *et al.* 2006). Currently the pathognomonic neuropathological lesions of

Alzheimer's disease are utilized to further classify and characterize the disease. Although the medical and scientific literature gyrates around the idea that these pathognomonic lesions of Alzheimer's disease are etiological, we, as well as many others, believe that these are end-stage epiphenomena (Raina *et al.* 1999; Castellani *et al.* 2001; Joseph *et al.* 2001; Nunomura *et al.* 2001; Rottkamp *et al.* 2001; Smith *et al.* 2002; Castellani *et al.* 2004; Lee *et al.* 2005; Castellani *et al.* 2006; Castellani *et al.* 2006; Nunomura *et al.* 2006; Castellani *et al.* 2007). Nevertheless, adequate neuropathological diagnosis in the setting of neurodegenerative disease includes quantification of Amyloid B plaques, neurofibrillary tangles, and inclusions as well as clinico pathological correlation with the disease phenotype (Hyman 1997; Hyman *et al.* 1997). Amyloid plaques in the first documented case of Alzheimer's disease were regarded as milliary foci as precipitates in Bielschowsky preparations. Now we know that these can also be seen in cognitively unremarkable individuals (Wilkins *et al.* 1969; Schorer 1985; Berrios 1990; Gold *et al.* 2001).

4.1. Gross Neuropathology and Histopathology

The most common gross findings in the brain of patients with Alzheimer's disease includes widening of the cerebral sulci with concomitant thinning of the gyri leading to a diffuse cortical atrophy. Compensatory ventricular dilation is often encountered. The cortical atrophy usually is more prominent in the hippocampus and medial temporal structures. The parietal and frontal lobes can also be affected, but usually there is sparing of the occipital lobe.

Neuropathological histological evaluation in a possible Alzheimer's disease brain involves the quantification of the hallmark lesions such as Amyloid (senile) plaques, neurofibrillary tangles and neuronal loss, as well as the extent of their presence. Additional lesions commonly encountered include Amyloid angiopathy and neuritic plaques. Amyloid beta plaques are deposits of Amyloid precursor protein (APP) fragments, a normal neuronal product codified in chromosome 21q21. These plaques are frequently surrounded by astrocytic and microglial proliferations. In addition, there might also be degenerating neurons and neurofibrillary tangles at the periphery.

Neurofibrillary tangles and neuropil threads are inclusions formed by hyperphosphorylated tau protein triplets. Neuronal loss in conjunction with neurofibrillary pathology can be seen in multiple locations such as the limbic nuclei, the amygdala, the neocortex, the locus cerelus and the nucleus basalis of Meynert. Neurofibrillary tangles often outlive neurons and are therefore left behind in the neuropil after the neuron dies, a phenomenon known as ghost tangle.

Amyloid angiopathy is the accumulation of Amyloid beta which is often present in the occipital cortex. It can lead to dysphoric angiopathy which is extension of these Amyloid deposits to the surrounding cerebellar parenchyma. It is considered a risk factor for strokes and can independently contribute to dementia in Alzheimer's disease patients (Kumar-Singh 2008).

The Consortium to Establish a Registry for Alzheimer's disease (CERAD) criterion is most commonly used due to its inclusion of clinical phenotype. Braak and Braak's criterion, more commonly used in the research setting, evaluates the density and distribution of neurofibrillary tangles by location in the entorhinal cortex, limbic cortex, and neocortex in increasing order of cognitive impairment. Multivariate analysis of a series of autopsies of elderly subjects revealed significant correlations between psychosocial status and both the CERAD criteria and Braak staging (Gold *et al.* 2001).

Indeed, it is now customary to view Amyloid plaques in Alzheimer disease as primary etiological, neurotoxic lesions and, hence, removing them (e.g., by immunotherapy) is believed to lead to clinical improvement (Castellani *et al.* 2006).

Amyloid plaques are an important diagnostic tool but a consequence rather than a cause in Alzheimer's disease (Castellani *et al.* 2006). Amyloid production appears to be a neuronal response to oxidative stress, therefore explaining the decreased amount of oxidative stress seen in end stage neurons (Yan *et al.* 1995). Amyloid has been proven to be a strong antioxidant of the likes of superoxide dismutase (Nunomura *et al.* 1999; Cuajungco *et al.* 2000; Huang *et al.* 2000; Nunomura *et al.* 2001; Castellani *et al.* 2006; Nunomura *et al.* 2006).

5. Oxidative Stress in Alzheimer's Disease

Evidence of oxidative stress in neurodegenerative diseases is extensive (Castellani *et al.* 1995; Castellani *et al.* 1996; Gerst *et al.* 1999; Guentchev *et al.* 2002; Hartzler *et al.* 2002; Kikuchi *et al.* 2002; Perry *et al.* 2002; Perry *et al.* 2002; Perry *et al.* 2003; Perry *et al.* 2003; Castellani *et al.* 2004; Ghanbari *et al.* 2004; Moreira *et al.* 2005; Petersen *et al.* 2005; Nunomura *et al.* 2006; Castellani *et al.* 2007; Gonzalez-Cuyar *et al.* 2007; Nunomura *et al.* 2007). Decreased glucose metabolism and mitochondrial abnormalities seen in Alzheimer's disease have been implicated in the origination of deleterious reactive oxygen species (Ishii *et al.* 1997; Castellani *et al.* 2002; Perry *et al.* 2003; Castellani *et al.* 2004). Oxidative phosphorylation produces superoxide radicals secondary to electron transport which are kept within the mitochondria (Castellani *et al.* 2002; Castellani *et al.* 2004). Mitochondrial abnormalities are mainly associated with enzyme activity deficiencies involving the enzymes of the electron transport chain, including cytochrome C oxidase (Yates *et al.* 1990; Mastrogiacomo *et al.* 1993; Simonian *et al.* 1994). In the presence of the redox active metals, increased cytoplasmic hydrogen peroxide may increase the production of reactive oxygen species. Metals active in antioxidant activity and prevention of propagation of reactive oxygen species, such as zinc, have been reported to be decreased in the Alzheimer's brain. Evidence of this is the implementation of Vitamin E supplementation as an antioxidant alternative in the treatment of Alzheimer's disease (Adelman 1997; Butterfield *et al.* 1999; Yatin *et al.* 2000; Sung *et al.* 2004; Boothby *et al.* 2005).

As molecules indicative of the oxidative stress are taking a more significant role, they have been measured by molecules such as 8-hydroxyguanosine (8OHG) and nitrotyrosine adduct formation (Nunomura *et al.* 2001; Kikuchi *et al.* 2002; Petersen *et al.* 2005). Nunomura et al (Nunomura *et al.* 2001) correlated the relationship of levels of 8OGH and the clinical phenotypes and concluded that indeed oxidative stress is an early event in neurodegeneration in patients with Down's syndrome as well as familial and sporadic

Alzheimer's disease (Nunomura *et al.* 1999; Nunomura *et al.* 2001; Castellani *et al.* 2006; Nunomura *et al.* 2006). Findings by Nunomura et al (Nunomura *et al.* 2001) demonstrate that oxidative stress is highest early in the disease and reduces with disease progression. Whereas end stage changes such as Amyloid deposition are associated with less quantitative oxidative damage. These authors suggest that oxidative stress is hallmark of the early stages of Alzheimer's disease and that its decrease is proportional to the formation of end stage lesions.

Bleomycin hydrolase BH is an intracellular endopeptidase expressed in oxidative environments and is up regulated in commonly vulnerable neuronal regions but not in neurons with end stage changes such as neurofibrillary pathology. This further demonstrates a role of oxidative stress early in the pathogenesis of Alzheimer's disease (Raina *et al.* 1999).

These authors suggest that perhaps these lesions, which for many decades have been regarded as the etiologic hallmarks of Alzheimer's disease, are but compensatory buffers that reduce intraneuonal oxidative damage, therefore, being epiphenomenal and not pathogenic of the disease (Nunomura *et al.* 2001).

6. Historical Perspective on the Role of Carbohydrates and Amyloidosis

Studies that attempt to clarify the role of carbohydrates in the pathophysiology of neurodegenerative disease are documented as far back as 1854 when Virchow introduced the term corpora amylacea to describe the microscopic intracellular lesions in the central nervous system of patients with Amyloidosis (Rottkamp *et al.* 2001; Smith *et al.* 2002; Castellani *et al.* 2005; Castellani *et al.* 2007). Five years later, Friedreich and Kehule utilized the term Amyloid to describe amorphous extracellular accumulations. Virchow's interpretation of the biochemical properties of corpora amylacea, iodine reactivity, was an accurate conclusion. Friedreich and Kekule's interpretation was also correct as they refered to what now is known as Amyloidosis, as an insoluble compound with green birefringence when Congo Red stained and polarized (Rottkamp *et al.* 2001; Smith *et al.* 2002; Castellani *et al.* 2005; Castellani *et al.* 2007). Friedereich and Kekule disagreed with Virchow's assessment due to the fact that they were talking about different compounds.

7. Metabolism and the Hexosamine Pathway

As we have mentioned, the search for the role of polysaccharides in the pathogenesis of neurodegenerative disease exceeds the century mark (Castellani *et al.* 2007). Bilateral temporoparietal and medial temporal hypoperfusion with subsequent decreased oxygen metabolism has been demonstrated by positron emission tomography (PET) and single photon emission computed tomography (SPECT) in Alzheimer's disease (AD) patients (Kumar *et al.* 1991; Ishii *et al.* 1996; Ishii *et al.* 1996; Ishii *et al.* 1997; Ishii *et al.* 1998; Nunomura *et al.* 2001). Additionally, there are reports that indicate that the cerebellar cortex is also affected by reduced oxygen tensions (Kumar *et al.* 1991; Ishii *et al.* 1996; Ishii *et al.* 1996; Ishii *et al.* 1997). Involvement of these areas is supported by histopathological examination. Hypoperfusion leads to decreased glucose metabolism by area which correlates with cognitive impairments (Ishii *et al.* 1997). Perhaps, downregulation of gene expression of oxidative phosphorylation in neuronal mitochondria accounts for the decreased glucose utilization in Alzheimer's disease patients (Jagust *et al.* 1991; Piert *et al.* 1996; Rapoport *et al.* 1996). Impaired glucose metabolism is also evidenced by the decreased concentrations of glucose transporters 1 and 3 (GLUT1) and (GLUT3) in different areas of the cerebral cortices of Alzheimer's disease patients (Simpson *et al.* 1994; Simpson *et al.* 1994). Several lines of evidence exist which implicate impaired glucose metabolism in Alzheimer's disease. These include reduced deoxyglucose utilization in PET scans, altered cortical glucose metabolism in Tg2576 mice, and decreased number of glucose transporters which shift the balance of available glucose, thus permitting the synthesis of chitin in the brain (Ishii *et al.* 1997; Ishii *et al.* 1998; Cuajungco *et al.* 2000; Nunomura *et al.* 2001; Niwa *et al.* 2002; Niwa *et al.* 2002).

Hyperglycemia has been proven to more than double hexosamine pathway activity in endothelial cells and subsequently increases Sp1 *O*-linked GlcNAc (Brownlee 2001). Activation of the hexosamine pathway leads to synthesis of glucosamine polymers in a manner similar to that described for Diabetes Mellitus (Brownlee 2000; Brownlee 2001; Niwa *et al.* 2002). Impaired glucose metabolism leads to intracellular hyperglycemia which shunts glucose into the hexosamine pathway, as suggested by the diabetic model (Brownlee 2001). In this pathway, substrates for proteoglycan synthesis as well as the formation of O-linked glycoproteins are produced after conversion of fructose-6-phosphate to glucosamine-6-phosphate by glutamine: fructose-6-phosphate aminotransferase. Therefore it is possible that upregulation of the hexosamine pathway leads to synthesis of glucosamine polymers (Castellani *et al.* 2005). Some authors have hypothesized that the intracellular hyperglycemia and increased glucosamine levels are secondary to hexosamine pathway activation due to impaired glucose metabolism and might have glucose and glucosamine polymers as end products.

Glucosamine polymers, the building blocks of chitin and chito saccharides, are synthesized from glucose by means of fructose and fructose 6 phosphate. High levels of both glucose and glucosamine are found in Alzheimer's disease at the cellular level secondary to existent activation of the hexosamine pathway, in turn, secondary to impaired glucose metabolism (Berenson *et al.* 1969; Brownlee 2001; Castellani *et al.* 2005). The combination of impairment in glucose metabolism and functional enzymes perhaps increases synthesis of glucose and glucosamine, polymers of starch and chitin respectively (Berenson *et al.* 1969; Brownlee 2001; Castellani *et al.* 2005; Castellani *et al.* 2007).

Characteristic hallmarks of cerebral histopathology, such as endothelial and circumferential arterial Amyloid deposition leading to increased thickness of the capillaries, might be one of the causes of impaired glucose influx transport across the blood brain barrier (Piert *et al.* 1996). Additionally, hexose transporter protein and hexokinase activity (leading to decreased glucose phosphorylation) is decreased in the microvasculature of Alzheimer's disease (Iwangoff *et al.* 1980; Iwangoff *et al.* 1980; Friedland *et al.* 1989; Sorbi *et al.* 1990; Jagust *et al.* 1991; Ober *et al.* 1991)

8. Chitin

Chitin is a glycopolymer that is well represented in several taxonomical kingdoms. It is most notable for its integral part in the exoskeleton of arthropods. It is a long polymer of beta 1, 4 linked N- Acetylglucosaminidase C8H13O5N) n (Glaser *et al.* 1957; Bakkers *et al.* 1997; Harris *et al.* 2000; Roncero 2002; Bulik *et al.* 2003; Banks *et al.* 2005; Bowman *et al.* 2006). The increased strength is attributed to the acetylamine groups increased hydrogen bonding. Its content is variable within species and is reported to compose approximately 2% of the fungal cell wall by dry weight (Klis 1994; Klis *et al.* 2002; Bowman *et al.* 2006). However, in some species, such as Aspergillus, it can account for as much as 20%. In such instances the chitin polymers impart tensile strength. The integrity of the cell wall is necessary for adequate homeostasis and proper environmental interaction (Cabib *et al.* 1975; Cabib 1987) Some fungi such as Saccharomyces cervesiae have three chitin synthetases: Chs1p, 2p, and (Roncero 2002; Bowman *et al.* 2006), and some have up to seven as seen in Aspergillus fumigatus (Roncero 2002). Chitin synthesis is an energy requiring process in which chitin synthetase catalyzes the transportation of N-acetyl glucosamine from uridine diphosphate (UDP N acetylglucosaminil transferase) and utilizes uridine diphosphate (UDP)-activated monomer as the sugar donor (Glaser *et al.* 1957). Chitin synthetase catalyzes the chitin polymer elongation by vectorial synthesis in such a way that the nascent chains are extruded through the cell wall (Bowman *et al.* 2006). Chitin synthetase inhibitors produce cell death (Glaser *et al.* 1957; Ling *et al.* 2004; Banks *et al.* 2005). In fungi, chitin microfibrills are composed from inter chain bonding of polymers.

No definitive mammalian chitin synthetase has been documented, however the pathogenic role of these oligosaccharides in vertebrates has been reported (Semino *et al.* 1996; Bakkers *et al.* 1997). Chi3L1, a chitin-like protein, binds chitin oligosaccharides in murine astrocytes and glioma lines (Tanwar *et al.* 2002). Hyaluronan synthase- 1 (HAS1) has been shown to convert activated glucosamine to chito-oligosaccharides *in vitro* using murine HAS1 gene product (Xu *et al.* 2002). Chitin, as well as chitosan, is utilized in the enhancement of the inflammatory roles of macrophages and neutrophils (Ueno *et al.* 2001; Ueno *et al.* 2001; Ueno *et al.* 2001). In such an instance, chitin might modulate cerebral microglial activation which has been documented in neurodegenerative diseases (Bakkers *et al.* 1997; Wyss-Coray *et al.* 2002; Wyss-Coray *et al.* 2002). CS-like genes have been reported in other evolutionary groups, such as insects, bacteria, protozoa and even vertebrates (Bulawa *et al.* 1991; Semino *et al.* 1996; Gagou *et al.* 2002). Additionally, there is evidence that chitin associated proteins, in both murine and human cell lines provide a possible pathogenic role to chitin-like proteins.

9. Chitin in the Alzheimer Brain

Given the documented evidence of deranged glucose metabolism in the Alzheimer's disease brain, studies have been conducted to elucidate whether hexosamine pathway activation and subsequent glucosamine polymer (chitin) deposition occur in a manner similar to that in Diabetes mellitus. In previous studies by Castellani et al (Castellani *et al.* 2005),

Hippocampal samples of Alzheimer's disease patients harboring A431E presenilin 1 mutation, as well as sporadic AD subjects were stained with Calcofluor fluorescence histochemistry, Amyloid β and phosphorylated tau protein (AT8) Immunohistochemistry. Calcofluor is a fluorochrome that becomes excited upon exposure to ultraviolet light. *In vivo* it exhibits strong affinity for chitin as it interacts with the β1-4 linkages that make up the chitin polymer (Garcia-Zapien *et al.* 1999; Klis *et al.* 2002; Castellani RJ 2004; Castellani *et al.* 2005; Castellani *et al.* 2007). It has been demonstrated that this is a convenient technique to determine chitin composition in tissues, as chitin is a linear polymer composed of glucosamine monomers linked β1-4 bonds (Klis *et al.* 2002). Calcofluor fluorescence histochemistry had robust staining of Amyloid angiopathy (Figure 1A), Amyloid plaques (Figure 1B), and also stained neurofibrillary tangles and cerebral vasculature (Castellani *et al.* 2005). Moreover, Calcofluor stained plaques were co-localized with Amyloid β Immunohistochemistry. The authors also subjected the tissue to chitinase treatment, thus degrading chitin to chitobiose and subsequently treated it with β N acetylglucosaminidase. Interestingly, these tissues had diminished Calcofluor fluorescence, therefore suggesting that chitin-like polysaccharides, including chitobiose, form an integral part of commonly encountered pathognomonic lesions of Alzheimer's disease such as senile plaques and Amyloid angiopathy (Castellani *et al.* 2005; Castellani *et al.* 2007). Because chitin and chitin-like polysaccharides are encountered in Amyloid plaques, several questions come to mind. For example, commercial forms of chitin stain with Congo Red histochemistry as well as have apple green birefringence when exposed to polarized light like Amyloid β. This raises the issue of how much of Amyloid β histochemical properties are imparted by chitin. Another property that both substances share is their relative insolubility. It is unknown if this is imparted by chitin, and therefore further studies should be conducted in order to answer this question. These properties are thought to be imparted to Amyloid β by its protein conformation (Fraser *et al.* 1992).

Another important issue involving the cerebral accumulation of chitin is studying exactly what role, if any, does it play. Several reports in the literature document that Amyloid β plaques may, in fact, be protective rather than pathologic elements in Alzheimer's disease (Mucke 2000; Wyss-Coray *et al.* 2000; Wyss-Coray *et al.* 2000; Castellani *et al.* 2001; Joseph *et al.* 2001; Nunomura *et al.* 2001; Castellani *et al.* 2006; Castellani *et al.* 2006; Nunomura *et al.* 2006; Castellani *et al.* 2007). Protective functions of Amyloid β take place by the binding of metallic ions while also decreasing free radical neurotoxicity (Rottkamp *et al.* 2001; Smith *et al.* 2002). It can therefore be hypothesized that perhaps chitin may also function as a protective mechanism by functioning as a primer for the deposition of Amyloid β (Castellani RJ 2004; Castellani *et al.* 2005; Castellani *et al.* 2007). Calcofluor fluorescence of β Amyloid wool plaques lacks Amyloid's customary ultrastructural fibril-like arrangement, which is also seen with Amyloid β immunohistochemistry. Potentially this can further assert to the fact that chitin deposition may precede and aid the process of Amyloid β accumulation.

Amyloid produced by light chain Amyloidosis is associated with highly sulfated glycosaminoglycans (GAG's) such as heparin sulfate (Bitter *et al.* 1965; Dalferes *et al.* 1967; Kumar *et al.* 1967; Kumar *et al.* 1967; Pennock 1968; Pennock *et al.* 1968); cardiac Amyloidosis is associated with hyaluronic acid (Clausen *et al.* 1964; Berenson *et al.* 1969; Dalferes *et al.* 1969). The study of the association of Amyloid deposits and glycosaminoglycan composition in end stage disease raises the question of whether these glycosaminoglycan deposits occur concomitantly with Amyloid deposits or as secondary phenomena.

It has been documented that highly sulfated GAG's such as keratin sulfate and heparan sulfate are synchronous with the deposition of Amyloid protein as an integral process in the pathogenesis of Amyloidosis (Garcia-Zapen *et al.* 1999; Castellani *et al.* 2004) Amyloid AA fibrils and heparan sulfate proteoglycans are closely associated as evidenced by co-localization of heparan sulfate with Amyloid β and in neuritic plaques and cerebral vasculature. Again, studies suggest that chitin and proteoglycans have a recruitment effect and thus promote Amyloid deposition (Snow *et al.* 1987; Snow *et al.* 1987; Szumanska *et al.* 1987; Snow *et al.* 1988). If this is accurate, the mechanism by which it is attained is unknown, and further studies should be conducted to elucidate this role.

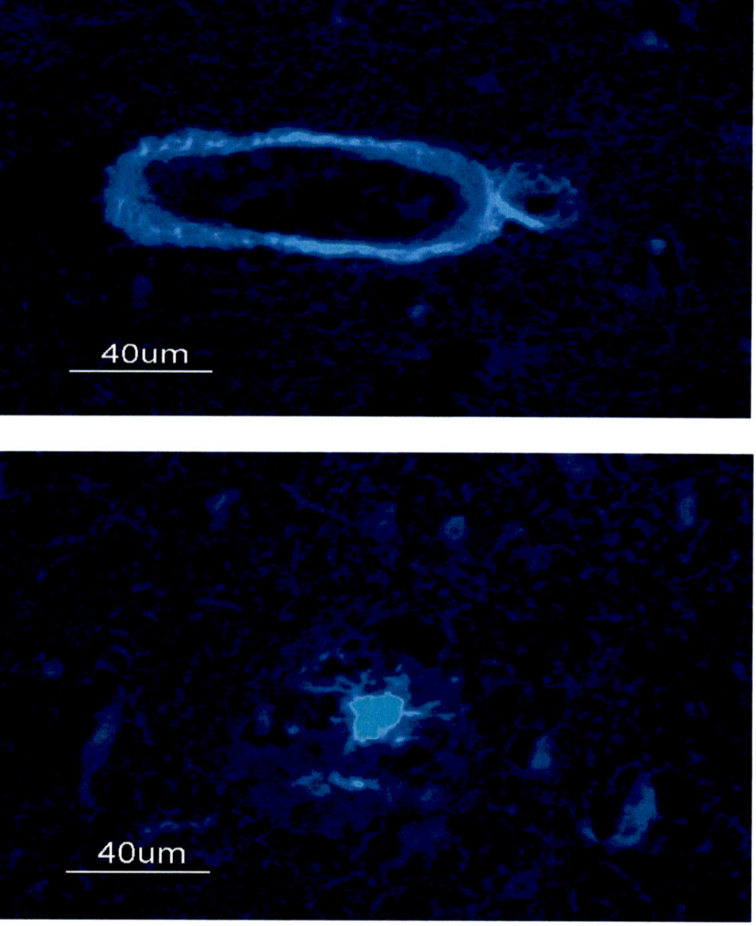

Figure 1 Calcofluor histochemistry of brain tissue from a patient with Alzheimer's disease demonstrates intense labeling of a blood vessel affected by Amyloid angiopathy (Figure 1a) and an Amyloid plaque (figure 1b). The labeling co-localizes with labeling for Amyloid-beta immunohistochemistry.
1.Castellani, R.J., et al., Chitin-like polysaccharides in Alzheimer's disease brains. Curr Alzheimer Res, 2005. 2: p. 419-23.

Conclusion

In conclusion, we have discussed the occurrence of chitin and chitin-like polysaccharides in Amyloid plaques and Amyloid angiopathy in both sporadic and familial Alzheimer's disease patients. This co-localization was performed with Calcofluor histochemistry co localization with Amyloid β immunohistochemistry. Therefore we demonstrate that chitin is an important component of the pathognomonic lesions of Alzheimer's disease. Chitin and Amyloid share several chemical properties such as insolubility. It is unclear if these properties may be influenced by the presence of Chitin within the Amyloid plaque. Our group has suggested that chitin and chitin like polysaccharides provide scaffolding for Amyloid accumulation. Since it has been described that Amyloid may in fact act as a neuroprotective agent against oxidative stress in, perhaps chitin is the first step in priming this neuroprotective action. The implications of these data in terms of the understanding of the disease as well as in treatment advances are considerable.

Acknowledgments

The authors wish to thank Krista J. Szafranski MS, PA for her critical review of the manuscript and helpful suggestions.

References:

Adelman, A. Selegiline and vitamin E in Alzheimer's disease. *J Fam Pract.* 1997; 45: 98-100.

Bakkers J, Semino CE, Stroband H, Kijne JW, Robbins PW, Spaink HP. An important developmental role for oligosaccharides during early embryogenesis of cyprinid fish. *Proc Natl Acad Sci . U S* A 1997; 94: 7982-6.

Banks IR, Specht CA, Donlin MJ, Gerik KJ, Levitz SM, Lodge JK. A chitin synthase and its regulator protein are critical for chitosan production and growth of the fungal pathogen Cryptococcus neoformans. *Eukaryot Cell* 2005; 4: 1902-12.

Berenson G, Dalferes ER Jr, Ruiz H, Radhakrishnamurthy B. Changes of mucopolysaccharides in the heart involved by amyloidosis. *Am J Cardiol.* 1996; 24: 358-64.

Berrios GE. Alzheimer's disease a conceptual history. *International Journal of Geriatric Psychiatry* 1990; 5: 355-365.

Bitter T, Muir H. Mucopolysaccharides in amyloidosis. *Lancet* 1965; 1(7389): 819.

Boothby LA, Doering PL.Vitamin C and vitamin E for Alzheimer's disease. *Ann Pharmacother* 2005; 39: 2073-80.

Bowman SM, Free SJ. The structure and synthesis of the fungal cell wall. *Bioessays* 2006; 28: 799-808.

Brownlee M. Negative consequences of glycation. *Metabolism* 2000; 49 (2 Suppl 1): 9-13.

Brownlee M. Biochemistry and molecular cell biology of diabetic complications. *Nature* 2001; 414: 813-20.

Bulawa CE, Wasco W. Chitin and nodulation. *Nature* 1991; 353: 710.

Bulik DA, Olczak M, Lucero HA, Osmond BC, Robbins PW, Specht CA. Chitin synthesis in Saccharomyces cerevisiae in response to supplementation of growth medium with glucosamine and cell wall stress. *Eukaryot Cell* 2003; 2: 886-900.

Butterfield DA, Koppal T, Subramaniam R, Yatin S. Vitamin E as an antioxidant/free radical scavenger against amyloid beta-peptide-induced oxidative stress in neocortical synaptosomal membranes and hippocampal neurons in culture: insights into Alzheimer's disease. *Rev Neurosci.* 1999; 10: 141-9.

Cabib E. The synthesis and degradation of chitin. *Adv Enzymol Relat Areas Mol Biol.* 1987; 59: 59-101.

Cabib E, Bowers B. Timing and function of chitin synthesis in yeast. *J Bacteriol.* 1975; 124: 1586-93.

Castellani RJ, Common R, Perry G, Ghetti B, Smith MA. Calcofluor stains amyloid-beta deposits. *Journal of Neuropathology and Experimental Neurology* 2004; 63: 524.

Castellani RJ, Webber KM, Moriera PI, Lee H, Casadesus G, Honda K, Zhu X, Perry G, Simth MA. Thinking outside the box in Alzheimer's disease treatment. *Development of Therapeutic Agents.* In Press.

Castellani RJ, Perry G, Smith MA. The role of novel chitin-like polysaccharides in Alzheimer disease. *Neurotox Res.* 2007; 12: 269-74.

Castellani RJ, Lee HG, Perry G, Smith MA. Antioxidant protection and neurodegenerative disease: the role of amyloid-beta and tau. *Am J Alzheimers Dis. Other Demen.* 2006; 21: 126-30.

Castellani RJ, Lee HG, Zhu X, Nunomura A, Perry G, Smith MA. Neuropathology of Alzheimer disease: pathognomonic but not pathogenic. *Acta Neuropathol.* 2006; 111: 503-9.

Castellani RJ, Honda K, Zhu X, Cash AD, Nunomura A, Perry G, Smith MA. Contribution of redox-active iron and copper to oxidative damage in Alzheimer disease. *Ageing Res Rev.* 2004; 3: 319-26.

Castellani RJ, Smith MA, Perry G, Friedland RP. Cerebral amyloid angiopathy: major contributor or decorative response to Alzheimer's disease pathogenesis. *Neurobiol Aging* 2004; 25: 599-602; *discussion* 603-4.

Castellani RJ, Moreira PI, Liu G, Dobson J, Perry G, Smith MA, Zhu X. Iron: the Redox-active center of oxidative stress in Alzheimer disease. *Neurochem Res.* 2007; 32: 1640-5.

Castellani RJ, Harris PL, Sayre LM, Fujii J, Taniguchi N, Vitek MP, Founds H, Atwood CS, Perry G, Smith MA. Active glycation in neurofibrillary pathology of Alzheimer disease: N(epsilon)-(carboxymethyl) lysine and hexitol-lysine. *Free Radic Biol Med.* 2001; 31: 175-80.

Castellani RJ, Siedlak SL, Fortino AE, Perry G, Ghetti B, Smith MA. Chitin-like polysaccharides in Alzheimer's disease brains. *Curr Alzheimer Res.* 2005; 2: 419-23.

Castellani RJ, Zhu X, Lee HG, Moreira PI, Perry G, Smith MA. Neuropathology and treatment of Alzheimer disease: did we lose the forest for the trees? *Expert Rev Neurother.* 2007; 7: 473-85.

Castellani R, Hirai K, Aliev G, Drew KL, Nunomura A, Takeda A, Cash AD, Obrenovich ME, Perry G, Smith MA. Role of mitochondrial dysfunction in Alzheimer's disease. *J. Neurosci Res.* 2002; 70: 357-60.

Castellani R, Smith MA, Richey PL, Perry G. Glycoxidation and oxidative stress in Parkinson disease and diffuse Lewy body disease. *Brain Res.* 1996; 37: 195-200.

Castellani R, Smith MA, Richey PL, Kalaria R, Gambetti P, Perry G. Evidence for oxidative stress in Pick disease and corticobasal degeneration. *Brain Res.* 1995; 696: 268-71.

Clausen J, Christensen HE. Paraproteins and Acid Mucopolysaccharides in Primary Amyloidosis. Biochemical and Histologic Studies of Four Human Cases of Primary Amyloidosis. *Acta Pathol Microbiol Scand.* 1964; 60: 493-511.

Cowburn RF, Popescu BO, Ankarcrona M, Dehvari N, Cedazo-Minguez A. Presenilin-mediated signal transduction. *Physiol Behav.* 2007; 92: 93-7.

Cuajungco MP, Goldstein LE, Nunomura A. Smith MA, Lim JT, Atwood CS, Huang X, Farrag YW, Perry G, Bush AI. Evidence that the beta-amyloid plaques of Alzheimer's disease represent the redox-silencing and entombment of abeta by zinc. *J Biol Chem.* 2000; 275: 19439-42.

Dalferes ER Jr, Radhakrishnamurthy B, Berenson GS. Acid mucopolysaccharides of amyloid tissue. *Arch Biochem Biophys.* 1967; 118: 284-91.

Dalferes ER Jr, Ruiz H, Radhakrishnamurthy B, Rigdon RH, Berenson GS. Acid mucopolysaccharides in the liver of Pekin duck with amyloidosis. *Proc Soc Exp Biol Med.* 1969; 131: 1382-5.

De Strooper B, Saftig P, Craessaerts K, Vanderstichele H, Guhde G, Annaert W, Von Figura K, Van Leuven F. Deficiency of presenilin-1 inhibits the normal cleavage of amyloid precursor protein. *Nature* 1968; 391: 387-90.

Desai AK, Grossberg GT. Diagnosis and treatment of Alzheimer's disease. *Neurology* 2005; 64 (12 Suppl 3): S34-9.

Ernst RL, Hay JW. The US economic and social costs of Alzheimer's disease revisited. *Am J Public Health* 1994; 84: 1261-4.

Farlow MR, Cummings JL. Effective pharmacologic management of Alzheimer's disease. *Am J Med.* 2007; 120: 388-97.

Fraser PE, Nguyen JT, Chin DT, Kirschner DA. Effects of sulfate ions on Alzheimer beta/A4 peptide assemblies: implications for amyloid fibril-proteoglycan interactions. *J Neurochem.* 1992; 59: 1531-40.

Friedland RP, Jagust WJ, Huesman RH, Koss E, Knittel B, Mathis CA, Ober BA, Mazoyer BM, Budinger TF. Regional cerebral glucose transport and utilization in Alzheimer's disease. *Neurology* 1989; 39: 1427-34.

Gagou ME, Kapsetaki M, Turberg A, Kafetzopoulos D. Stage-specific expression of the chitin synthase DmeChSA and DmeChSB genes during the onset of Drosophila metamorphosis. *Insect Biochem Mol Biol.* 2002; 32: 141-6.

Garcia-Marin V, Garcia-Lopez P, Freire M. Cajal's contributions to the study of Alzheimer's disease. *J. Alzheimers Dis.* 2007; 12: 161-74.

Garcia-Zapien AG, Gonzalez-Robles A, Mora-Galindo J. Congo red effect on cyst viability and cell wall structure of encysting. Entamoeba invadens. *Arch Med Res.* 1999; 30: 106-15.

Gerst JL, Siedlak SL, Nunomura A, Castellani R, Perry G, Smith MA. Role of oxidative stress in frontotemporal dementia. *Dement Geriatr Cogn Disord.* 1999; 10 (Suppl 1): 85-7.

Ghanbari HA, Ghanbari K, Harris PL, Jones PK, Kubat Z, Castellani RJ, Wolozin BL, Smith MA, Perry G. Oxidative damage in cultured human olfactory neurons from Alzheimer's disease patients. *Aging Cell* 2004; 3: 41-4.

Glaser L, Brown DH. The synthesis of chitin in cell-free extracts of Neurospora crassa. *J Biol Chem.* 1957; 228: 729-42.

Glenner GG, Wong CW. Alzheimer's disease: initial report of the purification and characterization of a novel cerebrovascular amyloid protein. *Biochem Biophys Res. Commun.* 1984; 120: 885-90.

Goedert M, Ghetti B. Alois Alzheimer: his life and times. *Brain Pathol.* 2007; 17: 57-62.

Gold G, Kovari E, Corte G, Herrmann FR, Canuto A, Bussiere T, Hof PR, Bouras C, Giannakopoulos P. Clinical validity of A beta-protein deposition staging in brain aging and Alzheimer disease. *J Neuropathol Exp Neurol.* 2001; 60: 946-52.

Gonzalez-Cuyar LF, Hunter B, Harris PL, Perry G, Smith MA, Castellani RJ. Cerebrotendinous xanthomatosis: case report with evidence of oxidative stress. *Redox Rep.* 2007; 12: 119-24.

Guentchev M, Siedlak SL, Jarius C, Tagliavini F, Castellani RJ, Perry G, Smith MA, Budka H. Oxidative damage to nucleic acids in human prion disease. *Neurobiol Dis.* 2002; 9: 275-81.

Harris MT, Lai K, Arnold K, Martinez HF, Specht CA, Fuhrman JA. Chitin synthase in the filarial parasite, Brugia malayi. *Mol Biochem Parasitol.* 2000; 111: 351-62.

Hartzler AW, Zhu X, Siedlak SL, Castellani RJ, Avila J, Perry G, Smith MA. The p38 pathway is activated in Pick disease and progressive supranuclear palsy: a mechanistic link between mitogenic pathways, oxidative stress, and tau. *Neurobiol Aging* 2002; 23: 855-9.

Hebert LE, Scherr PA, Bienias JL, Bennett DA, Evans DA. Alzheimer disease in the US population: prevalence estimates using the 2000 census. *Arch Neurol.* 2003; 60(8): 1119-22.

Huang X, Cuajungco MP, Atwood CS, Moir RD, Tanzi RE, Bush AI. Alzheimer's disease, beta-amyloid protein and zinc. *J Nutr.* 2000; 130(5S Suppl): 1488S-92S.

Hyman BT. The neuropathological diagnosis of Alzheimer's disease: clinical-pathological studies. *Neurobiol Aging.* 1997; 18(4 Suppl): S27-32.

Hyman BT, Trojanowski JQ. Consensus recommendations for the postmortem diagnosis of Alzheimer disease from the National Institute on Aging and the Reagan Institute Working Group on diagnostic criteria for the neuropathological assessment of Alzheimer disease. J. Neuropathol. *Exp Neurol.* 1997; 56: 1095-7.

Ishii K, Kitagaki H, Kono M, Mori E. Decreased medial temporal oxygen metabolism in Alzheimer's disease shown by PET. *J Nucl Med.* 1996; 37: 1159-65.

Ishii K, Sasaki M, Kitagaki H, Sakamoto S, Yamaji S. Maeda K. Regional difference in cerebral blood flow and oxidative metabolism in human cortex. *J Nucl Med.* 1996; 37: 1086-8.

Ishii K, Sasaki M, Kitagaki H, Yamaji S, Sakamoto S, Matsuda K, Mori E. Reduction of cerebellar glucose metabolism in advanced Alzheimer's disease. *J Nucl Med.* 1997; 38: 925-8.

Ishii K, Imamura T, Sasaki M, Yamaji S, Sakamoto S, Kitagaki H, Hashimoto M, Hirono N, Shimomura T, Mori E. Regional cerebral glucose metabolism in dementia with Lewy bodies and Alzheimer's disease. *Neurology* 1998; 51: 125-30.

Iwangoff P, Enz A, Armbruster R, Emmenegger H, Pataki A, Sandoz P. The influence of aging, post-mortem delay until isolation of the tissue and duration of agony on some glycolytic enzymes in human autoptic brain tissue. *Aktuelle Gerontol.* 1980; 10: 203.

Iwangoff P, Armbruster R, Enz A, Meier-Ruge W. Glycolytic enzymes from human autoptic brain cortex: normal aged and demented cases. *Mech Ageing Dev.* 1980; 14: 203-9.

Jagust WJ, Seab JP, Huesman RH, Valk PE, Mathis CA, Reed BR, Coxson PG, Budinger TF. Diminished glucose transport in Alzheimer's disease: dynamic PET studies. *J Cereb Blood Flow Metab.* 1991; 11: 323-30.

Janus C, Pearson J, McLaurin J, Mathews PM, Jiang Y, Schmidt SD, Chishti MA, Horne P, Heslin D, French J, Mount HT, Nixon RA, Mercken M, Bergeron C, Fraser PE, St George-Hyslop P, Westaway D. A beta peptide immunization reduces behavioural impairment and plaques in a model of Alzheimer's disease. *Nature* 2000; 408: 979-82.

Joseph J, Shukitt-Hale B, Denisova NA, Martin A, Perry G, Smith MA. Copernicus revisited: amyloid beta in Alzheimer's disease. *Neurobiol Aging.* 2001; 22: 131-46.

Kaplan M, Henderson AR. Solomon Carter Fuller, M.D. (1872-1953): American pioneer in Alzheimer's disease research. *J Hist Neurosci.* 2000; 9: 250-61.

Kikuchi A, Takeda A, Onodera H, Kimpara T, Hisanaga K, Sato N, Nunomura A, Castellani RJ, Perry G, Smith MA, Itoyama Y. Systemic increase of oxidative nucleic acid damage in Parkinson's disease and multiple system atrophy. *Neurobiol Dis.* 2002; 9: 244-8.

Klis FM. Review: cell wall assembly in yeast. *Yeast* 1994; 10: 851-69.

Klis FM, Mol P, Hellingwerf K, Brul S. Dynamics of cell wall structure in Saccharomyces cerevisiae. *FEMS Microbiol Rev.* 2002; 26: 239-56.

Kumar A, Schapiro MB, Grady C, Haxby JV, Wagner E, Salerno JA, Friedland RP, Rapoport SI. High-resolution PET studies in Alzheimer's disease. *Neuropsychopharmacology* 1991; 4: 35-46.

Kumar V, Berenson GS, Ruiz H, Dalferes ER Jr, Strong JP. Acid mucopolysaccharides of human aorta. 2. Variations with atherosclerotic involvement. *J Atheroscler Res.* 1967; 7: 583-90.

Kumar V, Berenson GS, Ruiz H, Dalferes ER Jr, Strong JP. Acid mucopolysaccharides of human aorta. 1. Variations with maturation. *J. Atheroscler Res.* 1967; 7: 573-81.

Kumar-Singh S. Cerebral amyloid angiopathy: pathogenetic mechanisms and link to dense amyloid plaques. *Genes Brain Behav.* 2008; 7 (Suppl 1): 67-82.

Larson EB, Shadlen MF, Wang L, McCormick WC, Bowen JD, Teri L, Kukull WA. Survival after initial diagnosis of Alzheimer disease. *Ann Intern Med.* 2004; 140: 501-9.

Lee HG, Castellani RJ, Zhu X, Perry G, Smith MA. Amyloid-beta in Alzheimer's disease: the horse or the cart? Pathogenic or protective? *Int J Exp Pathol.* 2005; 86: 133-8.

Ling H, Recklies AD. The chitinase 3-like protein human cartilage glycoprotein 39 inhibits cellular responses to the inflammatory cytokines interleukin-1 and tumour necrosis factor-alpha. *Biochem J.* 2004; 380(Pt 3): 651-9.

Mastrogiacomo F, Bergeron C, Kish SJ. Brain alpha-ketoglutarate dehydrogenase complex activity in Alzheimer's disease. *J Neurochem* 1993; 61: 2007-14.

Maurer K, Volk S, Gerbaldo H. Auguste D and Alzheimer's disease. *Lancet* 1967; 349: 1546-9.

Moller HJ, Graeber MB. The case described by Alois Alzheimer in 1911. Historical and conceptual perspectives based on the clinical record and neurohistological sections. *Eur Arch Psychiatry Clin Neurosci.* 1998; 248: 111-22.

Moreira PI, Smith MA, Zhu X, Nunomura A, Castellani RJ, Perry G. Oxidative stress and neurodegeneration. *Ann NY Acad Sci.* 2005; 1043: 545-52.

Morgan D, Diamond DM, Gottschall PE, Ugen KE, Dickey C, Hardy J, Duff K, Jantzen P, DiCarlo G, Wilcock D, Connor K, Hatcher J, Hope C, Gordon M, Arendash GW. A beta peptide vaccination prevents memory loss in an animal model of Alzheimer's disease. *Nature* 2000; 408: 982-5.

Mucke HA. Advances in Alzheimer therapy - sixth International Stockholm/Springfield Symposium. *Drugs* 2000; 3: 730-6.

Mucke L, Masliah E, Yu GQ, Mallory M, Rockenstein EM, Tatsuno G, Hu K, Kholodenko D, Johnson-Wood K, McConlogue L. High-level neuronal expression of abeta 1-42 in wild-type human amyloid protein precursor transgenic mice: synaptotoxicity without plaque formation. *J Neurosci.* 2000; 20: 4050-8.

Nicoll JA, Wilkinson D, Holmes C, Steart P, Markham H, Weller RO. Neuropathology of human Alzheimer disease after immunization with amyloid-beta peptide: a case report. *Nat Med* 2003; 9: 448-52.

Niwa K, Kazama K, Younkin L, Younkin SG, Carlson GA, Iadecola C. Cerebrovascular autoregulation is profoundly impaired in mice overexpressing amyloid precursor protein. *Am J Physiol Heart Circ Physiol.* 2002; 283: H315-23.

Niwa K, Kazama K, Younkin SG, Carlson GA, Iadecola C. Alterations in cerebral blood flow and glucose utilization in mice overexpressing the amyloid precursor protein. *Neurobiol Dis.* 2002; 9: 61-8.

Nunomura A, Perry G, Aliev G, Hirai K, Takeda A, Balraj EK, Jones PK, Ghanbari H, Wataya T, Shimohama S, Chiba S, Atwood CS, Petersen RB, Smith MA. Oxidative damage is the earliest event in Alzheimer disease. *J Neuropathol Exp Neurol.* 2001; 60: 759-67.

Nunomura A, Perry G, Hirai K, Aliev G, Takeda A, Chiba S, Smith MA. Neuronal RNA oxidation in Alzheimer's disease and Down's syndrome. *Ann NY Acad Sci.* 1999; 893: 362-4.

Nunomura A, Moreira PI, Lee HG, Zhu X, Castellani RJ, Smith MA, Perry G. Neuronal death and survival under oxidative stress in Alzheimer and Parkinson diseases. *CNS Neurol Disord Drug Targets* 2007; 6: 411-23.

Nunomura A, Castellani RJ, Lee HG, Moreira PI, Zhu X, Perry G, Smith MA. Neuropathology in Alzheimer's disease: awaking from a hundred-year-old dream. *Sci. Aging Knowledge Environ.* 2006; 2006: pe10.

Nunomura A, Castellani RJ, Zhu X, Moreira PI, Perry G, Smith MA. Involvement of oxidative stress in Alzheimer disease. *J Neuropathol Exp Neurol.* 2006; 65: 631-41.

Ober BA, Jagust WJ, Koss E, Delis DC, Friedland RP. Visuoconstructive performance and regional cerebral glucose metabolism in Alzheimer's disease. *J Clin Exp Neuropsychol.* 1991; 13: 752-72.

Pennock CA. Association of acid mucopolysaccharides with isolated amyloid fibrils. *Nature* 1968; 217(5130): 753-4.

Pennock CA, Burns J, Massarella G. Histochemical investigation of acid mucosubstances in secondary amyloidosis. *J Cli Pathol.* 1968; 21: 578-81.

Perry G, Nunomura A, Hirai K, Zhu X, Perez M, Avila J, Castellani RJ, Atwood CS, Aliev G, Sayre LM, Takeda A, Smith MA. Is oxidative damage the fundamental pathogenic mechanism of Alzheimer's and other neurodegenerative diseases? *Free Radic Biol Med.* 2002; 33: 1475-9.

Perry G, Taddeo MA, Nunomura A, Zhu X, Zenteno-Savin T, Drew KL, Shimohama S, Avila J, Castellani RJ, Smith MA. Comparative biology and pathology of oxidative stress in Alzheimer and other neurodegenerative diseases: beyond damage and response. *Comp Biochem Physiol C Toxicol Pharmacol.* 2002; 133: 507-13.

Perry G, Taddeo MA, Petersen RB, Castellani RJ, Harris PL, Siedlak SL, Cash AD, Liu Q, Nunomura A, Atwood CS, Smith MA. Adventiously-bound redox active iron and copper are at the center of oxidative damage in Alzheimer disease. *Biometals* 2003; 16: 77-81.

Perry G, Castellani RJ, Smith MA, Harris PL, Kubat Z, Ghanbari K, Jones PK, Cordone G, Tabaton M, Wolozin B, Ghanbari H. Oxidative damage in the olfactory system in Alzheimer's disease. *Acta Neuropathol.* 2003; 106: 552-6.

Petersen RB, Siedlak SL, Lee HG, Kim YS, Nunomura A, Tagliavini F, Ghetti B, Cras P, Moreira PI, Castellani RJ, Guentchev M, Budka H, Ironside JW, Gambetti P, Smith MA, Perry G. Redox metals and oxidative abnormalities in human prion diseases. *Acta Neuropathol.* 2005; 110: 232-8.

Piert M, Koeppe RA, Giordani B, Berent S, Kuhl DE. Diminished glucose transport and phosphorylation in Alzheimer's disease determined by dynamic FDG-PET. *J Nucl Med.* 1996; 37: 201-8.

Priller C, Bauer T, Mitteregger G, Krebs B, Kretzschmar HA, Herms J. Synapse formation and function is modulated by the amyloid precursor protein. *J Neurosci.* 2006; 26(27): 7212-21.

Qiu C, De Ronchi D, Fratiglioni L. The epidemiology of the dementias: an update. *Curr Opin Psychiatry* 2007; 20: 380-5.

Raina AK, Takeda A, Nunomura A, Perry G, Smith MA. Genetic evidence for oxidative stress in Alzheimer's disease. *Neuroreport* 1999; 10: 1355-7.

Rapoport SI, Hatanpaa K, Brady DR, Chandrasekaran K. Brain energy metabolism, cognitive function and down-regulated oxidative phosphorylation in Alzheimer disease. *Neurodegeneration* 1996; 5: 473-6.

Raschetti R, Albanese E, Vanacore N, Maggini M. Cholinesterase inhibitors in mild cognitive impairment: a systematic review of randomised trials. *PLoS Med.* 2007; 4: e338.

Roncero C. The genetic complexity of chitin synthesis in fungi. *Curr Genet.* 2002; 41: 367-78.

Rottkamp CA, Raina AK, Zhu X, Gaier E, Bush AI, Atwood CS, Chevion M, Perry G, Smith MA. Redox-active iron mediates amyloid-beta toxicity. *Free Radic Bio Med.* 2001; 30: 447-50.

Schorer CE. Historical essay: Kraepelin's description of Alzheimer's disease. *Int. J. Aging. Hum. Dev.* 1985; 21: 235-8.

Semino CE, Specht CA, Raimondi A, Robbins PW. Homologs of the Xenopus developmental gene DG42 are present in zebrafish and mouse and are involved in the synthesis of Nod-like chitin oligosaccharides during early embryogenesis. *Proc Natl Acad Sci U S A* 1996; 93: 4548-53.

Simonian NA, Hyman BT. Functional alterations in Alzheimer's disease: selective loss of mitochondrial-encoded cytochrome oxidase mRNA in the hippocampal formation. *J Neuropathol Exp Neurol.* 1994; 53: 508-12.

Simpson IA, Davies P. Reduced glucose transporter concentrations in brains of patients with Alzheimer's disease. *Ann Neurol.* 1994; 36: 800-1.

Simpson IA, Chundu KR, Davies-Hill T, Honer WG, Davies P. Decreased concentrations of GLUT1 and GLUT3 glucose transporters in the brains of patients with Alzheimer's disease. *Ann Neurol.* 1994; 35: 546-51.

Small DH, Cappai R. Alois Alzheimer and Alzheimer's disease: a centennial perspective. *J Neurochem.* 2006; 99: 708-10.

Smith MA, Casadesus G, Joseph JA, Perry G. Amyloid-beta and tau serve antioxidant functions in the aging and Alzheimer brain. *Free Radic Biol Med.* 2002; 33: 1194-9.

Snow AD, Mar H, Nochlin D, Kimata K, Kato M, Suzuki S, Hassell J, Wight TN. The presence of heparan sulfate proteoglycans in the neuritic plaques and congophilic angiopathy in Alzheimer's disease. *Am J Pathol.* 1988; 133: 456-63.

Snow AD, Willmer J, Kisilevsky R. A close ultrastructural relationship between sulfated proteoglycans and AA amyloid fibrils. *Lab Invest* 1987; 57: 687-98.

Snow AD, Kisilevsky R, Stephens C, Anastassiades T. Characterization of tissue and plasma glycosaminoglycans during experimental AA amyloidosis and acute inflammation. Qualitative and quantitative analysis. *Lab Invest.* 1987; 56: 665-75.

Solomon B. Immunological Approaches for Amyloid-beta Clearance Toward Treatment for Alzheimer's Disease. *Rejuvenation Res.* 2008.

Sorbi S, Mortilla M, Piacentini S, Tonini S, Amaducci L. Altered hexokinase activity in skin cultured fibroblasts and leukocytes from Alzheimer's disease patients. *Neurosci Lett.* 1990; 117: 165-8.

Sung S, Yao Y, Uryu K, Yang H, Lee VM, Trojanowski JQ, Pratico D. Early vitamin E supplementation in young but not aged mice reduces Abeta levels and amyloid deposition in a transgenic model of Alzheimer's disease. *FASEB J.* 2004; 18: 323-5.

Szumanska G, Vorbrodt AW, Mandybur TI, Wisniewski HM. Lectin histochemistry of plaques and tangles in Alzheimer's disease. *Acta Neuropathol.* 1987; 73: 1-11.

Tanwar MK, Gilbert MR, Holland EC. Gene expression microarray analysis reveals YKL-40 to be a potential serum marker for malignant character in human glioma. *Cancer Res.* 2002; 62: 4364-8.

Tanzi RE. The synaptic Abeta hypothesis of Alzheimer disease. *Nat Neurosci.* 2005; 8(8): 977-9.

Tator C, Bray G, Morin D. The CBANCH report--the burden of neurological diseases, disorders, and injuries in Canada. *Can J Neurol Sci.* 2007; 34: 268-9.

Tschape JA, Hartmann T. Therapeutic perspectives in Alzheimer's disease. *Recent Patents CNS Drug Discov.* 2006; 1: 119-27.

Ueno H, Nakamura F, Murakami M, Okumura M, Kadosawa T, Fujinag T. Evaluation effects of chitosan for the extracellular matrix production by fibroblasts and the growth factors production by macrophages. *Biomaterials* 2001; 22: 2125-30.

Ueno H, Murakami M, Okumura M, Kadosawa T, Uede T, Fujinaga T. Chitosan accelerates the production of osteopontin from polymorphonuclear leukocytes. *Biomaterials* 2001; 22: 1667-73.

Ueno H, Mori T, Fujinaga T. Topical formulations and wound healing applications of chitosan. *Adv Drug Deliv. Rev.* 2001; 52: 105-15.

Vassar R, Bennett BD, Babu-Khan S, Kahn S, Mendiaz EA, Denis P, Teplow DB, Ross S, Amarante P, Loeloff R, Luo Y, Fisher S, Fuller J, Edenson S, Lile J, Jarosinski MA, Biere AL, Curran E, Burgess T, Louis JC, Collins F, Treanor J, Rogers G, Citron M. Beta-secretase cleavage of Alzheimer's amyloid precursor protein by the transmembrane aspartic protease BACE. *Science* 1999;286: 735-41.

Walsh DM, Selkoe DJ. A beta oligomers - a decade of discovery. *J Neurochem.* 2007; 101: 1172-84.

Webber KM, Perry G, Smith MA, Casadesus G. The contribution of luteinizing hormone to Alzheimer disease pathogenesis. *Clin Med Res.* 2007; 5: 177-83.

Wilkins RH, Brody IA. Alzheimer's disease. *Arch Neurol.* 1969; 21: 109-10.

Wyss-Coray T, Mucke L. Inflammation in neurodegenerative disease--a double-edged sword. *Neuron* 2002; 35: 419-32.

Wyss-Coray T, Lin C, Sanan DA, Mucke L, Masliah E. Chronic overproduction of transforming growth factor-beta1 by astrocytes promotes Alzheimer's disease-like microvascular degeneration in transgenic mice. *Am J Pathol.* 2000; 156: 139-50.

Wyss-Coray T, Lin C, von Euw D, Masliah E, Mucke L, Lacombe P. Alzheimer's disease-like cerebrovascular pathology in transforming growth factor-beta 1 transgenic mice and functional metabolic correlates. *Ann N Y Acad Sci.* 2000; 903: 317-23.

Wyss-Coray T, Yan F, Lin AH, Lambris JD, Alexander JJ, Quigg RJ, Masliah E. Prominent neurodegeneration and increased plaque formation in complement-inhibited Alzheimer's mice. *Proc Natl Acad Sci U S A.* 2002; 99: 10837-42.

Xu H, Ito T, Tawada A, Maeda H, Yamanokuchi H, Isahara K, Yoshida K, Uchiyama Y, Asari A. Effect of hyaluronan oligosaccharides on the expression of heat shock protein 72. *J Biol Chem.* 2002; 277: 17308-14.

Yan SD, Yan SF, Chen X, Fu J, Chen M, Kuppusamy P, Smith MA, Perry G, Godman GC, Nawroth P et al. Non-enzymatically glycated tau in Alzheimer's disease induces neuronal oxidant stress resulting in cytokine gene expression and release of amyloid beta-peptide. *Nat Med.* 1995; 1: 693-9.

Yates CM, Butterworth J, Tennant MC, Gordon A. Enzyme activities in relation to pH and lactate in postmortem brain in Alzheimer-type and other dementias. *J Neurochem.* 1990; 55: 1624-30.

Yatin SM, Varadarajan S, Butterfield DA. Vitamin E Prevents Alzheimer's Amyloid beta-Peptide (1-42)-Induced Neuronal Protein Oxidation and Reactive Oxygen Species Production. *J Alzheimers Dis.* 2000; 2: 123-131.

Zilka N, Novak M. The tangled story of Alois Alzheimer. *Bratisl Lek Listy* 2006; 107: 343-5.

In: Binomium Chitin-Chitinase: Recent Issues
Editor: Salvatore Musumeci and Maurizio G. Paoletti
ISBN 978-1-60692-339-9

Chapter XII

Chitin and Chitinase in Anticancer Research

Xing Qing Pan[13]
College of Pharmacy, the Ohio State University,
500, West 12th Ave. Columbus, OH 43210, USA

Abstract

Cancer is a serious disease of human beings. So far satisfactory treatment and method of prevention are lacking. Many papers about the action of chitin, chitosan, and chitinase against cancer have been published. In this chapter, we are trying to collect related information from different areas, and the emphasis will be on the potential roles of chitin and chitinase in anticancer therapy. Much of the date reported in those papers is preliminary, many of the studies even had been done before the chitinase in human and mammalian animals being proven, and some of the discussions do need to have further direct evidences. Indeed, for further development in this important area, systematic studies are urgently needed. However, the results so far obtained are suggesting that development of low toxic anticancer treatment and preventive method from the study of chitin and chitinase is possible.

1. Introduction

Many cancer treatments are highly toxic and clinical results are often poor. New methods of treatment and prevention are required. On the other side, cancer regression occurred after some special bacterial infections has been recorded, although the reason unknown. Interestingly, those bacteria, which caused the infections in people whose cancer regressed, are strong chitinase producers, and many of them have been used for the commercial chitinase preparation.

13 E-mail: pan.67@osu.edu or pan_xingpan@yahoo.com.

It has also been reported that the injection of cancerous mice with chitin or water-soluble short chain chitin oligosacharides can induce inhibition of the cancer growth, while those materials do not show anticancer effects when they are directly added into cancer cell culture medium.

The special relationship between cancer growth in animal model and fungi infection, especially the involvement of N-acetyl-chitohexaose suggests that chitinase which produced by activated host animal macrophages may be involved in these anticancer actions.

Furthermore, the substantial anti-bladder cancer effect of BCG treatment, and the result of the activation of macrophages by Bacillus Calmette-Guerin also suggest the importance of macrophages in these anticancer actions.

Commercial bacterial chitinases and a recombinant chitinase from plant have been found to be able to selectively lyses cancer cells but not normal cells in culture, and the lyses of solid tumor and the cure of the cancer carrying mice after injection of commercial bacterial chitinase productions have also been reported.

In this chapter, we are going to discuss this information and related it to the role of chitinase in the damaging of cancer cells. The potential possibility of developing low toxic anticancer treatment and prevention through the activation of human own defense system will also be discussed.

2. Cancer Regression Induced by Special Bacterial Infections

Spontaneous tumor regression after bacterial or fungal infection has been reported in the history (Lin *et al.* 2004). The inoculation of bacteria into cancer patients also has been reported to be partly successful as an anticancer procedure (Ryan *et al.* 2005). People commonly consider the bacterial infection is related to antibody or other components of the immune systems, but in this chapter we are going to propose that the possible mechanism of this cancer regression is directly, or at least mainly related to the actions of chitinase and protease, which produced by those bacteria, or by the activated host macrophages.

2.1. Infection with the Bacteria Streptococcus and Serratia

The use of bacteria to treat cancer patient was reported by Dr. Willian Coley early in nineteen century, although at that time the mechanism was unknown. He noticed that a sarcoma patient with metastasis unexpectedly recovered after a serious *Stretococcus pyogenes* infection. As a result of this observation, he inoculated cancer patients with live *Streptococcus pyogenes* and *Serratia marcescens*. The reaction to the infection was strong, sometimes even difficult to control, but the cancer of the patients did show regression (Coley, 1893). He also used a heat-treated *Streptococci pyrogenes* and *Serratia marcescens* mixture as a vaccine to treat cancer patients. A typically example was an inoperable sarcoma patient, with sarcoma in the abdominal wall, pelvis and bladder. This patient recovered completely after the treatment, and the patient died from heart attack 26 years later. Since then, this treatment was widely used on other cancer patients with lymphomas, melanomas and myelomas (Hoption, 2002).

Looking back today, this method possibly would not be acceptable to most of the cancer patients, because of the fever, chills, inflammation and other side effects, but it did give us evidences that bacterial infection and unknown mechanism in our body can cure the cancer.

Noticeably, both *Streptococcus pyogenes* and *Serratia marcescens* are strong chitinase producers. Both of them can grow with the use of natural form chitin in the shells of crab or shrimp as their only carbon resource. *S. marcescens* produces a number of characteristically different chitinases, such as Chi A, Chi B, Chi C_1 and Chi C_2 etc., all of them belong to chitinase family 18. Only recently, it has been reported that bacteria also produce some family 19 chitinases (Suzuki *et al.* 2002). Family 18 chitinases and family 19 chitinases are quite different in their processing of the chitin digestion. The family 18 chitinases use a substrate-assisted catalysis mechanism, while family 19 chitinases use a general acid-and-base mechanism (Kawase *et al.* 2006). Some family 18 chitinases need other factors to join in their digestion of the natural chitin, such as "chitin-binding protein", such as CBP21, to bind onto chitin and change the natural chitin structure first, and then the chitinase can react to the exposed chitin molecules (Vaaje-Kolstad *et al.* 2005). The complication is that the levels of the "chitin-binding protein" dependence of different chitinases are different. For example, the activity of chitinase G is totally depending on the presence of the "chitin-binding protein", while the activity of chitinase B and C are partly dependent, and chitinase A is only slightly dependents.

Besides, in natural situations, chitin molecules commonly associate or bind with proteins and some other materials, and then the presence of some proteases may also be needed for chitinase to react on chitin. It has been found that *Streptomyces griseus* produces a special kind of "chitin-binding protease". This chitin-binding protease contains a chitin binding domain and an active serine protease center. With the help of this kind of "chitin-binding protease", the digestion of the natural chitin by these bacterial chitinases becomes much easier (Sidhu *et al.* 1994). Similar "chitin-binding protein" and "chitin-binding protease" also found in *Alteromonas sp.* (Miyamoto *et al.* 2002). So, it is possibly not only a single case.

In addition, Radwan *et al.* (1994) reported that the chitinase produced by bacteria *Streptomyces olivaceoviridis* originally was a 92 kDa protein. At its N-terminal, there was a 22 kDa sized serine protease domain. This 92 kDa chitinase protein cut off itself and formed a 70 kDa sized chitinase and a 22 kDa sized protease. In this way, the protease and chitinase always presented at same location and same time. We do not really know whether similar process also happens to other bacterial chitinases or not. But in many cases, when a family 18 chitinase reacts with natural chitin, proteases commonly present together.

Because most bacterial chitinases and our human macrophage chitinases are belonging to family 18, when we consider their action against cancer or some other things else, we should know the action needs other factors. Commonly, recombinant bacterial or human chitinase which produced by gene expression in *Escherichia coli* or other suitable cells may show low activity or even no activity on natural chitins. However, after combined with suitable chitin-binding protein and proteases, the activity of recombinant chitinase can be re-built up. But, most of the plant chitinases, which belong to chitinase family 19, they react to natural chitin directly because of the much simpler and direct mechanism (Roberts *et al.* 1988).

Another situation, which confusing us quite often is that in many chitinase activity assay, we use commercially synthesized poly-N-acetylglucosamine or chemically treated chitin, such as glycol chitin, fluoresce labeled short chain chitin or colloidal chitin etc. as substrates. We need to realize that the treatment of chitin, even with only β-mercaptol-ethanol or hot SDS, will denature the protein or other parts of the natural chitin materials, and then the exposure of the chitin molecule become possible to be digested by the chitinase.

Commonly, the protein of chitinase consists of a chitin-digestive center domain, a chitin-binding domain, and a hinge domain in between. At the time when we use human synthesized short chitin substrates, the chitin-digestive center can show its activity directly, no matter actually the chitinase has lost its chitin-binding domain in their structure or not. But, in the reaction to natural chitin, the chitin-binding domain is necessary, and as we mentioned above, that for family 18 chitinases the reaction to the natural chitin even require more. Then, to the natural chitin situation, those chitinase with no chitin-binding domain will show no activity, although in the test tube with human made short chain chitin kind of substrates, they do show chitinase activity. For example, the human macrophage cell 39 kDa chitotriosidase shows high digestive activity with artificial substrates, but in the natural situation, such as antifungal action, also possibly in the anticancer function, it can only show no activity. Therefore, tests with the use of artificial chitin substrates, the positive result just only tell us that there is a chitinase active center, but not to tell if this chitinase is functional to the natural chitin or not (Renkema *et al.* 1997).

2.2. Infection by Clostridium

Clostridium is an anaerobic, mesophilic chitinolytic bacterium, it can use natural chitin in the shells of crab or shrimp as their only carbon resource to grow and produce hydrogen gas as one of their productions (Evyernie, *et al.* 2000). In 1813, Vautier noticed that a cancer patient recovered from the cancer after infection of *Clostridium perfringen*. Busch (1868), and Fehleisen (1883) also reported separately that this bacterium selectively targeted solid tumors and growing specially inside the solid tumor tissue, at there it induced damage of the tumor (Pawelek, *et al.* 2003). Mose *et al.* (1964) injected the spores of a nonpathogenic strain of *C. butyricum* M-55 into the patients intravenously as a treatment of solid Ehrlich carcinomas. This treatment induced lyses of the tumor tissue. Grick *et al.* (1979) used a similar method to treat melanomas in mice; they found it worked fine, and if it combined with local irradiation and hyperthermia, significant enhancement in the survival rate could be obtained.

It has been noticed, that after intravenous injection of spores of the clostridium, the bacterial cells could only grow in the necrotic areas of the tumors. Oncolysis occurred only in the tissue of large tumors, not in the normal tissues, which surround the tumor. Besides, it was also noticed that the oncolysis did not occur in small solid tumors. Also, sometimes cancer cells survived in the rim area of the solid tumor (Lambin, *et al.* 1998). Thus, it was kind of complicated, and which was not understandable at that time.

At present time, we have known that the blood vessels in solid tumors are broken, and there are holes in the solid tumor blood vessel walls. The spores of clostridium just like the nano-particles since they are small. They can penetrate through the holes of the broken blood vessels of the cancer tissue and enter the tumor tissue (Pan, *et al.* 2004). In the necrotic areas, the oxygen supply is low, which permits the anaerobic bacterium to grow and produce chitinase and protease as well as other productions locally (we will discuss the anticancer function of chitinase later in this chapter). In the outer rim areas of solid tumor tissue, the oxygen supply is higher, which comes from the tumor blood vessel system and also the oxygen diffusion from the surrounding normal tissue. Thus, in the rim areas the anaerobic bacteria cannot grow very well and they produce less chitinase and protease. So, under this situation, some of the cancer cells may survive, and start re-growing after the bacteria are eliminated. In small solid tumors, the blood vessel systems are just built up, they are not yet broken, and therefore the spores of the clostridium have no way to get into the small tumor tissue. Besides, in the small size solid tumor, the oxygen penetration from surrounding normal tissue is more, which also inhibits the growth of the anaerobic bacteria. Thus, here we can see that the anticancer function of the spore injection treatment is closely related to the pattern of the bacterial growth, and possibly related to bacterial activity and the amount of its production.

Dang *et al.* (2001) combined the spore injection with the injection of the blood vessel damaging drugs, such as Dolastatin-10 etc. When the blood vessels in the solid tumor collapsed by the anti-blood vessel drug, more holes formed in the blood vessel walls. Meanwhile, the oxygen supply from the blood stream became lesser; those all benefited the growth of the anaerobic bacteria. Thus the combination improved the anticancer efficiency.

In addition of produce chitinase, Clostridia also produce a number of proteases. It would be meaningful to check the roles of chitinase and proteases, as well as the joint action of chitinase and protease in the lyses of the cancer tissue. Nevertheless, the side production, hydrogen gas, probably will be a cumber for the use of this bacterium. Besides, after the tumor damaged, the release of large amount of lysed tumor tissue and the bacterial cells will be a potential problem in some cases too.

2.3. The Stimulation of *Bacille Calmette Guerin* (BCG)

The use of *Bacille Calmette Guerin* has had substantial success in the treatment of early stage bladder surface cancer (Bevers, *et al.* 2004).

Distefano *et al.* (1983) pointed out that in the anticancer action of BCG, two materials were involved, one of them possibly was the H_2O_2, and another one was an unknown protein. Seya et al (2001) believed that it was the cell wall of BCG induced an innate immune response, which acted on the cancer.

Bucana et al (1976) transplanted hepatocarcinoma cells into syngeneic strain guinea pigs, and then injected living BCG cells into the tumor. After that they collected the macrophages from the cured guinea pigs and observed them in culture. They found that the BCG activated macrophages showed obviously different behavior from the regular macrophages, which obtained from the normal guinea pigs with no BCG treatment. The BCG activated macrophages had a great tendency to spread on glass, and these elongated cells often had a webbed appearance. When these activated macrophage cells met tumor cells, they attached to the tumor cell surface and extended their cytoplasm containing granules on to the tumor cell surface, or even transported the granules into the cytoplasm of the cancer cells. Actually, in 1974, Hibbs also reported that BCG activated macrophage cells transferred a number of granules into tumor cells, which is not done by the macrophage cells obtained from normal mice.

In a review, Patard et al (2003) suggested to use of viable BCG cells to treat cancer, he believed the viable BCG cell likely be able to give better anticancer results. Noticeably, Bacillus is also a chitinase producer, and it has been reported that the chitinase that produced by Bacillus processes antifungal function (Chang *et al.* 2007). But, like it happened to the other bacteria, after high degree purification, this chitinase lost its antifungal function (Melen'tiev *et al.* 2001), which is suggesting that this chitinase probably is a family 18 chitinase, the activity against natural chitin required other factors to join together to act on natural chitin materials (we will discuss the relationship of antifungal function and anticancer function again in part **3**).

Adams (1980) did a set of impressive observations on the anticancer activity of macrophages in vitro, these macrophages were collected from the peritoneal cavities of C57BL/6J mice after intraperitoneal injection of thioglycollate broth, concanavalin A (for mimic inflammatory), or viable BCG cells. He cultured these macrophages in serum free Nearen and Tytell Gibco medium for 2 days, and then the conditioned medium samples were used for observation of their anticancer activity. It was found that only the medium conditioned by the macrophages, which were collected from BCG stimulated mice showed strong anticancer activity against six different kinds of mouse cancer cell lines. However, this conditioned medium showed no effect against cells of two normal cell lines. The macrophages collected from thioglycollate or Con A stimulated mice showed very low or no effect on the cancer cells, neither on normal cells. Very interestingly, when they added protease inhibitors, such as 2 x 10^{-3} M diisopropyl-fluorophosphate (DPF) to the medium conditioned by BCG activated macrophages, inhibited the activity of the medium against cancer cells by about one third. But, further increasing the concentration of DPF did not further decrease the activity of the conditioned medium against cancer cells. In another paper published by the same research group, they confirmed that there was an unknown cancer cell specific cytolytic material in the conditioned medium; to it they named "CF", which was produced by BCG stimulated mouse macrophages at the same time when the activated macrophages were producing protease. When the anticancer activity of the protease inhibited by DPF, the anticancer activity of CF still showed, this was not inhibited by DPF. Further more, they proved that CF was a protein, which was synthesized by the activated macrophages after the BCG stimulation. If heated the macrophages at 56^{0} C for one hour, or a protein synthesis inhibitor cycloheximide was added into the Neuman and Tytell culture Medium during the two days incubation, the CF no longer produced.

3. Fungal Infections and the Inhibition of the Cancer Cells

3.1. Infection of *Aspergillus Fumigatus*

Aspergillus fumigatus is a saprophytic fungus; it synthesizes its own chitin and chitinase and can be grown in culture medium that contains 1% chitin as its sole carbon source (Escott, *et al.* 1998).

Intravenous injection of *A. fumigatus* into guinea pigs induced an increasing chitinase activity in the serum and tissues, while the activity of β-hexosaminidase increased very little. The activity of lysosomal α-mannosidase, β-galactosidase and β-glusosidase showed no increase. It was proved that this chitinase came from the host animal not from the fungi. The authors speculated that the mammalian chitinase was a defense mechanism against chitin-containing pathogens (Overdijk *et al.* 1999).

Repentigny *et al.* reported (1993), that *A. fumigatus* cell wall contains α- and β- (1, 3)-glucans, chitin, galatomannan and β- (1, 3), and (1, 4)-glucan, and these compounds stimulated the host animal immune system, and which mediated by macrophages.

Noticeably, Tzankov *et al.* (2001) reported that an adult acute myelogenous leukemia (AML) patient underwent a spontaneous remission after a severe invasive pulmonary aspergillosis infection. The 60-year-old female patient was only given treatment with antibiotics and granulocyte colony stimulating factor (GCSF).

3.2. Infection of *Candida Albicans*

Candida albicans is a serious agent of infection. Approximately 80-90% of the *C. albicans* cell wall is carbohydrate, including β-1, 3-glucan, β-1, 6-glucan, chitin, and polymer of mannose (mannan). All of them covalently associated with proteins. Chitin is a minor component of the wall (0.6-9 %). *Candida albicans* produces chitinase, which is inhibited by allosamidin (Dickinson, *et al.* 1989).

Similar to what occurred following the *A. fumigatus* injection in guinea pigs, when newborn children were infected with *C. albicans*, the human chitotriosidase activity in the plasma and urine increased, after the infection was cured, the chitotriosidase level returned to normal (Labadaridis, *et al.* 1998).

Interestingly, mice injected with chitin or chitosan intraperitoneally, became resistant to *C. albicans* infection. If the challenge of *C. albicans* was through intravenous injection, the chitin treated mice showed higher resistance, and the number of polymorphonuclear leukocytes in blood increased obviously. However, if the challenge of *C. albicans* was by intraperitoneal injection, then the chitosan treated mice showed higher resistance to the infection. This time the number of activated peritoneal exudates cells was larger in chitosan treated mice (Suzuki, *et al.* 1984).

Kobayashi *et al.* noticed (1990) that the mice which carried big sarcoma tumors showed a distinct ability to resist *C. albicans* infection, but those mice which bearing small tumor did not. However, after these small tumors carrying mice received an injection of N-acetylchitohexaose, they also became resistant *C. albicans* infection. More important was that after N-acetylchitohexaose injection, the mice not only built up the resistance to *C. albicans* infection, and the tumor growth was also partly inhibited.

Repentigny *et al.* proved (1993) that the activated macrophage was the key to the inhibition of *C. albicans* infection. In tumor bearing mice, an unknown macrophage-activating factor showed up, which let the macrophages became sensitive to N-acetylchitohexaose, But, in normal mice, without sarcoma, N-acetylchitohexaose injection alone did not protect the mice from *C. albicans* infection.

Actually, Robinette *et al.* (1975) and Robert *et al.* (1976) already reported that Lewis lung carcinoma bearing mice showed resistance to *C. albicans* infection. They pointed out that the macrophages were the source of resistance, while neutrophils and mononuclear phagocytic cells might be also involved. Gamma interferon and macrophage colony stimulating factor enhanced the phagocytic activity, while tumor necrosis factor, interleukin-1 (IL-1), IL-3 and IL-4 did not (Torres *et al.* 1997).

Similar to Adams, who observed the BCG activated macrophages conditioned medium against cancer cells, as it was mentioned in part **2.3.** Suzuki *et al* (1987) observed the candidacidal activity of the conditioned medium from chitin activated microphage cells. Those macrophages were collected from mice that had been injected with chitin intraperitoneally. They cultured the chitin activated macrophages in RPMI 1640 medium with 10% fetal bovine serum for 24 hours, and then the conditioned medium were tasted for candidacidal activity. They found that the candidacidal activity of the conditioned medium was increased two fold higher than that of the medium conditioned by control macrophage cells, which obtained from normal mice without chitin treatment. Interestingly, in this observation, Suzuki also found that the addition of the protease inhibitor diisopropylfluorophosphate (DFP) at a concentration over 0.1 mM caused a partial inhibition of the candidacidal activity. Also, further increase the DFP concentration up to 1.0 mM, did not enhance the inhibition more. These results suggested that protease was also involved in the candidacidal activity of the macrophages, and there were something else unknown also produced by the activated macrophage s at same time.

In the anticancer observation of Adams, an anticancer protein CF was produced by BCG activated macrophage cells together with protease, and here in Sazuki's candidacidal observation, protease and "something unknown" were produced by chitin-activated macrophages. These observations together with other related phenomenon possibly support the speculation that the protein CF detected in Adams experiment is the same thing, which brought the cancidacidal action in Sazuki's candidacidal observation. Unfortunately, so far we still do not have direct evidences to say so, although at present time we have known that the activated macrophages are capable of producing both chitinase and protease, which actually is an important part of our innate immune defending system.

Stevens *et al.* (2000) expressed two 50 kDa sized recombinant human chitotriosidases in Chinese hamster ovary cells with their gene DNA obtained from human gene library. They found that with the use of fluorescence poly-N-acetylglucosamine as a substrate, these two recombinant chitotriosidases showed high level of chitinase activity, but both of them showed no anti-fungal activity. However, when these recombinant human chitotriosidases were used in combination with chemical anti-fungal drugs, a synergistic function showed up. These results suggested that that possibly some thing, such as "chitin-binding protein" or protease, are not presenting, therefore the 50 kDa recombinant chititriosidases could not show their activity against the natural chitin in fungal cells, although these chitinase even still have their chitin binding domain. In the situation of combination with anti-fungal drugs, the chitin compounds of the fungal cells possibly were no longer normal, which allowed the recombinant chitotriosides to react on, and then the synergistic antifungal function showed up. But as the same, the direct evidences are also not available yet.

In plants, chitinase 19 family chitinases are the main defending system for inhibition of fungi invading. Different to bacterial and mammalian macrophage family 18 chitinases, the plant family 19 chitinases react to natural chitin directly with different mechanism (Kawase *et al.* 2006). However it was proved, that the antibody of plant chitinase inhibited the chitinase activity, also blocked the anti-fungal defending activity (Schlumbaum *et al.* 1986). In part **6.2.** We will see the recombinant plant chitinase TYchi directly showed anticancer activity in culture (Xu *et al.* 2008).

4. Inhibition of Cancer Cells by Chitin, Chitosan, and Poly N-Acetylglucosamine Injection

While chitin type materials were being tested against fungal infection in animal models, these carbohydrate compounds were also tested for anticancer function (Rheinnecker, *et al.* 1996). Results showed that the injections of chitin or chitosan or short oligosaccharides prevented the growth of transplanted cancer cells in mice. Suzuki *et al.* (1986) observed the anticancer effectiveness of the polysaccharides from tetra-N-acetylchititetraose to hepta-N-acetylchitoheptaose. Their results showed that the N-acetyl-chitohexaose (hexamer of chitin oligosaccharide) and Chitohexaose (hexamer of chitooligosaccharide) were the most effective compounds. Intravenous injection of 100 mg/kg of these compounds for five times inhibited transplanted sarcoma 180 growing in all the mice. So as to the MM-40 and Meth-A solid tumors (Tsukada *et al.* 1990). In 2003, the same group further reported that intraperitoneal injection of chitin or chitosan showed strong stimulation on resistance of *Listeris monocytogenes* and *Pseudomonas aeruginosa* infection, through the activation of the peritoneal exudate cells (Okawa *et al.* 2003).

Qin et al (2002) also reported that the intraperitoneally injection of water-soluble short chain chitin or chitosan compounds inhibited sarcoma 180 growth in mice. The maximum inhibitory rate was 64.2%. However, in their observation, the results showed that if the molecule size of the compound was bigger than hexamer, the anticancer activity was better.

In Roby's report (1987), that in plant melon, 12-24 hours after the injection of chitin kind oligosaccharide, the chitinase activity increased 3-5 times.

Tokura et al (1999) reported that the natural big chitin molecules had a rigid crystalline structure, established by inter- and intra-residual hydrogen bonds between acetyl amino groups and hydroxyl groups. This kind of natural chitin did not directly activate peritoneal macrophages, but 70% deacetylated chitosan, CM-chitin and hydroxyethyl-chitin did activate the macrophages after they were injected intraperitioneally to the mice.

Lately, a lot of researches are using chitosan as a carrier to transport anticancer drugs or genes to cancer cells (Yao, *et al.* 2007). Chitosan is a chitin like compound; it also processes some chitin activity (Tanioka, *et al.* 1995). Since chitosan is also a big molecule of polyamine compound, it will disturb the cell membrane (Fischer, D, *et al.* 2003). The multiple free NH_2 groups in a big molecule allow chitosan to associate with different materials and works as a carrier. When the size of chitosan particles is small enough along 100 nm in diameter, this nano-particle of chitosan can penetrate into solid tumor tissue through the holes of the broken blood vessel in solid tumor tissue (Pan *et al.* 2004). Thus, chitosan particles can be used as a carrier to transport nucleic acids, proteins or other materials such as drugs into the solid tumors.

5. Some Other Kinds of Materials Can Also Stimulate the Innate Immune System and Activate Macrophages

Muramyl acid (3-O-D-lactyl ether of D-glucosamine) and its dipeptide derivative N-acetylmuramyl-L-alanyl-D-isoglutamineis are important components in mycobacterial cells, which have been used in Freund's adjuvant. Human blood monocytes or macrophages commonly are not cytotoxic to tumorigenic cells in vitro, but after exposure to muramyl tripeptide-PE containing liposomes, macrophages will be activated and behaving like BCG activated macrophage cells. Those activated macrophage cells rapidly contact tumor cells, damage tumor cell membrane, and possibly transfer some granule form materials into cytoplasm of the tumor cells (Bucana *et al.* 1983).

Mannan is a compound of the yeast cell wall, its mode of action is not clear, but it has been reported capable of activate macrophages and inhibit the growth of tumor cells (Hashimoto *et al.* 1983). B-glucans and related compounds have been used in animal research, and in cancer patient treatment (Demir, *et al.* 2007). The mechanism of the action possibly is also related to macrophage activation.

Bacterial lipopolysacchrides and interferon-γ have been used in culture to activate macrophages. Haewix *et al.* (1992) pointed out that only the mature macrophages produced a high molecular weight protein in culture medium after stimulation, but the freshly isolated human peripheral blood lymphocytes did not. Russell *et al.* (1977) reported that macrophages obtained from cancer bearing mice commonly did not show obvious anticancer activity, but actually the growth of the cancer did give some stimulations to macrophages, although it was not enough to make macrophage cells to form tumoricidal. Russell *et al.* found that if stimulate those macrophages which collected directly from solid tumor tissue with very low level (such as 1-10 ng/ml) purified bacterial lipopolysaccharide (LPS), these stimulated macrophages became strong anticancer cells. The anticancer activation occurred in 4 hours. But, to normal macrophages which obtained from normal mice, much higher concentrations of LPS (such as 100 μg) and longer times (6-8 hours) were needed to show a low level of anticancer activity.

6. The Anticancer Activity of Chitinases

6.1. The Anticancer Activity of Commercial Bacterial Chitinases

Radiotherapy and surgery are the main methods for cancer patient treatment at present time. Although more than 1500 anticancer agents have been studied, and over 500 chemical compounds have been tried in clinics, to date cancer chemotherapy have not yet been entirely satisfactory. The high toxicity is the most common problem in cancer chemotherapy. Thus, study on the systems in cancer cells, which are different to the normal cells possibly will be a better way for further developing. Based on this idea, a number of materials were tested in vitro and in cancer bearing mice. Since many cancer cells contained unusual higher level of N-glucosamine, therefore chitinase and related reagents were tested. The results were encouraging; among the tested materials the selective anticancer function of chitinase was most attractive (Pan *et al.* 2001 and 2005).

6.1.1. Selective Anticancer Cell Activity of Commercial Bacterial Chitinases Sample in Culture

Human breast cancer MCF-7 cells were cultured in pH 7.4, D-MEM/F-2 medium contained with 10% preheated fetal bovine serum and antibiotics. Human oral squamous carcinoma KB cells, mouse sacoma cancer 24JK cells and mouse melanoma B16 cancer cells were cultured in RPMI 1640 medium system contained with 10% preheated newborn calf serum and antibiotics. It was found, that 0.50 units/ml concentration of bacterial chitinase (Sigma C6137, produced by *Streptomyces griseus*) selectively lysed MCF-7 cells, KB cells, 24JK cells and B16 cells in 12 hours. However, under the same condition, the normal B6D2F1 mouse spleen cells showed still healthy and unchanged (Pan *et al.* 2005).

In 72 hours observation system, the cell activity was assayed by MTT methods (Mlynarcyuh *et al.* 2001), it was found that 0.125 unit/ml chitinase inhibited the cancer cell activity. When checked with methods of microscopy or electronmicroscopy, the cancer cells showed various levels of morphological change if the concentration of chitinase over 0.25 unit/ml. Figure 1 shows the results of a MTT assay with KB cells and the normal mouse spleen cells, under different concentrations of chitinase.

6.1.2. The Anticancer Activity of Commercial Bacterial Chitinases in Mice

Intra-tumor injection of 5 units of bacterial chitinase in saline or pH 7.4 PBS solution, induced lyses of human breast cancer, human lung cancer, human colon cancer and human melanoma xenografts in SCID mice. The solid tumors became soft and the color of the tumor turned from pink to dark brown. The tumor size contracted within couple of days. If the solid tumor size was bigger than 0.3-0.5 cm^3, second injection or a third injection might be needed. Finally, the solid tumor became a black crust (see Picture 2). This crust shed off about ten days later. If some alive tumor cells still left there, the solid tumor would grow up again.

Thus, the dosage of chitinase should be given enough to completely destroy the tumor. Commonly for tumor sizes of 0.5 cm^3 or less, two injections were enough.

Three Sigma chitinase samples were tried. Sigma C1525 and C7809 both produced by *Serratia mercescens*, and C6137 was produced by *Streptomyces griseus*. Their anticancer results were similar.

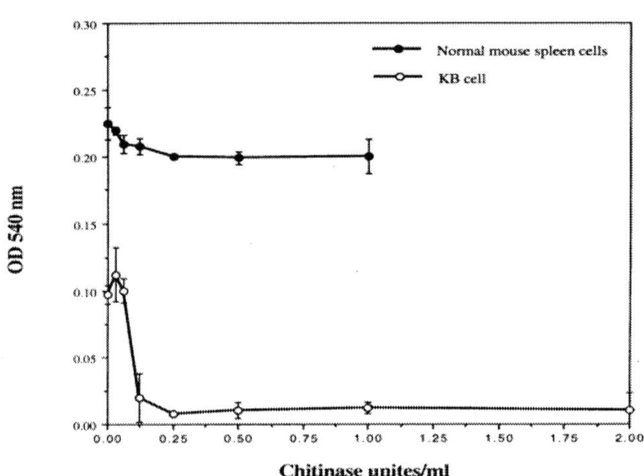

Figure 1. Comparison of MTT assay results of chitinase treated human oral squamous carcinoma KB cells and B6D2F1 normal spleen cells. Cells were cultured in RPMI-1640 medium with 10% preheated newborn calf serum and antibiotics, also with different concentrations of chitinase (Sigma C1525). The cell samples were kept in a regular 37° C, 5% CO2 incubator for 72 hours. Then a MTT assay was run following the literature (Mlynarczuh *et al.* 2001).

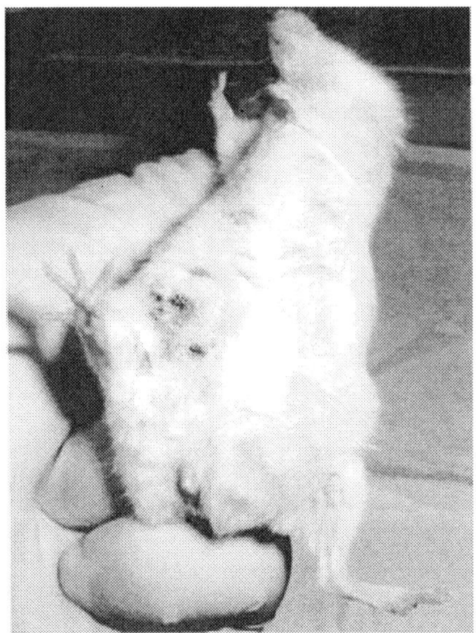

Figure 2. Human colon adenocarcinoma xenograft in SCID mouse. On the left side of the mouse is the untreated control solid tumor. On the right side of the mouse is the tumor, one week after chitinase 5 units x 2 treated.

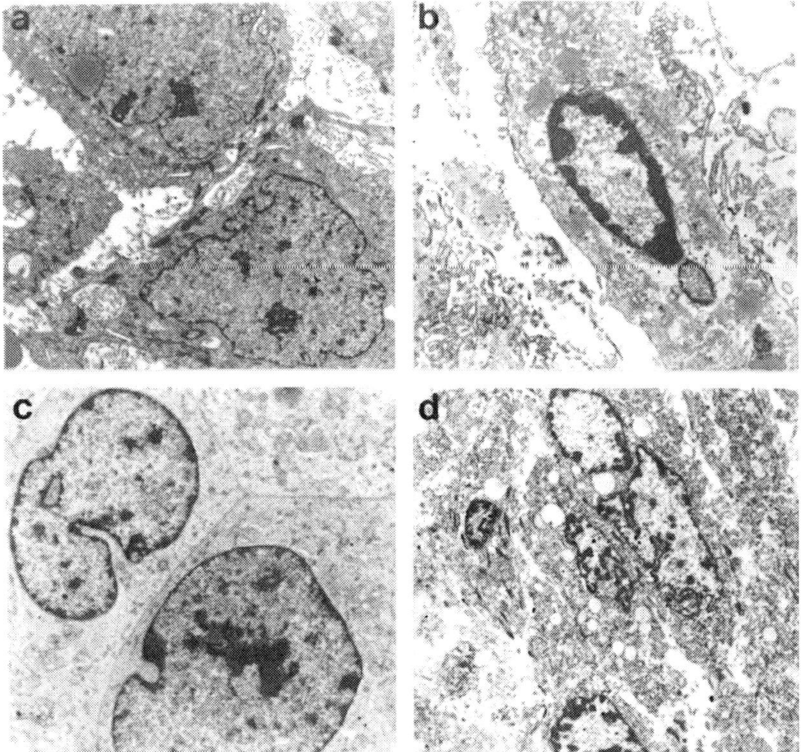

Figure 3. Transmission electronic microscopic observations on chitinase treated human cancer xenograft tissues in SCID mice. (a). Human lung cancer xenograft L1-5 tissue control. (b) Human lung cancer xenograft L1-5 tissue, 24 hours after chitinase 5 unites x 1 intratumoral injection. (c) Human melanoma xenograft M2-2 tissue control. (d). Human melanoma xenograft M2-2 tissue, 24 hours after chitinase 5 unites x 1 intratumor injection.

6.2. Anticancer Function of Plant Chitinase

Xu et al (2008) obtained a recombinant chitinase TYchi from plant *Trichosanthes Kirilowii* Maximowiczanti, it possessed chitinase activity in cell culture, also possessed RNA N-glycosidase activity in a rabbit reticulocyte cell lysate system. In *T. kinilowii* plants the expression of TYchi was a defense reaction, induced by fungal pathogen infection. However, in RPMI-1640 plus fetal bovine serum system, recombinant TYchi showed strong toxicity to the histocyte lymphoma U937 cell at IC_{50} 54 μg/ml, and to choriocarcinoma JAR cells at IC_{50} 73 μg/ml. In a cell-free rabbit reticulocyte lysate system, TYchi inhibited protein synthesis with the IC_{50} at approximately 5 nM.

7. Discussion

At the time we observed the anticancer function of the bacterial chitinase, we also tested some related reagents for comparison under similar condition. Intratumor injection of chitosanase (Sigma C9830) showed good anticancer activity, which was very similar to that of the chitinases. But, lysosome (Sigma L9283), lysozyme (Sigma L6876), lipase (Sigma L3126), esterase (Sigma E0887), type IV DNase (Sigma D5025), and RNase A (Sigma R6513) did not show antitumor activity even at high dosage. Β-N-Acetylglucosaminidase (Sigma A2264) showed very high toxicity to the mice, possibly it reacted to all the cells with no selectivity. But, β-N-Acetyl-hexosaminidase (Sigma A7708) (Woynarowska *et al.* 1992) and β-glucuronidase (Sigma G0251, G7351) showed anticancer effect against human lung cancer xenograft at the concentration lower than the concentration at which the toxicity showed. Recombinant β-N-Acetyl-hexosaminidase (obtained from company V-LBS, catalog number R-1010) also showed anticancer activity. Different proteases such as chymotrypsin (Sigma C4129), trypsin (Sigma T1426) etc. did show some levels of ability to lyse tumor tissue. Collagenase (Sigma C0130) was toxic, which caused death of mice before any anticancer effect was showing. Elastase (Sigma E0127) and proteinase K (Sigma P2308) showed good anticancer function at milligram level, both induced tumor tissue lyses.

Sanders *et al.* (2007) observed the mucolytic activity of chitinase, and they found recombinant *Serrana marcescens* Chitinase A showed no effect and Chitinase B showed a low effect on human cystic fibrosis patient's sputum. But, when Chitinase B mixed with a low concentration of proteinase K, the lytic activity increased obviously, although the same concentration of proteinase K alone only showed low activity. They also found that the commercial Sigma chitinase, which produced by *Serratia marcescens*, showed obviously high mucolytic activity, but there was also protease activity showed in this commercial chitinase production. The commercial chitinase obtained from *Serratia marcescens* culture is a mixture of several chitinases. It is possible these preparation samples contained protease impurities. In addition of producing chitinases, bacteria also produce chitin-binding protease or a number of other protease (Sidhu *et al.* 1994). Some bacterial chitinase even contain protease domain in their protein structure (Radwan *et al.* 1994). Anyway, compare to the anticancer activity of protenase mentioned above, we have to admit that the anticancer activity of commercial chitinase samples may include the actions of those protease impurities. It is possible that because of the presence of the protease impurity, let the anticancer activity of these family 18 chitinase showed stronger. It is also possible, that the commercial chitinase samples may contain low level of chitin-binding protein, such as CBP21 etc. (Vaaje-Kolstad *et al.* 2005), although so far we do not know yet. However, since the plant family 19 chitinase react system is simpler (Schlumbaum *et al.* 1986; Robert *et al.* 1988), the recombinant plant chitinase still keeps its anticancer activity directly (Xu *et al.* 2008).

In this chapter, we tried to put a number of results, which obtained from different areas together. Indeed, it will be much better if we can provide with some direct evidences to support our discussion. Unfortunately, although we have been trying to propose doing so, so far we have not obtained opportunity yet. However, if we compare the results reported in those papers, especially in some cases, such as, *Serratia marcescens* is exactly the bacterium, which has been used to treat cancer patients in history, and now it has been used for the commercial chitinase preparation, while this chitinase sample showed anticancer activity in culture and in mice. In addition to the information from related areas, we believe that it may be reasonable to make such a hypothesis.

If this hypothesis is correct, then three different kinds of chitinase possibly have showed their anticancer activity. One is the bacterial chitinase, together with protease. The second is the chitinase, which produced by the activated macrophages, this will be the most important part of the study in future, since it comes human own defending system, and good for cancer treatment as well as cancer prevention. The third one is the family 19 plant chitinases, which can react to cancer cells directly. Obviously, to each of them, systematic studies are urgently needed.

Macrophage cells naturally have two main jobs, one is protecting host animal body against pathogen invaders, and another one is regulating the tissue remodeling. Commonly, cancer cells do not produce materials, which stimulate macrophages enough to recognize them as enemy cells. So, in cancer tissue, macrophages actually are helping tumors to grow and re-organization (Pollard, 2004). People have tried to stimulate the innate immune system with INF-γ, IL-2 and GM-CSF injections, but only obtained limited success. Possibly, without some other proper stimulations such as chitin kind compounds or other related materials, and let them work together, only INF-γ, IL-2 or GM-CSF injection was not enough (Eijk *et al.* 2005; Hill *et al.* 2002). A lot of work has been done on how to stimulate the Th-1 innate immune response, and a good result has been obtained in the study of leishmaniasis. After proper stimulation, the susceptible mice turned to be resistant to the infection (Bretscher, 1992) that is just what we want in the cancer treatment and prevention.

In the report of Xu *et al.* (2008), chitinase inhibited the RNA N-glycosidase system at much lower concentration. On the other side, Bucana (1976), Hibbs (1974), and other scientists have reported that the activated macrophage can directly transfer granules into cancer cells. If granules really can be transported from the activated macrophage cells into the cytoplasm of the cancer cells, then chitinase possibly be able to get into the cytoplasm of the cancer cells. In this case, chitinase could react to the RNA N-glycosidase system directly, and then the killing of the cancer cell will be much more effective.

So far we still do not really know what kind of targets on cancer cells are attacked by the chitinase yet, including the cooperation of the proteases. As we know that cell tumorigenesis is accompanied with the changes of gene expression and post-translational modifications, which resulted in the changes of the cellular phenotype and membrane composition. The common changes in tumor cells have showed in the high molecular weight glycoproteins, proteoglycans and glycosphingolipids, cancer cells contain several times more glycosaminoglycans than that of the normal cells, and the structures of cancer related structures are also different from those of the normal cells. The glucosamine rich polymers such as cell surface mucin, heparan and hyaluronan etc. are playing important roles in the life and function of cells (Iozzo, 1985, Dannis, 1992, Ohyama, 2008, Stern, 2008). These materials directly or indirectly related to N-acetyl-glucosamine and proteins. Although their structure is not really a chitin, but we will not be surprised to see that chitinase and protease, or they join together to act on to these structures, and then kill the cancer cells.

Lately, Ujita *et al.* (2003) reported the recombinant chitin-binding domain of human macrophage chitinase is specific for chitin, not glucan, xylan or mannan. But this binding site did bind to hyaluronan and hybrid type N-linked oligosaccharide chains, while chondroitin not. Also, the binding was only inhibited by N-acetylglucosamine or di-N-acetylchiobiose. These results are supporting the hypothesis that chitinase possibly cross-reacting to those N-acetylglucosamine containing polymers on the cancer surface.

Conclusion

The potential values of chitin, chitosan and chitinase in cancer prevention and treatment have been discussed in this chapter. Although some parts are still not yet clear, and much information is still lacking, but the possibility of using related compounds to activate the macrophages, stimulate the innate immune system, and let human own defending system to selectively kill cancer cells is apparent. The use of microbes or their chitinases, and the plant family 19 chitinase in the treatment of cancer is also important subjects to be studied further.

Reference

Adam DO, Kao K, Farb R, Pizzo SV. Effector mechanisms of cytolytically activated macrophages. *J Immunol.* 1980; 124: 293-300.

Bevers RF, Kurth KH, Schamhart DH. Role of urothelial cells in BCG immunotherapy for superficial bladder cancer. *Br J Cancer*. 2004; 91: 607-612.

Bhattacharya D, Nagpure A, Gupta RK. Bacterial chitinases: prpperties and potential. *Critical Rew Biotechnol*. 2007; 27: 21-28.

Bretscher PA, Wei G, Menon JN, Ohmannt HB. Establishment of stable, cell-mediated immunity that makes "susceptible" mice resistant to leishmania major. *Science.* 1992; 257: 539-542.

Bucana C, Hoyer LC, Hobbs B, Breesman S, McDaniel M, Hanna MG. Morphological evidence for the translocation of lysomal organelles from cytotoxic macrophages into the cytoplasm of tumor target cells. *Cancer Res.* 1976; 36: 4444-4458.

Bucana CD, Hoyer LC, Schroit AJ, Kleinerman E, Fidler IJ. Ultrastructural studies of the interaction between liposome-activated human blood monocytes and allogeneic tumor cells in vitro. *Am J Pathol.* 1983; 112: 101-111.

Chang WT, Chen YC, Jao CL. Antifungal activity and enhancement of plant growth by Bacillus cereus grown on shellfish chitin wastes. *Bioresource Technol.* 2007; 98: 1224-1230.

Coley WB. The treatment of malignant tumors by repeated inoculalations of erysiplos with a report of ten original cases. *Am J Med Sci.* 1893; 105: 487-511.

Dang LH, Bettegowda C, Huso DL, Kinzler KW, Vogelstein B. Combination bacteriolytic therapy for the treatment of experimental tumors. *PNAS.* 2001; 98: 15155-15160.

Dannis JW. Changes in glycosylation associated with malignant transformation and tumor progression. In M. Fukuda (Eds.), Cell surface carbojudrates and cell development. 1992; Pp.161-194. London, CRC Press.

Demir G, Klein HO., Molinas NM, Tuzuner N. Beta glucan induces proliferation and activation of monocytes in peripheral blood of patients with advanced breast cancer. *Int Immunopharmacol*. 2007; 7: 113-116.

Dickinson K, Keer V, Hitchcock CA, Adams DJ. Chitinase activity from Candida albicans and its inhibition by allosamidin. *J General Microbiol.* 1989; 135: 1417-1421.

Distefano JF, Beck G, Zucker S. Mechanism of BCG-activated macrophage-induced tumor cell cytotoxicity: evidence for both oxygendependent and independent mechanisms. *Int. Arch Allergy Appl Immunol.* 1983; 70 : 252-260.

Eijk MV, Roomen CPAA, Renkema GH, Bussink AP, Andrews L, Blommaart EFC, Sugar A, Verhoeven AJ, Boot RG, Serts JMFG. Characterization of human phagocyte-drived chitotriosidase, a compount of innate immunity. *Int Immunol.* 2005; 17: 1505-1512

Evyernie D, Yamazaki S, Morimoto K, Karita S, Kimura T, Sakka K, Ohmiya K. Identification and charcterization of Clostridium paraputrificum M-21, a chitinolytic, mesophilic and hydrogen-producing bacterium. *J Bioscience Bioengineering.* 2000; 89: 596-601.

Fischer D, Li Y, Ahlemeyer B, Krieglstein J, Kissel T. In vitro cytotoxicity testing of polycations: influence of polymer structure on cell viability and hemolysis. *Biomayerials* 2003; 24: 1121-1131.

Grick D, Dietzel F, Ruster I. Further progress with oncolysis due to local high frequency hyperthermia, local x-irradiation and apathogenic clostridia. *J Microw Power.* 1979; 14: 163-166.

Haewix S, Andreesen R, Ferber E, Schwamberger G. Human macrophages secrete a tumoricidal activity distinct from tumor necrosis factor-α and reactive nitrogen intermediates. *Res Immunol.* 1992; 143: 89-94.

Hashimoto K, Okawa Y, Suzuki K, Okura Y, Suzuki S, Suzuki M. Antitumor activity of acidic mannan fraction from bakers'yeast. *J Pharm Dyn.* 1983; 6: 668-676.

Hibbs JB. Heterocytolysis by macrophages activated by Bacillus Calmette-Guerin: lysosome exocytosis into tumor cells. *Science.* 1974; 184: 468-471.

Hill HC, Conway TF, Sabel MS, Jong YS, Mathiowitz E, Bankert RB, Egilmez NK. Cancer immunotherapy with interleukin 12 and granulocyte-macrophage colony-stimulating factor-encapsulated microspheres: Coinduction of innate and adaptive antitumor immunity and cure of disseminated disease. *Cancer Res.* 2002; 62: 7254-7263.

Hoption CSA, van Netten JP, van Netten C, Golver DW. Spontaneous regression: a hidden treasure buried in time. *Med Hypotheses.* 2002; 58: 115-119.

Iozzo RV. Proteoglycans: structure, function, and role in neoplasia. *Lab Investigatio*n. 1985; 53: 373-396.

Kawase T, Yokokawa S, Saito A, Fujii T, Nikaidou N, Miyashita K, Watanabe T. Comparison of enzymatic and antifungal properties between family 18 and 19 chitinase from S.coelicolor A3 (2). *Biosci Biotechnol Biochem.* 2006; 70: 988-998.

Kobayashi M, Watanabe T, Suzuki S, Suzuki M. Effect of N-acetylchitohexaose against Candida albicans infection of tumor-bearing mice. *Microbiol Immunol.* 1990; 34: 413-426.

Labadaridis J, Dimitriou E, Costalos C, Serts J, Weely SV, Koopman WED, Michelakakis H. Serial chitotriosidase activity estimations in neonatal systemic candidiasis. *Acta Pediatr.* 1998; 87: 605-606.

Lambin P, Theys J, Landuyt W, Rijken P, van der Kogel A, van der Schueren E, Hodgkiss R, Fowler L. Colonisation of Clostridium in the body is restricted bto hypoxic and necrotic areas of tumor. *Anaerobe.* 1998; 4: 183-188.

Lin J, Lin E, Nemunaitis J. Bacteria in the treatment of cancer. *Curr Opin Mol Therapeutics.* 2004; 6: 629-639.

Melen'tiev AI, Aktuganov GE, Galimzianova NF. The role of chitinase in the antifungal activity of Bacillus sp.739. *Mikrobiologiia.* 2001; 70: 636-641.

Miyamoto K, Nukui E, Itoh H, Sato T, Kobayashi T, Imala C, Watanabe E, Inamori Y, Tsujibo H. Molecular analysis of the gene encoding a nivel chitin-binding protease from Alteromonas sp. StrainO-7 and its role in the chitinolyse system. *J Bacterol.* 2002; 184: 1865-1872.

Mlynarczuh I, Hoser G, Grzela T, Stoklosa T, Wojcik C, Malejczyk J, Jakobisiak M. Augmentred pro-spoptotic effects of TRAIL and proteasome inhibitor in human promonocytic leukemic U937 cells. *Anticancer Res.* 2001; 21: 1237-1240.

Mose JR, Mose G. Oncolysis by clostridia. I. Activity os Clostridium butyricum (M-55) and other non-pathogenic clostridia against the Ehrlich carcinoma. *Cancer Res.* 1964; 24: 212-216.

Ohyama C. Glycosylation in bladder cancer. *Int J Clin Oncol.* 2008; 13: 308-313.

Okawa Y, Kobayashi M. Suzuki S, Suzuki M. Comparative study of protective effects of chitin, chitosan, and N-acetyl chitohexaose against *Pseudomonas aeruginosa* and *Listeria monocytogenes* infections in mice. *Biol Pharm Bull.* 2003; 26: 902-904.

Overdijk B, Steijn GJ, Odds FC. Distribution of chitinase in guinea pig tissues and increases in levels of this enzyme after systemic infection with Aspergillus fumigatus. *Microbiology.* 1999; 145: 259-269.

Pan XQ, Hardy J, Lee R, Shih CC, Wang HQ. Chitinase selectively attacks tumor cells and cures cancer transplanted model mice. *Minerva Medica.* 2001; 92: Suppl 1: 127-128.

Pan XQ, Lee RJ, Ratnam M. Penetration into solid tumor tissue of fluorescent latex microspheras: a mimic of liposome particles. *Anticancer Res.* 2004; 24: 3005-3008.

Pan XQ, Shih CC, Harday J. Chitinase induces lyses of MCF-7 cells in culture and of human breast cancer xenograft B11-2 in SCID mice. *Anticancer Res.* 2005; 25: 343-346.

Patard JJ, Rodriguez A, Lobel B. The current status of intravesical therapy for superficial bladder cancer. *Current Opin Urol.* 2003; 13: 357-362.

Pawelek JM, Low KB, Bermudes D. Bacteria as tumor-targeting vectors. *Lancet Oncol.* 2003; 4: 548-556.

Pollard JW. Tumor-educated macrophages promote tumor progression and metastasis. *Nature Rev Cancer.* 2004; 4: 71-78.

Qin C, Du Y, Xiao L, Li Z, Gao X. Enzymic preparation of water-soluble chitosan and their antitumor activity. *Int J Biol Macromolecul.* 2002; 31: 111-117.

Radwan HH, Plattner HJ, Menge U, Diekmann H. The 92-kDa chitinase from Streptomyces olivaceoviridis contains a lysine-C endoproteinase at its N-terminus. *FEMS Microbio Lett.* 1994; 120: 31-35.

Renkema GH, Boot RG, Strjland A, Donker KWV, Berg M, Muijers AO, Aert JM. Synthesis, sorting and precessing into distinct isoforms of human macrophage chitoriosidase. *Eur J Biochem.* 1997; 244: 279-285.

Repentigny L, Petitbois S, Boushira M, Michaliszyn E, Senechal S, Gendron N, Montplaisir S. Acquired immunity in experimental murine Aspergillosis is mediated by macrophages. *Infect Immunity.* 1993; 61: 3791-3802.

Rheinnecker M, Hardt C, Ilag LL, Kufer P, Gruber R, Hoess A, Lupas A, Rottenberger C, Pluckthum A, Pack P. Multivalent antibody fragments with high functional affinity for a tumor associated carbohydrate antigen. *J Immunol.* 1996; 157: 2989-2997.

Robert JN, Dowvid PK, Richard LT. Subversion of host defense mechanisms by murine tumors I. A circulation factor that suppresses macrophage-mediated resistance to infection. *J Exp Med.* 1976; 143: 559-573.

Roberts WK, Selitrennikoff CP. Plant and bacterial chitinase differ in antifungal activity. *J Ceneral Microbiol.* 1988; 134: 169-176.

Robinette EH, Mardon DH. Brief communication: delayed lethal response to Candida albicans infection in mice bearing the lewis lung carcinoma. *J Nat Cancer Inst.* 1975; 55: 731-733.

Roby D, Gaddelle A, Toppan A. Chitin pligosaccharides as elicitators of chitinase activity in melon plants. *Biochem Biophys Res Communications* 1987; 143: 885-892.

Russell SW, Doe WF, McIntosh AT. Functional characterization of a stable noncytolytic stage of macrophage activity in tumors. *J Exp Med.* 1977; 146: 1511-1520.

Ryan RM, Green J, Lewis CE. Use of bacteria in anti-cancer therapies. *Bio Essays* 2005; 28: 84-94.

Sanders NN, Eijsink VG, Pangaart VD, Neerven JV, Simons PJ, Smedt De, Demeester J. Mucolytic activity of bacterial human chitinases. *Biochim Biophys Acta.* 2007; 1770: 839-846.

Schlumbaum A, Mauch F, Vogeli U, Boller T. Plant chitinases are potent inhibitors of fungal growth. *Nature.* 1986; 324: 365-367.

Seya T, Matsumoto M, Tsuji S, Begum NA, Nomura M, Azuma I. Hayashi A, Toyoshima K. Two receptor theory in innate immune activation: studies on the receptors for bacillus culmet guillen-cell wall skeleton. *Arch Immunol Ther Exp.* (Warsz). 2001; 49 Suppl1: S13-21.

Sidhu SS, Kalmar GB, Willis LG, Borgford TJ. *Streptomyces griseus* proteinase C. *J Biol Chem.* 1994; 269: 20167-20171.

Stern R. Association between cancer and "acid mucopolysaccharides": an old concept comes of age, finally. *Saminars Cancer Biol.* 2008; 18: 238-243.

Stevens DA, Mikami T, Okawa Y, Toko A, Suzuki S, Suzuki, M. Recombinant human chitinase. *Curr Opin Anti-infect Invest Drug.* 2002; 2: 399-404.

Suzuki K, Okawa Y, Hashimoto K, Suzuki S, Suzuki M. Protecting effect of chitin and chitosan on experimentally induced murine candidiasis. *Microbiol Immunol.* 1984; 28: 903-912.

Suzuki K, Okawa Y, Suzuki S, Suzuki M. Candidacidal effect of peritoneal exudates cells in mice administered with chitin or chitosan: the role of serine protease on the mechanisn of oxygen-independent candidacidal effect. *Microbiol Immunol.* 1987; 31: 375-379.

Suzuki K, Sugawara N, Suzuki M, Uchiyama T, Katouno F, Nikaidou N, Watanabe T. Chitinases A, B, and C1 of *Serratia marcescens* 2170 produced by recombinant *Escherichia coli*: Enzymatic properties and synergism on chitin degradation. *Biosci Biotechnol Biochem.* 2002; 66: 1075-1082.

Suzuki K, Tokowa Y, Suzuki S, Suzuki M. Effects of N-acetylchito-oligosaccharides on activation of phagocytes. *Microbiol Immunol.* 1986; 30: 777-787.

Tanioka SI, Tanigawa T, Tanaka Y. Activation of macrophage by chitin and chitosan. *Kichin Kitosan Kenkyu.* 1995; 1: 108-109.

Tokura S, Tamura H, Azuma I. Immunological aspects of chitin and chitin derivatives administered to animals. Chitin and Chitinases. Jolles, R. and Muzzarelli, R.A.A. (Eds). 1999; Birkhauser Verlag Basel, Switzerlan.

Torres AV, Balish E. Macrophages in resistance to candidiasis. *Microbiol Mole Biol Rev.* 1997; 61: 170-192.

Tsukada K, Matsumoto T, Aizawa K, Tokoro A, Naruse RS, Suzuki S. Antimetastatic and growth-inhibitory effects of N-acetylchitohexaose in mice bearing lewis lung carcinoma. *Jpn J Cancer Res.* 1990; 81: 259-265.

Tzankov A, Ludescher C, Duba HC, Steinlechner M, Knapp R, Schmid T, Grunewald K, Gastl G, Stauder R. Spontaneous remission in a secondary acute myelogenous leukaemia following invasive pulmonary aspergillosis. *Ann Hematol.* 2001; 80: 423-425.

Ujita M, Sakai K, Hamazaki K, Yoneda M, Isomura S, Hara A. Carbohydrate binding specificity of the recombinant chitin-binding domain of human macrophage chitinase. *Biosci Biotechnol Biochem.* 2003; 67: 2402-2407.

Vaaje-Kolstad G, Horn S, Aalten DMF, Synstad B, Eijsink VGH. The non-catalytic chitin-binding protein CBP21 from Serratia marcescens is essential for chitin degradation. *J Biol Chem.* 2005; 280: 28492-28497.

Woynarowska B, Wikiel H, Sharma M, Carpenter N, Fleet GWJ, Bernaki PJ. Inhibition of human overian carcinoma cell- and hexosaminінidase- mediated degradation of extracellular matrix by sugar analogs. *Anticancer Res.* 1992; 12: 161-166.

Xu L, Wang Y, Wang L, Gao Y, Ann C. TYchi, a novel chitinase with RNA N-glycosidase and anti-tumor activity. *Fronities in Bioaci.* 2008; 13: 3127-3135.

Yao O, Hou SX, Zhang X, Zhao G, Gou XJ, You JZ. Preparation and characterization of biotinylated chitosan nanoparticle. *Yao Xue Xue Bao.* 2007; 42: 557-561.

In: Binomium Chitin-Chitinase: Recent Issues ISBN 978-1-60692-339-9
Editor: Salvatore Musumeci and Maurizio G. Paoletti

Chapter XIII

YKL-40 in Inflammation, Tissue Remodeling and Cancer

Julia S. Johansen[14]
Department of Rheumatology, Herlev Hospital,
University of Copenhagen, Denmark

Abstract

YKL-40 (also named Chitinase-3-like-1, CHI3L1) is a 40 kDa heparin-, chitin- and collagen-binding glycoprotein without chitinase activity and a member of "mammalian chitinase-like proteins". The YKL-40 gene is located on chromosome 1q32.1, has a size of 7948 base pairs and contains 10 exons. The crystallographic structure for YKL-40 is known, but cellular receptors are not identified. High YKL-40 mRNA and protein expressions are found in human embryonic and fetal cells, macrophages during late state of differentiation, macrophages in inflammed synovial membrane, atheromatous plaques, arteritic vessels, alveolar macrophages in inflamed lung tissue, microglia/macrophages from central nervous system, and in tumor-associated macrophages, neutrophils, mast cells, arthritic chondrocytes, differentiated vascular smooth muscle cells, fibroblast-like synovial cells, endothelial cells and by several types of cancer cells. The YKL-40 gene and protein are overexpressed compared to normal tissues in glioblastoma, melanoma, squamous cell carcinoma and many types of adenocarcinoma.

The exact biological functions of YKL-40 are unknown. YKL-40 is a growth factor for fibroblasts and chondrocytes, modulates the rate of type I collagen fibril formation, acts synergistically with IGF-1, is regulated by TNFα and IL-6, requires sustained activation of NF-kappaB, initiates MAP kinase and PI-3K signalling cascades leading to the phosphorylation of ERK-1/2 MAP kinase and protein kinase B (AKT)-mediated signalling cascades, which are associated with control of mitogenesis. YKL-40 may play a role in inflammation and the innate immune response, enhances bacterial adhesion to colonic

14 Correspondence: Julia S. Johansen, Department of Rheumatology Q107, Herlev Hospital, University of Copenhagen, Herlev Ringvej 75, DK-2730 Herlev, Denmark. Phone: (45) 44884243; FAX: (45) 44884214; E-mail: julia.johansen@post3.tele.dk.

epithelial cells and has a role in cancer cell proliferation, differentiation, metastasis potential, protects the cells from undergoing apoptosis, stimulates angiogenesis, and has an effect on extracellular tissue remodeling surrounding the tumour, although *in vivo* proof of this is yet to be obtained.

Plasma levels of YKL-40 are elevated compared to healthy subjects in patients with acute inflammation (e.g. pneumonia, endotoxaemia, hepatitis) or chronic inflammation (e.g. rheumatoid arthritis, inflammatory bowel disease, asthma, sarcoidosis, type II diabetes, coronary artery disease) and in patients with liver fibrosis. Plasma YKL-40 levels are also elevated in some patients with primary or metastatic cancer and may be useful as an independent "prognosticator" of survival, a predictor of treatment response, and in monitoring cancer recurrence/progression after treatment. Unfortunately, most of these studies are small and retrospective. Recently, two large studies suggest that plasma YKL-40 may have a value in screening for colorectal cancer.

In the future, more research on the function of YKL-40 is needed and large prospective, longitudinal clinical studies should be performed to determine if plasma YKL-40 levels have a clinical value as a biomarker in patients with inflammation, tissue remodeling, fibrosis and cancer.

1. Introduction

During the last few years there has been a growing interest in the glycoprotein protein YKL-40. From it first discovery in 1985 by Millis *et al.* only 29 papers were published about the protein in the following 15 years, but since 2000 the number of publications have increased to a little more than 250. Most of these publications are related to plasma YKL-40 as a biomarker. The present Chapter will focus on human YKL-40 and plasma concentrations of YKL-40 as a potential biomarker in cancer patients and in patients with diseases characterized by inflammation, tissue remodeling and fibrosis.

2. YKL-40 Protein and Gene

YKL-40 (Johansen *et al.* 1992) is also named human cartilage glycoprotein-39 (HC-gp39) (Hakala *et al.* 1993), 38-kDa heparin-binding glycoprotein (gp38k) (Shackelton *et al.* 1995), chitinase-3-like-1 (CHI3L1) (Rehli *et al.* 1997), and chondrex (Harvey *et al.* 1998). YKL-40 is a phylogenetically highly conserved glycoprotein and a member of "Family 18 chitolectins" (Bussink *et al.* 2007; Funkhouser *et al.* 2007; Zaheer-ul-Haq *et al.* 2007). YKL-40 is described in human (Hakala *et al.* 1993), chimpanzee, pig (Shackelton *et al.* 1995), cow, goat (Mohanty *et al.* 2003), sheep, guinea pig (De Ceuninck *et al.* 2001), rat and mouse (Morrison *et al.* 1994). Furthermore, the fruit fly *Drosophila melanogaster* (Kirkpatrick *et al.* 1995; Kawamura *et al.* 1999), the mosquito *Anopheles gambiae*, the zebra fish *Danio rerio*, the pacific oyster *Crassostrea gigas* (Badariotti *et al.* 2006) and the nematode *Caenorhabditis elegans* have multiple putative YKL-40-like proteins (National Center for Biotechnology Information (NCBI)).

The YKL-40 protein contains a single polypeptide chain of 383 amino acids, has a molecular mass of 40 kDa (Hakala *et al.* 1993) and an isoelectric point of 7.6 (Renkema *et al.*

1998). The crystallographic structure for human YKL-40 is known (Fusetti *et al.* 2003; Houston *et al.* 2003) and display the typical fold of family 18 glycosyl hydrolases (Henrissat *et al.* 1997).

The YKL-40 gene is located on chromosome 1q32.1, has a size of 7948 base pairs and consists of 10 exons (Rehli *et al.* 1997, 2003). Recently, two splice forms of the YKL-40 gene are reported; isoform 1 (containing exon 1-10) and isoform 2 (in which exon 8 has been spliced out) (Johansen *et al.* 2007b).

3. YKL-40 Expression in Non-Malignant Cells and Tissues

Immunohistochemical studies have demonstrated that YKL-40 is strongly expressed in all germ layers of human embryos and subsequently expressed in human fetal tissues of ecto-, meso- and endoderm (Johansen *et al.* 2007b). At the cellular level YKL-40 protein expression is high in embryonic and fetal tissues characterized by rapid proliferation and marked differentiation, and in tissues undergoing morphogenetic changes (Johansen *et al.* 2007b). Interestingly, recent studies also found that YKL-40 is produced by human embryonic stem cells and their progenitors (Johansen *et al.* 2007d). In adult human normal tissue high YKL-40 expression is observed in cells with a high cellular activity (Ringsholt *et al.* 2007).

YKL-40 is expressed *in vitro* by macrophages during late state of differentiation (Renkema *et al.* 1998; Rehli *et al.* 1997 and 2003), and *in vivo* by tumour-associated macrophages (Junker *et al.* 2005), by macrophages in inflamed synovial tissue (Baeten *et al.* 2000; Volck *et al.* 2001), by macrophages and giant cells in arteritic vessels (Johansen *et al.* 1999) and in sarcoid lesions of patients with pulmonary sarcoidosis (Johansen *et al.* 2005). Macrophages in atherosclerotic plaques express YKL-40, particularly macrophages that have infiltrated deeper into the lesion, and the highest YKL-40 mRNA expression is found in macrophages in the early atherosclerotic lesion (Boot *et al.* 1999). Recently, YKL-40 protein expression is shown in macrophages and subepithelial cells in bronchial-biopsy specimens and in macrophages and neutrophils in cytospin of broncho-alveolar lavage from patients with asthma (Chupp *et al.* 2007). Furthermore YKL-40 protein expression is found in microglia/macrophages from primary cultures of human fetal brain (Bonneh-Barkay *et al.* 2008). YKL-40 is also located in the specific granules of neutrophils (Volck *et al.* 1998) and in mast cells (Ringsholt *et al.* 2007; Roslind *et al.* 2008). In patients with rheumatoid arthritis, YKL-40 protein is expressed by CD16+ monocytes with a dim expression of CD14 (Baeten *et al.* 2000). This CD14+/CD16+ phenotype can differentiate from classic CD14++ monocytes by maturation *in vitro* and is considered as pro-inflammatory with properties of tissue macrophages and are a source of TNFα (Belge *et al.* 2002). Furthermore arthritic chondrocytes (Hakala *et al.* 1993; Johansen *et al.* 2001; Volck *et al.* 2001), differentiated vascular smooth muscle cells (Millis *et al.* 1985, 1986; Malinda *et al.* 1999; Nishikawa *et al.* 2003), endothelial cells (Johansen *et al.* 1999), fibroblast-like synovial cells (Nyirkos *et al.* 1990; Hakala *et al.* 1992; Dasuri *et al.* 2004) and colonic epithelial cells (Mizoguchi 2006) produce YKL-40.

Increased YKL-40 protein expression is found in fibrotic liver tissue from patients with alcoholic liver disease and chronic hepatitis C virus infection (Johansen *et al.* 1997, 2000). Hepatocytes did not express YKL-40 protein and no expression was found in normal liver tissue except in mesenchymal structures of the portal tract. It was not possible by the immunohistochemical methods to discriminate the extent to which the YKL-40 protein expression was intracellular, and if hepatic stellate cells, leucocytes and macrophages expressed YKL-40 (Johansen *et al.* 1997, 2000). I biopsies with chronic active hepatitis C virus YKL-40 protein expression was found in areas with piecemeal necrosis, but not in lymphocytes.

YKL-40 is overexpressed, compared to normal tissue, in hippocampus tissue from patients with schizophrenia (Chung *et al.* 2003), and in brain tissue from patients with Alzheimer disease (Colton *et al.* 2006). Furthermore, YKL-40 is expressed in normal human neural retina and retinal pigment epithelium-choroid complex, and upregulated in pathological human exudative age-related macular degeneration, in glaucoma and in experimental murine choroidal neovascular membranes (Sharon *et al.* 2002; Lo *et al.* 2003; Miyahara *et al.* 2003; Rakic *et al.* 2003; Liton *et al.* 2005, 2006; Gonzales *et al.* 2006).

4. The Function of YKL-40 in Non-Malignant Diseases

Mechanistically very little is known about the function of YKL-40, and cellular receptors mediating the biological effects of YKL-40 are not identified. Several possible functions are suggested (Figure 1).

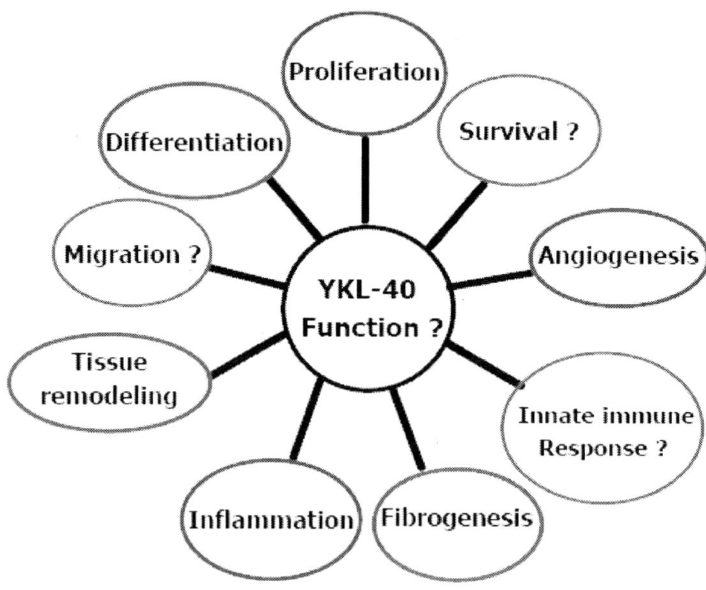

Figure 1. The exact functions of YKL-40 are unknown. Green circles represent potential functions based on 1-2 studies. Red circles represent potential functions based on more than 2 studies.

YKL-40 is a proliferation factor for fibroblasts, synovial cells and chondrocytes (De Ceuninck *et al.* 2001; Recklies *et al.* 2002) and acts synergistically with IGF-1 (Recklies *et al.* 2002). Studies of embryonic and fetal cells (Johansen *et al.* 2007d), macrophages (Rehli *et al.* 1997, 2003; Renkema *et al.* 1998; Krause *et al.* 1996; Hashimoto *et al.* 1999; Suzuki *et al.* 2000) and fetal chondrocytes (Benz *et al.* 2002; Stokes *et al.* 2002; Imabayashi *et al.* 2003; Johansen *et al.* 2007) indicate that YKL-40 is a proliferation and differentiation marker. Furthermore, YKL-40 contributes to chondrocyte differentiation by inducing the transcription factor SOX9 and type II collagen expressions, and the induction of SOX9 depends on ERK1/2 and PI3K activities, but not on p38 and JNK MAPK (Jacques *et al.* 2007).

YKL-40 is regulated in chondrocytes by TNFα (Ling *et al.* 2004; Recklies *et al.* 2005) and requires sustained activation of NF-κB (Recklies *et al.* 2005). This important transcription factor controls cell survival by regulating cell proliferation, growth arrest and death (Balkwill *et al.* 2004). YKL-40 initiates MAP kinase and PI-3K signaling cascades in fibroblasts leading to phosphorylation of both the extracellular signal-regulated kinase (ERK)-1/2 MAP kinase and protein kinase B (AKT)-mediated signalling cascades, which are associated with the control of mitogenesis (Recklies *et al.* 2002, 2005; Ling *et al.* 2004). Stimulation of human articular chondrocytes or skin fibroblasts with IL-1 or TNFα in the presence of YKL-40 result in reduction of both p38 and SAPK/JNK phosphorylation, and YKL-40 suppresses the cytokine-induced secretion of several metalloproteinases and the chemokine IL-8 (Ling *et al.* 2004). This suggest that YKL-40 may play a protective role in inflammatory environments, limiting degradation of the extracellular matrix and thereby controlling tissue remodeling (Recklies *et al.* 2005). IL-6 may also regulate YKL-40 in humans (personal observation) and IFNγ strongly induces YKL-40 mRNA in human peripheral blood-derived macrophages, while dexamethasone has inhibitory effects (Kzhyshkowska *et al.* 2006).

YKL-40 binds collagen type I, II and III and modulates the rate of type I collagen fibril formation (Bigg *et al.* 2006). Furthermore, YKL-40 binds chitin, but has no chitinase activity (Hakala *et al.* 1993; Renkema *et al.* 1998; Houston *et al.* 2003) due to amino acid substitution in the active site of chitinases (Hakala *et al.* 1993; Renkema *et al.* 1998). Vertebrates in an embryonic stage use short chito-oligosaccharides as primers for the synthesis of hyaluronan (Meyer *et al.* 1996; Semino *et al.* 1996; Varki 1996). YKL-40 has heparin and hyaluronan binding motifs (Fusetti *et al.* 2003) and may bind to cell surface receptors such as heparin sulphate proteoglycans and may recognize hyaluronan or its precursor as a substrate in the extracellular matrix and interfere with the synthesis and local concentrations of hyaluronan (Fusetti *et al.* 2003). Is this correct, YKL-40 may consequently influence the effects of high hyaluronan in tissues, e.g. the extent of cell adhesion and migration during the tissue remodeling processes that take place during metastasis, inflammation, fibrosis, and atherogenesis (Toole 2004; Fjeldstad *et al.* 2005).

YKL-40 modulates vascular endothelial cell morphology by promoting the formation of branching tubules and migration and adhesion of vascular smooth muscle cells (Shackelton *et al.* 1995; Millis *et al.* 1985, 1986; Malinda *et al.* 1999; Nishikawa *et al.* 2003) suggesting a role in angiogenesis.

YKL-40 may be an anti-apoptotic protein, since it initiates PI-3K signaling pathways and phosphorylation of AKT (Recklies *et al.* 2002, 2005; Ling *et al.* 2004), which are associated with cell survival. YKL-40 is called the "breast regression protein (Brp-39)" (Morison *et al.* 1994), because it is induced in mice mammary epithelial cells a few days after weaning. Mammary involution involves programmed cell death. It is hypothesized that YKL-40 utilizes a chitin oligosaccharide binding ability while participating in the various signal transduction pathways that leads to apoptosis of the regressing cells and that YKL-40 is a protective signaling factor that determines which cells are to survive the drastic tissue remodeling that occurs during involution (Mohanty *et al.* 2003).

Several studies have demonstrated that YKL-40 is a candidate autoantigen in rheumatoid arthritis. YKL-40 derived peptides, which bind with high affinity to the rheumatoid arthritis-associated HLA-DR1 and DR4, are recognized by peripheral T cells from patients with rheumatoid arthritis, and these T cells showed a proliferative response to YKL-40 peptides (Verheijden *et al.* 1997; Cope *et al.* 1999; Vos *et al.* 2000). HLA-DM-dependent presentation pathway is involved in the presentation of autoantigenic YKL-40 epitopes (Patil *et al.* 2001). Furthermore HLA-DR/YKL-40$^{263-275}$ complexes are expressed by dendritic cells in rheumaotid arthritis synovium and associated with histologic features of follicular synovitis and is specific for rheumatoid arthritis (Steenbakkers *et al.* 2003; Baeten *et al.* 2004). YKL-40 induced a chronic relapsing arthritis in BALB/c mice, which clinically and histologically resembles rheumatoid arthritis, and this arthritis could be delayed and suppressed by intranasal administration of YKL-40 prior to immunization (Verheijden *et al.* 1997; Joosten *et al.* 2000). Nasal tolerization with YKL-40 had a beneficial effect on both inflammation and tissue destruction in collagen-induced arthritis in mice (Joosten *et al.* 2000). A small randomized, double blind, placebo controlled phase I/II study of patients with rheumatoid arthritis treated with intravenous infusions of a soluble complex of native HLA-DR4 (β*0401) complexed to a 1311 Da peptide corresponding to amino acid residues 263-275 of YKL-40 demonstrated that this treatment led to T cell inactivation and immunologic tolerance in patients with persistent rheumatoid arthritis and 67% of the patients who received the highest dose had a clinical response after 5 infusions compared with 14% in the placebo treated group (Kavanaugh *et al.* 2003). These findings suggest that YKL-40 may play a fundamental role in the pathophysiology of rheumatoid arthritis, and that immunological tolerance of the protein may control disease activity in these patients.

Recently, it is suggested that YKL-40 has a role in the innate immune response based on studies of patients with asthma (Chupp *et al.* 2007; Dickey *et al.* 2007). Furthermore, during lentiviral encephalitis, YKL-40 may interfere with the biological activity of basic fibroblast growth factor and other heparin-binding growth factors and chemokines that can affect neuronal function or survival (Bonneh-Barkay *et al.* 2008; Kolson 2008). Interestingly, it has also been shown that YKL-40 plays a pathogenetic role in colitis, presumably by enhancing the adhesion and invasion of bacteria on/into colonic epithelial cells, and leads to exacerbations of intestinal inflammation (Mizoguchi 2006, 2007; Kawada *et al.* 2008) (see also Chapter XXI in this book by Mizoguchi and Kawada). These observations suggest that YKL-40 may play different roles depending upon the cell types under inflammatory conditions.

The level of expression of YKL-40 has previously been shown to be genetically regulated. An association is found between schizophrenia and haplotypes within the promoter region of the gene encoding for YKL-40 (Zhao *et al.* 2007; Yang *et al.* 2008), suggesting a functional mechanism for involvement of YKL-40 in schizophrenia susceptibility. Whereas there are no difference in haplotype frequencies between patients with sarcoidosis and controls

(Kruit *et al.* 2007). A single nucleotide polymorphism (SNP)-329 G/A in the YKL-40 gene was shown to significantly influence plasma YKL-40 levels in healthy subjects, but not in patients with sarcoidosis (Kruit *et al.* 2007). A promoter SNP-131C/G in the YKL-40 gene was associated with elevated plasma YKL-40 levels, asthma, bronchial hyper responsiveness and pulmonary function in the Hutterites (Ober *et al.* 2008). However, genetic variants of YKL-40 in the general population have not been described yet.

5. Measurement of Plasma YKL-40

YKL-40 concentrations in human serum or EDTA plasma can be determined by a commercial two-site, sandwich-type enzyme-linked immunoassay (ELISA) (Quidel, Santa Clara, CA, USA) (Harvey *et al.* 1998). This assay uses streptavidin-coated microplate wells, a biotinylated-Fab monoclonal mouse antibody against human YKL-40 (capture antibody) and an alkaline phosphatase-labeled polyclonal rabbit antibody against human YKL-40 (detection antibody). Bound enzyme activity is detected with p-nitrophenyl phosphate as substrate. The ELISA is finished within 4 hours and involves three 1-hour incubation steps, is carried out at room temperature and does require sample dilution if the concentration of YKL-40 in the sample is very high. The sensitivity of the ELISA is 20 μg/l (detection limit 8 μg/l) and the recovery 102%. The intra- and inter-assay coefficients of variation are <5.0% and <10%, respectively.

This ELISA can also be used for determination of YKL-40 levels in bronchoalveolar-lavage fluid (Kuepper *et al.* 2008), cerebrospinal fluid (Østergaard *et al.* 2002; Bonneh-Barkay *et al.* 2008; Kacira *et al.* 2008), synovial fluid levels (Johansen *et al.* 1996; Volck *et al.* 2001) and conditioned medium from human cell cultures (Johansen *et al.* 2001; Junker *et al.* 2005b). Plasma YKL-40 levels from baboons (Mahaney *et al.* 1998) and cynomolgus macaques (Register *et al.* 2001) and cerebrospinal fluid levels of YKL-40 in pigtailed macaques (Bonneh-Barkay *et al.* 2008) can also be determined using this ELISA, whereas there is no cross-reactivity with YKL-40 from cow, pig, rabbit, mouse and rat.

Several factors must be considered when handling blood samples for the measurement of YKL-40. The time interval between drawing of blood and centrifugation of blood stored at room temperature must be less than 3 hours for serum and 8 hours for EDTA plasma samples. Otherwise significant and not disease related elevations of YKL-40 are found in the serum and EDTA plasma samples left on the clot for a longer time when compared with YKL-40 concentrations in serum and EDTA plasma samples centrifuged within 1 hour after venipuncture. If the blood is stored at 4°C before centrifugation YKL-40 concentrations are stable in serum for 24 hours and in EDTA plasma for 72 hours (Høgdall *et al.* 2000). Degranulation of neutrophils with release of YKL-40 from the specific granules is the most likely explanation for this time dependent increase in YKL-40 concentrations in serum and EDTA plasma. Repetitive freezing and thawing of serum samples up to 9 times have no effect on serum YKL-40 levels (Johansen *et al.* 1993; Harvey *et al.* 1998; Høgdall *et al.* 2000). YKL-40 concentrations in serum are stable in samples stored up to 5 days at room temperature (Johansen *et al.* 1993), up to 9 days at 4°C (Harvey *et al.* 1998), and at -80°C for at least 10 years (personal observation). YKL-40 concentrations in corresponding serum and

EDTA plasma samples are correlated (rho = 0.98, p<0.001), but YKL-40 is significantly higher in serum compared to EDTA plasma with a YKL-40 serum/EDTA plasma ratio of 1.4 (Johansen *et al.* 1993; Høgdall *et al.* 2000). This is probably caused by a small release of YKL-40 from activated neutrophils during the coagulation process.

Serum YKL-40 levels are relatively stable in healthy subjects during a day, a month, a year and during a period of 3 years. The within subject coefficient of variation including variation over time and inter-assay was 28.8% and 30.2% over a period of 2 and 3 years, and the intraclass correlation coefficients were 72.4% and 72.2% indicating reasonable reliability of serum YKL-40 measurements. An estimated variation in serum YKL-40 within healthy subjects including inter-assay variation suggests that an increase of more than 109% or a decrease of more than 52% in serum YKL-40 could be considered as significant and not only a reflection of pre-analytical conditions, methodological and normal biologic variability (Johansen *et al.* 2008b).

In this Chapter there will be no discrimination between YKL-40 in serum and plasma EDTA samples, since the serum or plasma EDTA concentrations of YKL-40 in the patients were accordingly related to the serum or plasma EDTA concentrations of YKL-40 in healthy subjects.

6. Plasma YKL-40 in Healthy Subjects

Plasma YKL-40 increases with age in small studies of healthy subjects (Johansen *et al.* 1996, 2008b; Harvey *et al.* 1998), and confirmed in a recent large study of 8899 participants from the general population (Johansen *et al.* 2009). There is no difference in plasma YKL-40 between the two sexes. The median plasma YKL-40 level in healthy subjects is 40-50 μg/l and the upper 97.5 percentage limit is 168 μg/l in healthy subjects (Harvey *et al.* 1998; Johansen 2006a, 2008b, 2009). However, a truly "normal" level and "abnormal" level of plasma YKL-40 have not yet been determined.

7. Plasma YKL-40 in Patients with Diseases Characterized by Inflammation, Tissue Remodeling and Fibrosis

YKL-40 can be regarded as an acute phase protein, since its plasma concentration increases by more than 25% following an inflammatory stimulus, and plasma YKL-40 has been suggested as a potential biomarker of acute and chronic inflammation (Johansen 2006a), including systemic low-grade inflammation (Johansen *et al.* 2008c). In contrast to the most widely used acute phase protein serum C-reactive protein (CRP) (Gabay *et al.* 1999; Kushner *et al.* 2006) that is produced in the liver by hepatocytes in response to high IL-6, YKL-40 is produced locally *in vivo* in tissues with inflammation by macrophages, mast cells and neutrophils.

Table 1. Plasma YKL-40 as a potential biomarker in patients with diseases characterized by inflammation, tissue remodeling, and fibrosis

Disease type	Retrospective studies	Prospective studies	Total number of patients	% of patients with elevated plasma YKL-40
Bacterial infections	2	0	169	75
Rheumatoid arthritis	9	3	797	13 - 49
Diabetes	4	0	367	not desribed
Asthma	2	0	110	not described
Sarcoidosis	1	0	63	not descibed
Systemic sclerosis	2	0	128	27 - 35
Hearth disease	4	0	4630	not described
Inflammatory bowel disease	3	1	668	11 - 69
Viral hepatitis	5	1	1330	35 - 80
Alcoholic liver disease	2	0	580	64 - 90

The percentage of patients with elevated plasma YKL-40 is calculated as the number of patients in the different studies with a plasma YKL-40 level above the age-adjusted 95th percentile of the plasma YKL-40 level in healthy subjects.

Several small and retrospective studies have determined plasma concentrations of YKL-40 in patients with diseases characterized by inflammation, increased extracellular tissue remodeling or ongoing fibrosis (Table 1). In many of these studies no or low correlations were found between serum CRP and plasma YKL-40, suggesting that these two biomarkers reflect different aspects of the inflammatory process, and that plasma YKL-40 provides independent information of inflammation. Only few of these studies included a large number of patients, but in these studies an elevated plasma YKL-40 was associated with poor prognosis and short survival in patients with alcoholic liver disease (Nøjgaard *et al.* 2003a) and in patients with coronary artery disease (Kastrup *et al.* 2009). In the following sections the different studies are described briefly.

Patients with *Streptococcus pneumoniae* pneumonia or bacteremia have 8 - 10 fold higher plasma YKL-40 compared to healthy subjects (Nordenbæk *et al.* 1999, Kronborg *et al.* 2002). It is found that plasma or cerebrospinal fluid concentration of YKL-40 to some extent reflect the severity and prognosis of a bacterial infection, and plasma YKL-40 may add to the information of serum CRP in patients with severe acute bacterial infections (Nordenbæk *et al.* 1999; Kronborg *et al.* 2002; Østergaard *et al.* 2002). Human endotoxaemia, which is followed by increased plasma TNFα and IL-6 levels, increases plasma YKL-40 (Johansen *et al.* 2005). A large prospective longitudinal study is ongoing to evaluate if plasma YKL-40 provides better information of prognosis in patients with sepsis compared to serum CRP. Recently, it has been found in patients with immunodeficiency virus encephalitis that YKL-40 levels and viral load in the cerebrospinal fluid was correlated (Bonneh-Barkay *et al.* 2008).

Increased levels of YKL-40 are found in synovial fluid and plasma of patients with rheumatoid arthritis and active disease compared to inactive patients or healthy subjects (Johansen *et al.* 1993, 1999, 2001; Harvey *et al.* 1998, 2000; Vos *et al.* 2000; Combe *et al.* 2001; Peltomaa *et al.* 2001; Matsumoto *et al.* 2001; Volck *et al.* 2001; Knudsen *et al.* 2006, 2008). Plasma YKL-40 is correlated with clinical parameters of disease activity and serum CRP, and in a few studies with progression of joint destruction (Johansen *et al.* 1999, 2001). However, none of these studies found that plasma YKL-40 provides better information of disease activity and prognosis in patients with rheumatoid arthritis compared to serum CRP.

Recently it has been shown that plasma YKL-40 levels were high in some patients with asthma compared to healthy subjects, and that plasma YKL-40 correlated with the severity of asthma measured by clinical variables, including FEV1 and with the thickness of the subepithelial basement membrane in biopsy specimens of the lung (Chupp *et al.* 2007). A small pilot study showed thereafter, that YKL-40 concentrations in plasma and bronchoalveolar-lavage-fluid increased significantly 24 hours after allergen challenge in patients with allergic asthma (Kuepper *et al.* 2008). The levels of YKL-40 in bronchoalveolar-lavage fluid were also correlated with eosinophil counts 24 hours after allergen challenge (Kuepper *et al.* 2008). It is suggested that plasma YKL-40 may contribute to the stratification of the underlying severity of disease among patients with asthma, independent of disease activity (Dickey 2007).

Plasma YKL-40 is elevated in patients with pulmonary sarcoidosis compared to controls and correlated with serum angiotensin-converting enzyme (Johansen *et al.* 2005). Patients with systemic sclerosis and elevated plasma YKL-40 have a poor prognosis and these patients died more often due to extensive interstitial or vascular fibrosing processes (Montagna *et al.* 2003; Nordenbæk *et al.* 2005). Patients with giant cell arteritis have elevated plasma YKL-40 at time of diagnosis, but during treatment with glucocorticoids plasma YKL-40 was not related to disease activity and serum CRP (Johansen *et al.* 1999).

Patients with type 2 diabetes have higher plasma YKL-40 levels compared to subjects with normal glucose tolerance (Harvey *et al.* 1998; Rathcke *et al.* 2006; Nielsen *et al.* 2009), and plasma YKL-40 is related to insulin resistance (Rathcke *et al.* 2006), fasting plasma glucose, and plasma IL-6 (Nielsen *et al.* 2009), but not to serum CRP (Rathcke *et al.* 2006) or obesity (Nielsen *et al.* 2009). Recently, it has been shown that patients with type 1 diabetes have eleveated plasma YKL-40 compared to healthy subjects, and that increasing plasma YKL-40 was independently associated with increasing albuminuria in patients with type 1 diabetes (Rathcke *et al.* 2009).

Plasma YKL-40 is increased up to 7 fold in patients after an acute myocardial infarction (Nørgaard *et al.* 2008; Wang *et al.* 2008) and is associated with the number of diseased vessels assessed by coronary angiography (Kucur *et al.* 2007). Serum YKL-40 is also correlated with creatine kinase fraction B in non-thrombolyzed patients with acute myocardial infarction (Nørgaard *et al.* 2008). A large study of patients with stable coronary artery disease has recently shown that high plasma YKL-40 is associated with poor prognosis (Kastrup *et al.* 2009). It is unknown if circulating YKL-40 may reflect the total burden of coronary atherosclerosis or may identify a high-risk atherosclerosis phenotype with ongoing inflammation and severe atherosclerotic plaque formation.

Plasma YKL-40 correlated with plasma IL-6 in 80-year old women and men, but only with serum TNFα and CRP among 80-year old women, and high plasma YKL-40 was associated with a low CD4:CD8 cell ratio. High plasma YKL-40 in 80-year old subjects were associated with all-cause mortality and independent of potential confounders (sex, smoking, BMI, chronic disease and anti-inflammatory medicine) (Johansen *et al.* 2008c).

The role of YKL-40 in inflammatory bowel diseases are decribed in more details in Chapter XXI in this book by Mizoguchi and Kawada. Elevated plasma YKL-40 levels are found in a some patients with inflammatory bowel disease (Vos *et al.* 2000b; Koutroubakis *et al.* 2003; Punzi *et al.* 2003; Vind *et al.* 2003). In patients with ulcerative colitis plasma YKL-40 correlated with serum CRP levels and a disease activity score (Koutroubakis *et al.* 2003; Vind *et al.* 2003). Low correlations were found in patients with Crohn's disease between plasma YKL-40, CRP and Harvey-Bradshaw score (Vind *et al.* 2003), but in another study of patients with Crohn's disease plasma YKL-40 correlated with serum CRP and Crohn's Disease Activity Index score (Koutroubakis *et al.* 2003). Patients with inflammatory bowel disease and joint involvement had higher plasma YKL-40 than patients without joint involvement (Punzi *et al.* 2003). The subgroup of patients with inflammatory bowel disease with elevated plasma YKL-40 may not only have intestinal inflammation or arthropathy; patiens with Crohn's disease and stenotic disease had higher plasma YKL-40 than patients with non-stenotic disease and plasma YKL-40 was independent of other clinical parameters (Koutroubakis *et al.* 2003; Erzin *et al.* 2008). Plasma YKL-40 may therefore reflect ongoing fibrogenesis and could be a risk factor in patients with Crohn's disease. This needs to be tested in larger prospective studies of patients with inflammatory bowel disease.

In 8899 subjects from the general population increasing plasma levels of YKL-40 are found with increasing alcohol intake (Johansen *et al.* 2009). In studies of patients with alcoholic liver disease elevated plasma YKL-40 levels are related to both liver fibrosis and inflammation in the liver (Johansen *et al.* 1997, 2000; Nøjgaard *et al.* 2003a). Patients with alcoholic liver disease and very high plasma YKL-40 have shorter survival than patients with normal plasma YKL-40 (Nørgaard *et al.* 2003a). Plasma YKL-40 is elevated in most patients with moderate to severe liver fibrosis and cirrhosis, independently of disease etiology, and may provide new information of ongoing fibrogenesis in the liver (Johansen *et al.* 1997, 2000; Tran *et al.* 2000; Nøjgaard *et al.* 2003a, 2003b; Nunes *et al.* 2005; Kamal *et al.* 2006; Alexander *et al.* 2007; Esmat *et al.* 2007; Fontana *et al.* 2008; Berres *et al.* 2009). Although plasma YKL-40 could predict cirrhosis (Ishak 5/6) in a large group of patients with chronic hepatitis C, plasma YKL-40 was not included in the final 3-variable model consisting of serum hyaluronan, TIMP-1 and platelet count (Fontana *et al.* 2008). In children, plasma YKL-40 could not differentiate patients with advanced liver fibrosis from those with mild fibrosis (Lebensztejn *et al.* 2007). Large prospective studies of patients with liver diseases are needed to determine if patients with slight liver fibrosis and high plasma YKL-40 are at risk of developing cirrhosis, and if plasma YKL-40 in combination with other biomarkers of liver fibrosis (e.g. hyaluronan, TIMP-I and procollagen type III) can predict the severity of liver fibrosis and will be useful in monitoring patients with liver fibrosis. Plasma YKL-40 may also be useful to monitor in patients with liver diseases during anti-fibrotic or anti-viral therapy.

Mechanistically, we do not know the function of YKL-40 in liver diseases and if it has a role in the pathogenesis of liver fibrosis/cirrhosis. YKL-40 is a growth factor of fibroblast, but it is unknown if YKL-40 stimulates the hepatic stellate cells and their production of collagen. Reducing the extracellular matrix production by activated hepatic stellate cells is crucial in preventing liver fibrosis. If YKL-40 has a role in development of liver fibrosis then inhibition of YKL-40 production or blocking of YKL-40 activity in patients with alcoholic liver disease or hepatitis C or B virus may be a valuable method to inhibit the development of liver fibrosis. It has recently been shown that a promoter SNP-131C/G in the YKL-40 gene determines YKL-40 plasma levels and is associated with the severity of hepatitis C virus induced liver fibrosis (Berres *et al.* 2009), suggesting a functional role of YKL-40 in liver fibrogenesis.

8. YKL-40 Gene and Protein Expression in Cancer Cells

YKL-40 gene mutations and polymorphism in cancer patients have not been described but studies are ongoing. A search of the YKL-40 gene in the "Expressed Sequence Tags" database (dbEST, NCBI) shows that YKL-40 mRNA is found in several types of cancer: breast, cervix, colon, esophagus, gastrointestinal, germ cell, glioblastoma, head and neck, kidney, leukaemia, liver, lung, lymphoma, melanoma, ovary, pancreas, skin, and uterus. Although numerous cell lines have been studied the secretion of YKL-40 protein *in vitro* has only been reported in a few human cancer cell lines; the MG63 osteosarcoma cell line (Johansen *et al.* 1992), the U87 glioblastoma cell line (Junker 2005a, Saidi *et al.* 2008), the glioma cell lines U1242MG, U343MG and U1231MG (Krona *et al.* 2007), a few myeloma cell lines (Mylin *et al.* 2006) and some tumor cell lines that originate from immature cells of the monocytic differentiation lineage (Kirkpatrick *et al.* 1997; Rehli *et al.* 2003; Verhoeckx *et al.* 2004).

Microarray gene analyses have shown that the YKL-40 gene is overexpressed, compared to normal tissue, in human papillary thyroid carcinoma (Huang *et al.* 2001), glioblastomas (Lal *et al.* 1999; Markert *et al.* 2001; Tanwar *et al.* 2002; Shostak *et al.* 2003; Colin *et al.* 2005; Nigro *et al.* 2005; Philips *et al.* 2006; Kroes *et al.* 2007; Ducray *et al.* 2008; Saidi *et al.* 2008), astrocytomas (Krona *et al.* 2007), and extracellular myxoid chondrosarcoma (Sjögren *et al.* 2003). YKL-40 protein expression in glioblastoma tissue assessed by immunohistochemistry is a biomarker of genetic subtype, therapeutic response, prognosis, and a differential diagnostic marker for histological subtypes of gliomas (Nutt *et al.* 2005; Pelloski *et al.* 2005, 2007; Philips *et al.* 2006; Rousseau *et al.* 2006). The YKL-40 gene is overexpressed in gliomas with EGFR amplification (Ducray *et al.* 2008), and other found that high YKL-40 protein expression in glioblastoma was associated with loss of chromosome

10q but not with amplification of the EGF-receptor (Pelloski *et al.* 2005). High YKL-40 gene- and protein expression in glioblastoma tissue are associated with poor radiation response and short time to disease progression and death (Nutt *et al.* 2005; Pelloski *et al.* 2005, 2007; Saidi *et al.* 2008). Pair-wise combinations of markers identified epidermal growth factor receptor variant III (EGFRvIII) and YKL-40 as prognostically important, and patients with EGFRvIII-negative/YKL-40-negative tumours had the best prognosis (Pelloski *et al.* 2007). In tumour cells of breast carcinomas YKL-40 protein is overexpressed compared to normal breast tissue (Kim *et al.* 2007; Qin *et al.* 2007; Roslind *et al.* 2007, 2008b) and was in a small study a predictor of short disease-free survival (Kim *et al.* 2007). This however, could not be confirmed in a large study of patients with primary breast cancer (Roslind *et al.* 2008b). In tumor cells of hepatocarcinoma, YKL-40 protein is overexpressed compared to normal liver tissue (Lau *et al.* 2006), and the YKL-40 gene and protein expressions are increased in peritumoral non-neoplastic pancreatic tissue compared to normal pancreatic tissue (Fukushima *et al.* 2005).

9. The Function of YKL-40 in Cancer

The biological functions of YKL-40 in cancer are unknown, and very few studies have evaluated the function of YKL-40 in cancer. It is suggested that YKL-40 plays a role in proliferation and differentiation of malignant cells, protects the cells from undergoing apoptosis, stimulates angiogenesis, has an effect on extracellular tissue remodeling and stimulates fibroblast activity/proliferation surrounding the cancer cells, although *in vivo* proofs of these hypotheses are yet to be obtained (Johansen *et al.* 2006a, 2006b, 2007c). Membrane receptors mediating the biological effects of YKL-40 are not known. YKL-40 induced activation of intracellular signal-transduction pathways suggests that YKL-40 interacts with one or several signalling components on the cell membrane (Recklies *et al.* 2002, 2005; Ling *et al.* 2004).

A possible role of YKL-40 in the malignant phenotype as a cellular survival factor in an adverse microenvironment is supported by the observation that up-regulated YKL-40 expression is found in human glioblastoma cells following stress stimuli such as hypoxia, serum depletion, ionizing radiation and chemotherapy (Junker *et al.* 2005b). The response in YKL-40 expression was late, 24 hours to 72 hours after stimuli, indicating that YKL-40 is a secondary response downstream of other mechanisms. Astrocytes transfected with YKL-40 have increased resistance to serum depletion and radiation as well as increased invasion potential (Nigro *et al.* 2005). *In vivo*, in glioblastoma cells, a positive association between YKL-40 and activated Akt1 pathways and MAPK intermediates has been reported (Pelloski *et al.* 2007). It was hypothesized that YKL-40 as a secreted protein may serve as an extracellular signal, inducing increased downstream activity of Ras (Pelloski *et al.* 2006), or may be a surrogate measurement of Ras/PI3-K activation (Pelloski *et al.* 2007). Transfection of the human glioblastoma cell line U87 with short-interfering RNA against vascular endothelial growth factor (VEGF-A) and implantation on a chick chorio-allantoic membrane resulted in a strongly up-regulation (8.48 fold) of the YKL-40 gene, suggesting a role of YKL-40 in regulating response of cancer cells to hypoxia (Saidi *et al.* 2008).

It has also been shown that *in vivo* tumour growth of the human U87 glioblastoma cells as xenografts on nude mice was delayed by treatment with monoclonal antibodies against human YKL-40 (Junker *et al.* 2007). This suggests that YKL-40 may be a potential cancer target.

The stroma around the periphery of solid tumours has several similarities with granulation tissue such as that found in wound-healing or inflammation, and regulates essential aspects of tumour proliferation, cell death, progression, matrix remodeling and angiogenesis, and subsequently promotes tumour growth and progression of metastatic disease (Dvorak 1986; Balkwill *et al.* 2001, 2004, 2005; Coussens *et al.* 2002; Declerck *et al.* 2004; Moss *et al.* 2005; de Visser *et al.* 2006; Tlsty *et al.* 2006; Lin *et al.* 2007). YKL-40 is a proliferation factor of fibroblasts (De Ceuninck *et al.* 2001; Recklies *et al.* 2002). One could speculate that YKL-40 secreted by cancer cells and inflammatory cells (macrophages, mast cells and neutrophils) surrounding and/or infiltrating the tumour may play a role in proliferation, activation and differentiation of the fibroblasts/myofibroblasts surrounding the tumour. YKL-40 could thereby influence development of the prominent desmoplastic stroma seen in both primary cancer and metastatic sites. This phenomenon, termed stromal reaction, includes activation of fibroblasts and myofibroblasts transformation, inflammation, enhanced secretion of cytokines, matrix proteins and metalloproteinases, and neovasculation. All of which play a role in cancer development and spread, affecting the proliferation, differentiation, invasion or regression of cancer cells, particularly in cancers of epithelial origin (Gregoire *et al.* 1995; Bissell *et al.* 2001; Kenny *et al.* 2003; Dranoff *et al.* 2004; Tlsty *et al.* 2006).

Finally, YKL-40 is syntesized by vascular smooth muscle cells (Millis *et al.* 1985, 1986; Shackelton *et al.* 1995; Malinda *et al.* 1999; Nishikawa *et al.* 2003), stimulates migration of endothelial cells (Malinda *et al.* 1999), and promotes vascular smooth muscle cell attachment, spreading and migration (Nishikawa *et al.* 2003) suggesting a function in angiogenesis and thereby playing a role in the growth of the tumour (Hillen *et al.* 2007).

10. Plasma YKL-40 Levels in Patients with Cancer

Several studies of patients with solid tumours have demonstrated that plasma YKL-40 is elevated (defined as > the 95th confidence limit of plasma YKL-40 in healthy age-matched subjects) and related to tumour stage in some patients with primary or metastatic cancer of the breast (Johansen *et al.* 1995; Jensen *et al.* 2003; Johansen *et al.* 2003), colorectal (Cintin *et al.* 1999; Cintin *et al.* 2002), ovary (Høgdall *et al.* 2003; Dehn *et al.* 2003; Dupont *et al.* 2004; Grønlund *et al.* 2006; Fredriksson *et al.* 2008), small cell lung (Johansen *et al.* 2004), prostate (Brasso *et al.* 2006; Johansen *et al.* 2007a; Kucur *et al.* 2008), kidney (Geertsen *et al.* 2003), pancreas (Fukushima *et al.* 2005; Fredriksson *et al.* 2008), endometrial (Diefenbach *et al.* 2007), cervical (Johansen *et al.* 2006c; Mitsuhashi A *et al.* 2009), head and neck (Roslind *et al.* 2008a), glioblastoma (Tanwar *et al.* 2002; Hormigo *et al.* 2006), and melanoma (Schmidt *et al.* 2006a, 2006b; Jensen *et al.* 2007). Plasma YKL-40 is also elevated in patients with acute myeloid leukemia (Bergmann *et al.* 2005), multiple myeloma (Mylin *et al.* 2006, 2008), and Hodgkin lymphoma (Biggar *et al.* 2008) (Table 2). High plasma YKL-40 in patients with all these different types of solid tumours and hematologic malignancies is an independent prognostic biomarker of short recurrence or progression-free interval and short overall survival. This was found in patients with local or metastatic cancer, at time of first cancer diagnosis and at time of relapse, and plasma YKL-40 was independent of other prognosticators when tested in multivariate Cox analysis (e.g. estrogen receptor status, serum HER2, CEA, CA125 and LDH) (Johansen *et al.* 2006a, 2006b and 2007c).

Table 2. Plasma YKL-40 as a potential biomarker of poor prognosis in cancer patients

Cancer type	Retrospective studies	Prospective studies	Total number of patients	% of patients with elevated plasma YKL-40
Breast	3	0	425	9 - 61
Colorectal	1	2	1027	16 - 44
Endometrial	1	0	34	76
Ovarian	4	0	252	55 - 72
Prostate	3	0	246	43
Kidney	1	0	58	83
Small cell lung	1	0	131	22 - 40
Pancreas	2	0	68	56
Cervival	0	2	257	20 - 100
Head and neck	1	0	173	53
Glioblastoma	1	1	208	57 - 72
Melanoma	3	0	575	13 - 45
Acute misleid leukemia	1	0	77	52
Multiple myeloma	2	0	116	29 - 56
Hodgkin lymphoma	1	0	470	Elevated 3.6 fold

The percentage of patients with elevated plasma YKL-40 is calculated as a the number of patients in the different studies with a plasma YKL-40 level above the age-adjusted 95th percentile of the plasma YKL-40 level in healthy subjects. The percentage reflect differences according to the stage of the cancer disease in the different studies.

A high plasma YKL-40 in patients with first recurrence of breast cancer predicted less responsiveness to anthracycline therapy (Jensen *et al.* 2003), and high plasma YKL-40 was a predictor of second-line chemoresistance in patients with ovarian cancer (Grønlund *et al.* 2006).

Five studies have suggested that plasma YKL-40 may be useful for monitoring disease recurrence and progression in cancer patients after treatment (Cintin *et al.* 2002; Schmidt *et al.* 2006b; Hormigo *et al.* 2006; Johansen *et al.* 2007a; Roslind *et al.* 2008a). Plasma YKL-40 decreased after curative operation for colorectal cancer, and patients with elevated plasma YKL-40 six months after operation had shorter recurrence-free interval and overall survival (Cintin *et al.* 2002). Furthermore, elevated plasma YKL-40 during the later follow-up period of these patients operated for colorectal cancer increased the risk of recurrence and death (Cintin *et al.* 2002). In patients operated for stage I and II melanoma an association was found between plasma YKL-40 during follow-up and recurrence-free survival and overall survival (Schmidt *et al.* 2006b). In patients operated for high-grade gliomas plasma YKL-40 during follow-up was lower in patients with no radiographic evidence of disease compared to patients with signs of disease, and high plasma YKL-40 during follow-up was associated with short survival (Hormigo *et al.* 2006). In patients with prostate cancer treated with endocrine therapy elevated plasma YKL-40 during treatment was related to short survival (Johansen *et al.* 2007a). During follow-up of patients with squamous cell carcinoma of the head and neck (TNM stage III and IV) a high plasma YKL-40 after radiotherapy predicted poorer overall survival within 6 months (Roslind *et al.* 2008a).

Until recently there was no studies regarding plasma YKL-40 as a biomarker for screening of cancer. We have recently determined plasma YKL-40 in a prospective cohort study of 8899 subjects (aged 20-95 years) from the Danish general population, the Copenhagen City Heart Study, followed for 11 years for cancer incidence and for 14 years for death. We found that elevated plasma YKL-40 predicted a 3.4 fold increased risk of gastrointestinal cancer in the general population and decreased survival after any cancer diagnosis (Johansen *et al.* 2009). Furthermore, we have measured plasma YKL-40 in a prospective, population based study of 4987 subjects (aged 18-97 years) referred to endoscopy due to symptoms or other risk factors for colorectal cancer. 303 subjects were diagnosed with colorectal cancer and multivariate logistic regression analysis including plasma YKL-40, age, sex, BMI, smoking, alcohol intake and co-morbidity demonstrated that plasma YKL-40 independently predicted colorectal cancer (Johansen *et al.* 2008a). Both studies suggest that plasma YKL-40 may be useful in the assessment of risk for colorectal cancer.

11. Plasma YKL-40 - A Potential Cancer Biomarker?

A "Biomarker" (Biological marker) can be defined as "A characteristic that is objectively measured and evaluated as an indicator of normal biological processes, pathogenic processes, or pharmacological responses to a therapeutic intervention" (Atkinson 2001). The use of plasma YKL-40 has not received FDA approval for use as a biomarker in patients with cancer or any other disease. Most of the clinical studies of plasma YKL-40 are small and retrospective, and the definitive value of plasma YKL-40 as a valuable biomarker of inflammation, tissue remodeling, liver fibrosis and cancer is yet to be determined. It is still unknown if knowledge of the plasma YKL-40 level in an individual patient is so reliable that it can be used to make clinical decisions that will improve outcome of the patient. Generally, elevated plasma YKL-40 levels are not found in healthy subjects, and high plasma YKL-40 levels can reflect several types of disease reviewed briefly in this chapter. It will therefore be necessary to consider co-morbidity and repeated measurements when evaluating plasma YKL-40 as a biomarker for any type of disease. Furthermore, a low plasma YKL-40 level cannot rule out cancer or other diseases, since plasma YKL-40 does not have very high sensitivity.

The term "tumour marker" embraces a spectrum of molecules of widely divergent characteristics, but sharing an association with malignancy that facilitates their application in the clinical detection (diagnosis, screening) and management (monitoring, prognosis) of cancer patients (Werner *et al.* 1993). Table 3 gives the present results of plasma YKL-40 as a cancer biomarker. YKL-40 is neither organ nor tumour specific, but plasma YKL-40 levels may be useful as a "prognosticator" and in monitoring of cancer patients, and may also have a role in screening. Elevated plasma YKL-40 levels are found in a subgroup of patients with 15 different types of cancer, and the highest plasma YKL-40 levels are found in patients with metastatic cancer and with the poorest prognosis. Furthermore, plasma YKL-40 provides independent information of survival. The potential values of plasma YKL-40 as a biomarker in monitoring and screening of cancer need more studies, and its value in combinations with other biomarkers has to be determined.

Table 3. Is plasma YKL-40 a cancer biomarker?

Tumour specific? - produced exclusively by specific cancer cells	No
High specificity? - absent in healthy or benign disease	No
High sensitivity? - present frequently in the target cancer	No
Useful for screening? - is detectable in early stage subclinical disease	?
Prognosticator? - reflects prognosis in an individual cancer patient	Yes
Predictor? - reflects treatment response in an individual cancer patient	?

In order to propose guidelines on how promising tumour markers progress from the laboratory into the clinic, Hayes and colleagues (1996, 1998) introduced the "Tumor Marker Utility Grading System". According to this system, plasma YKL-40 is on the "Utility scale +" or "Utility scale +/-", and a number of validation requirements have to be fulfilled before plasma YKL-40 can be considered to have reached "Level of Evidence I", whereupon clinical implementation is feasible. Most of the plasma YKL-40 biomarker studies are on

"Level of Evidence III", defined as retrospective studies where samples are not originally collected with the intent of testing the value (e.g. prognostic value).

There are therefore limitations to the conclusions that can be made from the present studies of plasma YKL-40 as a biomarker. According to the "Tumor Marker Utility Grading System" guidelines, the next step would be to launch an appropriate prospective study where the benefit of using plasma YKL-40 levels in the clinical decision-making process is assessed.

Endpoints should include overall survival, disease-free survival, quality of life and cost-effectiveness. The study could be designed either as a single, highly-powered, prospective, controlled study with the primary objective of testing plasma YKL-40 level as a "prognosticator" or a similar prospective study where the primary goal could be the testing of a therapeutic hypothesis and secondly testing plasma YKL-40 as a biomarker. Finally "Level of Evidence II" studies are either 1) highly-powered prospective studies specifically addressing the issue of the utility of the biomarker or 2) an overview or meta-analysis of studies, each of which have a lower level of evidence.

Conclusion

It has been challenging and rewarding to work with the strange and unique protein YKL-40, and to investigate the potential of plasma YKL-40 as a biomarker in human diseases. It has also been exciting to see the increasing interest in YKL-40. There is reason to be optimistic that plasma YKL-40 may have a place in the routine clinical management of a number of human diseases, since plasma YKL-40 may provide new information compared to rutinely used biomarkers.

However, YKL-40 is more than a biomarker, and the study of YKL-40 has just started and many questions regarding this protein remain to be answered. The complete biological functions of YKL-40 are unclear and its role in cancer development and the mechanisms by which it reflects cancer aggressiveness and cancer progression are poorly understood. It is not known if YKL-40 has a receptor. The present studies suggest that YKL-40 has a role in both embryonic/fetal growth and in pathological growth like cancer, has metastatic potential, and probably also has a function in inflammation and tissue remodeling processes and in pathological conditions leading to fibrosis. The mechanisms by which stimuli lead to increased expression and synthesis of YKL-40 are unknown and deserve intensive studies. It remains to be determined whether YKL-40 is related to the innate immune responses and the autoimmune responses underlying autoimmune diseases. A major issue to explore is the question if YKL-40 could become a target for the development of new cancer therapeutics and in the treatment of inflammatory diseases. Therefore much more research on YKL-40 is needed.

References

Alexander J, Tung BY, Croghan A, Kowdley KV. Effect of iron depletion on serum markers of fibrogenesis, oxidative stress and serum liver enzymes in chronic hepatitis C: results of a pilot study. *Liver Int.* 2007; 27: 268-73.

Arion D, Unger T, Lewis DA, Levitt P, Mirnics K. Molecular evidence for increased expression of genes related to immune and chaperone function in the prefrontal cortex in schizophrenia. *Biol Psychiatry.* 2007; 62: 711-21.

Atkinson AJ, Jr. Biomarkers and surrogate endpoints: Preferred definitions and conceptual framework. *Clin Pharmacol Ther.* 2001; 69: 89-95.

Badariotti F, Kypriotou M, Lelong C, Dubos MP, Renard E, Galera P, Favrel P. The phylogenetically conserved molluscan chitinase-like protein 1 (cg-clp1), homologue of human hc-gp39, stimulates proliferation and regulates synthesis of extracellular matrix components of mammalian chondrocytes. *J Biol Chem.* 2006; 281: 29583-96.

Baeten D, Boots AMH, Steenbakkers PGA, Elewaut D, Bos E, Verheijden GFM, Verbruggen G, Miltenburg AMM, Rijnderes AWM, Veys EM, de Keyser F. Human cartilage gp-39+, CD16+ monocytes in peripheral blood and synovium.Correlation with joint destruction in rheumatoid arthritis. *Arthritis Rheum.* 2000; 43: 1233-43.

Balkwill F, Mantovani A. Inflammation and cancer: back to Virchow? *Lancet.* 2001; 357: 539-45.

Balkwill F, Coussens LM. An inflammatory link. *Nature.* 2004; 431: 405-6.

Balkwill F, Charles KA, Mantovani A. Smoldering and polarized inflammation in the initiation and promotion of malignant disease. *Cancer Cell.* 2005; 7: 211-7.

Belge KU, Dayyani F, Horelt A, Siedlar M, Frankenberger M, Frankenberger B, Espevik T, Ziegler-Heitbrock L. The proinflammatory CD14+CD16+DR++ monocytes are a major source of TNF. *J Immunol.* 2002; 168: 3536-42.

Benz K, Breit S, Lukoschek M, Mau H, Richter W. Molecular analysis of expansion, differentiation, and growth factor treatment of human chondrocytes identifies differentiation markers and growth-related genes. *Biochem Biophys Res Commun.* 2002; 293: 284-92.

Bergmann OJ, Johansen JS, Klausen TW, Mylin AK, Kristensen JS, Kjeldsen E, Johnsen HE. High serum concentration of YKL-40 is associated with short survival in patients with acute myeloid leukemia. *Clin Cancer Res.* 2005; 11: 8644-52.

Berres ML, Papen S, Pauels K, Schmitz P, Zaldivar MM, Hellerbrand C, Mueller T, Berg T, Weiskirchen R, Trautwein C, Wasmuth HE. A functional variation in CHI3L1 is associated with severity of liver fibrosis and YKL-40 serum levels in chronic hepatitis C infection. *J Hepatol.* 2009; in press.

Biggar RJ, Johansen JS, Smedby KE, Rostgaard K, Chang ET, Adami H-O, Glimelius B, Molin D, Hamilton-Dutoit S, Melbye M, Hjalgrim H. Serum YKL-40 and IL-6 levels in Hodgkin lymphoma. *Clin Cancer Res.* 2008; 14: 6974-9.

Bigg HF, Wait R, Rowan AD, Cawston TE. The mammalian chitinase-like lectin, YKL-40, binds specifically to type I collagen fibril formation. *J Biol Chem.* 2006; 281: 21082-95.

Bissel MJ, Radisky D. Putting tumours in context. *Nat Rev Cancer.* 2001; 1: 46-54.

Bonneh-Barkay D, Bissel SJ, Wang G, Fish KN, Nicholl GCB, Darko SW, Medina-Flores R, Murphey-Corb M, Rajakumar PA, Nyaundi J, Mellors JW, Bowser R, Wiley CA. YKL-40, a marker of simian immunodeficiency virus encephalitis, modulates the biological activity of basic fibroblast growth factor. *Am J Pathol* 2008; 173: 130-43.

Boot RG, van Achterberg TAE, van Aken BE, Renkema GH, Jacobs MJHM, Aerts JMFG, de Vries CJM. Strong induction of members of the chitinase family of proteins in atherosclerosis. Chitotriosidase and human cartilage gp-39 expressed in lesion macrophages. *Arterioscler Thromb Vasc Biol.* 1999; 19: 687-94.

Brasso K, Christensen IJ, Johansen JS, Teisner B, Garnero P, Price PA, Iversen P. Prognostic value of PINP, bone alkaline phosphatase, CTX-I, and YKL-40 in patients with metastatic prostate carcinoma. *Prostate.* 2006; 66: 503-13.

Bussink AP, Speijer D, Aerts JMFG, Boot RG. Evolution of mammalian chitinase(-like) members of family 18 glycosyl hydrolases. *Genetics.* 2007; 177: 959-70.

Chung C, Tallerico T, Seeman P. Schizophrenia hippocampus has elevated expression of chondrex glycoprotein gene. *Synapse.* 2003; 50: 29-34.

Chupp GL, Lee CG, Jarjour N, Shim YM, Holm CT, He S, Dziura JD, Reed J, Coyle AJ, Kiener P, Cullen M, Grandsaigne M, Dombret MC, Aubier M, Pretolani M, Elisa JA. A chitinase-like protein in the lung and circulation of patients with severe asthma. *N Engl J Med.* 2007; 357: 2016-27.

Cintin C, Johansen JS, Christensen IJ, Price PA, Sørensen S, Nielsen HJ. Serum YKL-40 and colorectal cancer. *Br J Cancer.* 1999; 79: 1494-9.

Cintin C, Johansen JS, Christensen IJ, Price PA, Sørensen S, Nielsen HJ. High serum YKL-40 level after surgery for colorectal carcinoma is related to short survival. *Cancer.* 2002; 95: 267-74.

Colin C, Baeza N, Bartoli C, Fina F, Eudes N, Nanni I, martin PM, Ouafik L, Figarella-Branger D. Identification of genes differentially expressed in glioblastoma versus pilocytic astrocytoma using suppression subtractive hybridization. *Oncogene.* 2006; 25: 2818-29.

Colton CA, Mott RT, Sharpe H, Xu Q, van Nostrand WE, Vitek MP. Expression profiles for macrophage alternative activaiton genes in AD and in mouse models of AD. *J Neuroinflam.* 2006; 3: 1-12.

Combe B, Dougados M, Goupille P, Cantagrel A, Eliaou JF, Sibilia J, Meyer O, Sany J, Daures JP, Dubois A. Prognostic factors for radiographic damage in early rheumatoid arthritis: a multiparameter prospective study. *Arthritis Rheum.* 2001; 44: 1736-43.

Cope AP, Patel SD, Hall F, Congia M, Hubers HAJM, Verheijden GF, Boots AMH, Menon R, Trucco M, Rijnders AMW, Sønderstrup G. T cell responses to a human cartilage autoantigen in the context of rheumatoid arthritis-associated and nonassociated HLA-DR4 alleles. *Arthritis Rheum.* 1999; 42: 1497-1507.

Coussens LM, Werb Z. Inflammation and cancer. *Nature.* 2002; 420; 860-7.

Dasuri K, Antonovici M, Chen K, Wong K, Standing K, Ens W, El-Gabalawy H, Wilkins JA. The synovial proteome: analysis of fibroblast-like synoviocytes. *Arthritis Res Ther.* 2004; 6: R161-8.

De Ceuninck F, Gaufillier S, Bonnaud A, Sabatini M, Lesur C, Pastoureau P. YKL-40 (cartilage gp-39) induces proliferative events in cultured chondrocytes and synoviocytes and increases glycosaminoglycan synthesis in chondrocytes. *Biochem Biophys Res Commun.* 2001; 285: 926-31.

De Visser KE, Eichten A, Coussens LM. Paradoxical roles of the immune system during cancer development. *Nat Rev Cancer.* 2006; 6: 24-37.

DeClerck YA, Mercurio AM, Stack MS, Chapman HA, Zutter MM, Muschel RJ, Raz A, Matrisian LM, Sloane BF, Noel A, Hendrix MJ, Coussens L, Padarathsingh M. Proteases, extracellular matrix, and cancer. *Am J Pathol.* 2004; 164: 1131-9.

Dehn H, Høgdall EVS, Johansen JS, Price PA, Jørgensen M, Engelholm SAA, Høgdall CK. Plasma YKL-40, as a prognostic tumor marker in recurrent ovarian cancer. *Acta Obstet Gynecol Scand.* 2003; 82: 287-93.

Dickey BF. Editorial. Exoskeletons and exhalation. *N Engl J Med.* 2007; 357: 2082-4.

Diefenbach CS, Shah Z, Iasonos A, Barakat RR, Levine DA, Aghajanian C, Sabbatini P, Hensley ML, Konner J, Tew W, Spriggs D, Fleisher M, Thaler H, Dupont J. Preoperative serum YKL-40 is a marker for detection and prognosis of endometrial cancer. *Gynecol Oncol.* 2007; 104: 435-42.

Dranoff G. Cytokines in cancer pathogenesis and cancer therapy. *Nature Reviews.* 2004; 4: 11-22.

Ducray F, Idbaih A, de Reynies A, Bieche I, Thillet J, Mokhtari K, Lair S, Marie Y, Paris S, Vidaud M, Hong-Xuan K, Delattre O, Delattre JY, Sanson M. Anaplastic oligodendrogliomas with 1p19q codeletion have a proneural gene expression profile. *Molecular Cancer.* 2008; 7: 41.

Dupont J, Tanwar MK, Thaler HT, Fleisher M, Kauff N, Hensley ML, Sabbatini P, Anderson S, Aghajanian C, Holland EC, Spriggs DR. Early detection and prognosis of ovarian cancer using serum YKL-40. *J Clin Oncol.* 2004; 22: 3330-9.

Dvorak HF. Tumors: wounds that do not heal. Similarities between tumor stroma generation and wound healing. *N Engl J Med.* 1986; 315: 1650-9.

Erzin Y, Uzun H, Karatas A, Celik AF. Serum YKL-40 as a marker of disease activity and stricture formation in patiens with Crohn's disease. *J Gastroenterol Hepatol.* 2008; 23: e357-62.

Esmat G, Metwally M, Zalata KR, Gadalla S, Abdel-Hamid M, Abouzied A, Shaheen AA, El-Raziky M, Khatab H, El-Kafrawy S, Mikhail N, Magder LS, Afdhal NH, Strickland GT. Evaluation of serum biomarkers of fibrosis and injury in Egyptian patients with chronic hepatitis C. *J Hepatol.* 2007; 46: 620-7.

Fjeldstad K, Kolset SO. Decreasing the metastatic potential in cancers – targeting the heparan sulfate proteoglycans. *Curr Drug Targets.* 2005; 6: 665-82.

Fredriksson S, Horecka J, Brustugun OT, Schlingemann J, Koong AC, Tibshirani R, Davis RW. Multiplexed proximity ligation assays to profile putative plasma biomarkers relevant to pancreatic and ovarian cancer. *Clin Chem.* 2008; 54: 582-9.

Funkhouser JD, Aronson NN. Chitinase family GH18: evolutionary insights from the genomic history of a diverse protein family. *BMC Evol Biol.* 2007; 7: 96.

Fukushima N, Koopmann J, Sato N, Prasad N, Carvalho R, Leach SD, Hruban RH, Goggins M. Gene expression alterations in the non-neoplastic parenchyma adjacent to infiltrating pancreatic ductal adenocarcinoma. *Mod Pathol.* 2005; 18: 779-87.

Fusetti F, Pijning T, Kalk KH, Bos E, Dijkstra BW. Crystal structure and carbohydrate binding properties of the human cartilage glycoprotein-39. *J Biol Chem.* 2003; 278: 37753-60.

Fontana RJ, Goodman ZD, Dienstag JL, Bonkovsky HL, Naishadham D, Sterling RK, Su GL, Ghosh M, Wright EC, HALT-C Trial Group. Relationship of serum fibrosis markers with liver fibrosis stage and collagen content in patients with advanced chronic hepatitis C. *Hepatology.* 2008; 47: 789-98.

Gabay C, Kushner I. Acute-phase proteins and other systemic responses to inflammation. *N Engl J Med.* 1999; 340: 448-54.

Geertsen PF, Johansen JS, von der Maase H, Jensen BV, Price PA. High pretreatment serum level of YKL-40 is related to short survival in patients with advanced renal cell carcinoma treated with high-dose continuous intravenous infusion of interleukin-2. ASCO 2004 *Ann Meet Proc.* 2003; 22: 399 (Abstract 1603).

Gonzalez P, Epstein DL, Luna C, Liton PB. Characterization of free-floating spheres from human trabecular meschwork (HTM) cell culture in vitro. *Exp Eye Res.* 2006; 82: 959-67.

Gregoire M, Lieubeau B. The role of fibroblasts in tumor behavior. *Cancer Met Rev.* 1995; 14: 339-50.

Grønlund B, Høgdall EVS, Christensen IJ, Johansen JS, Nørgaard-Pedersen B, Engelholm SA, Høgdall C. Pre-treatment prediction of chemoresistance in second-line chemotherapy of ovarian carcinoma: value of serological tumor marker determination (tetranectin, YKL-40, CASA, CA125). *Int J Biol Markers.* 2006; 21: 141-8.

Hakala BE, White C, Recklies AD. Human cartilage gp-39, a major secretory product of articular chondrocytes and synovial cells, is a mammalian member of a chitinase protein family. *J Biol Chem.* 1993; 268: 25803-10.

Harvey S, Weisman M, O'Dell J, Scott T, Krusemeier M, Visor J, Swindlehurst C. Chondrex: new marker of joint disease. *Clin Chem.* 1998; 44: 509-16.

Harvey S, Whaley J, Eberhardt K. The relationship between serum levels of YKL-40 and disease progression in patients with early rheumatoid arthritis. *Scand J Rheumatol.* 2000; 29: 391-3.

Hashimoto S, Suzuki T, Dong HY, Yamazaki N, Matsushima K. Serial analysis of gene expression in human monocytes and macrophages. *Blood* 1999; 94: 837-44.

Hayes DF, Bast RC, Desch CE, Fritsche H Jr., Kemeny NE, Jessup JM., Locker GY, Macdonald RG, Mennel RG, Norton L, Ravdin P, Taube S, Winn RJ. Tumor Marker Utility Grading System: a framework to evaluate clinical utility of tumor markers. *J Natl Cancer Inst.* 1996; 88: 1456-66.

Hayes DF. Determination of clinical utility of tumor markers: a tumor marker utility grading system. *Recent Results Cancer Res.* 1998; 152: 71-85.

Henrissat B, Davies G. Structural and sequence-based classification of glycoside hydrolases. *Curr Opin Struct Biol.* 1997; 7: 637-44.

Hillen F, Griffioen AW. Tumour vascularization: sprouting angiogenesis and beyond. *Cancer Metastasis Rev.* 2007; 26: 489-502.

Hormigo A, Gu B, Karimi S, Riedel E, Panageas KS, Edgar MA, Tanwar MK, Rao JS, Fleisher M, DeAngelis LM, Holland EC. YKL-40 and matrix metalloproteinase-9 as potential serum biomarkers for patients with high-grade gliomas. *Clin Cancer Res*. 2006; 12: 5698-704.

Houston DR, Recklies AD, Krupa JC, van Aalten DMF. Structure and ligand-induced conformational change of the 39-kDa glycoprotein from human articular chondrocytes. *J Biol Chem*. 2003; 278: 30206-12.

Huang Y, Prasad M, Lemon WJ, Hampel H, Wright FA, Kornacker K, LiVolsi V, Frankel W, Kloos RT, Eng C, Pellegata NS, de la Chapelle A. Gene expression in papillary thyroid carcinoma reveals highly consistent profiles. *Proc Natl Acad Sci USA*. 2001; 98: 15044-9.

Høgdall EVS, Johansen JS, Kjaer SK, Price PA, Blaakjaer J, Høgdall CK. Stability of YKL-40 concentration in blood samples. *Scand J Clin Lab Invest*. 2000; 60: 247-52.

Høgdall EVS, Johansen JS, Kjaer SK, Price PA, Christensen L, Blaakaer J, Bock JE, Glud E, Høgdall CK. High plasma YKL-40 level in patients with ovarian cancer stage III is related to shorter survival. *Oncol Rep*. 2003; 10: 1535-8.

Imabayashi H, Mori T, Gojo S, Kioyono T, Sugiyama T, Irie R, Isoga T, Hata J, Toyama Y, Umezawa A. Redifferentiation of dedifferentiated chondrocytes and chondrogenesis of human bone marrow stromal cells via chondrosphere formation with expression profiling by large-scale cDNA analysis. *Exp Cell Res*. 2003; 288: 35-50.

Jacques C, Recklies AD, Levy A, Berenbaum F. HC-gp39 contributes to chondrocyte differentiation by inducing SOX9 and type II collagen expressions. *Osteoarthritis Cartilage*. 2007; 15: 138-46.

Jensen BV, Johansen JS, Price PA. High levels of serum HER-2/neu and YKL-40 independently reflect aggressiveness of metastatic breast cancer. *Clin Cancer Res*. 2003; 9: 501-12.

Jensen BV, Qvortrup C, Christensen IJ, Pfeiffer P, Nicolaisen N, Nielsen S, Johansen JS. Association of biomarkers of inflammation and YKL-40 with survival of patients with metastatic colorectal cancer. Proceedings of 2008 Gastrointestinal Cancer Symposium, Orlando, Florida. 2008; (Abstract 323).

Jensen M, Johansen JS, Christensen IJ, Schmidt H, Nørgaard PH, Bastholt L. Serum YKL-40 is elevated in a subgroup of patients after surgery for stage IIb-III melanoma. ASCO 2007 *Ann Meet Proc*. 2007; 25: 490 (Abstract 8574).

Johansen JS, Williamson MK, Rice JS, Price PA. Identification of proteins secreted by human osteoblastic cells in culture. *J Bone Miner Res*. 1992; 7: 501-12.

Johansen JS, Jensen HS, Price PA. A new biochemical marker for joint injury. Analysis of YKL-40 in serum and synovial fluid. *Br J Rheumatol*. 1993; 32: 949-55.

Johansen JS, Cintin C, Jørgensen M, Kamby C, Price PA. Serum YKL-40: a new potential marker of prognosis and location of metastases of patients with recurrent breast cancer. *Eur J Cancer* 1995; 31A: 1437-42.

Johansen JS, Hvolris J, Hansen M, Backer V, Lorenzen I, Price PA. Serum YKL-40 levels in healthy children and adults. Comparison with serum and synovial fluid levels of YKL-40 in patients with osteoarthritis or trauma of the knee joint. *Br J Rheumatol*. 1996; 35: 553-9.

Johansen JS, Møller S, Price PA, Bendtsen F, Junge J, Garbarsch C. Plasma YKL-40: a new potential marker of fibrosis in patients with alcoholic cirrhosis? *Scand J Gastroenterol.* 1997; 32: 582-90.

Johansen JS, Baslund B, Garbarsch C, Hansen M, Stoltenberg M, Lorenzen I, Price PA. YKL-40 in giant cells and macrophages from patients with giant cell arteritis. *Arthritis Rheum.* 1999a; 42: 2624-30.

Johansen JS, Stoltenberg M, Hansen M, Florescu A, Hørslev-Petersen K, Lorenzen I, Price PA. Serum YKL-40 concentrations in patients with rheumatoid arthritis: relation to disease activity. *Rheumatology* 1999b; 38: 618-26.

Johansen JS, Christoffersen P, Møller S, Price PA, Henriksen JH, Garbarsch C, Bendtsen F. Serum YKL-40 is increased in patients with hepatic fibrosis. *J Hepatol.* 2000; 32: 911-20.

Johansen JS, Kirwan JR, Price PA, Sharif M. Serum YKL-40 concentrations in patients with early rheumatoid arthritis: relation to joint destruction. *Scand J Rheumatol.* 2001a; 30: 297-304.

Johansen JS, Olee T, Price PA, Hashimoto S, Ochs RL, Lotz M. Regulation of YKL-40 production by human articular chondrocytes. *Arthritis Rheum.* 2001b; 44: 826-37.

Johansen JS, Christensen IJ, Riisbro R, Greenall M, Han C, Price PA, Smith K, Brünner N, Harris AL. High serum YKL-40 levels in patients with primary breast cancer is related to short recurrence free survival. *Breast Cancer Res Treat.* 2003; 80: 15-21.

Johansen JS, Drivsholm L, Price PA, Christensen IJ. High serum YKL-40 level in patients with small cell lung cancer is related to early death. *Lung Cancer.* 2004; 46: 333-40.

Johansen JS, Krabbe K, Møller K, Pedersen BK. Circulating YKL-40 levels during human endotoxaemia. *Clin Exp Immunol.* 2005a; 140: 343-8.

Johansen JS, Milman N, Hansen M, Garbarsch C, Price PA, Graudal N. Increased serum YKL-40 in patients with pulmonary sarcoidosis. A potential marker of disease activity? *Respiratory Medicine.* 2005b; 99: 396-402.

Johansen JS. Studies on serum YKL-40 as a biomarker in diseases with inflammation, tissue remodeling, fibrosis and cancer. *Dan Med Bull.* 2006a; 53: 172-209.

Johansen JS, Jensen BV, Roslind A, Nielsen D, Price PA. Review. Serum YKL-40, a new prognostic biomarker in cancer patients? *Cancer Epidemiol Biomarkers Prev.* 2006b; 15: 194-202.

Johansen JS, Roslind A, Palle C, Christensen IJ, Nielsen HJ, Price PA, Nielsen D, Mosgaard B. Serum YKL-40 levels in patients with cervical cancer are elevated compared to patients with cervical intraepithelial neoplasia and healthy controls. ASCO *Ann Meet Proc*. 2006c; 24: 267 (Abstract 5047).

Johansen JS, Brasso K, Iversen P, Teisner B, Garnero P, Price PA, Christensen IJ. Changes of biochemical markers of bone turnover and YKL-40 following hormonal treatment for metastatic prostate cancer are related to survival. *Clin Cancer Res.* 2007a; 13: 3244-9.

Johansen JS, Høyer PE, Larsen LA, Price PA, Møllgård K. YKL-40 protein expression in the early developing human musculoskeletal system. *J Histochem Cytochem.* 2007b; 55: 1213-28.

Johansen JS, Jensen BV, Roslind A, Price PA. Review. Is YKL-40 a new therapeutic target in cancer? *Expert Opin Ther Targets* 2007c; 11: 219-34.

Johansen JS, Møllgård K. YKL-40: Common and distinct in cancer and autoimmunity, frequent in early development and abundant in human embryonic stem cells. In: International Symposium Honoring Prof. David Naor, September 2-3, 2007, Abstract # 18. Ein Kerem, Jerusalem, 2007d.

Johansen JS, Christensen IJ, Price PA, Nielsen HJ, and the Danish Colorectal Cancer Study Group Serum YKL-40 in risk assessment for colorectal cancer. A population based, prospective study of 4987 subjects at risk of colorectal cancer. *ASCO Ann Meet Proc.* 2008a; 26: 212 (Abstract 4136).

Johansen JS, Lottenburger T, Nielsen HJ, Jensen JEB, Svendsen MN, Kollerup G, Christensen IJ. Diurnal, weekly, and long-time variation in serum concentrations of YKL-40 in healthy subjects. *Cancer Epidemio Biomarkers Prev.* 2008b; 17: 2603-8.

Johansen JS, Pedersen AN, Schroll M, Jørgensen T, Pedersen BK, Bruunsgaard H. High serum YKL-40 level in a cohort of 80-years old is associated with increased risk of all-cause mortality. *Clin Exp Immunol.* 2008c; 151: 260-6.

Johansen JS, Bojesen SE, Mylin AK, Frikke-Schmidt R, Price PA, Nordestgaard BG. Elevated plasma YKL-40 predicts increased risk of gastrointestinal cancer and decreased survival after any cancer diagnosis in the general population. *J Clin Oncol* 2009, in press.

Joosten LAB, Coenen de-Roo CJJ, Helsen MMA, Lubberts E, Boots AMH, van den Berg, WB, Miltenburg AMM. Induction of tolerance with intranasal administration of human cartilage gp-39 in DBA/1 mice. *Arthritis Rheum.* 2000; 43: 645-55.

Junker N, Johansen JS, Andersen CB, Kristjansen PEG. Expression of YKL-40 by peritumoral macrophages in human small cell lung cancer. *Lung Cancer.* 2005a; 48: 223-31.

Junker N, Johansen JS, Hansen LT, Lund EL, Kristjansen PEG. Regulation of YKL-40 expression during genotoxic or microenvironmental stress in human glioblastoma cells. *Cancer Sci.* 2005b; 96: 183-190.

Junker N, Johansen JS, Kristjansen PEG, Price PA. Monoclonal antibody therapy against YKL-40 delays growth in glioblastoma xenografts. American Association for Cancer Research Annual Meeting: Proceedings 2007; Abstract nr. 4099.

Kacira T, Hanimoglu H, Kucur M, Sanus GZ, Kafadar AM, Tanriverdi T, Kaynar MY. Elevated cerebrospinal fluid and serum YKL-40 levels are not associated with symptomatic vasospasm in patients with aneurysmal subarachnoid haemorrhage. *J Clin Neuroscience.* 2008; 15: 1011-6.

Kamal SM, Turner B, He Q, Rasenack J, Bianchi L, Al Tawil A, Nooman A, Massoud M, Koziel MJ, Afdhal NH. Progression of fibrosis in hepatitis C with and without schistosomiasis: correlation with serum markers of fibrosis. *Hepatology.* 2006; 43: 771-9.

Kastrup J, Johansen JS, Winkel P, Hansen JF, Hildebrandt P, Jensen GB, Jespersen CM, Kjøller E, Kolmos HJ, Lind I, Nielsen H, Gluud C, and the CLARICOR Trial Group. High serum concentration of YKL-40 in patients with stable coronary artery disease is associated with increased risk of myocardial infarction, cardiovascular death and all-cause mortality. *Eur Heart J* 2009; in press.

Kavanaugh A, Genovese M, Baughman J, Kivitz A, Bulpitt K, Olsen N, Weisman M, Mattesson E, Furst D, van Vollenhoven R, Anderson J, Cohen S, Wei N, Meijerink J,

Jacobs C, Mocci S. Allele and antigen-specific treatment of rheumatoid arthritis: a double blind, placebo controlled phase 1 trial. *J Rheumatol.* 2003; 30: 449-54.

Kawada M, Chen CC, Arihiro A, Nagatani K, Watanabe T, Mizoguchi E. Chitinase 3-like-1 enhances bacterial adhesion to colonic epithelial cells through the interaction with bacterial chitin-binding protein. *Lab Invest.* 2008; 88: 883-95.

Kawamura K, Shibata T, Saget O, Peel D, Bryant PJ. A new family of growth factors produced by the fat body and active on Drosophila imaginal disc cells. *Development.* 1999; 126: 211-9.

Kelleher TB, Mehta SH, Bhaskar R, Sulkowski M, Astemborski J, Thomas, DL, Moore RE, Afdhal NH. Prediction of hepatic fibrosis in HIV/HCV co-infected patients using serum fibrosis markers: the SHASTA index. *J Hepatol.* 2005; 43: 78-84.

Kenny PA, Bissell MJ. Tumor reversion: correction of malignant behavior by microenvironmental cues. *Int J Cancer.* 2003; 107: 688-95.

Kim SH, Das K, Noreen S, Coffman F, Hameed M. Prognostic implications of immunohistochemically detected YKL-40 expression in breast cancer. *World J Surg Oncol.* 2007; 5: 17.

Kirkpatrick RB, Matico RE, McNulty DE, Strickler JE, Rosenberg M. An abundantly secreted glycoprotein from Drosophila melanogaster is related to mammalian secretory proteins produced in rheumatoid tissues and by activated macrophages. *Gene.* 1995; 153: 147-54.

Kirkpatrick RB, Emery JG, Connor JR, Dodds R, Lysko PG, Rosenberg M. Induction and expression of human cartilage glycoprotein 39 in rheumatoid inflammatory and peripheral blood monocyte-derived macrophages. *Exp Cell Res.* 1997; 237: 46-54.

Knudsen LS, Østergaard M, Baslund B, Narvestad E, Petersen J, Nielsen HJ, Ejbjerg BJ, Szkudlarek M, Johansen JS. Plasma IL-6, plasma VEGF and serum YKL-40: relationship with disease activity and radiographic progression in rheumatoid arthritis patients treated with infliximab and methotrexate. *Scand J Rheumatol.* 2006; 35: 489-91.

Knudsen LS, Klarlund M, Skjødt H Jensen T, Østergaard M, Jensen KE, Hansen MS, Hetland ML, Nielsen HJ, Johansen JS. Biomarkers of inflammation in patients with unclassified polyarthritis and early rheumatoid arthritis. Relationship to disease activity and radiographic outcome. *J Rheumatol.* 2008; 35: 1277-87.

Kolson DL. YKL-40: a candidate biomarker for simian immunodeficiency virus and human immunodeficiency virus encephalitis. *Am J Pathol.* 2008; 173: 25-29.

Krause SW, Rehli M, Kreutz M, Schwarzfisher L, Paulauski JD, Andreesen R. Differential screening identifies genetic markers of monocyte to macrophage maturation. *J Leukocyte Biol.* 1996; 60: 540-5.

Kroes RA, Dawson G, Moskal JR. Focused microarray analysis of glyco-gene expression in human glioblastomas. *J Neurochem.* 2007; 103: 14-24.

Krona A, Aman P, Orndal C, Josefsson A. Oncostatin M-induced genes in human astrocytomas. *Int J Oncol.* 2007; 31: 1457-63.

Kronborg G, Østergaard, Weis N, Nielsen H, Obel N, Pedersen SS, Price PA, Johansen JS. Serum level of YKL-40 is elevated in patients with Streptococcus pneumoniae bacteremia and is associated with the outcome of the disease. *Scand J Infect Dis.* 2002; 34: 323-6.

Kruit A, Grutters JC, Ruven HJT, van Moorsel CC, van den Bosch JM. A CHI3L1 gene polymorphism is associated with serum levels of YKL-40, a novel sarcoidosis marker. *Respir Med.* 2007; 101: 1563-71.

Koutroubakis IE, Petinaki E, Dimoulios P, Vardas E, Roussomoustakaki M, Maniatis AN, Kouroumalis EA. Increased serum levels of YKL-40 in patients with inflammatory bowel disease. *Int J Colorectal Dis.* 2003; 18: 254-9.

Kucur M, Isman FK, Karadag B, Vural VA, Tavsanoglu S. Serum YKL-40 levels in patients with coronary artery disease. *Coron Artery Dis.* 2007; 18: 391-6.

Kucur M, Isman FK, Balci C, Onal B, Hacibekiroglu M, Ozkan F. Serum YKL-40 levels and chitotriosidase activity as potential biomarkers in primary prostate cancer and benign prostatic hyperplasia. *Urol Oncol.* 2008; 26: 47-52.

Kuepper M, Bratke K, Virchow JC. Chitinase-like protein and asthma. *N Engl J Med.* 2008; 358: 1073-5.

Kushner I, Rzewnicki D, Samols D. What does minor elevation of C-reactive protein signify? *Am J Med.* 2006; 119: 166.e17-28.

Kzhyshkowska J, Mamidi S, Gratchev A, Kremmer E, Schmuttermaier C, Krusell L, Haus G, Utikal J, Schledzewski K, Scholtze J, Goerdt S. Novel stabilin-1 interacting chitinase-like protein (SI-CLP) is up-regulated in alternatively activated macrophages and secreted via lysosomal pathway. *Blood.* 2006; 107: 3221-8.

Lal A, Lash AE, Altschul SF, Velculescu V, Zhang L, McLendon RE, Marra MA, Prange C, Morin PJ, Polyak K, Papadopoulos N, Vogelstein B, Kinzler KW, Struasberg RL, Riggins GJ. A public database for gene expression in human cancers. *Cancer Res.* 1999; 59: 5403-7.

Lau SH, Sham JST, Xie D, Tzang CH, Tang D, Ma N, Hu L, Wang Y, Wen JM, Xiao G, Zhang WM, Lau GK, Yang M, Guan XY. Clusterin plays an important role in hepatocellular carcinoma metastasis. *Oncogene.* 2006; 25: 1242-50.

Lebensztejn DM, Skiba E, Werpachowska I, Sobaniec-Lotowska ME, Kaczmarksi M. Serum level of YKL-40 does not predict advanced liver fibrosis in children with chronic hepatitis B. *Adv Med Sci.* 2007; 52: 120-4.

Ling H, Recklies AD. The chitinase 3-like protein human cartilage glycoproein 39 inhibits cellular responses to the inflammatory cytokines interleukin-1 and tumour necrosis factor-alpha. *Biochem J.* 2004; 380: 651-9.

Liton PB, Liu X, Stamer WD, Challa P, Epstein DL, Gonzalez, P. Specific targeting of gene expression to a subset of human trabecular meschwork cells using the chitinase 3-like 1 promoter. *Invest Ophtahlmol Vis Sci.* 2005; 46: 183-90.

Liton PB, Luna C, Challa P, Epstein DL, Gonzalez P. Genome-wide expression profile of human trabecular meshwork cultured cells, nonglaucomatous and primary open angle glaucoma tissue. Molecular. *Vision.* 2006; 12: 774-90.

Lin WW, Karin M. A cytokine-mediated link between immunity, inflammation, and cancer. *J Clin Invest.* 2007; 117: 1175-83.

Lo WR, Rowlette LL, Caballero M, Yang P, Hernandez MR, Borras T. Tissue differential microarray analysis of dexamethasone induction reveals potential mechanisms of steroid glaucoma. *Invest Ophthalmol Vis Sci.* 2003; 44: 473-85.

Mahaney MC, Czerwinski SA, Rogers J. Genetic effects on serum levels of human cartilage glycoprotein-39 (YKL-40) in a pedigreed baboon model for age-related changes and pathology in bone. *Am Soc Bone Miner Res.* 1998; Abstract no SA121.

Malinda KM, Ponce L, Kleinman HK, Shackelton LM, Millis AJT. Gp38k, a protein synthesized by vascular smooth muscle cells, stimulates directional migration of human umbilical vein endothelial cells. *Exp Cell Res.* 1999; 250: 168-73.

Markert JM, Fuller CM, Gillespie GY, Bubien JK, McLena LA, Hong RL, Lee K, Gullans SR, Mapstone TB, Benos DJ. Differential gene expression profiling in human brain tumors. *Physiol Genomics.* 2001; 5: 21-33.

Matsumoto T, Tsurumoto T. Serum YKL-40 levels in rheumatoid arthritis: correlations between clinical and laboratory parameters. *Clin Exp Rheumatol.* 2001; 19: 655-60.

Meyer MF, Kreil G. Cells expressing the DG42 gene from early *Xenopus* embryos synthesize hyaluronan. *Proc Natl Acad Sci USA.* 1996; 93: 4543-7.

Millis AJT, Hoyle M, Reich E, Mann DM. Isolation and characterization of a Mr = 38,000 protein from differentiating smooth muscle cells. *J Biol Chem.* 1985; 260: 3754-61.

Millis AJT, Hoyle M, Kent L. In vitro expression of a 38,000 dalton heparin-binding glycoprotein by morphologically differentiated smooth muscle cells. *J Cell Physiol.* 1986; 127: 366-72.

Mitsuhashi A, Matsui H, Usui H, Nagai Y, Tate S, Unno Y, Hirashiki K, Seki K, Shozu M. Serum YKL-40 as a marker for cervical adenocarcinoma. *Ann Oncol.* 2009; in press.

Miyahara T, Kikuchi T, Akimoto M, Kurokawa T, Shibuki H, Yoshimura N. Gene microarray analysis of experimental glaucomatous retina from cynomologous monkey. *Invest Ophthalmol Vis Sci.* 2003; 44: 4347-56.

Mizoguchi E. Chitinase 3-like-1 exacerbates intestinal inflammation by enhancing bacterial adhesion and invasion in colonic epithelial cells. *Gastroenterology.* 2006; 130: 398-411.

Mizoguchi E, Mizoguchi A. Is the sugar always sweet in intestinal inflammation? *Immunol Res.* 2007; 37: 47-60.

Mohanty AK, Singh G, Paramasivam M, Saravanan K, Jabeen T, Sharma S, Yadav S, Kaur P, Kumar P, Srinivasan A, Singh TP. Crystal structure of a novel regulatory 40 kDa mammary gland protein (MGP-40) secreted during involution. *J Biol Chem.* 2003; 278: 14451-60.

Montagna GL, D'Angelo S, Valentini G. Cross-sectional evaluation of YKL-40 serum concentrations in patients with systemic sclerosis. Relationship with clinical and serological aspects of disease. *J Rheumatol.* 2003; 30: 2147-51.

Morrison BW, Leder P. Neu and ras initiate murine mammary tumors that share genetic markers generally absent in c-myc and int-2-initiated tumors. *Oncogene.* 1994; 9: 3417-26.

Moss SF, Blaser MJ. Mechanisms of disease: inflammation and the origins of cancer. Review. *Nature Clin Practice.* 2005; 2: 90-7.

Mylin AK, Rasmussen T, Johansen JS, Knudsen LM, Nørgaard PH, Lenhoff S, Dahl IMS, Johnsen HE; the Nordic Myeloma Study Group. Serum YKL-40 concentrations in newly diagnosed multiple myeloma patients and YKL-40 expression in malignant plasma cells. *Eur J Hematol.* 2006; 77: 416-24.

Mylin AK, Abildgaard N, Johansen JS, Andersen NF, Heickendorff L, Standal T, Gimsing P, Knudsen LM. High serum YKL-40 concentration is associated with severe bone disease in newly diagnosed multiple myeloma patients. *Eur J Haematol.* 2008; 80: 310-7.

Nielsen AR, Erikstrup C, Johansen JS, Fischer CP, Plomgaard P, Krogh-Madsen R, Taudorf S, Mortensen OH, Petersen AMW, Lindegaard B, Pedersen BK. Plasma YKL-40 – a BMI-independent marker of type 2 diabetes. *Diabetes* 2009, in press.

Nigro JM, Misra A, Zhang L, Smirnov I, Colman H, Griffin C, Ozburn N, Chen M, Pan E, Koul D, Yung WK, Feuerstein BG, Aldape KD. Integrated array-comparative genomic hybridization and expression array profiles identify clinically relevant molecular subtypes of glioblastoma. *Cancer Res.* 2005; 65: 1678-86.

Nishikawa KC, Millis AJT. gp38k (CHI3L1) is a novel adhesion and migration factor for vascular cells. *Exp Cell Res.* 2003; 287: 79-87.

Nordenbæk C, Johansen JS, Junker P, Borregaard N, Sørensen O, Price PA. YKL-40, a matrix protein of specific granules in neutrophils, is elevated in serum of patients with community-acquired pneumonia requiring hospitalization. *J Infect Dis.* 1999; 180: 1722-6.

Nordenbæk C, Johansen JS, Halberg P, Wiik A, Garbarsch C, Ullman S, Price PA, Jacobsen S. High serum levels of YKL-40 in patients with systemic sclerosis are associated with pulmonary involvement. *Scand J Rheumatol.* 2005; 34: 293-7.

Nunes D, Fleming C, Offner G, O'Brien M, Tumilty S, Fix O, Heeren T, Koziel M, Graham C, Craven DE, Stuver S, Horsburgh CR Jr. HIV infection does not affect the performance of noninvasive markers of fibrosis for the diagnosis of hepatitis C virus-related liver disease. *J Acquir Immune Defic Syndr.* 2005; 40: 538-44.

Nutt CL, Betensky RA, Brower MA, Batchelor TT, Louis DN, Stemmer-Rachamimow AO. YKL-40 is a differential diagnostic marker for histologic subtypes of high-grade gliomas. *Clin Cancer Res.* 2005; 11: 2258-64.

Nyirkos P, Golds EE. Human synovial cells secrete a 39 kDa protein similar to a bovine mammary protein expressed during the non-lactating period. *Biochem J.* 1990; 268: 265-8.

Nøjgaard C, Johansen JS, Christensen E, Skovgaard LT, Price PA, Becker U and The EMALD Group Serum levels of YKL-40 and PIIINP as prognostic markers in patients with alcoholic liver disease. *J Hepatol.* 2003a; 39: 179-86.

Nøjgaard C, Johansen JS, Krarup HB, Holten-Andersen M, Møller A, Bendtsen F and the Danish Viral Hepatitis Study Group. Effect of antiviral therapy on markers of fibrogenesis in patients with chronic hepatitis C. *Scand J Gastroenterol.* 2003b; 38: 659-5.

Nøjgaard C, Høst NB, Christensen IJ, Poulsen SH, Egstrup K, Price PA, Johansen JS. Serum levels of YKL-40 increases in patients with acute myocardial infarction. *Coron Artery Dis.* 2008; 19: 257-63

Ober C, Tan Z, Sun Y, Possick JD, Pan L, Nicolae R, Radford S, Parry RR, Heinzmann A, Deichmann KA, Lester LA, Gern JE, Lemanske RF Jr, Nicolae DL, Elias JA, Chupp GL. Effect of variation in CHI3L1 on serum YKL-40 level, risk of asthma, and lung function. *N Engl J Med.* 2008; 358: 1682-91.

Patil NS, Hall FC, Drover S, Spurrell DR, Bos E, Cope AP, Sonderstrup G, Mellins ED. Autoantigenic HCgp39 epitopes are presented by the HLA-DM-dependent presentation pathway in human B cells. *J Immunol.* 2001; 166: 33-41.

Pelloski CE, Mahajan A, Maor M, Chang EL, Woo S, Gilbert M, Colman H, Yang H, Ledoux A, Blair H, Passe S, Jenkins RB, Aldape KD. YKL-40 expression is associated with poorer response to radiation and shorter overall survival in glioblastoma. *Clin Cancer Res.* 2005; 11: 3326-34.

Pelloski CE, Lin E, Zhang L, Yung WK, Colman H, Liu JL, Woo SY, Heimberger AB, Suki D, Prados M, Chang S, Barker FG 3rd, Fuller GN, Aldape KD. Prognostic associations of activated mitogen-activated protein kinase and akt pathways in glioblastoma. *Clin Cancer Res.* 2006; 12: 3935-41.

Pelloski CE, Ballman KV, Furth AF, Zhang L, Lin E, Sulman EP, Bhat K, McDonald JM, Yung WK, Colman H, Woo SY, Heimberger AB, Suki D, Prados MD, Chang SM, Barker FG 2nd, Buckner JC, James CD, Aldape K. Epidermal growth factor receptor variant III status defines clinically distinct subtypes of glioblastoma. *J Clin Oncol.* 2007; 25: 2288-94.

Peltomaa R, Paimela L, Harvey S, Helve T, Leirisalo-Repo M. Increased level of YKL-40 in sera from patients with early rheumatoid arthritis: a new marker for disease activity. *Rheumatol Int.* 2001; 20: 192-6.

Phillips HS, Kharbanda S, Chen R, Forrest WF, Soriano RH, Wu TD, Misra A, Nigro JM, Colman H, Soroceanu L, Williams PM, Modrusan Z, Feuerstein BG, Aldape K. Molecular subclasses of high-grade glioma predict porgnosis, delineate a pettern of disease progressin, and resemble stages in neurogenesis. *Cancer Cell.* 2006; 9: 157-73.

Punzi L, Podswiadek M, D'Inca R, Zaninotto M, Bernardi D, Plebani M, Sturniolo GC. Serum human cartilage glycoprotein 39 as a marker of arthritis associated with inflammatory bowel disease. *Ann Rheum Dis.* 2003; 62: 1224-6.

Qin W, Zhu W, Schlatter L, Miick R, Loy TS, Atasoy U, Hewett JE, Sauter ER. Increased expression of the inflammatory protein YKL-40 in precancers of the breast. *Int J Cancer.* 2007; 121: 1536-42.

Rakic JM, Lambert V, Deprez M, Foidart JM, Noël A, Munaut C. Estrogens reduce the expression of YKL-40 in the retina: implications for eye and joint diseases. *Invest Ophthalmol Vis Sci.* 2003; 44: 1740-6.

Rathcke CN, Johansen JS, Vestergaard H. YKL-40, a biomarker of inflammation, is elevated in patients with type 2 diabetes and is related to insulin resistance. *Inflamm Res.* 2006; 55: 53-9.

Rathcke CN, Persson F, Tarnow L, Rossing P, Vestergaard H. YKL-40, a marker of inflammation and endothelial dysfunction, is elevated in patients with type 1 diabetes and increases with levels of albuminuria. *Diabetes Care.* 2009; in press.

Recklies AD, White C, Ling H. The chitinase 3-like protein human cartilage 39 (HC-gp39) stimulates proliferation of human connective-tissue cells and activates both extracellular signal-regulated kinase-and protein kinase B-mediated signalling pathways. *Biochem J.* 2002; 365: 119-6.

Recklies AD, Ling H, White C, Bernier SM. Inflammatory cytokines induce production of CHI3L1 by articular chondrocytes. *J Biol Chem.* 2005; 280: 41213-21.

Register TC, Carlson CS, Adams MR. Serum YKL-40 is associated with osteoarthritis and atherosclerosis in nonhuman primates. *Clin Chem.* 2001; 47: 2159-61.

Rehli M, Krause SW, Andreesen R. Molecular characterization of the gene for human cartilage gp-39 (CHI3L1), a member of the chitinase protein family and marker for late stages of macrophage differentiation. *Genomics.* 1997; 43: 221-5.

Rehli M, Niller HH, Ammon C, Langmann S, Schwarz-Fisher L, Andreesen R, Krause SW. Transcriptional regulation of CHI3L1, a marker gene for late stages of macrophage differentiation. *J Biol Chem.* 2003; 278: 44058-67.

Renkema GH, Boot GR, Au FL, Donker-Koopman WE, Strijland A, Muijsers AO, Hrebicek M, Aerts JMFG. Chitotriosidase, a chitinase, and the 39-kDa human cartilage glycoprotein, a chitin-binding lectin, are homologues of family 18 glycosyl hydrolases secreted by human macrophages. *Eur J Biochem.* 1998; 251: 504-9.

Ringsholt M, Høgdall EVS, Johansen JS, Price PA, Christensen LH. YKL-40 protein expression in normal human tissues - an immunohistochemical study. *J Mol Histol.* 2007; 38: 33-43.

Roslind A, Johansen JS, Junker N, Nielsen DL, Dzaferi H, Price PA, Balslev E. YKL-40 expression in benign and malignant lesions of the breast: A methodologic study. Appl. Immunohisto. *Mol Morphol.* 2007; 15: 371-81.

Roslind A, Johansen JS, Christensen IJ, Kiss K, Balslev E, Nielsen DL, Bentzen J, Price PA, Andersen E. High serum levels of YKL-40 in patients with squamous cell carcinoma of the head and neck are associated with short survival. *Int J Cancer.* 2008a; 122: 857-63.

Roslind A, Knoop AS, Jensen M, Johansen JS, Nielsen DL, Price PA, Balslev E. YKL-40 protein expression is not a prognostic marker in patients with primary breast cancer. *Breast Cancer Res Treat.* 2008b; 112: 275-85.

Rousseau A, Nutt CL, Betensky RA, Iafrate AJ, Han M, Ligon KL, Rowitch DH, Louis DN. Expression of oligodendroglial and astrocytic lineage markers in diffuse gliomas: use of YKL-40, ApoE, ASCL1, and NKX2-2. *J Neuropathol Exp Neurol.* 2006; 65: 1149-56.

Saidi A, Javerzat S, Bellahcene A, de Vos J, Bello L, Castronovo V, Deprez M, Loiseau H, Bikfalvi A, Hagedorn M. Experimental anti-angiogenesis causes upregulation of genes associated with poor survival in glioblastoma. *Int J Cancer.* 2008; 122: 2187-98.

Schmidt H, Johansen JS, Gehl J, Geertsen PF, Fode K, and von der Maase H. Elevated serum level of YKL-40 is an independent prognostic factor for poor survival in patients with metastatic melanoma. *Cancer.* 2006a; 106: 1130-9.

Schmidt H, Johansen JS, Sjoegren P, Christensen IJ, Sørensen BS, Fode K, Larsen J, von der Maase H. Serum YKL-40 predicts relapse-free and overall survival in patients with American Joint Committee on Cancer stage I and II melanoma. *J Clin Oncol.* 2006b; 24: 798-804.

Semino CE, Specht CA, Raimondi A, Robbins PW. Homologs of the *Xenopus* developmental gene DG42 are present in zebrafish and mouse and are involved in the synthesis of Nod-like chitin oligosaccharides during early embryogenesis. *Proc Natl Acad Sci USA.* 1996; 93: 4548-53.

Shackelton LM, Mann DM, Millis AJT. Identification of a 38-kDa heparin-binding glycoprotein (gp38k) in differentiating vascular smooth muscle cells as a member of a group of proteins associated with tissue remodeling. *J Biol Chem.* 1995; 270: 13076-83.

Sharon D, Blackshaw S, Cepko CL, Dryja TP. Profile of the genes expressed in the human peripheral retina, macula, and retinal pigment epithelium determined through serial analysis of gene expression (SAGE). *PNAS.* 2002; 99: 315-20.

Shostak K, Labunskyy V, Dmitrenko V, Malisheva T, Shamayev M, Rozumenko V, Zozulya Y, Zehetner G, Kavsan V. HC gp-39 gene is upregulated in glioblastoma. *Cancer Lett.* 2003; 198: 203-10.

Sjögren H, Meis-Kindblom JM, Örndal C, Bergh P, Ptaszynski K, Åman P, Kindblom LG, Stenman G. Studies on the molecular pathogenesis of extraskeletal myxoid chondrosarcoma-cytogenetic, molecular genetic, and cDNA microarray analyses. *Am J Pathol.* 2003; 162: 781-92.

Steenbakkers PGA, Baeten D, Rovers E, Veys EM, Rijnders AWM, Meijerink J, Keyser FD, Boots AMH. Localization of MHC Class II/human cartilage glycoprotein-39 complexes in synovia of rheumatoid arthritis patients using complex-specific monoclonal antibodies. *J Immunol.* 2003; 170: 5719-27.

Stokes DG, Liu G, Coimbra IB, Piera-Velazquez S, Crowl RM, Jimenez SA. Assessment of the gene expression profile of differentiated and dedifferentiated human fetal chondrocytes by microarray analysis. *Arthritis Rheum.* 2002; 46: 404-19.

Suzuki T, Hashimoto S, Toyoda N, Nagai S, Yamazaki N, Dong HY, Skakai J, Yamashita T, Nukiwa T, Matsushima K. Comprehensive gene expression profile of LPS-stimulated human monocytes by SAGE. *Blood.* 2000; 96: 2584-91.

Tanwar MK, Gilbert MR, Holland EC. Gene expression microarray analysis reveals YKL-40 to be a potential serum marker for malignant character in human glioma. *Cancer Res.* 2002; 62: 4364-8.

Tlsty TD, Coussens LM. Tumor stroma and regulation of cancer development. *Annu Rev Pathol Mech Dis.* 2006; 1: 119-50.

Toole BP. Hyaluronan: from extracellular glue to pericellular cue. *Nat Rev Cancer.* 2004; 4: 528-39.

Tran A, Benzaken S, Saint-Paul MC, Guzman-Granier E, Hastier P, Pradier C, Barjoan EM, Demuth N, Longo F, Rampal P. Chondrex (YKL-40), a potential new serum fibrosis marker in patients with alcoholic liver disease. *Eur J Gastroenterol Hepatol.* 2000; 12: 989-93.

Tsark EC, Wang W, Teng YC, Arkfeld D, Dodge GR, Kovats S. Differential MHC class II-mediated presentation of rheumatoid arthritis autoantigens by human dendritic cells and macrophages. *J Immunol.* 2002; 169: 6625-33.

Werner M, Fraser C, Silverberg M. Clinical utility and validation of emerging biochemical markers for mammary adenocarcinoma. *Clin Chem.* 1993; 39; 2386-96.

Varki A. Does DG42 synthesize hyaluronan or chitin? A controversy about oligosaccharides in vertebrate development. *Proc Natl Acad Sci USA.* 1996; 93: 4523-5.

Verheiden GFM, Rijnders AWM, Bos E, Coenen-de ROO CJJ, van Staveren CJ, Miltenburg, AMM, Meijerink JH, Elewaut D, Keyser F, Veys E, Boots AMH. Human cartilage glycoprotein-39 as a candidate autoantigen in rheumatoid arthritis. *Arthritis Rheum.* 1997; 40: 1115-25.

Verhoeckx KCM, Bijlsma S, De Groene EM, Witkamp RF, van der Greed J, Rodenburg RJT. A combination of proteomics, principal component analysis and transcriptomics is a powerful tool for the identification of biomarkes for macrophage maturation in the U937 cell line. *Proteomics.* 2004; 4: 1014-28.

Vind I, Johansen JS, Price PA, Munkholm P. Serum YKL-40, a potential new marker of disease activity in patients with inflammatory bowel disease. *Scand J Gastroenterol.* 2003; 38: 599-605.

Volck B, Price PA, Johansen JS, Sørensen O, Benfield T, Calafat J, Nielsen HJ, Borregaard N. YKL-40, a mammalian member of the chitinase family, is a matrix protein of specific granules in human neutrophils. *Proc Assoc Am Phys.* 1998; 110: 351-60.

Volck B, Johansen JS, Stoltenberg M, Garbarsch C, Price PA, Østergaard M, Østergaard K, Løvgreen-Nielsen P, Sonne-Holm S, Lorenzen I. Studies on YKL-40 in knee joints of patients with rheumatoid arthritis and osteoarthritis. Involvement of YKL-40 in the joint pathology. *Osteoarthritis Cartilage*. 2001; 9: 20314.

Vos K, Miltenburg AMM, van Meijgaarden KE, van den Heuvel M, Elferink DG, van Galen PJM, van Hogezand RA, van Vliet-Daskalopoulou E, Ottenhoff THM, Breedveld FC, Boots AMH, de Vries RRP. Cellular immune response to human cartilage glycoprotein-39 (HC gp-39)-derived peptides in rheumatoid arthritis and other inflammatory conditions. *Rheumatology.* 2000a; 39: 1326-31.

Vos K, Steenbakkers P, Miltenburg AMM, Bos E, van den Heuvel MW, van Hogezand RA, de Vries RRP, Breedveld FC, Boots AMH. Raised human cartilage glycoprotein-39 plasma levels in patients with rheumatoid arthritis and other inflammatory conditions. *Ann Rheum Dis.* 2000b; 59: 544-8.

Wang Y, Ripa RS, Johansen JS, Gabrielsen A, Steinbrüchel DA, Friis T, Bindslev L, Haack-Sørensen M, Jørgensen E, Kastrup J. YKL-40 a new biomarker in patients with acute coronary syndrome or stable coronary artery disease. *Scand Card J.* 2008: 42: 295-302.

Yang MS, Morris DW, Donohoe G, Kenny E, O'Dushalaine CT, Schwaiger S, Nangle JM, Clarke S, Scully P, Quinn J, Meagher D, Baldwin P, Crumlish N, O'Callaghan E, Waddington JL, Gill M, Corvin A. Chitinase-3-like 1 (CHI3L1) gene and schizophrenia: genetic association and a potential functional mechanism. *Biol Psychiatry.* 2008; 64: 98-103.

Zaheer-ul-Haq, Dala P, Aronson NN, Madura JD. Family 18 chitolectins: comparison of MGP40 and HUMGP39. *Biochem Biophys Res Commun.* 2007; 359: 221-6.

Zhao X, Tang R, Gao B, Shi Y, Zhou J, Guo S, Zhang J, Wang Y, Tang W, Meng J, Li S, Wang H, Ma G, Lin C, Xiao Y, Feng G, Lin Z, Zhu S, Xing Y, Sang H, St Clair D, He L. Functional variants in the promoter region of Chitinase 3-Like 1 (CHI3L1) and susceptibility to schizophrenia. *Am J Hum Genet.* 2007; 80: 12-8.

Zheng M, Cai WM, Zhao JK, Zhu SM, Liu RH. Determination of serum levels of YKL-40 and hyaluronic acid in patients with hepatic fibrosis due to schistosomiasis japonica and appraisal of their clinical value. *Acta Trop.* 2005; 96: 148-52.

Østergaard C, Johansen JS, Benfield T, Price PA, Lundgren JD. YKL-40 is elevated in cerebrospinal fluid from patients with purulent meningitis. *Clin Diagn Lab Immun.* 2002; 9: 598-604.

In: Binomium Chitin-Chitinase: Recent Issues ISBN 978-1-60692-339-9
Editor: Salvatore Musumeci and Maurizio G. Paoletti

Chapter XIV

Plasma Chitotriosidase Activity in Beta-Thalassemia

Rita Barone
Centre for Inherited Metabolic Disorders, Department of Pediatrics, University of Catania and Institute of Chemistry and Technology of Polymers, National Research Council (CNR)- Catania-Italy.

Abstract

Chitotriosidase, a functional chitinase secreted by activated macrophages, is extremely increased in plasma of patients with Gaucher disease (GD) or beta-glucocerebrosidase deficiency. GD is a lysosomal storage disorder characterized by blocked catabolism of glucosylceramide (GC), a metabolic intermediate derived from the cellular turnover of membrane gangliosides and globosides. The primary cell type affected in GD is the macrophage (Gaucher cell) where the presence of GC and other sphingolipids at non-physiological concentrations is thought to interfere with other biochemical pathways outside the lysosome, leading to cell dysfunction including reticuloendothelial expansion and macrophage activation. In 1999, we investigated if plasma chitotriosidase levels are increased in patients with beta-thalassemia, an haematological disorder characterized by the genetic defect of beta-globin chains synthesis and resulting in unproductive erythropoiesis and enormous expansion of the reticuloendothelial system. We found that plasma chitotriosidase activity was increased to a variable extent in a group of patients with beta-thalassemia major including those treated with the intense transfusion regimen and iron chelation therapy. We suggested that the increased chitotriosidase production in beta-thalassemia might reflect macrophage activation probably related to the intracellular iron overload and storage of erythrocytes membrane break-down products. After this initial description, chitotriosidase evaluation in patients with beta-thalassemia has been the object of further clinical work up. We will review here the present knowledge on chitotriosidase in beta-thalassemia with the aim to describe the role and significance of human chitinase increase in this haematological disorder.

1. Introduction

Chitotriosidase (ChT) enzyme is produced in large amounts by the lipid-laden macrophages in Gaucher disease (GD) (Hollak *et al.*1994). GD is a lysosomal glycolipid storage disorder caused by recessive inherited mutations in the acid β-glucosidase gene. This defect leads to a deficiency in the glucocerebrosidase enzyme and accumulation in macrophages of glucosylceramide, a metabolic intermediate derived from the cellular turnover of membrane gangliosides and globosides (Beutler and Grabowski 2001). In particular, senescent blood cells that are rich in glucosylceramide and its precursors follow endocytic uptake by macrophages. The lysosomal apparatus of macrophages may become stressed with glucosylceramide and overloaded. Gaucher cells (GC) release in patient's plasma numerous compounds including cytokines, chemokines and hydrolases some of which contribute to the various tissue damages (Beutler and Grabowski 2001; Bussink *et al.* 2006). Two main isoforms of ChT of approximately 50 and 39 kDa, respectively, have been described. Cultured macrophages predominantly produce and secrete the 50 kDa isoform (Renkema *et al.* 1995; 1997). Post-translational modifications further enhance the molecular heterogeneity of ChT by the existence of isoforms that differ in their glycosylation status. In this regard, O-linked glycosylation has been documented in chitinases from other species (Fuhrman *et al.* 1995). Renkema *et al.* (1997) described different ChT activities in pI fractionated medium of culture of macrophages. These different activities were dependent also on glycosylation as demonstrated by their modification after neuraminidase treatment (Rosalki *et al.* 1985).

The existence of different ChT isoforms is confirmed by proteomic analyses of GD plasma by 2-DE (Quintana *et al.* 2006). In this study, modifications of ChT spot profile on neuraminidase treatment, indicated lost of sialic acid molecules linked to their O-glycans. Molecular heterogeneity of ChT enhances biological diversity of this glycoprotein and it is coherent with multiple involvement of ChT in different physiological conditions and diseases (Barone *et al.* 2007).

2. Chitotriosidase and Beta-Thalassemia

In 1971 Zaino *et al.* described GC or Gaucher-like cells in the spleen and bone marrow of patient with thalassemia major (Figure 1).

Ultrastructure of GC in beta-thalassemia shows intracytoplasmic tubules and phagocytosis of mature and immature erythrocytes. Enrichment of monohexosyl ceramide is found in splenic tissue from thalassemia patients. Erythrophagocytosis is a common feature of both Gaucher disease and thalassemia and it appears that the impaired catabolism of erythrocytes beta-thalassemia may give rise to the increased glucocerebroside.

The homozygous beta-thalassemias are a group of genetically inherited hemoglobin (Hb) disorders of beta-globin chains synthesis characterized by unproductive erythropoiesis due to intramedullary hemolysis. According to the degree of anemia, two main forms, sharing a common basic molecular mechanism, are distinguished: thalassemia major (TM) and thalassemia intermedia (TI).

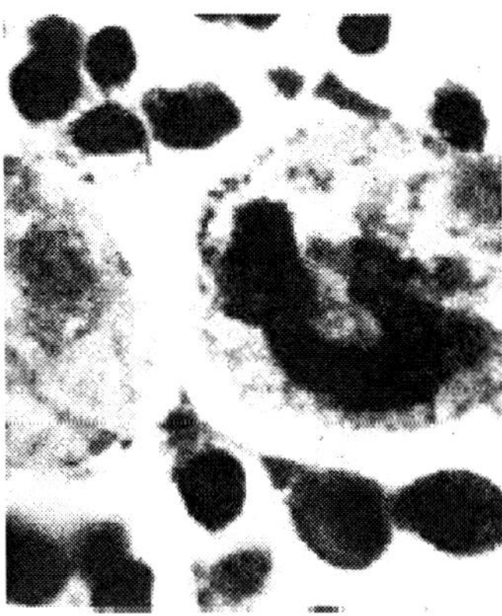

Figure 1. Light-microscopy of a lipid-laden macrophage (Gaucher cell) in the spleen tissue of a patient with beta-thalassemia (Adapted from Zaino *et al.* 1971).

The severity of the clinical phenotype differentiates the two forms. TM usually presents as a severe anemia requiring life-long transfusion therapy for survival. In TM patients the enormous expansion of the reticuloendothelial system results in an elevated blood consumptions which in turn leads to a progressive iron overload and the potential for impaired endocrine, cardiac and hepatic function (Weatherall *et al.* 2001). The improvement in life expectancy of beta-thalassemia patients achieved during the past few decades by virtue of therapeutic advances including the appropriate use of iron chelators, has motivated investigators' interest in a better understanding of the clinical consequences of this genetic defect.

In this regard, Barone *et al.* (1999) investigated plasma ChT levels in Sicilian beta-thalassemia patients with respect to clinical and laboratory parameters reflecting the status of the disease. Seventy-two patients with TM with age ranging from 4 to 29 years were studied at the Thalassemic Center of Catania University . All patients underwent an intense transfusion regimen in order to maintain their baseline Hb level above 9.5 g/dl and were treated with subcutaneous desferrioxamine infusion (40-50 mg/Kg, 3-6 days a week). Fourteen subjects underwent splenectomy. For comparison, plasma ChT levels were measured in 25 TI patients aging 4-37 years which were only occasionally transfused and have never received iron chelation therapy. Moreover, results of ChT levels in plasma of TM and TI patients were compared to those obtained in healthy subjects (n: 85) and in patients with GD type I (n: 8). Thirteen out of 72 (18%) of TM patients have increased plasma ChT activity including 7 subjects with ChT levels as high at that observed in GD which means at least 100-fold elevated than the median normal values.

TM patients with highest ChT levels (median: 4173; range: 1589-9282 nmol/h/ml) had significantly higher mean serum ferritin levels (< 3000 ng/ml), higher serum GPT level and higher urinary excretion compared to TM patients with normal or moderately increase ChT. The increase of plasma ChT activity was not a finding of TI as all our studied patients had ChT activity in the normal range with the exception of 3 subjects with an increase of almost 5 to 15 times that of control subjects (Figure 2).

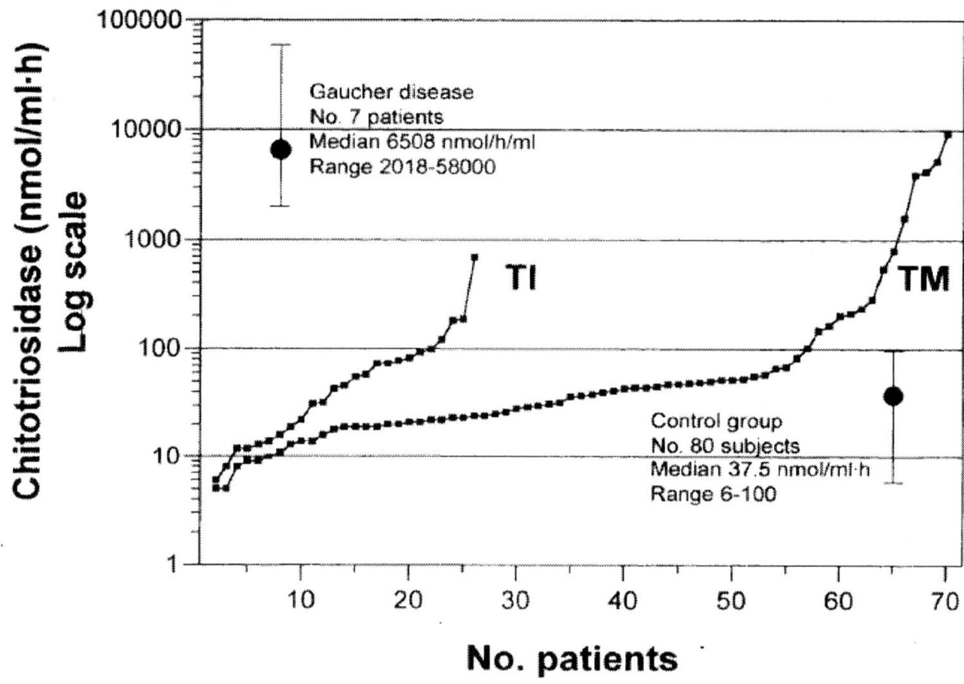

Figure 2. Plasma ChT levels (nmol/h/ml) in TM, TI, GD and controls. (Adapted from Barone *et al.* 1999).

Based on the above mentioned results, ChT evaluation has been recommended as a useful marker of disease status in TM (Barone *et al.* 1999).

The existence of patients with TM and increased ChT was confirmed in a comparative study between Sicilian and Sardinian TM patients (Barone *et al.* 2001). Plasma ChT activity was found most frequently elevated among Sardinian (48.4%) than Sicilian (17.1%) patients.

Three groups of patients might be detected among Sicilian and Sardinian patients, according to their ChT levels (table 1). The groups with the highest ChT levels (>1000 nmol/h/ml) also had significantly higher mean serum ferritin levels.

Differences in the rate and extent of ChT increase between Sicilian and Sardinian patients with TM might be related to their differences in β-thalassemia molecular bases which in different way might influence macrophage activation and hence, ChT overproduction.

In the Sardinian population, the most common type of TM is the β^0 variety, which is characterized by the absent production of β-globin chains, with the absolute prevalence of the codon β-39 nonsense mutation, which accounts for 95.7% of the β-thalassemia chromosomes (Rosatelli *et al.* 1992). On the other hand, a wide genetic heterogeneity with five more common mutations underlies β-thalassemia in Sicily (Schilirò *et al.* 1995). In addition, differences in environmental features between these studied populations including past malarial endemicity (Clegg and Weatherall, 1999), and their interaction with 24 bp duplication in the ChT gene responsible for ChT deficiency (Boot *et al.* 1998), might play a role in the light of recently discovered differences of ChT activity among different Mediterranean areas (Malaguarnera *et al.* 2003).

Table 1. Clinical parameters in Sardinian and Sicilian TM patients (Adapted from Barone *et al.* 2001)

	Sardinian patients (n = 64)	Sicilian patients (n = 70)
Age (years)	29 (20–40)	21 (2.5–41)
Age at diagnosis (months)	8 (3–18)	12 (1–84)
Age at the first transfusion (months)	9 (3–18)	12 (2–84)
Gender (m/f)	30/34	27/43
Splenectomy	12 (18.7%)	14 (20%)
Pre-transfusional Hb level (g/dl)	10.5 ± 0.44	9.59 ± 0.58
Blood consumption (ml/kg/year)	156 ± 30	161.4 ± 33.81
Serum ferritin (ng/ml) (normal range: 22–322)	4500 (1000–19,000)	2000 (57–10,000)
Chronic hepatitis	26 (40.6%)	25 (35.7%)

The significance of ChT evaluation in the clinical assessment of β-thalassemia was investigated in a cohort of Israeli patients (Altarescu *et al.* 2002). Eleven out of 30 (37%) adult TM patients and 1 of 14 TI subjects, had slight to moderate increase of plasma ChT levels. Interestingly, a significant correlation between ChT levels and both ferritin levels and number of blood transfusion per year was found in this study which confirmed the pivotal role of iron overload and macrophage activation in leading to ChT overproduction in TM. Iron overload results in lysosomal instability due to mechanical injury as well as to iron mediated peroxidative injury of lysosomal membrane (Michelakakis *et al.* 1997). Iron loading *in vivo* leads to ChT overproduction in different rat tissues (Dimitriou *et al.* 2000). Notably, in Altarescu study (2002) all children with β-thalassemia had normal ChT activity, suggesting that a certain degree of iron accumulation might be required before an increase in ChT levels is induced as well as a long-lasting macrophage damage also related to the uninterrupted schedule of transfusions and erythrocyte membrane breakdown products accumulation.

These pathomechanisms are supported also by the occurrence of lower serum ferritin and plasma chitotriosidase levels in transplanted β-thalassemia patients with respect to the transfused thalassemia patients as reported in Figure 3. (Maccarone *et al.* 2001). Allogeneic hematopoietic cell transplantation (HCT) provides correction of the genetic defect in β-thalassemia hence it represents the only cure proved for these disorders. The normal plasma ChT levels in transplanted β-thalassemia patients suggests that HCT might be effective in reversing hematological abnormalities and reducing macrophage activation.

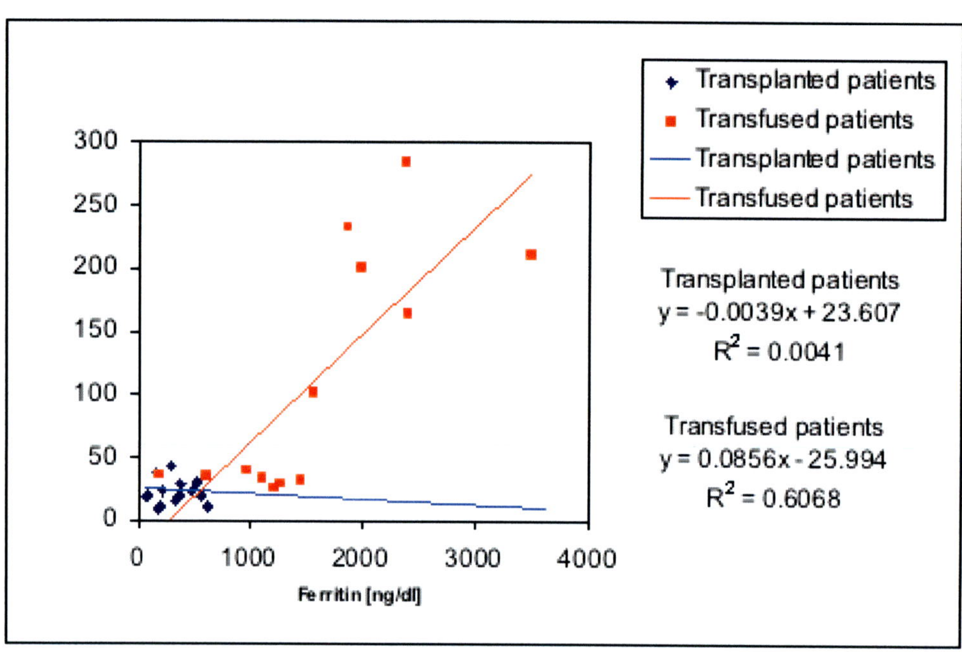

Figure 3. Plasma ChT activity and serum ferritin levels in TM patients treated with HCT and in TM patients underwent intense transfusion regimen (Adapted from Maccarone *et al.* 2001).

More recently, Dimitriou *et al.* (2005) detected a significant increase of plasma CCL18/PARC chemokine (Boot *et al.* 2004) in Greek patients with β-thalassemia. In this cohort it was also found that 11/36 TM patients had significantly higher plasma ChT levels than normal individuals. CCL18/PARC had a positive correlation to ferritin and chitotriosidase levels and a negative correlation to iron chelation therapy (Dimitriou *et al.* 2005) then supporting in the causative role of iron overload in ChT overproduction in β-thalassemia.

A comparison among reported ChT levels in different cohorts of β-thalassemia patients leads to observe some differences related to the proportion of patients with increased ChT levels (Barone *et al.* 1999, 2001; Altarescu *et al.* 2002; Dimitriou *et al.* 2005) and also to the measured absolute values of plasma ChT in the same patients. Likewise beta-thalassemia, the increase of plasma ChT levels may be observed only in a percentage of patients with lysosomal storage disorders different from GD and including Krabbe disease, Fabry disease, GM1-gangliosidosis, Morquio B disease, Niemann-Pick B/C, cholesteryl ester storage disease, Wolman disease, fucosidosis, galactosialidosis (Guo *et al.* 1995; Vedder *et al.* 2006; Barone *et al.* 2007).

Increased serum ferritin and erythrocyte superoxide dismutase (SOD) levels and oxidative damage biomarker F_2–isoprostanases (Matayatsuk C *et al.* 2007) in TM patients are consistent with oxidative damage to targets such as lipids in thalassemia. Future studies to assess possible correlation of ChT and oxidative damage biomarkers are envisaged to establish antioxidants therapy in this hematological disorder.

Conclusions

The increase of plasma ChT activity in β-thalassemia is not an universal finding as in GD. In addition, almost 6% of patients in the general population is homozygous for the 24 bp duplication in the ChT gene, which abolishes plasma ChT activity hampering the possible application of ChT evaluation in the assessment of disease severity and response to treatment of β-thalassemia patients. However, likewise in lysosomal diseases and other infectious or immunological disorders, the detection of increased ChT levels in β-thalassemia suggests the occurrence of an immunological response and macrophage activation most probably related to chronic accumulation of lipids and iron in phagocytes. The constant finding of a positive correlation between plasma ChT activity and serum ferritin levels in all studied cohorts of TM patients, indicate that there may be a role for monitoring ChT in β-thalassemia.

Acknowledgments

A special thank to the patients and their families and to Prof. S. Musumeci who allowed to expand my interest in chitotriosidase evaluation from the lysosomal storage disorders to other different diseases. The skillful technical collaboration of Mr. Giuseppe Rapicavoli is gratefully acknowledged.

This study is in memory of Prof. Felicia Di Gregorio, head of the thalassemia center at the Department of Pediatrics, University of Catania.

References

Altarescu G, Rudensky B, Abrahamov A, Goldfarb A, Rund D, Zimran A, Elstein D. Plasma chitotriosidase activity in patients with beta-thalassemia. *Am J Hematol.* 2002; 71: 7 –10.

Barone R, Di Gregorio F, Romeo MA, Schiliro G, Pavone L. Plasma chitotriosidase activity in patients with beta-thalassemia. *Blood Cells Mol Dis.* 1998; 25: 2 – 9.

Barone R, Simpore J, Malaguarnera L, Pignatelli S, Musumeci S Plasma chitotriosidase activity in acute Plasmodium falciparum malaria. *Clin Chim Acta* 2003; 331: 79-85.

Barone R, Sotgiu S, Musumeci S. Plasma chitotriosidase in health and pathology. *Clin Lab* 2007; *53:321-333.*

Beutler E, Grabowski GA in: Scriver CR, Beaud*et al.* Sly WS, Valle D (Eds.), The Metabolic and Molecular Basis of Inherited Disease, 8th ed., McGraw-Hill, New York, 2001, pp. 3635–68.

Boot RG, Renkema GH, Verhoek M, Strijland A, Bliek J, de Meulemeester TMAMO, Mannens MMAM, Aerts JMFG. The human chitotriosidase gene, nature of inherited enzyme deficiency. *J Biol Chem.* 1998; 273:25680–25685.

Boot RG, Verhoek M, de Fost M, Hollak CEM, Maas M, Bleijlevens B, van Breemen MJ, van Meurs M, Boven LA, Laman JD, Moran MT, Cox TM, Aerts JMFG. Marked elevation of the chemokine CCL18/PARC in Gaucher disease: a novel surrogate marker for assessing therapeutic intervention. *Blood* 2004; 103: 33–39.

Bussink AP, van Eijk M, Renkema GH, Aerts JM, Boot RG. The biology of the Gaucher cell: the cradle of human chitinases. *Int Rev Cytol.* 2006; *252*:*171-128.*

Chakraborty D; Bhattacharyya, M. Antioxidant defense status of red blood cells of patients with β-thalassemia.. *Clin Chim Acta* 2001; 305:123–129.

Cighetti G, Duca L, Bortone L, Sala S, Nava I, Fiorelli G, Cappellini, MD. Oxidative status and malondialdehyde in betathalassemia patients. *Eur J Clin Invest.* 2002; 32: 55–60.

Clegg JB, Weatherall DJ. Thalassemia and malaria: new insights into an old problem. *Proc Assoc Am Physicians.* 1999;111:278–82.

Dimitriou E, Kairis M, Sarafidou J, Michelakakis H. Iron overload and kidney lysosomes. *Biochim Biophys Acta.* 2000 15; *1501:138-148.*

Dimitriou E, Verhoek M, Altun S, Karabatsos F, Moraitou M, Youssef J, Boot R, Sarafidou J, Karagiorga M, Aerts H, Michelakakis H. Elevated plasma chemokine CCL18/PARC in beta-thalassemia. *Blood Cells Mol Dis. 2005;35:3282-31.*

Fuhrman JA, Lee J, Dalamagas D. Structure and function of a family of chitinase isozymes from Brugian microfilariae. *Exp Parasitol.* 1995; 80: 672–680.

Guo Y, He W, Boer AM, Wevers RA, de Bruijn AM, Groener JE, Hollak CE, Aerts JM, Galjaard H, van Diggelen OP. Elevated plasma chitotriosidase activity in various lysosomal storage diseases. *J Inherit Metab Dis.* 1995; 18: 717-722.

Hollak CEM, van Weely S, van Oers MHJ, Aerts JMFG. Marked elevation of plasma chitotriosidase activity, a novel hallmark of Gaucher disease. *J Clin Invest.* 1994; 93: 491–502.

Maccarone C, Pizzarelli G, Barone R, Musumeci S. Plasma chitotriosidase activity is a marker of recovery in transplanted patients affected by beta-thalassemia major. *Acta Haematol. 2001; 105:109-110.*

Malaguarnera L, Simporè J, Prodi DA, Angius A, Sassu A, Persico I, Barone R, Musumeci S. A 24-bp duplication in exon 10 of human chitotriosidase gene from the sub-Saharan to the Mediterranean area: role of parasitic diseases and environmental conditions. *Genes Immun.* 2003;4:*570-574.*

Michelakakis H, Dimitriou E, Georgakis H, Karabatsos F, Fragodimitri C, Saraphidou J, Premetis E, Karagiorga-Lagana M. Iron overload and urinary lysosomal enzyme levels in beta-thalassemia major. *Eur J Pediatr.* 1997; 156: 602-604.

Quintana L, Monasterio A, Escuredo K, del Amo J, Alfonso P, Elortza F, Santa Cruz S, Simón L, Martínez A, Giraldo P, Pocoví M, Castrillo JL. Identification of chitotriosidase isoforms in plasma of Gaucher disease patients by two dimensional gel electrophoresis. *Biochim Biophys Acta 2006; 1764: 1292-1298.*

Renkema GH, Boot RG, Muijsers AO, Donker-Koopman WE, Aerts JMFG. Purification and characterization of human chitotriosidase, a novel member of the chitinase family of proteins. *J Biol Chem.* 1995; 270: 2198–2202.

Renkema GH, Boot RG, Strijland A, Donker-Koopman WE, Van den Berg M, Muijsers AO, Aerts JMFG. Synthesis, sorting, and processing into distinct isoforms of human macrophage chitotriosidase. *Eur J Biochem.* 1997; 244:279–285.

Rosalki SB,Ying Foo A. Incubation with neuraminidase and affinity electrophoresis with wheat-germ lectin compared for separating and quantifying alkaline phosphatase isoenzymes in plasma. *Clin Chem.* 1985; 31: 1198–1200.

Weatherall DJ, Clegg JB, Higgs DR, Wood WG. The hemoglobinopathies. in: Scriver CR, Beaud*et al.* Sly WS, Valle D, Childs B, Kinzler KW, Vogestein B. 8th ed. The Metabolic and Molecular

Bases of Inherited Disease, vol. III, McGraw Hill, New York, 2001, pp. 4571– 4637.

Naithani, R; Chandra, J.; Bhattacharjee, J.; Verma, P.; Narayan, S. Peroxidative stress and antioxidant enzymes in children with beta-thalassemia major. *Pediatr Blood Cancer* 2006; 46:780–785.

Walter P B, Fung EB, Killilea DW, Jiang Q, Hudes M, Madden J, Porter J, Evans P, Vichinsky E, Harmatz P. Oxidative stress and inflammation in iron-overloaded patients with β-thalassemia or sickle cell disease. *Br J Haematol.* 2006; 135: 254–263.

Rosatelli MC, Dozy A, Faa V, Meloni A, Sardu R, Saba L. Molecular characterization of b-thalassemia in the Sardinian population. *Am J Hum Genet* 1992;50:422–6.

Schilirò G, Di Gregorio F, Samperi P, Mirabile E, Liang R, Curuk MA, Ye Z, Huisman TH. Genetic heterogeneity of beta-thalassemia in Southeast Sicily. *Am J Hematol.* 1995; 48: 5–11.

Vedder AC, Cox-Brinkman J, Hollak CE, Linthorst GE, Groener JE, Helmond MT, Scheij S, Aerts JM. Plasma chitotriosidase in male Fabry patients: A marker for monitoring lipid-laden macrophages and their correction by enzyme replacement therapy. *Mol Genet Metab.* 2006 Jun 7;

Zaino EC, Rossi MB, Pham TD, Azar HA. Gaucher's cells in thalassemia. *Blood* 1971; *38: 457-462.*

In: Binomium Chitin-Chitinase: Recent Issues ISBN 978-1-60692-339-9
Editor: Salvatore Musumeci and Maurizio G. Paoletti

Chapter XV

Chitotriosidase in *Plasmodium*, *Anopheles* and Human Interaction

Maria Musumeci [1]***, Andrea Giansanti*** [2] ***and Salvatore Musumeci*** [3]
[1]Department of Hematology, Oncology and Molecular Medicine,
National Institute of Health, Rome, Italy
[2]Department of Physics, "La Sapienza" University of Rome, Italy.
[3]Department of Neurosciences and Mother and Child Sciences, University of Sassari and Institute of Biomolecular Chemistry, National Research Council (CNR), Li Punti (SS), Italy

Abstract

High levels of plasma chitotriosidase (Chit) are a marker of macrophage activation in several infectious pathologies and, in particular, in human malaria. *Plasmodium falciparum (P. falciparum)*, during its maturative cycle in the midgut of the *Anopheles* mosquito, secretes a specific chitinase enabling it to cross chitin-containing peritrophic membrane (PM) which surrounds the blood meal. This represents a necessary step in the migration of the parasite from the midgut to the salivary glands of malaria's vector. The cooperation between human Chit and the chitinase produced by *P. falciparum* in attacking the peritrophic membranes in the Anopheles midgut has been recently demonstrated by *in vivo* experiments, and seems to favour the trasmissibility of human malaria in African sub-Saharan regions. Optical microscopy (OP) showed that the formation of the PM was completed after 16 h in the posterior midgut of *Anopheles* already fed with healthy donor bloods. In contrast, PM formation was partly conserved after 16 h, when mosquitoes were fed with malaria and Gaucher patient blood, but the PM appeared clearly damaged at 20 and 24 hours. In addition, the PM formation was almost completely inhibited in the midgut of *Anopheles* fed with *P. falciparum* chitinase enriched blood. These alterations in the PM formation were confirmed by Transmission Electronic Microscopy (TEM). This functional homology between human Chit and *P. falciparum* chitinase was confirmed also by computational methods. A simple sequence analysis method, potentially useful to assess fine textual closeness in families of homologous proteins, was applied to a set of chitinases from mammals and *plasmodia*. This analysis confirmed the clustering and the phylogenetic relationships obtained with well known alignment methods, but also showed that the sequences of chitinases from different malaria hosts and from different malaria parasites are strictly correlated. This correlation confirms a functional homology among chitinases, which is seen as a condition for the spreading of the different forms of malaria. From this perspective, one can get insight into the origins of malaria, and its genetic or pharmacological control.

1. Introduction

Chitin is the scaffold of arthropods' exoskeleton, fungal cell walls, nematode egg shells and sheaths of some filarial nematodes. Chitinases have a basic role in the defense of organisms against pathogenic parasites (Van Alten *et al.* 2001; Synstad *et al.* 2004). Chitotriosidase (Chit) is a human chitinase, produced by macrophage cells (Boot *et al.* 1998), is relatively abundant in mononuclear and polymorphonuclear leukocytes (Bouzas *et al.* 2003). High levels of Chit activity have been found in the plasma of individuals affected by various diseases (Hollak *et al.* 1994; Barone *et al.* 1999; Boot *et al.* 2005). Common denominator in all the situations, where a plasma Chit increase was reported, is the intense macrophage activation (Barone *et al.* 2007). Some years ago, we described an increase of Chit in patients affected by acute and/or severe malaria induced by *P. falciparum* (Barone *et al.* 2003), but no evidence we found that the elevated plasma Chit had an impact on the severity and outcome of malaria, and no study has compared genotypes with disease status. Its concentration has been found up to 2 orders of magnitude larger in plasma of patients suffering from Gaucher disease; a rare genetic disorder caused by a mutation in the glucocerebrosidase gene (Giraldo *et al.* 2001, Aerts and Boot 2008 in this Book). This enzyme is active against colloidal chitin and is inhibited by the chitinase inhibitor allosamidin.

2. Plasma Chitotriosidase and Malaria

There is a strict correlation between ferritin and Chit in malaria at favourable evolution, while this correlation is absent in severe malaria (Musumeci *et al.* 2006), suggesting that an accelerated red cell turnover, associated to macrophage accumulation of red cell membrane degradation products, may induce macrophages to produce Chit, resulting in an increase of its plasma level proportional to ferritin (Bouzas *et al.* 2003).

Moreover, in parts of Sub-Saharan Africa, where the malaria infection is endemic, the wild type allele of CHIT is prevalent and the frequence of this allele in the world follows the diffusion of past malaria endemia with some exceptions. (Chien *et al.* 2005; Piras *et al.* 2007).

Recently in Papua New Guinea, where the malaria is also endemic, an association analysis showed no significant difference between CHIT genotypes and the presence of *P. falciparum* or *P. vivax* malaria or other parasitic diseases (Hall *et al.* 2007). This observation put in discussion the relationship between CHIT expression and malaria, but focus differences in the frequency of mutated CHIT1 allele not always associate to selective mechanism (Chien *et al.* 2005; Piras *et al.* 2007).

Indeed, an increased plasma Chit seem actually may serve to increase the trasmissibility of malaria, by inhibiting the formation of the peritrophic membrane (PM) in the anopheline gut.

3. Chitinase in The *Plasmodium/Anopheles* Relationship

Anopheline mosquitoes, as most insects, produce a peritrophic membrane (PM) permeable to gastric juice that separates the blood meal from stomach epithelium. Diffusion of hydrolytic enzymes from the midgut epithelium across the PM is essential for blood digestion and survival of mosquitoes. In principle, a PM absence could permit to digestive enzymes a quickly access to the food bolus thus resulting in faster digestion and consequently, better absorption of nutrients. Nevertheless, the PM's main task is the protection of the epithelium cells from hard/wrinkled undigested particles or foreign organisms, including pathogens. Thus, during evolution, the slight disadvantage caused by a small decrease in the digestion rate has been compensated by the higher protection afforded by the PM (Villalon *et al.* 2003). The chitinase activity represent a necessary step in the migration of parasite from the midgut to the salivary glands of the intermediate host (Figure 1).

Billingsley and Rudin, 1992 observed that PM thickening partially inhibits development of *P. falciparum*, thus representing a physical barrier that parasites have to overcome by secreting a specific chitinase. In fact *P. falciparum* produces a chitinase able to digest the PM and essential to cross this physical barrier in a maturative phase idoneous to complete the asexual cycle. The oocyte matures after this step and reaches the salivary glands as sporozoites (see Figure 1).

4. *In Vitro* Experiments of Chit/*P. Falciparum* Chitinase Interaction

We demonstratesd that human Chit from different sources is active on the Anopheline PM, probably enhancing the penetration action of malaria parasites through the mosquito midgut.

Optical microscopy morphological examination of stomach sections of untreated samples 1, 2, 16 h after blood feeding, revealed a detectable PM (figure 2 A). PM thickness increased at 20 h and decreased at 24 h (figure 2 B-C). The PM has been recognized as a fibrous and laminated structure that encloses and surrounds the entire blood meal, just above the epithelium. Heme precipitates and matrix bound heme are clearly visible between the erythrocytes, at 20 h.

TEM observation confirmed a perfectly maintained PM structure, showing an electron dense layer that separates the blood meal (bolus of erythrocytes) from the microvilli (figure 3 A). During the blood digestion, the PM internal portion became filled by heme precipitates (figure 3 B-C) even though it maintained both its compact structure and barrier function.

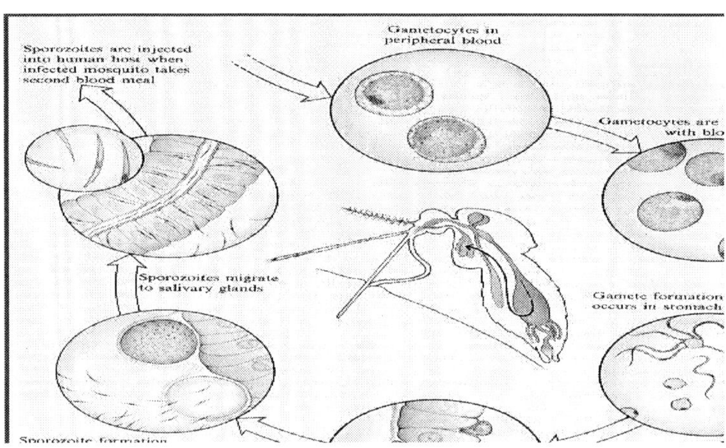

Figure 1. Sexual cycle of P. *falciparum* in the stomach of Anopheles from gamete formation, fertilization, oocyst formation, sporozoite formation and migration to salivary glands.

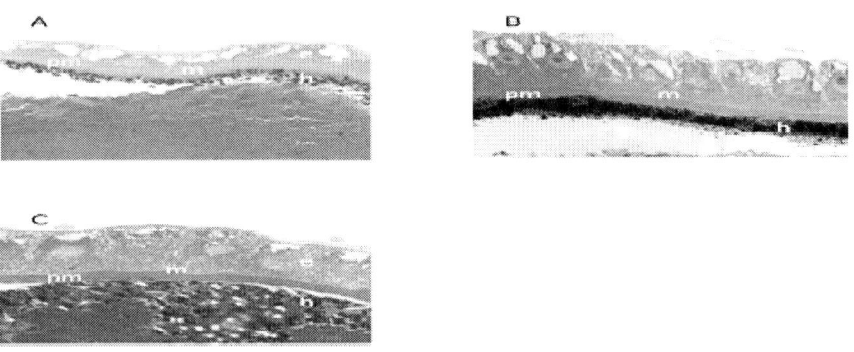

Figure 2. Morphological examination of the midgut from Anopheles fed with blood of a healthy donor reveals a fully developed PM by 16 h (Figure 2 A), whose thickness increases at 20 h and decreases at 24 h (Figure 2 B and C). (Adapted from Di Luca *et al.* 2006).

Both OM (not shown) and TEM data, concerning midgut from *Anopheles* fed with blood from malaria patients, indicated delayed and damaged PM formation, according to Chit concentration (0.68 and 0.95 mU/ml of Chit activity, respectively).

Also, in *Anopheles* fed with blood of patients with Gaucher disease (5.2 and 10.4 mU/ml of Chit activity respectively), the PM thickness was markedly reduced. In Figure 3 the midgut section of mosquitoes fed with blood from the patient with the lower Chit activity (5.2 mU/ml) is shown. PM appears filled by products of digestion since 16 h after blood meal (Figure 3 A). A PM thin fibrillar portion divided the blood products from microvilli, during the first two digestion phases (Figure 3 A, B). At 24 h, in some areas heme precipitates are found in contact with the microvilli layer (Figure 3 C).

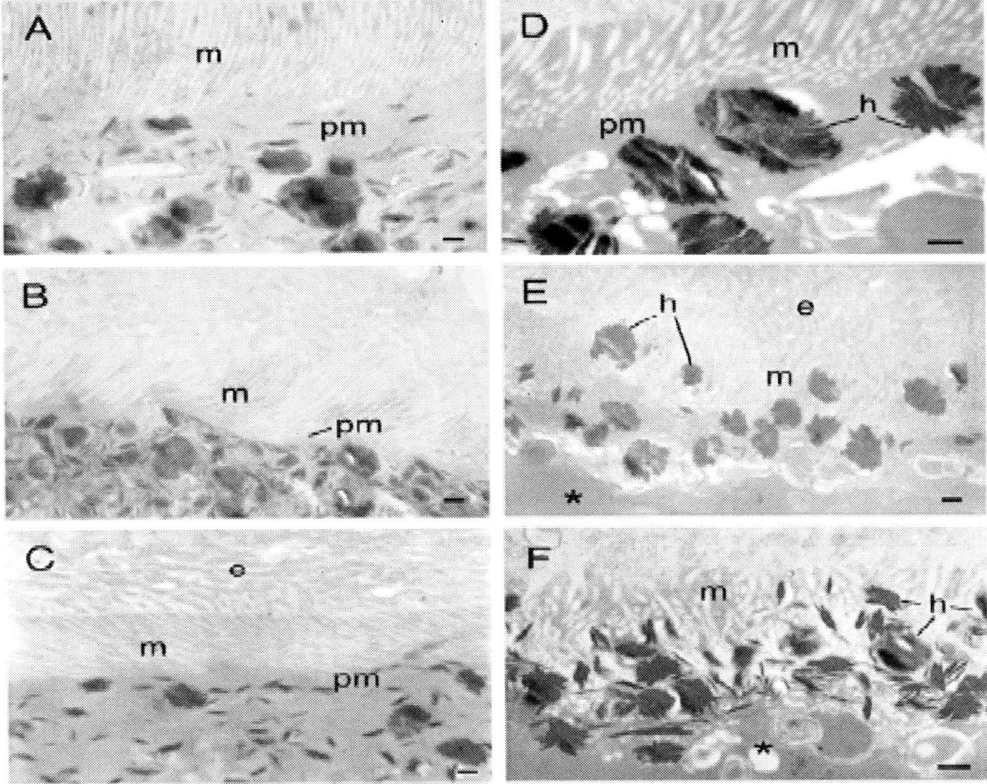

Figure 3: TEM micrographs of midgut sections from Anopheles stephensi female mosquitoes at 16h (A-D), 20h (B-E) and 24h (C-F) after blood feeding. Scale bars = 500nm A-B-C: mosquitoes fed with blood of Gaucher disease patient containing 5.2 mU/ml of Chit. PM is completely filled by digestion products since 16h from blood feeding. A thin fibrillar internal portion of PM divides the blood products from microvilli (m) during the first two digestion phases. At 24 h from feeding, heme precipitates (h) directly contact microvilli (m) of epithelial cells (e). D-E-F: mosquitoes fed with blood containing 2000mU/ml of commercial chitinase from *P. falciparum*. A thin fibrillar, external portion of PM, is interposed between the bloodmeal (asterisk) only in a few areas at 16h after feeding (D). The products of blood digestion including heme precipitates (h) are clearly visible among the microvilli (m) and inside the epithelial cells (e) since the 20h. (Adapted from Di Luca *et al.* 2006)

When the blood samples were enriched with the commercial chitinase from *P. falciparum* (100, 500 and 2000 mU/ml of PfCHTactivity, respectively), the PM appears clearly disrupted and the heme precipitates reached the epithelium. These alterations were clearly observed by TEM at 16, 20 and 24 h. Figure 3 (D-E-F) shows midgut sections of the sample obtained adding 2000 mU/ml of PfCHT1. Since 16 h, PM appears as a thin fibrillar structure acting as a barrier only in a few areas (Figure 3 D). Subsequently, at 20 and 24 h after bloodmeal, PM seems disorganised and completely permeable to blood digestion products, clearly visible among the microvilli and inside the epithelial cells (Figure 3 E-F).

In Figure 4 the progression or regression of the PM in presence of Chit, as measured on the microscopic section of stomach (MP/epithelium) is graphically reported: the PM formation was measurable after 16 h, but clearly reduced at 20-24 h in a degree proportional to Chit contents. These alterations were clearly documented by TEM at 16, 20 and 24 hours (see figure 3).

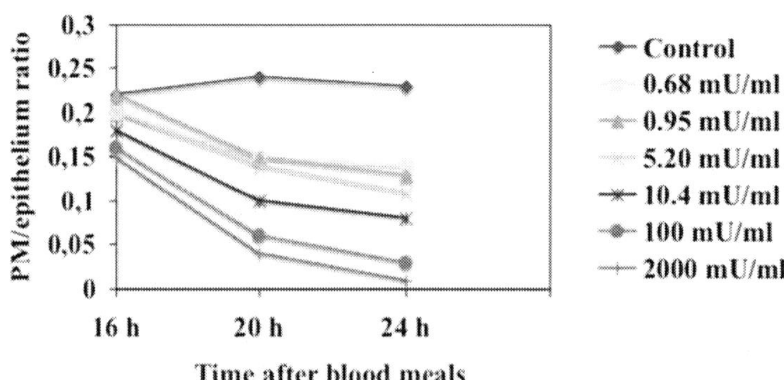

Figure 4 The PM/epithelium ratios were measured on the microscopic section of stomach of Anopheles fed with blood containing different concentration of chitinase (Adapted from Di Luca *et al.* 2006).

Now most of heme bound to the PM is associated with enormous number of small electron dense granules and the residual PM is full of erythrocytes and heme precipitates. Also if the PM structurally could appear conserved, the heme precipitate were clearly visible in mosquitoes which were feeded with blood containing 5.2 mU/ml from Gaucher patient and 2000mU/ml of chitinase from *P. falciparum* at different magnitudes (figure 3). Moreover, on the endoperitrophic side, some of these granules seem to dissociate from the PM and leak into the lumen. This observation is important since it could open the way to develop of other study protocols aimed to block the midgut phase, since the heme is toxic both for *P. falciparum* and mosquitoes.

Therefore, *P. falciparum* chitinase could be a potential target for blocking transmission of *P. falciparum* from human host to the mosquito vector (Shahabuddin and Kaslow, 1993). There it would probably be no need to fully inhibit the chitinase activity of the parasite, as partial inhibition (e.g. 50%) could be enough to prevent the *P. falciparum* migrating inside the mosquitoes. Such innovative drugs could also use specific differences between the intermediate and the human host to achieve the selective release of such inhibitors in the mosquito. Another way could be to stimulate chitinase activity in malaria patients, favouring the PM damage in the mosquitoes. If chitinase activity is stimulated in the human host, the change for the parasite to cross the PM apparently should be facilitated, but considering the image of electronic microscopy, where a lot of heme is generated by the digestion of Hb, heme precipitates could represent an obstacle for the *Plasmodium* survival (Shahabuddin *et al.* 1993). If chloroquine or artemisine linking free heme, inhibits the formation of hemozoin, the free heme (actually is heme-QHS complex) should kill the malaria parasite in the stomach of *Anopheles* (Pandey *et al.* 1999; Taramelli *et al.* 1999). This hypothesis is conflictuing with previous affermation that an increased plasma Chit actually may serve to increase the transmissibility of malaria, by inhibiting the formation of peritrophic membrane (PM) in the *Anopheles*, but it remains a possibility to be proved in further future study.

5. Chitinase Control with Inhibitors

Some years ago, experimental data showed that incorporation of exogenous chitinases into a blood meal could prevent PM formation (Huber *et al.* 1991; Filho *et al.* 2002). Moreover, the inhibition of the endogenous mosquitoes' midgut chitinase with allosamidin, an inhibitor of chitinase, delays PM disassembling at the beginning of blood digestion (Shen and Jacobs-Lorena, 1997).

In 1993 Shahabuddin *et al.* suggested the use of allosamidin, or specific antibodies, to reduce the malaria infection rate, thus hampering the development of malaria parasites in their vector. Unfortunately, this transmission blocking strategy did not result in a practical application, but remained a mere scientific exercitation (Kaslow, 1993). In the light of the discovery of chitinases in humans, the possibility to block the transmissibility with chitinase inhibition, with either allosamidin or specific antibodies against Chit, seems even further removed from practical application. In fact, human Chit and plasmodium chitinase are structurally and functionally very similar, since they belong to the same 18 glycosyl hydrolases family. For this reason, selective *P. falciparum* chitinase (PfCHT1) inhibition (Langer and Vinetz, 2001) appears difficult to obtain, without inhibiting the plasma Chit.

On the contrary, our findings support the hypothesis that human plasma Chit may influence the *P. falciparum*/human host relationship, thus helping the parasite to cross the PM and reach the midgut epithelium. Different levels of human Chit have been reported in healthy populations living in areas where malaria was/is endemic. It is known that 5-6 % of healthy Caucasian individuals are deficient in Chit. This is due to homozygous duplication of 24-base pairs in exon 10 of the CHIT gene (Boot *et al.* 1998), while 35-45% of Caucasian individuals are heterozygotes for this mutation. Therefore, a higher level of plasma Chit is found in healthy individuals from Africa, Sardinia and Sicily, proportional to the allele frequency in these populations (Malaguarnera *et al.* 2003), (see Table I).

Because of *P. falciparum* penetration through the Anopheles midgut depends on PM disruption, we may hypothesize that elevated Chit activity levels in plasma of African malaria patients could help *P. falciparum* to complete its life cycle in the vector, thus contributing to the malaria maintenance in endemic areas.

Table I. Allele frequencies of wild and mutated alleles of *CHIT* in population of different continents in relation to past and actual malaria endemicity (Adapted from Malaguarnera *et al.* 2003)

References	Country or Region	Number studied subjects	Past Malaria	To day	Allele Frequencies	
					wild	mutated
Boot *et al.* 1998	Israel	68	++	(-)	0.77	0.23
Boot *et al.* 1998	Holland	171	+	(-)	0.76	0.24
Canudas *et al.* 2001	Spain	116	+	(-)	0.74	0.26
Choi *et al.* 2001	South India	216	+++	+++	0.56	0.44
Malaguarnera *et al.* 2003	Benin	100	++++	++++	1	0
Malaguarnera *et al.* 2003	Burkina Faso	100	++++	++++	0.98	0.02
Malaguarnera *et al.* 2003	Sardinia	107	+++	(-)	0.79	0.21
Malaguarnera *et al.* 2003	Sicily	100	++	(-)	0.73	0.27
Chien *et al.* 2005	Taiwan	82	+++	(-)	0.42	0.58
Piras *et al.* 2007	Sardinia (>400 m)	136	-	(-)	0.886	0.114
Piras *et al.* 2007	Sardinia (0-200 m)	128	+++	(-)	0.754	0.246
Hall *et al.* 2008	Papua New Guinea	693	+++	+++	0.749	0.251

On the contrary, in European regions where Chit became redundant for the expansion of the inactive mutated CHIT allele, it could be possible that the mutant genotype has been a contributing factor in the rapid eradication of malaria, together with the use of DDT and land reclamation.

It is well-known that a genetic protection from malaria is conferred by heterozygotes for HbS, thus reducing the circulating asexual parasitemia (Luzzatto *et al.* 1970). On the other hand, HbS carriers show a compensatory effect on the transmissibility of parasite from humans to mosquitoes. Robert *et al.* 1996 observed that the percentage of Anopheles infected by HbAS gametocyte carriers was higher compared to mosquitoes fed with HbAA carrier blood. This increase of parasite transmissibility could depend on higher Chit levels in the HbAS carrier blood, due to an increased red cell turnover by macrophage cells and a Chit macrophage overproduction in African population (Malaguarnera *et al.* 2003). In Sardinia and Sicily, a similar phenomenon could also have occurred, induced by the different prevalence of β-thalassemia or G6PDH deficiency. In conclusion, an adaptive strategy of the malaria parasite, based on a selective advantage in Africa (Vinetz *et al.* 1999), could be hypothesized from our data. This could have developed a significant epidemiological effect by increasing the global level of malaria transmission in countries such as Burkina Faso, through genetic characteristics of a population which maintains a high Chit production in activated macrophages (Malaguarnera *et al.* 2003).

6. Fuctional Homology between Human and *P. Falciparum* Chitotriosidases

To support the hypothesis of an interaction between *P. falciparum* chitinase (PfCHT1) and human plasma Chit we conducted a computational analysis of chitinase sequences that introduces the concept of functional homology, distinct from structural homology. In fact a simple sequence analysis method, potentially useful to assess fine contextual closeness in families of homologous proteins, was applied to a set of 14 chitinases from mammals and plasmodia (see Table II).

Table II. The first column lists the National Center for Biotechnology Information (NCBI) gene identification numbers of the various chitinases. In the second column the organisms are identified. The third column lists the labels used in the present work. It has to be noted that for *P. gallinaceum* there are two different chitinase genes: cht1 and cht2, as reported in Li *et al.* 2005. (Adapted from Giansanti *et al.* 2007)

	Gene id.	Gene	Organism	Label
1	gi|73909055|	Cht1	[*Homo sapiens*]	Homo_sapi
2	gi|58037265|	Cht1	[*Mus musculus*]	Mus_muscu
3	gi|62659337|	Cht1	[*Rattus norvegicus*]	Rattus_n1
4	gi|55589088|	Cht1	[*Pan troglodytes*]	Pan_trogl
5	gi|39598848|	Cht1	[*Rattus norvegicus*]	Rattus_n2
6	gi|23509192|	Cht1	[*P. falciparum* 3D7]	PfCHT1
7	gi|61661212|	Cht2	[*P.gallinaceum*]	PgCHT2
8	gi|61661214|	Cht1	[*P. reichenowi*]	PrCHT1
9	gi|7530424|	Cht1	[*P. gallinaceum*]	PgCHT1
10	gi|14275849|	Cht1	[*P. berghei*]	PbCHT1
11	gi|37590968|	Cht1	[*P. yoelii*]	PyCHT1
12	gi|37591029|	Cht1	[*P. chabaudi*]	PcCHT1
13	gi|37590964|	Cht1	[*P. vivax*]	PvCHT1
14	See (Li et al 2005)	Cht1	[*P. knowlesi*]	PkCHT1

Chitinases were aligned into highly conserved blocks, which comprise residues relevant for the catalysis in human Chit and the high degree of conservation in these key residues strongly suggests that most of the *Plasmodium* and mammalian chitinases should have very similar hydrolytic pockets. An almost perfect superposition of models of the active site of PfCHT1, PgCHT1 and human Chit has been already shown in Vinetz *et al.* 1999. Of course, the affinity (free energy of binding) against the common substrate can be modulated, in different chitinases, by i) fine structural details of the hydrolytic pocket; ii) the presence or absence of chitin binding domains; iii) the presence of shared sequence motifs more or less in the proximity of the essential residues, as suggested by the co-occurrence matrix. It is also worth mentioning that mammalian acidic chitinase (AMCase), which has an active site of the same chitinase type 18 family, is active only in very acidic environments (pH in the range 1.5-2.0), whereas Chit is active at a milder acidity, around pH 5.2 (Chou *et al.* 2006).

However Bussing *et al* 2008 demonstrated that His187 conserved residue is responsible for the acidic optimum in mouse, while the acidic activity of human AMCase is not as pronounced as that of mouse AMCase, despite the presence of His 187. This effect suggests a pH dependent modulation of the reaction mechanism that is unique to AMCases. In fact as the surface potential of mouse AMCase is significantly lower , the loss of structural stability at very low pH may limit the extreme acidic activity of human AMCase. The catalytic efficiency of this class of molecules, surely determined by the steric complementarity of the active site to substrates, can be modulated by patterns of amino acidic substitutions close to the hydrolytic pocket.

The method of co-occurrences we used in this study proves to be a valid support to the "3Dcoffee" multiple alignment of the sequences and offers insight into possible functional homologies between otherwise distantly related chitinases (Clamp *et al.* 2004; Tcoffee@igs). Moreover, the method proposed here is able to detect a significant correlation between the sequences of chitinases in malaria parasite/host pairs. In particular, going back to the dark grey area in Table 3 of co-occurrences values, it has to be noted that the significant score of 4 is shared by the co-occurrences between the chitinases of: *Homo sapiens* and *P. falciparum*; *Rattus n_1* and *P. berghei* ; *Pan troglodytes* and *P. reichenowi*. This consideration must be related to the recent experimental result pointing to a functional homology and cooperation of human and plasmodial chitinases in digesting the peritrophic membrane (Di Luca *et al.* 2006; Di Luca *et al.* 2006). At a macroscopic level, the essential point to be stressed is that *Anopheles* mosquitoes living in an infected area have a high probability of ingesting blood from subjects with acute malaria and with high levels of expressed Chit. If human Chit is in turn highly active against the chitin of the PM, then it can facilitate ookinetes in the traversing of the midgut wall of the *Anopheles* mosquito increasing the transmissibility of malaria. This derives from the sequence and functional homologies between *P. falciparum* chitinase and human Chit.

However the functional homology recognized here, suggests caution in adopting strategies to control malaria through the use of inhibitors of the *P. falciparum* chitinase. Clearly, a high affinity inhibitor of the plasmodial chitinase might well be an effective inhibitor of human Chit too, interfering with its physiological role. However homozygous mutants in European and other countries have no obvious signs of compromised immunity or increased susceptibility to microbial infection.

This analysis confirms the clustering and the phylogenetic relationships obtained with well known alignment methods, but also showed that the sequences of chitinases from different malaria hosts and from different malaria parasites are strictly correlated.

Our study provides a rationale that a cooperation between the host's and *P. falciparum* chitinases in favouring the spread of malaria could exist. Chit secretion, elevated in African subjects living in precarious and poor hygienic status condition, reflects the genetics of individuals who maintain high the wild type CHIT gene frequency (observed in the 99% of the population) as a consequence of environmental factors. High levels of active Chit contribute to the persistence of endemic malaria through the functional homology against the PM of *P. falciparum* chitinase and human Chit, documented by "in vitro" observations and by computational analysis (Di Luca *et al.* 2006; Giansanti *et al.* 2007).

Table III. The element i-j of the co-occurrence matrix counts the number of statistically significant k-motifs common to chitinase sequences i-th and j-th. Elements on the principal diagonal have been set to zero. Different colors highlight different groups of organisms. In light grey: mammals (upper left box); orthologous chitinases of *P. falciparum, P. gallinaceum and P. reichenowi* (see Li *et al.* 2005) (center box); chitinases of the other plasmodia (lower right box). The relation between mammalian and plasmodial chitinases is shown in dark grey (white digits). (Adapted from Giansanti *et al.* 2007)

Label	N.	1	2	3	4	5	6	7	8	9	10	11	12	13	14
Homo_sapi	1	0	132	124	344	50	4	0	4	2	1	1	1	1	1
Mus_Muscu	2	132	0	171	132	32	3	1	3	2	1	1	1	1	1
Rattus_n1	3	124	171	0	130	42	3	1	3	3	4	3	3	2	3
Pan_trogl	4	344	132	130	0	56	4	0	4	1	2	1	1	2	1
Rattus_n2	5	50	32	42	56	0	2	3	2	2	2	2	1	1	1
PfCHT1	6	4	3	3	4	2	0	89	288	11	14	15	19	13	19
PgCHT2	7	0	1	1	0	3	89	0	92	8	11	11	11	12	11
PrCHT1	8	4	3	3	4	2	288	92	0	12	14	15	19	13	19
PgCHT1	9	2	2	3	1	2	11	8	12	0	126	123	93	16 7	93
PbCHT1	10	1	1	4	2	2	14	11	14	126	0	327	0	13 0	209
PyCHT1	11	1	1	3	1	2	15	11	15	123	327	0	212	12 9	212
PcCHT1	12	1	1	3	1	1	19	11	19	93	209	212	0	10 6	282
PvCHT1	13	1	1	2	2	1	13	12	13	167	130	129	106	0	106
PkCHT1	14	1	1	3	1	1	19	11	19	93	209	212	282	10 6	0

At the end the functional homology between the chitinases of *Homo sapiens* and *Pan troglodytes* with those of their respective malaria parasites is interesting and worth to be further analyzed. It is well known that the infectivity of *P. reichenowi,* parasite of chimpanzees against man is limited and viceversa *P. falciparum* hardly infects chimpanzees (Martin *et al.* 2005). The above observation may have relevance for the interesting scholarly debate about placing the origins of malaria in Africa at the recent era of the onset of agriculture (around 10,000 years ago) or, at least, at the time of the separation of man from the hominids (around 8 Myears ago) (Pennisi, 2001).

Table IV. List of shared 4-motifs among mammalian and plasmodial chitinases. Each row corresponds to entries of scores 4 and 3 in the dark grey box from table III. (Adapted from Giansanti *et al.* 2007)

Matrix Element	Shared 4-motifs
1/9: CHIT-PfCHT1	FDGL, NPRE, SFDG, KYSF
2/6: Mus_Muscu/PfCHT1	FDGL, GFDG SSDN,
2/8: Mus_Muscu/PrCHT1	FDGL, GFDG, SSDN,
3/6: Rattus_n1/PfCHT1	FDGL, GFDG, SLSS
3/8: Rattus_n1/PrCHT1	FDGL, GFDG, SLSS
3/9: Rattus_n1/PgCHT1	PGLV, VDKI ,EQEV
3/10:Rattus_n1/PbCHT1	ASGK, HVDA, SKAI, TFVN
3/11: Rattus_n1/PyCHT1	HVDA, SKAI, TFVN
3/12: Rattus_n1/PcCHT1	NNHQ, SKAI, TFVN
3/14: Rattus_n1/PkCHT1	SGPP, SGVY, TFVN, NDEL
4/6: Pan_trogl/PfCHT1	FDGL, NPRE, SFDG, KYSF
4/8: Pan_Trogl/PrCHT1	FDGL, NPRE, SFDG, KYSF
5/7: Rattus_n2/PgCHT2	DDIN, GFDG, LNKA

The hypothesis of a parallel functional homology between the chitinases of chimpanzee with that of *P. reichenowi* and between the human Chit and *P. falciparum* chitinase as a favourable condition for the spreading of the disease, points to a very old mechanism at the origin of the malaria epidemics (Escalante *et al.* 2005). The conservated sequences among chimpanzees and humans (Gianfrancesco and Musumeci, 2004) support the basis to analyse the relation among these enzymes and their substrates as the results of a selective pressure.

From this perspective, one can get insight into the origins of malaria, on its genetic or pharmacological control.

References

Barone R, Di Gregorio F, Romeo MA, Schilirò G, Pavone L. Plasma chitotriosidase activity in patients with beta-Thalassemia. *Blood Cells Molecules and Diseases* 1999; 21:1-8.

Barone R, Simpore J, Malaguarnera L, Pignatelli S, Musumeci S. Plasma chitotriosidase activity in acute *Plasmodium falciparum* malaria. *Clin Chim Acta*. 2003; 331:79-85.

Barone R, Sotgiu S, Musumeci S. Plasma chitotriosidase in health and pathology. *Clin Lab.* 2007; 53:321-33.

Billingsley PF, Rudin W. The role of the mosquito peritrophic membrane in blood meal digestion and infectivity of *Plasmodium* species. *J Parasitol.* 1992; 78:430-40

Boot RG, Rankema GH, Verhoek M, Strijland A, Bliek J, de Meulemeester TM, Mannens MM, Aert JM. The human chitotriosidase gene. Nature of inherited enzyme deficiency. *J Biol Chem.* 1998; 273:25680-5.

Boot RG, Bussink AP, Verhoek M, de Boer PA, Moorman AF, Aerts JM. Marked differences in tissue-specific expression of chitinases in mouse and man. *J Histochem Cytochem.* 2005; 53:1283-1292.

Bouzas L, Carlos Guinarte J, Carlos Tutor J. Chitotriosidase activity in plasma and mononuclear and polymorphonuclear leukocyte populations. *J Clin Lab Anal.* 2003; 17:271-275.

Bussink AP, Vreede J, Aerts JM, Boot RG. A single histidine residue modulates enzymatic activity in acidic mammalian chitinase. *FEBS Lett.* 2008; 582:931-5.

Chien YH, Chen JH, Hwu WL. Plasma chitotriosidase activity and malaria. *Clin Chim Acta.* 2005; 353:215.

Choi EH, Zimmerman PA, Foster CB, Zhu S, Kumaraswami V, Nutman TB, Chanock SJ. Genetic polymorphisms in molecules of innate immunity and susceptibility to infection with *Wuchereria bancrofti* in South India. *Genes Immun.* 2001; 2:248-253.

Chou YT, Yao S, Czerwinski R, Fleming M, Krybaev R, Xuan D, Zhou H, Brooks J, Fitz L, Strand J, Presman E, Lin L, Aulabaugh A, Huang X. Kinetic characterization of recombinant human acidic mammalian chitinase. *Biochemistry* 2006; 45:4444-54.

Clamp M, Cuff J, Searle SM, Barton GJ. The Jalview Java Alignment Editor. *Bioinformatics* 2004; 12: 426-427.

Di Luca M, Romi R, Severini F, Toma L, Musumeci M, Fausto AM, Mazzini M, Gambellini G, Musumeci S. Human chitotriosidase helps *Plasmodium falciparum* in the Anopheles midgut. *J Vector Borne Dis.* 2006; 43:144-6

Di Luca M, Romi R, Severini F, Toma L, Musumeci M, Fausto AM, Mazzini M, Gambellini G, Musumeci S. High levels of human chitotriosidase hinder the formation of peritrophic membrane in anopheline vectors. *Parasitol Res.* 2006; 100: 1033-9.

Escalante A, Cornejo OE, Freeland DE, Poe AC, Durrego E, Collins WE, Lal AA. A monkey's tale: The origin of Plasmodium vivax as a human malaria parasite. *Proc Natl Acad Sci. USA* 2005; 102:1980-1985.

Filho BP, Lemos FJ, Secundino NF, Pascoa V, Pereira ST, Pimenta PF. Presence of chitinase and beta-N-acetylglucosaminidase in the Aedes aegypti. A chitinolytic system involving peritrophic matrix formation and degradation. *Insect Biochem Mol Biol.* 2002; 32:1723-9.

Gianfrancesco F, Musumeci S. The evolutionary conservation of the human chitotriosidase gene in rodents and primates. *Cytogenet Genome Res.* 2004;105:54-56.

Giansanti A, Bocchieri M, Rosato V, Musumeci S. A fine functional homology between chitinases from host and parasite is relevant for malaria transmissibility. *Parasitol Res.* 2007;101:639-45.

Giraldo P, Cenarlo A, Alfonso P, Perez-Calvo JI, Rubio-Felix D, Girald M, Pocovi M. Chitotriosidase genotype and plasma activity in patients with type 1 Gaucher's disease and their relatives (carrier and non-carriers) *Haematologica* 2001; 86:977-984

Hall AJ, Quinnell RJ, Raiko A, Lagog M, Siba P, Morroll S, Falcone FH. Chitotriosidase deficiency is not associated with human hookworm infection in a Papua New Guinean population. *Infect Genet Evol.* 2007; 7:743-7.

Hollak CEM, van Weely S, van Oers MHJ, Aerts JM. Elevated plasma chitotriosidase: a novel biochemical hallmark of Gaucher disease. *J Clin Invest.* 1994; 93:1288-1292.

Huber M, Cabib E, Miller LH. Malaria parasite chitinase and penetration of the mosquito peritrophic membrane. *Proc Natl Acad Sci U S A*. 1991; 88:2807-10.

Kaslow DC.Transmission-blocking immunity against malaria and other vector-borne diseases. *Curr Opin Immunol.* 1993; 5:557-65.

Langer RC, Vinetz JM. *Plasmodium* ookinete-secreted chitinase and parasite penetration of the mosquito peritrophic matrix. *Trends Parasitol.* 2001; 1:269-72.

Li F, Patra KP, Vinetz JM. An anti-Chitinase malaria transmission-blocking single-chain antibody as an effector molecule for creating a *Plasmodium falciparum*-refractory mosquito. *J Infect. Dis.* 2005; 192:878-887.

Luzzatto L, Nwachiku-Jarrett ES, Reddy S. Increased sickling of parasitized erythrocytes as mechanism of resistance against malaria in the sickle-cell trait. *Lancet* 1970; 1:319-321.

Malaguarnera L, Simpore J, Prodi DA, Angius A, Sassu A, Persico I, Barone R, Musumeci S. A 24-bp duplication in exon 10 of human chitotriosidase gene from the sub-Saharan to the Mediterranean area: role of parasitic diseases and environmental conditions. *Genes Immun.* 2003; 4:570-4.

Martin MJ, Rayner JC, Gagneux P, Barnwell JW, Varki A. Evolution of human-chimpanzee differences in malaria susceptibility: Relationship to human genetic loss of N-glycolylneuraminic acid. *Proc Natl Acad Sci. USA* 2005; 102:12819-12824.

Musumeci M, Simpore J, Barone R, Angius A, Musumeci S. Synchronic macrophage response and *Plasmodium falciparum* malaria. *J Vector Borne Dis.* 2006; 43:84-7.

Pandey AV, Tekwani BL, Singh RL, Chauhan VS. Artemisinin, an endoperoxide antimalarial, disrupts the hemoglobin catabolism and heme detoxification systems in malarial parasite.*J Biol Chem.* 1999; 274:19383-8.

Pennisi E. Malaria's Beginnings: On the heels of Hoes? Science 2001; 293:416-417

Piras I, Melis A, Ghiani ME, Falchi A, Luiselli D, Moral P, Varesi L, Calò CM, Vona G. Human CHIT1 gene distribution: new data from Mediterranean and European populations. *J Hum Genet.* 2007; 52:110-6.

Robert V, Tchuinkam T, Mulder B, Bodo JM, Verhave JP, Carnevale P, Nagel RL. Effect of the sickle cell trait status of gametocyte carriers of *Plasmodium falciparum* on infectivity to anophelines. *Am J Trop Med Hyg*. 1996; 54:111-3.

Shahabuddin M, Kaslow DC. Chitinase: a novel target for blocking parasite transmission? *Parasitol Today.* 1993; 9:252-5.

Shahabuddin M, Toyoshima T, Aikawa M, Kaslow DC. Transmission-blocking activity of a chitinase inhibitor and activation of malarial parasite chitinase by mosquito protease. *Proc Natl Acad Sci U S A*. 1993; 90:4266-70.

Shen Z, Jacobs-Lorena M. Characterization of a novel gut-specific chitinase gene from the human malaria vector *Anopheles gambiae*. *J Biol Chem.* 1997; 272:28895-900

Synstad S, Gaseidnes S, van Aalten DMF,Vriend G, Nielsen JE, Eijsink VGH. Mutational and computational analysis of the role of conserved residues in the active site of family 18 chitinase. *Eur J Biochem.* 2004; 271:253-262.

Taramelli D, Monti D, Basilico N, Parapini S, Omodeo-Salé F, Olliaro P. A fine balance between oxidised and reduced haem controls the survival of intraerythrocytic plasmodia. *Parassitologia.* 1999; 41:205-8.

"3DCoffee: Combining Protein Sequences and Structures within Multiple Sequence Alignments."O. O'Sullivan, K Suhrc, C. Abergel, D.G. Higgins, C. Notredame. *Journal of Molecular Biology*,Vol 340, pp385-395,2004

van Aalten DMF, Komander D, Synstad B, Gaseidnes S, Peter MG, Eijsink VGH. Structural insights into the catalytic mechanism of family 18 exochitinase. *Proc Natl Acad Sci USA* 2001; 98:8979-8984.

Villalon JM, Ghosh A, Jacobs-Lorena M. The peritrophic matrix limits the rate of digestion in adult *Anopheles s tephensi* and *Aedes aegypti* mosquitoes. *J Insect Physiol.* 2003; 49: 891-5.

Vinetz JM, Dave SK, Specht CA, Brameld KA, Xu B, Hayward R, Fidock DA. The chitinase PfCHT1 from the human malaria parasite *Plasmodium falciparum* lacks proenzyme anc chitin binding domains and displays unique substrate preferences. *Proc Natl Acad Sci USA* 1999; 96:14061-14066.

Vinetz JM. *Plasmodium* ookinete invasion of the mosquito midgut. *Curr Top Microbiol. Immunol.* 2005; 295:357-82.

In: Binomium Chitin-Chitinase: Recent Issues
Editor: Salvatore Musumeci and Maurizio G. Paoletti ISBN 978-1-60692-339-9

Chapter XVI

Role of AMCase in the Allergic and Non Allergic Ocular Pathologies

Maria Musumeci [1] ***and Salvatore Musumeci*** [2]
[1]Department of Hematology, Oncology and Molecular Medicine, National Institute of Health, Rome; Italy
[2]Department of Neurosciences and Mother and Child Sciences, University of Sassari and Institute of Biomolecular Chemistry, National Research Council (CNR), Li Punti (SS), Italy

Abstract

Chitin is abundant in the structural coatings of fungi, insects, and parasitic nematodes, but it is not produced in mammals. The host defense against chitin-containing pathogens includes production of chitinases. An acidic mammalian chitinase (AMCase) is produced in human epithelial cells of lower airways and conjunctiva via a Th2-specific, IL-13-dependent pathway and seems to be associated with asthma and allergic ocular pathologies. The understanding of the role of AMCase in allergic disease is only at its beginning and many issues open new possibilities for its control using specific inhibitors of AMCase activity or modulating its expression.

In patients with vernal keratoconjunctivitis (VKC) and with seasonal allergic conjunctivitis (SAC) the level of AMCase activity in the tears was found significantly elevated when compare to healthy controls and the highest levels were found in VKC. When RNA was extracted by conjunctival epithelial cells of these patients, quantitative Real Time PCR measurement confirmed that mRNA expression correlates with tear AMCase activity and the expression was significantly higher in VKC and SAC.

Also Receiver Operating Characteristic (ROC) analysis demonstrated that the sensitivity and specificity of AMCase measurement were 100 %, addressing the use of AMCase assay in the biochemical diagnosis of VKC and SAC.

Recent studies in rabbits, where a reactive uveitis was induced by LPS injection into the eye's anterior chamber, confirmed that increased AMCase activity was measurable in tears and that epithelial cells of conjunctiva express specific mRNA. A well as it was previously demonstrated in experimental model of mouse asthma, the inflammatory reaction induced by LPS was controlled by the chitinase inhibitor and steroid, instilled at 3 hr interval in conjunctival sacs.

In dry eye, another non allergic ocular pathology, an increased AMCase activity was documented and the specific mRNA expressed by epithelial conjunctival cells. In this pathology the eye inflammation can be ascribed to a common mechanism mediated by AMCase, via a Th2 specific, IL-13 dependent way. In synthesis, AMCase may be considered an important mediator in the pathogenesis of Th2 inflammation eye's diseases, suggesting its potential diagnostic and therapeutic utility.

1. Introduction

Chitin, the third most abundant polysaccharide in nature, is not expressed in mammals, but it is found in the structural coatings of fungi (Debono and Gordee, 1994), the exoskeleton of many arthropods (Neville *et at.* 1976) and parasitic nematodes (Araujo *et al.* 1993). These chitin coats provide protection for pathogens from harsh conditions inside the host and chitinases have a basic role in the defense of organisms against containing chitin parasites (Van Alten *et al.* 2001; Synstad *et al.* 2004). Paradoxically, although chitin and chitin synthase do not exist in mammals, members of glycosyl hydrolase (GH18) family, such as AMCase, chitotriosidase, YKL-40 and YKL-39, Ym1, oviduct-specific glycoprotein and stabilin-1-interacting chitinase like protein have recently been described (Kawada *et al.* 2007). Chitotriosidase and acidic mammalian chitinase (AMCase) posses chitinase enzymatic activity, whereas other mammalian chitinases do not possess this activity. AMCase is a 50 kDa protein that contains a 30-kDa N-terminal catalytic domain that can hydrolyse chitin. Characteristic of this chitinase is the resistance to acidic pH, which distinguishes AMCase from Chit, that has its optimum at pH 6 (Boot *et al.* 2001; Chou *et al.* 2006)

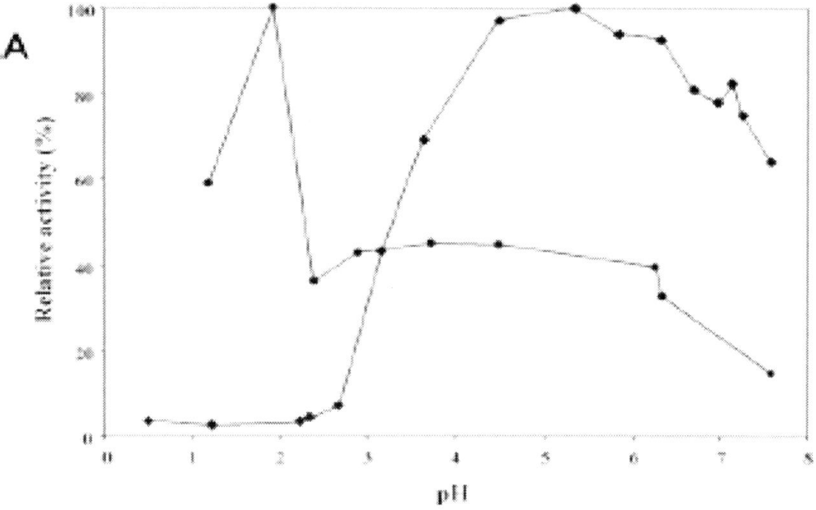

Figure 1. Relative activity (%) of Chitotriosidase and AMCase at differerent pH values (Adapted from Boot *et al.* 2001).

A role for AMCase in asthma pathophysiology has been suggested by Zhu *et al.* 2004, following the demonstration that AMCase expression increased in the lungs of ovalbumin-sensitized mice which developed airway hyperresponsiveness, compared to control animals. In this asthma model, AMCase is expressed in both airway epithelial cells and alveolar macrophages. Elevated expression of AMCase was also observed in lung tissue of asthmatic patients, when compared to normal subjects. Furthermore, Zhu *et al.* 2004 showed that AMCase does not directly induce Th2 cytokine response, but mediates the effector response of IL-13 (a cytokine produced by Th2 cells). In fact, IL-13 production was required for induction of AMCase, since IL-13-expressing mice had much higher levels of AMCase expression (mRNA) in lung epithelial cells compared to control animals. The role of AMCase in mediating the inflammatory response in asthma remains unclear. However, it seems that AMCase activity is required for the increased production of monocyte and macrophage chemokines MCP-1, MCP-2 and MIP-1β, the eosinophil chemoattractants eotaxin and eotaxin 2, and the neutrophil chemoattractant epithelial-derived neutrophil-activating protein 78 (ENA-78). Increased production of these chemoattractants would lead to the increase of inflammatory infiltrates observed in chronic asthma.

Interestingly, AMCase activity, suppresses the expression of the Th1 chemokines IP-10 and I-TAC, further altering the immune balance in favour of a Th2 response (Zhu *et al.* 2004). AMCase also possesses the ability to exacerbate local inflammation by the activation of IL-13 pathway and facilitating the production of chemical mediators, and has been proposed as a potential therapeutic target in the Th2-mediated inflammation (Donnelly and Barnes, 2004). In fact, the inhibition of AMCase activity with the chitinase inhibitor allosamidin (Sakuda *et al.* 1987) or administration of antisera against AMCase, decreased the number of inflammatory cells in the bronchoalveolar lavage (BAL) fluid of ovalbumin-sensitized and challenged mice and reduced the asthma symptoms (Zhu *et al.* 2004). Interestingly, Ramanathan *et al.* (2006) showed that AMCase mRNA was significantly more expressed in nasal mucosa of patients with severe sinus inflammation than in control subjects. Clearly this finding does not mean that actually there are parasites containing chitin in the nose causing sinusitis, but rather support the concept that severe and persistent sinusitis may be a consequence of a misplaced immune response against parasites that are not really present. Also the IL-13, already known to be increased in asthmatics, was found to be higher in those with nasal polyposis, without any predictive values differently from AMCase (Ramanathan *et al.* 2006).

The epithelial cells lining nasal and conjunctival surfaces play an important role as first responders of the immune system (Ramanathan *et al.* 2006). When they are stimulated fighting also non-existent parasites, they can produce chitinases, which represent a marker of innate immunity and of allergic response. More recently Chupp *et al.* 2007 found a large quantity of YKL-40, a chitinase-like protein that lacks enzymatic activity, in the serum and lungs of patients with severe asthma, which correlated with the grade of the disease thus representing a new marker of severity, supporting the previous observation of Zhu *et al.* 2004 in experimental asthma model. The role of AMCase in allergic diseases is suggestive of new possibilities for its control, using specific inhibitors of AMCase or modulating its expression. These observations allow us to hypothesize that AMCase could be also a potential key enzyme in the pathogenesis of allergic conjunctivitis (Figure 2).

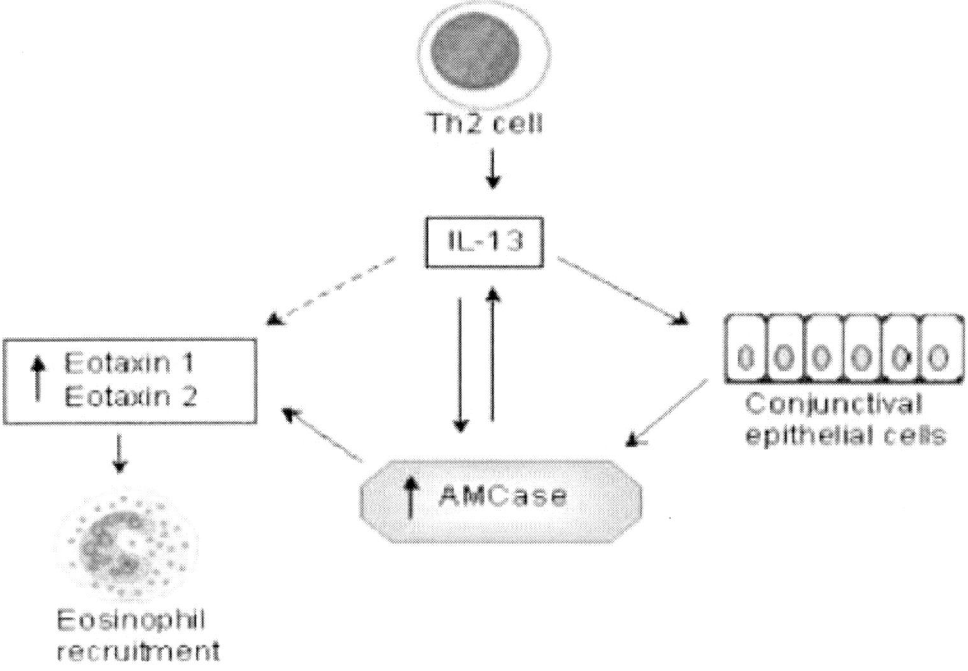

Figure 2. Model of AMCase control: Th2-driven allergic conjunctival response with IL-13 modulation of eotaxin 1-2 overexpression and eosinophil recruitment (Adapted from Musumeci *et al.* 2008).

2. Allergic Conjunctivitis

Allergic conjunctivitis is very frequent and many eye doctors know seasonal allergic conjunctivitis (SAC) or vernal conjunctivitis (VKC) as the same entity- i.e., caused by an allergic reaction involving mast cell activation via IgE. However VKC is more severe and often resistant to antiallergic and immunosuppressive treatment. VKC is a bilateral, recurrent inflammation of the conjunctiva that tends to occur in children. Its onset is in the spring and summer (i.e., vernal) - going into remission in the cooler months. There are 3 forms: palpebral, limbal and mixed, but they usually are considered to be different expressions of the same disease (Tuft *et al* 1989), .

Both VKC and SAC are a Th2-driven disease with a Th2 cytokine derived pattern, with an overexpression of eotaxins, RANTES, MCP-1, matrix metalloproteinases (MMPs)s (Ono, 2003; Metz *et al.* 1997). Moreover, Th2-associated cytokines and chemokines have been identified in the tears of patients with both VKC and season allergic conjunctivitis (SAC) (Ono, 2003; Metz *et al.* 1997).

Our research in VKC and SAC confirm the role of AMCase in human conjunctival allergic pathologies and the measurement of AMCase showed a sensitivity and specificity of 100 %, suggesting the possible use of AMCase assay in the biochemical diagnosis of VKC and SAC (Figure 3) (Musumeci *et al.* 2008).

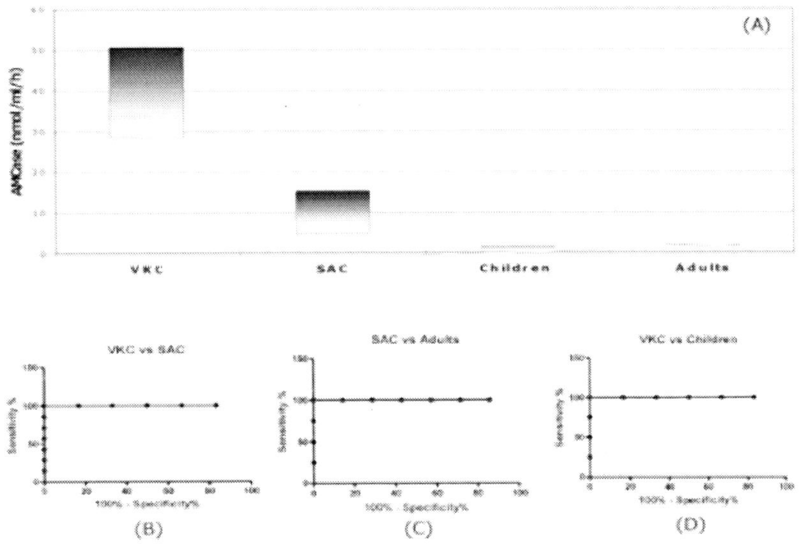

Figure 3. Distribution (panel A) and receiver-operator characteristic (ROC) analysis (panel B, C, D) of AMCase activities in VKC, SAC, adult and children controls (Adapted from Musumeci *et al.* 2008).

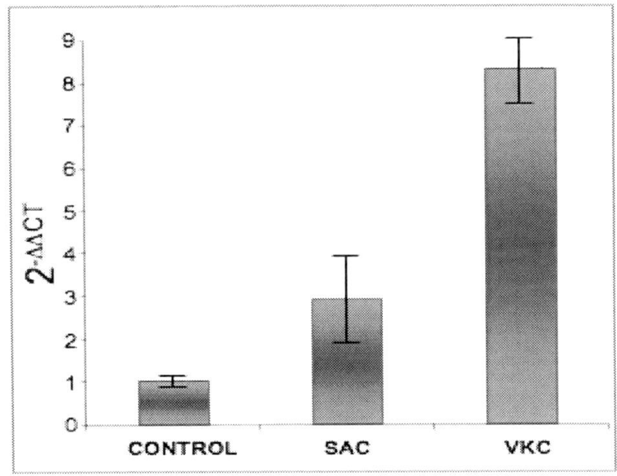

Figure 4. Detection of AMCase expression by quantitative real time PCR of RNA obtained from 8 Controls, 6 Vernal conjunctivitis (VKC), and 7 Season conjunctivitis (SC) patients. The differences among the level of 2-ΔΔCT were statistically significant with a $p < 0.0001$, comparable to that found analysing the AMCase activity. (Adapted from Musumeci *et al.* 2008)

The activity of AMCase was significantly increased in VKC and the exact nature of the chitinolytic activity in tears of allergic ocular pathologies was confirmed by the acid stability of this enzyme at pH 2, while the source of AMCAse from the conjunctival epithelial cells was confirmed by RT-PCR of RNA, extracted by conjunctival epithelial cells and obtained by cytology impression. The real time PCR measurement demonstrated a correlation between mRNA expression and tear AMCase activity (Figure 4).

3. Experimental Uveitis

A recent study in rabbits, where an experimental uveitis was induced injecting LPS into the anterior chamber of the eye, confirmed that the chitinolitic activity in tears collected at different time 0, 6, 24 h after injection, was due to AMCase (Bucolo *et al.* 2008). In fact uveitis makes up a group of heterogeneous diseases characterized by an intraocular inflammatory process that involves many complex immune pathways still not completely determined. It seems that Th1 system dominates intraocular inflammation and that on the ocular surface Th2 immune responses may also be present (Trinh *et al.* 2007). In fact in both uveitis and VKC, a CC chemokine receptor 4 (CCR4) related to the Th2 system was found to be significantly increased (Trinh *et al.* 2007). According to this observation, a hypothetical model of AMCase and IL-13 control of Th2 driven ocular surface inflammation has been proposed in Figure 5

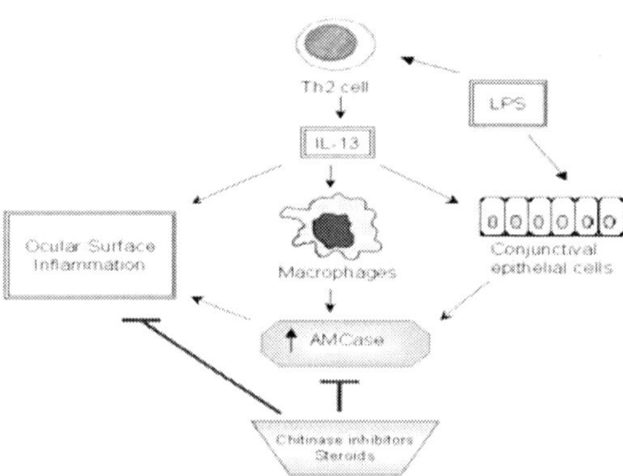

Figure 5. Hypothetical model of AMCase and IL-13 control of Th2-driven ocular surface inflammation in rabbit endotoxin induced uveitis (EIU): effect of chitinase inhibitors and steroids (Adapted from Bucolo *et al.* 2008).

Similarly to that found in asthma model of Zhu *et al.* 2004, the inflammatory reaction induced by LPS was controlled by instillation of chitinase inhibitors such as allosamidin and/or caffeine and dexamethasone.

Tear AMCase activity in rabbit treated with allosamidin, caffeine, dexamethasone and PBS has been measured and expressed as nmol/ml/h. Figure 6 shows the kinetics of AMCase activity in tears of rabbit treated with PBS before and after LPS injection. Before the induction of EIU with LPS injection the AMCase activity (basal) was very low (22.82 $\pm$ 1.07 nmol/ml/hr) in the tears. LPS injection caused a significant increase of AMCase activity, which reached at 6 h and 24 h values of 324.50 $\pm$ 30.0 and 328.40 $\pm$ 26.00 nmol/ml/hr respectively (see Figure 6).

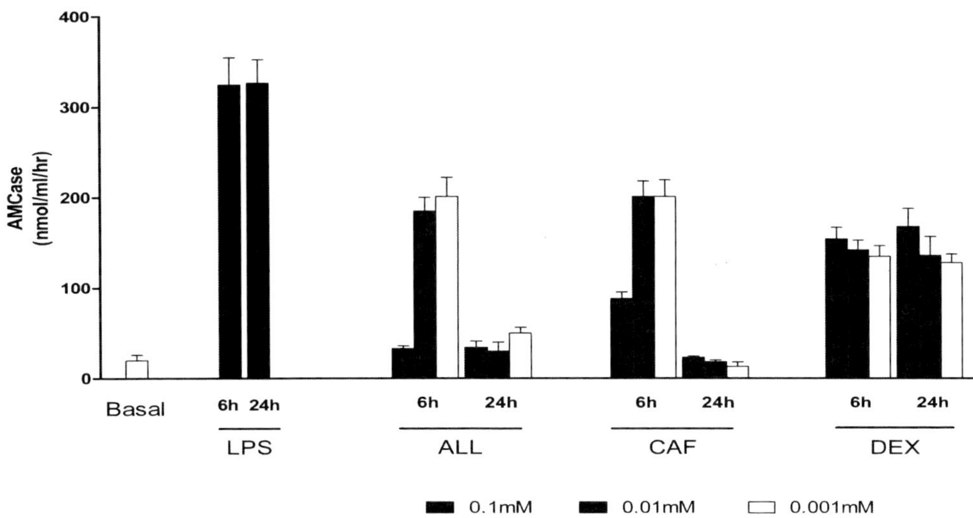

Figure 6. AMCase activity (mU/mg tear protein) before (basal) and at 6 h and 24 h after LPS intravitreal injection, at 6 and 24 h after conjunctival instillation of different concentration 0.1 mM, 0.01 mM and 0.001 mM of allosamidin (ALL), caffeine (CAF) and Dexamethasone (DEX) (Adapted from Bucolo *et al.* 2008).

The chitinolytic activity was not reduced decreasing the mixture reaction at pH 2 in the samples obtained at 6 and 24 h, confirming that this activity was characteristic of AMCase.

A significantly dose-related inhibition of AMCase activity has been observed at 6 h and 24 h, in allosamidin-treated group as well as in caffeine-treated group compared to the LPS group ($P < 0.0001$). Likewise, the group treated with dexamethasone showed a significant inhibition of AMCase activity that was not dose-dependent both at 6 h and 24 h ($P < 0.0001$). A statistical difference in terms of AMCase activity (Figure 6) between allosamidin and caffeine (33.00 ± 5.21 and 88.00 ± 7.30 nmol/ml/hr respectively, $P < 0.0001$), has been observed at 6 h with the highest concentration (0.1mM), but this difference was not significant with 0.01 and 0.001mM concentrations. With the concentration of 0.0001 mM the effect on the AMCase activity was comparable to that observed with PBS, and no significant differences were observed at 24 h among 0.0001 mM concentration in all responder treated groups (data not shown). The clinical evaluation of the drug treatments is summarized in figure 7.

Absence of inflammation (score 0) was observed in all groups before LPS injection, on the contrary a severe inflammatory response (clinical score = 3-4) was found at 6 h and 24 h after EIU induction. Topical instillation of allosamidin or caffeine, caused a remarkable reduction of ocular inflammation (clinical score = 2 and 1 for allosamidin and caffeine respectively) at all doses tested, except at the lowest (0.0001 mM) concentration. Also the group treated with dexamethasone showed an important decrease of ocular inflammation at all responder doses with a clinical score of 1. Further, the clinical effect with 0.0001 mM dexamethasone was not different from the group treated with PBS, where the score remained about 3-4 (data not shown).

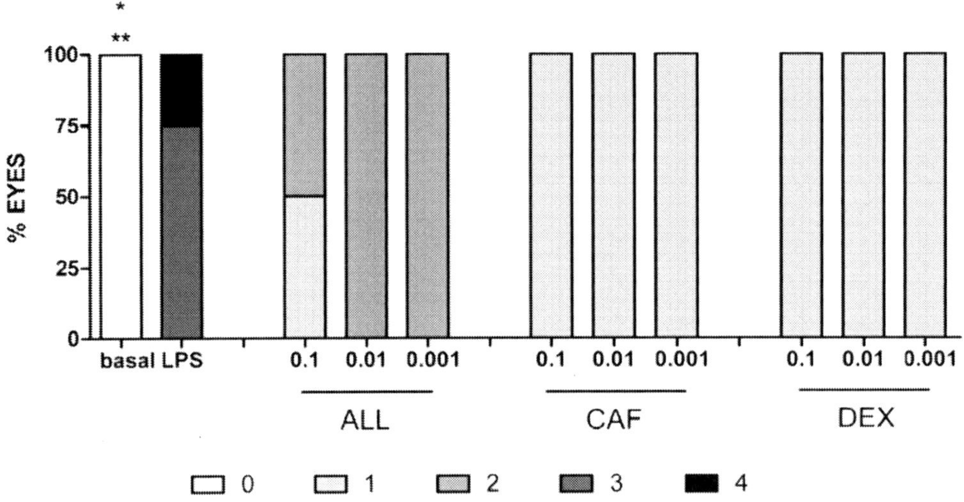

Figure 7. Clinical evaluation 24 h after EIU induction. Grading criteria: 0 = normal; 1 = discrete dilatation of iris and conjunctival vessels; 2 = moderate dilatation of iris and conjunctival vessels; 3 = intense iridal hyperemia with flare in the anterior chamber; 4 = intense iridal hyperemia with flare in the anterior chamber and presence of fibrinous exudates. *$p<0.001$ vs LPS and **$p<0.01$ vs 0.01mM and 0.001mM ALL by Dunn test after Kruskal-Wallis (Adapted from Bucolo *et al.* 2008).

In our previous work we demonstrated that the drugs used in this study, when instilled alone, did not show any ocular irritation. In this study caffeine was used because it is known to bind to the active site of chitinase in the same way as the well-known chitinase inhibitor allosamidin (Rao *et al.* 2005), but surprisingly, among its multiple mechanism of action, i.e. adenosine receptor antagonist, phosphodiesterase inhibitor, histone deacetylase inducer, caffeine can act via chitinase inhibition, and this mechanism over the others is probably associated to anti-inflammatory effect (Undem, 2006). The chitinase represents also a target for glucocorticoids, in fact recently Zhao *et al.* 2007 demonstrated that in mice sensitized and challenged with ovalbumin (OVA), dexamethasone down-regulated acidic mammalian chitinase (AMCase) in the bronchial tissue. Similarly, dexamethasone inhibited the AMCase expression in conjunctival epithelial cells in rabbit uveitis, where a significant reduction of AMCase activity was observed in the tears of rabbit with uveitis treated with the glucocorticoid. The source of AMCase from the conjunctival epithelial cells is sustained from the direct evidence in our previous study of a correlation among AMCase activity and expression of mRNA by conjunctival epithelial cells of SAC and VKC patients, obtained by impression cytology (Musumeci *et al.* 2008). In addition, the inhibition of ocular inflammation in the group treated with dexamethasone was significantly higher compared to the group treated with allosamidin, probably because corticosteroids act through multiple mechanisms (Undem, 2006; Zhao *et al* 2007). Interestingly, a similar effect was observed with caffeine that determined a significant inhibition of AMCase and a robust attenuation of clinical score. Also in this case the reason is likely due to the several mechanisms of action of this methylxanthine (Undem, 2006).

The exact function of AMCase in the conjunctival tissue is still object of speculations. The only consistent data come again from Ramanathan *et al.* 2006, who demonstrated by real time PCR the presence of specific mRNA in the tissue of recurrent nasal polyps and from our data in SAC and VKC, where AMCase mRNA was over-expressed by epithelial cells of conjuntival tissue and the magnitude of this expression was correlated with the clinical severity of the diseases (Musumeci *et al.* 2008).

Since chitin is not naturally present in mammals, the abnormal production of corresponding enzyme (i.e. AMCAse) in allergy and inflammatory diseases, and its role in the build up of mucus and other fluid or in the polyp formation in human chronic sinusitis opens the way to legitimate this enzyme as a target for drug therapies to block its action, or where possible, its production. This consideration seems to be in contrast with the observation of Reese *et al.* 2007, who demonstrated that the inoculation of chitin in mice determines a recruitment of inflammatory cells (eosinophils and basophils), characteristic of allergic response and that the preincubation of chitin with AMCase precluded this effect, raising the possibility that inherent deficits in human chitin degradation could underlie airway inflammation and favour allergic reaction. However, as Reese *et al.* 2007, correctly acknowledge, other studies suggest that chitin has a rather relevant role in immune response. Oral administration of chitin, for instance, has been shown to down modulate a murine model of allergic airway inflammation (Shibata *et al* 2000). Another study by Strong *et al.* 2002 demonstrated that direct application of chitin microparticles to the respiratory tract can alleviate allergic symptoms in a mouse model of allergy. In these studies, the importance of the route of administration (oral, nasal and parenteral respectively) seems to be a key point for the ability of chitin to induce a Th1 anti-allergic effect. or to induce a Th2-specific, IL-13 mechanism proposed in this chapter.

In addition the size of the chitin particles that people are exposed may result in different effects on biology and may expose some of the mystery of chitin and chitinases binomium. For example, depending on the size of chitin particles, these may or may not be phagocytized and displayed by dendritic cells to T-cells to stimulate an immune response. The size or polymerization of chitin could affect the chemoattractive potential of the molecule.

Nonetheless, our previous results in humans regarding ocular allergies and the data generated from the present study on experimental uveitis are very intriguing. We believe that all hypotheses generated at this point in the field are plausible and interesting to pursue (Burton *et al.* 2007). In fact, both hypotheses regarding AMCase's role in immuno inflammatory conditions may be true at the same time, taking into account that the site of inflammatory process as well as the administration route of the pharmacological tools are of primary importance. Chitin preparation used in murine model of allergy could mimic the effect of inhibitors of chitinase because of the high affinity with AMCase and support our observation in the rabbit model of EIU.

In this study we did not measure the level of IL-13, because kits for the measurement of IL-13 in rabbit were not available, but data reported by Zhu *et al.* 2002 support the correlation between AMCase expression in bronchial tissue and increase of IL-13 as a marker of Th2 response typical of allergic diseases with a sequential mechanism (Figure 2). Also in this study the inhibition of AMCase activity with the chitinase inhibitors, allosamidin and caffeine, determined in rabbits, injected by LPS, a decrease of inflammatory response as well as in the mice that over expressed IL-13 the inhibition of AMCase with allosamidin decreases also the inflammation, suggesting a modulatory effect of AMCase in the airways hyper-responsiveness (Zhu *et al.* 2002).

These results open the way to the role of chitinases in the mechanism of allergic eye inflammation or to consider inhibitor innovative target for treatment.

4. Dry Eye Syndrome

Other non allergic ocular inflammation may be associated with AMCase secretion in tears and conjunctival cell expression. Since the chitinase is thought to belong to innate immunity, in dry eye caused by alteration to lacrimal film with different mechanisms an increased AMCase expression could be demonstrated.

In fact dry eye is an ocular disease determined by a reduced tear film stability that can be characterized by a decreased aqueous component production or by an increased evaporation. This condition is accompanied by ocular discomfort symptoms such as burning, foreign body sensation, impaired vision and may interfere with patients' quality of life (International Dry Eye Workshop, 2007). Epidemiological studies have_estimated that it affects between 11 and 17% of the general population, with an increasing prevalence in older people (Moss *et al.* 2000; Lee *et al.* 2002); this number increases with the aging population, resulting in a significant decrease in the quality of life; it was found that, among people asking for ophthalmologic referral, 29% had a dry eye (Moss *et al.* 2004; Miljanovic *et al.* 2007). Dry eye is caused by an impaired function of the tear film that lubricates the ocular surface. Tear film is a complex structure where three main layers are recognizable: 1. The lipid layer contains oils secreted by the meibomian glands. The outer-most layer of the tear film coats the aqueous layer to provide a hydrophobic barrier that retards evaporation of tears; 2. The aqueous layer contains water and other substances such as proteins secreted by the lacrimal gland, it serves to promote spreading of the tear film, control of infectious agents and osmotic regulation; 3. The mucous layer contains mucin secreted by the conjunctival goblet cells, it provide a hydrophilic layer that allows for even distribution of the tear film, as well as mucus covering of the cornea.

Dry eye is caused when the tears produced by the eyes are insufficient to moisture and lubricate the ocular surface. Environmental factors can also play a role in its pathogenesis (Stern *et al.* 1998). Examples include dusty air, hot-dry or windy climate, or fumes like cigarette smoke which can evaporate tears much speedily or hamper their effectiveness (Altinors *et al.* 2006). Contact lens wear may also induce dry eye as the lens materials tend to absorb water and protein from the tear film and because a long term contact lens use reduces corneal sensitivity and lacrimal secretion (Epstein, 2006).

People who watch TV, use a laptop, or read for a longer duration may cause eye strain due to dry eye. This is due to straining of the eyes, reduced blinking and is often accompanied by altered meibomian gland secretion (Bergqvist and Knave, 1994; Fenga *et al.* 2007)). Inadequate sleep or insomnia can also cause dry eyes, as the ocular surface is overexposed and is apt to dry up faster. Finally, dry eyes are a frequent side effect of many systemic and topical drugs and medical conditions, such as Thyroid diseases, Parkinson's Disease, Sjögren's Syndrome (SS), and deficiency in vitamin A (O'Day and Horn, 2001).

Most women experience dry eyes as they enter menopause, due to the changes of the hormonal status (Versura and Campos, 2005).

Ocular discomfort and visual impairment are the possible consequence of the dry eye condition [29] and the standard treatment for dry eye treatment is the use of artificial tears, which is often accompanied by the use of anti-inflammatory eye drops, containing corticosteroid and/or cyclosporine A (Willen *et al.* 2008).

Recently, several reports associate the dry eye to an inflammatory condition of the ocular surface. Increased endogenous ICAM-1 expression and production was detected in conjunctival epithelial cells and accessory lacrimal tissues of dry eye patients. ICAM-1, synthesized by epithelial cells, may serve as a signaling molecule for the predisposition to ocular surface inflammation and facilitate the presentation of potential antigen by epithelial cells (Narayanan *et al.* 2006). This could serve as a clue to understand the pathogenetic mechanisms underlying the association among the dry eye manifestations and allergic ocular diseases (Stern *et al.* 2005).

Moreover, experimental dry eye in mouse stimulates expression and production of IL-1 alpha, IL-6, TNF-alpha, and MMP-9, and activates MAPK signaling pathways on the ocular surface, which could play an important role in the induction of those inflammatory mechanisms implicated in the pathogenesis of dry eye (Pflugfelder *et al.* 1999; Corrales *et al.* 2007).

There are different response in C57BL/6 mice, where desiccating stress significantly increased the concentrations of MIP-1alpha, MIP-1beta, IP-10, and MIG proteins in the corneal epithelium and conjunctiva, while in BALB/c mice the levels of MCP-3, eotaxin-1, and CCR3 transcripts increased in ocular tissues (Corrales *et al.* 2007). These two different responses demonstrate that some desiccating stress can induce specific patterns of Th-1 and Th-2 chemokines and their receptors in a strain-related fashion. In fact expression of Th-1 interleukin (IL)-1alpha, IL-6, and tumor necrosis factor (TNF)-alpha transcripts were higher in the corneal epithelium and conjunctiva of C57BL/6 mice and Th-2 cytokines IL-4 and IL-10 were significantly greater in BALB/c tears (Corrales *et al.* 2007). These observations obtained in the dry eye mouse model could be translated to human, where the continuous physical trauma induced by alteration of tear composition, activate an inflammation with the same mediator characteristic of Th-1 and Th-2 immune response (Stern *et al.* 2005).

If studies encourage further understanding of the intricate interactions of Th1 and Th2 cytokines in dry eye syndrome, the role of innate immunity was never considered sufficiently.

The epithelial cells lining nasal and conjunctival surfaces could play an important role as first responders of the immune system. They are capable of actively participating in immune reactions via expression of surface antigens, such as adhesion molecules, and synthesis of cytokines. This appears to be important in the pathophysiology of non-ocular allergic disorders (Hingorani *et al.* 1998). *Among other factors, it was shown that* they can produce chitinases (Musumeci *et al.* 2008). In fact it was demonstrated that the AMCase activity and the mRNA expression in the epithelial conjunctival cells of vernal keratoconjunctivitis (VKC) and season allergic conjunctivitis (SAC) was significantly increased, introducing a new diagnostic and therapeutic marker in the pathogenesis of allergic conjunctivitis.

In patients with meibomian gland disfunction (MGD) dry eyes and with secondary to Sjogren's (SS) dry eye and in healthy subjects (all adults), we collected tears to measure AMCase activity and conjunctival epithelial cells to measure AMCase expression.

Figure 8 shows the levels of AMCase activity and the quantity (pg) of mRNA in patients with MGD dry eyes, SS dry eye and healthy controls.

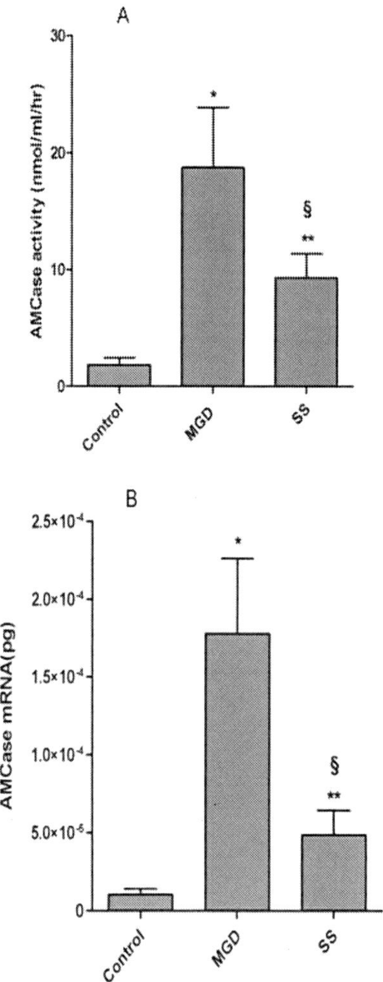

Figure 8. (A) tear AMCase activity (nmol/ml/h) and (B) quantity of AMCase mRNA (pg) in healty subject, MGD dry eye patients and primary SS dry eye patients. * $p<0.001$ vs control; ** $p<0.05$ vs control; § $p<0.001$ vs MGD group (Adapted from Musumeci *et al.* 2008).

AMCase activity was significantly higher in the group of MGD dry eyes compared to SS dry eye ($p<0.0001$), while in the healthy controls the AMCase was very low (data not shown). Real-time PCR analysis showed a consistent up-regulation of AMCase in the MGD dry eye patients with respect to SS dry eye and normal controls (Figure 8), providing data on its cellular origin. The quantitative real time evaluation of AMCase indicated a strong increase in MGD dry eye compared to control, whereas in SS dry eye the increase is minor, but still statistically significant. In the SS dry eye the pathogenetic mechanism is linked to an immune process which involves the ocular surface with a marked reduction of the aqueous phase of tears as a hallmark of the disease (aqueous deficient dry eye). In the MGD form, the pathogenetic mechanisms are different (evaporative dry eye): altered tear film is characterized by a primary alteration of the lipid component which can be accompanied by a modified bacterial flora on the conjunctiva (Seal *et al* 1982; Mahajan, 1983). This might induce a stronger innate response, as demonstrated by the higher AMCase expression level.

The results support the hypothesis that similar mechanisms associate the allergic ocular pathologies with dry eye. In Figure 9, a model combines the two pathologies at least in the pathogenetic mechanisms and suggests a common treatment with chitinase inhibitors or steroids.

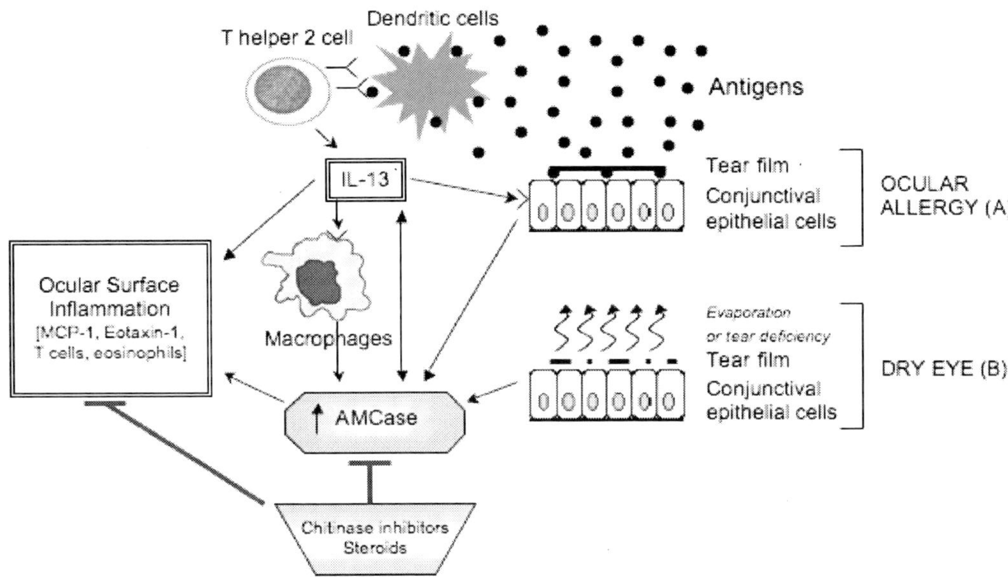

Figure 9. Model of AMCase-mediated ocular allergy and dry eye inflammation: (A) dendritic cells actively uptake antigens (Chitin ?) and present antigen to Th2 cell producing IL-13, which play a role to induce AMCase by conjunctival cells and macrophages, both expressing receptor (IL-13R) on their surface. Of note, AMCase stimulates MCP-1 and eotaxin-1 production, which induce the recruitment of T cells, eosinophils and macrophages, sustaining ocular surface inflammation; (B) alteration of tear film (e.g. excessive evaporation or tear deficiency) stimulates conjunctival cells to AMCase expression which induces ocular surface inflammation with the same mechanism. Chitinase inhibitors may suppress ocular surface inflammation blocking the AMCase enzymatic activity, while steroids inhibit its expression (Adapted from Musumeci *et al.* 2008).

The final consideration on the role of AMCAse in the ocular allergic and non allergic pathologies seem now supported by enzymatic AMCase activity in tears and mRNA expression in conjunctival cells. Our preliminary results confirm that the instillation of chitinase inhibitor in conjunctival sac both in rabbit model and in human allergic conjunctivitis and in dry eye syndromes reduce the clinical symptoms. Both inhibitor and steroids interrupt the pathogenetic mechanism inhibiting the enzymatic activity or blocking the molecular epression of AMCase.

The steroid and cyclosporine A use in dry eye and in VKC and SAC find their rationale in the new role of the AMCase in the pathogenesis of ocular inflammation, but the role of chitinase in the physiopathology of eye is not completely concluded.

In fact another chitinase, chitotriosidase, has been shown to be expressed in connection with lysozyme in the human lacrimal gland (Hall *et al.* 2007). This chitinase produced by macrophages and neutrophils points to a role in innate immunity (Van Eijk *et al.* 2005). Since this chitinase was considered more active in the control of chitin containing pathogens, the presence of measurable chitotriosidase activity in tears seems to support the eye protection in both human and mouse, where the chitotriosidase gene was evolutionary conserved (Gianfrancesco and Musumeci, 2004). In this context it is interesting to note that chitotriosidase, unlike bacterial chitinases, does not appear to have any mucolytic activity (Sanders *et al.* 2007), so the lacrimal film is not modified in its structure, maintaining the integrity of visual function.

The trigger function of innate immunity, conditioning the more complex adaptive immunity, deserves further study in pathologies where tissue inflammation is responsible for clinical manifestations.

References

Altinors DD, Akça S, Akova YA, Bilezikçi B, Goto E, Dogru M, Tsubota K. Smoking associated with damage to the lipid layer of the ocular surface. *Am J Ophthalmol.* 2006; 141:1016-1021.

Araujo AC, Souto-Padron T, de Souza W. Cytochemical localization of carbohydrate residues in microfilariae of Wuchereria bancrofti and Brugia malayi. *J Histochem Cytochem.* 1993; 41:571-78

Bergqvist UO, Knave BG. Eye discomfort and work with visual display terminals. *Scand J Work Environ Health* 1994; 20:27-33.

Boot RG, Blommaart EF, Swart E, Ghauharali-van der Vlugt K, Bijl N, Moe C, Place A, Aerts JM. Identification of a novel acidic mammalian chitinase distinct from chitotriosidase. *J Biol Chem.* 2001; 276: 6770-6778.

Bucolo C, Musumeci M, Maltese A, Drago F, Musumeci S. Effect of chitinase inhibitors on endotoxin-induced uveitis in rabbits. *Pharmacology Research* 2008; 57(3):247-52.

Burton OT, Zaccone P. The potential role of chitin in allergic reactions. *Trends Immunol.* 2007; 28:419-22.

Chou YT, Yao S, Czerwinski R, Fleming M, Krykbaev R, Xuan D, Zhou H, Brooks J, Fitz L, Strand J, Presman E, Lin L, Aulabaugh A, Huang X. Kinetic characterization of recombinant human acidic mammalian chitinase. *Biochemistry.* 2006; 45:4444-54.

Chupp GL, Lee CG, Jarjour N, Shim YM, Holm CT, He S, Dziura JD, Reed J, Coyle AJ, Kiener P, Cullen M, Grandsaigne M, Dombret MC, Aubier M, Pretolani M, Elias JA. A chitinase-like protein in the lung and circulation of patients with severe asthma. *N Engl J. Med.* 2007; 357:2016-27.

Corrales RM, Villarreal A, Farley W, Stern ME, Li DQ, Pflugfelder SC. Strain-related cytokine profiles on the murine ocular surface in response to desiccating stress. *Cornea.* 2007; 26:579-84.

Debono M, Gordee RS. Antibiotics that inhibit fungal cell wall development. *Annu Rev Microbiol.* 1994; 48:471-97

Donnelly LE, Barnes PJ. Acidic mammalian chitinase–a potential target for asthma theraphy. *Trends Pharmacol. Sci.* 2004; 25: 509-11

Epstein AB. Contact lens care products effect on corneal sensitivity and patient comfort. *Eye Contact Lens*. 2006; 32:128-32.

Fenga C, Aragona P, Cacciola A, Spinella R, Di Nola C, Ferreri F, Rania L. Meibomian gland dysfunction and ocular discomfort in video display terminal workers. *Eye.* 2008; 22:91-5.

Gianfrancesco F, Musumeci S. The evolutionary conservation of the human chitotriosidase gene in rodents and primates. *Cytogenet Genome Res.* 2004; 105: 54-6.

Hall AJ, Morroll S, Tighe P, Götz F, Falcone FH. Human chitotriosidase is expressed in the eye and lacrimal gland and has an antimicrobial spectrum different from lysozyme. *Microbes Infect.* 2008; 10:69-78..

Hingorani M, Calder VL, Buckley RJ, Lightman SL.The role of conjunctival epithelial cells in chronic ocular allergic disease. *Exp Eye Res*. 1998; 67:491-500.

Kawada M, Hachiya Y, Arihiro A, Mizoguchi E. Role of mammalian chitinases in inflammatory conditions. *Keio J Med.* 2007; 56:21-7

Neville AC, Parry DA, Woodhead-Galloway J. The chitin crystallite in arthropod cuticle. *J Cell Sci.* 1976; 21:73-82

Lee AJ, Lee J, Saw SM, Gazzard G, Koh D, Widjaja D, Tan DT. Prevalence and risk factors associated with dry eye symptoms: a population based study in Indonesia. *Br J Ophthalmol.* 2002; 86:1347-51.

Mahajan VM. Acute bacterial infections of the eye: their aetiology and treatment. *Br J Ophthalmol.* 1983; 67:191-4.

Metz DP, Hingorani M, Calder VL, Buckley RJ, Lightman SL. T-cell cytokines in chronic allergic eye disease *J Allergy Clin Immunol.* 1997; 100:817-24

Miljanovic B, Dana R, Sullivan DA, Schaumberg DA. Impact of dry eye syndrome on vision-related quality of life. *Am J Ophthalmol.* 2007; 143:409-15.

Moss SE, Klein R, Klein BE. Prevalence and risk factors for dry eye syndrome. *Arch Ophthalmol.* 2000; 118:1264-8.

Moss SE, Klein R, Klein BE. Incidence of dry eye in an older population. *Arch Ophthalmol.* 2004; 122:369-73.

Musumeci M, Maltese A, Bucolo C, Musumeci S. Chitinase levels in the tears of subjects with ocular allergies. *Cornea* 2008; 27:168-73

Narayanan S, Miller WL, McDermott AM. Conjunctival cytokine expression in symptomatic moderate dry eye subjects. *Invest Ophthalmol. Vis. Sci.* 2006; 47:2445-50.

O'Day DM, Horn JD. The Eye and Rheumatic Disease. In: Kelley WN editor. Textbook of Rheumatology, 6th ed. Philadelphia: Saunders, 2001: chapter 29.

Ono SJ. Vernal keratocojunctivitis: evidence for immunoglobulin E-dependent and immunoglobulin E-independent eosinophilia. *Clin Exp Allergy.* 2003; 33:279-81

Pflugfelder SC, Jones D, Ji Z, Afonso A, Monroy D. Altered cytokine balance in the tear fluid and conjunctiva of patients with Sjögren's syndrome keratoconjunctivitis sicca. *Curr Eye Res*. 1999; 19:201-11.

Ramanathan MJ, Lee WK, Lane AP. Increased expression of acidic mammalian chitinase in chronic rhinosinusitis with nasal polyps. *Am J Rhinol.* 2006; 20:330-35

Rao FV, Andersen OA, Vora KA, Demartino JA, van Aalten DM. Methylxanthine drugs are chitinase inhibitors: investigation of inhibition and binding modes. *Chem Biol.* 2005; 12:973-80.

Reese TA, Liang HE, Tager AM, Luster AD, Van Rooijen N, Voehringer D, Locksley RM. Chitin induces accumulation in tissue of innate immune cells associated with allergy. *Nature.* 2007; 447:92-6.

Sakuda S, Isogai A, Matsumoto S, Suzuki A. Search for microbial insect growth regulators. II. Allosamidin, a novel insect chitinase inhibitor. *J Antibiot (Tokyo).* 1987; 40:296-300

Sanders NN, Eijsink VG, van den Pangaart PS, Joost van Neerven RJ, Simons PJ, De Smedt SC, Demeester J. Mucolytic activity of bacterial and human chitinases. *Biochim Biophys Acta.* 2007; 1770:839-46.

Seal DV, Barrett SP, McGill JI. Aetiology and treatment of acute bacterial infection of the external eye. *Br J Ophthalmol.* 1982; 66:357-60.

Shibata Y, Foster LA, Bradfield JF, Myrvik QN. Oral administration of chitin down-regulates serum IgE levels and lung eosinophilia in the allergic mouse. *J Immunol.* 2000; 164:1314-21.

Stern ME, Beuerman RW, Fox RI, Gao J, Mircheff AK, Pflugelder SC. The pathology of dry eye: The interaction between the ocular surface and lacrimal glands. *Cornea* 1998; 17:584-589.

Stern ME, Siemasko KF, Gao J, Calonge M, Niederkorn JY, Pflugfelder SC. Evaluation of ocular surface inflammation in the presence of dry eye and allergic conjunctival disease. *Ocular Surf.* 2005; 3 (4 Suppl):S161-4.

Stern ME, Siemasko KF, Niederkorn JY. The Th1/Th2 paradigm in ocular allergy. *Curr. Opin Allergy Clin Immunol.* 2005; 5:446-50.

Strong P, Clark H, Reid K. Intranasal application of chitin microparticles down-regulates symptoms of allergic hypersensitivity to Dermatophagoides pteronyssinus and Aspergillus fumigatus in murine models of allergy. *Clin Exp Allergy.* 2002; 32:1794-800.

Synstad S, Gaseidnes S, van Aalten DMF,Vriend G, Nielsen JE, Eijsink VGH Mutational and computational analysis of the role of conserved residues in the active site of family 18 chitinase. *Eur J Biochem.* 2004; 271:253-262.

Trinh L, Brignole-Baudouin F, Raphaël M, Dupont-Monod S, Cassoux N, Lehoang P, Baudouin C. Th1 and Th2 responses on the ocular surface in uveitis identified by CCR4 and CCR5 conjunctival expression. *Am J Ophthalmol.* 2007; 144:580-5.

Tuft SJ, Dart JK, Kemeny M. Limbal vernal keratoconjunctivitis: clinical characteristics and immunoglobulin E expression compared with palpebral vernal. *Eye*. 1989; 3:420-7.

Undem BJ. Pharmacotheraphy of asthma. In: Brunton LL, Lazo JS, Parker KL, eds. Goodman and Gilman's The pharmacological basis of therapeutics. 11th ed. McGraw-Hill, 2006: 717-36

van Aalten DMF, Komander D, Synstad B, Gaseidnes S, Peter MG, Eijsink VGH. Structural insights into the catalytic mechanism of family 18 exochitinase. *Proc Natl Acad Sci USA* 2001; 98:8979-8984.

van Eijk M, van Roomen CP, Renkema GH, Bussink AP, Andrews L, Blommaart EF, Sugar A, Verhoeven AJ, Boot RG, Aerts JM. Characterization of human phagocyte-derived chitotriosidase, a component of innate immunity. *Int Immunol.* 2005; 17:1505-12.

Versura P, Campos EC. Menopause and dry eye. A possible relationship. *Gynecol Endocrinol.* 2005; 20:289-98.

Zhao J, Yeong LH, Wong WS. Dexamethasone alters bronchoalveolar lavage fluid proteome in a mouse asthma model. *Int Arch Allergy Immunol.* 2007; 142:219-29.

Zhu Z, Ma B, Zheng T, Homer RJ, Lee CG, Charo IF, Noble P, Elias JA. IL-13-induced chemokine responses in the lung: role of CCR2 in the pathogenesis of IL-13-induced inflammation and remodeling. *J Immunol.* 2002; 168:2953-62.

Zhu Z, Zheng T, Homer RJ *et al.* Acidic mammalian chitinase in asthmatic Th2 inflammation and IL-13 pathway activation. *Science.* 2004; 304:1678-82

The definition and classification of dry eye diseases: report of the definition and classification subcommittee of the International Dry Eye WorkShop Ocul. Surf. 2007; 5:75-92.

Willen CM, McGwin G, Liu B, Owsley C, Rosenstiel C. Efficacy of Cyclosporine 0.05% ophthalmic emulsion in contact lens wearers with dry eyes. *Eye Contact Lens.* 2008; 34:43-45.

In: Binomium Chitin-Chitinase: Recent Issues
Editor: Salvatore Musumeci and Maurizio G. Paoletti
ISBN 978-1-60692-339-9

Chapter XVII

Chitinases in the Immune Response

Maria Musumeci [1] ***and Salvatore Musumeci*** [2]
[1]Department of Hematology, Oncology and Molecular Medicine, National Institute of Health, Rome, Italy
[2]Department of Neurosciences and Mother and Child Sciences, University of Sassari and Institute of Biomolecular Chemistry, National Research Council (CNR), Li Punti (SS), Italy

Abstract

The mammalian family 18 chitinase members include different enzymes. The true enzymes which hydrolyze chitin are Chitotriosidase (Chit) and AMCase. The YKL-40, YKL-39, SI-CLP, oviductin and murine Ym1/2 are chitinase like proteins which have lost the hydrolytic activity. Several studies, in the last years, demonstrated the role of chitinases in the immunological response. The first human observation was that in Gaucher disease the lipid-laden macrophages are able to produce very high level of Chit in response to the presence of glucosylceramide and ceramide. Moreover clinical data showed also that Chit is higher in patients with *Plasmodium falciparum* malaria, expression of macrophage activation. Recent findings support the hypothesis that chitinases have a role in the innate immunity. Our research of some years ago demonstrated that the INF-gamma, TNF-alpha, LPS and Prolactin stimulate monocyte-derived macrophages to produce Chit, conditioning immune function. These results open a new view on the function of innate immunity, in the modulation of adaptive immune response and in allergic diseases. In fact AMCase has been found to be implicated in the Th2-mediated inflammations such as asthma, inflammatory bowel disease, chronic rhinosinusitis and eye pathologies. A recent study in mice suggested that the presence of chitin determines the accumulation of innate immune cells in tissues with allergy and that this mechanism could be abrogated by AMCase, concluding that chitinase may also have a regulatory mechanism in mounting the immune response. Moreover Chit has been associated to neurodegenerative diseases as shown by the studies of multiple sclerosis and Alzheimer disease, which make of Chit measurement in blood and in cephalorachidian liquid the most sensible parameters for follow up. The group of chitinase like proteins, expressed in several tissue and cells, YKL-40 and YKL-39 are implicated in autoimmune diseases as rheumatoid arthritis, where they are also involved in immune regulatory mechanisms. In this chapter we will try to illustrate the known mechanisms implicated in immunological response.

1. Introduction

Chitin, after cellulose and hemicellulose, is the third most abundant polysaccharide in nature. It is a component of the fungal cell walls, the exoskeleton of many arthropods and is also found in parasitic nematodes, while mammals do not contain chitin (Shahabuddin *et al.* 1993). The fundamental role of chitinases in lower life forms is prevalently in the defence against chitin contain pathogens (Elias *et al.* 2005), and in the remodelling of chitin structure of fungi (Taib *et al.* 2005). The absence of chitin and chitin synthase in all mammals including rodents and primates focalizes its biological role in health and pathology (Malaguarnera *et al.* 2006; Barone *et al.* 2007), where the chitinase sequences are prevalently conserved (Gianfrancesco *et al.* 2004).

The human family of chitinases is very interesting in pathology and it is characterized by the presence of a Glyco 18 domain. To this family appertain two true chitinases that are able to hydrolise chitin, named chitotriosidase (Chit) and acidic mammalian chitinase (AMCase), but other chinase like proteins, YKL-40, YKL-39, SI-CLP, oviductin and murine Ym1/2 proteins, have lost the enzymatic activity maintaining a high binding affinity for chitin (Kzhyshkowska *et al.* 2006).

2. Chitinase and Macrophages

The first issue in pathology about Chit, a phagocyte specific chitinase, come from the study of Gaucher disease, where it was purified from a patient spleen in two active isoform, 50 and 39 KDa, where the predominant isoform is 50 Kda (Renkema *et al.* 1995). Particularly in this pathology it is a valuable diagnostic tool to monitor the efficacy of enzymatic substitute therapy (Hollak *et al.* 1994). As in Gaucher disease an increased Chit activity was found also in beta-thalassemia major, where this increase of Chit reflects macrophage activation due to intracellular iron overload and all mechanisms typical of this pathology (Barone *et al.* 2001). Also in beta-thalassemia it was useful for monitoring the effect of bone marrow transplant (Maccarone *et al.* 2001). In *Plasmodium falciparum* malaria Chit activity was found high and it is a marker of macrophage activation [Barone *et al.* 2003; Musumeci *et al.* 2006). In Alzheimer disease, where the senile plaque, the strong oxidative stress and the great production of pro-inflammatory cytokines and chemokines suggest the important role of inflammatory response, a significant increase of plasma Chit activity was demonstrated (Di Rosa *et al.* 2006). The elevated expression is probably due to a strong macrophage activation or to pathogenic role of chitin like glucosamine polymers [Minagar *et al.* 2002; Castellani *et al.* 2005; Sotgiu *et al.* 2007) in this neurodegenerative disease. In addition, high Chit levels were found in stroke, where this increase confirms the relevant role of activated macrophages in brain in absence of infections [Palasik *et al.* 2005; Sotgiu *et al.* 2005), while in multiple sclerosis elevated activity of Chit in blood and cerebrospinal liquid, suggests an important role for macrophages in this pathology (Sotgiu *et al.* 2006; Sotgiu *et al.* 2007). The specific expression of Chit by phagocytic cells suggests also a

specific role of this enzyme in defence against chitin-contain pathogens. Our previous observation of high Chit activity in colostrums of African vs. Caucasian women suggested the presence of activated macrophages that correlated with the genetic characteristic of the African population (Malaguarnera *et al.* 2003) and sustain the immune function in the defence against parasites (Musumeci *et al.* 2005), especially in African Burkinabe women (Figure 1). Milk samples were collected in both Caucasian and African women using a standardized procedure for 3 consecutive days at 24-h intervals. The level of Chit in colostrum of African women was elevated in the first 24 h and this level progressively decreased on the 2nd day and 3rd day. The values for Chit were significantly lower (p-0.001) in colostrum of Caucasian women and this level rapidly decreased to a minimal level on the second and third days respectively

The role of Chit in the immune system was also demonstrated from the presence of high levels of this enzyme in neonates with fungal infections, moreover the subministration of Chit in mice showed a protective effect against *Aspergillus fumigatus* and *Candida albicans* infections [Labadaridis *et al.* 1998; Labadaridis *et al.* 2005).

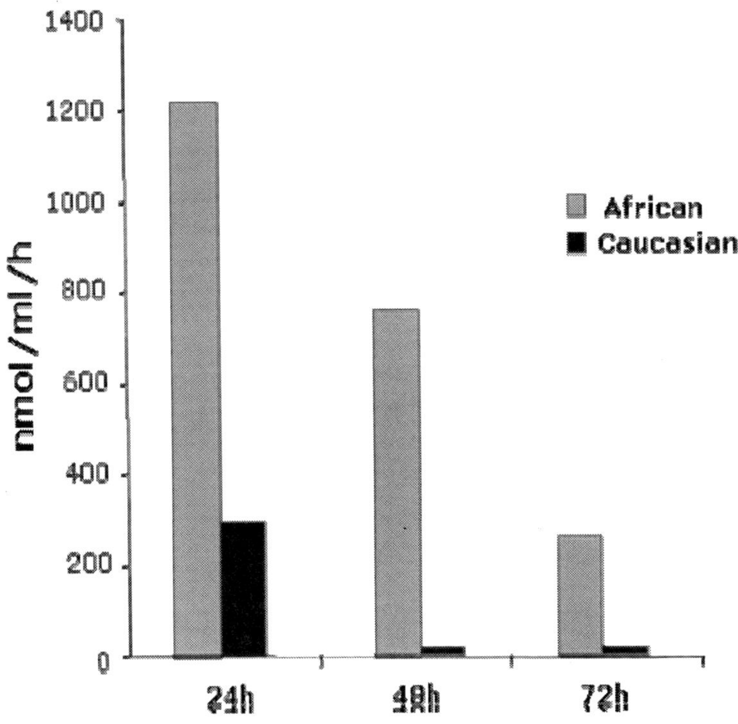

Figure 1. Levels of Chit in colostrum of African and Caucasian in the first 72 h. (Adapted from Musumeci *et al.* 2005).

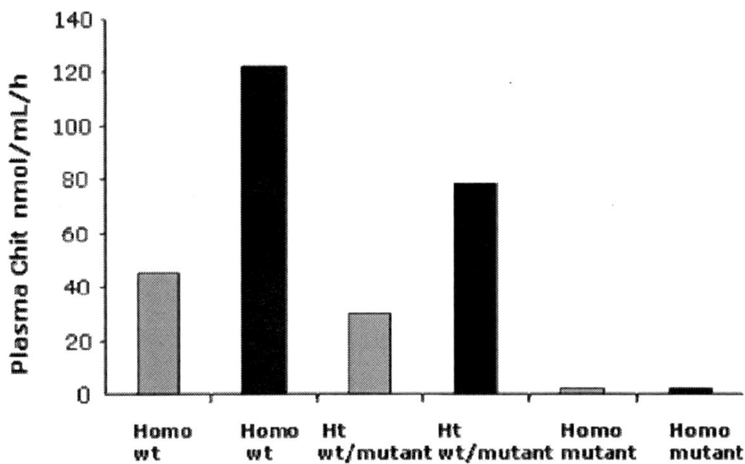

Figure 2. Plasma Chit levels (nmol/ml/h) in young Sicilian subjects from 22 to 52 years (grey) and in old Sicilian subjects from 60 to 90 years (black) divided for their genetic composition: homozygous (homo) for wild type allele; heterozygous (Ht) for wild type and mutant allele; homozygous (homo) for mutant allele. (Adapted from Barone *et al.* 2007).

Another contribution to this specific role come from a study which described a correlation between CHIT1 genotype HH variant, associated to a decreased Chit activity, and the susceptibility to filarial infections caused by *Wuchereria bancrofti* (Choi *et al.* 2001). A role of immunity in the pathogenesis of elderly alterations derives from an association between high level of Chit in plasma and aging, suggesting a continuous macrophage activation to remove oxidation products (Barone *et al.* 2007; Kurt *et al.* 2007) (Figure 2).

All of these observations suggest that chitinases are directly or indirectly involved in the immune response mediating an innate response or modulating an adaptive immune response. In fact another active chitinase, AMCase, was later discovered. It has a higher activity and stability at acidic pH and a specific expression in lung, gastrointestinal tract and recently in the conjunctival allergic pathology of eyes (Musumeci *et al.* 2008) and when dysregulated it plays an important role in allergic asthma and nasal poliposis (Zhu *et al.* 2004; Ramanathan *et al.* 2006), both diseases where an immune mechanism has been largely demonstrated. In fact the Th2 response was the protagonist of these disorders and it was associated to increased production of IL-4, IL-5, IL-13 and IgE and tissue eosinophilia [Relova *et al.* 2001; Chung, 2003). This response is typical also for parasites and other pathogens, but probably in some of these situations this response occurs also when the parasites are absent, introducing the concept of "ghost" parasite.

Recently it was shown that chitin alone induces the accumulation of IL-4 expressing innate immune cells in mice. Reese *et al.* 2007 confirmed previous research (Nair *et al.* 2006), utilizing mice infected with the migrating helminth, *Nippostrongylus brasilensis,* where the presence of AMCase and Ym1/2 was dependent on signal transducer and activator of transcription 6 (Stat6), which mediates worm expulsion in the lung tissue. In this study AMCase and Ym2 are expressed in lung as mRNA by day 3 but as protein by day 9. It is possible that the dependence of Stat6 assigns a role to chitin as a recognition element for tissue infiltration by immune cells as well as IL-4 and IL-13, introducing a new key role for chitin in the pathogenesis of allergic diseases. In fact to demonstrate this hypothesis, Reese *et al.* 2007 administrated chitin to the lung of knock-in IL-4 GFP transgenic mice to allow cells competent to produce IL-4 to be detected. This stimulus determined the recruitment of eosinophils and basophils to the lung that returned to basal level by day 9. The intraperitoneal injection demonstrated that the eosinophil accumulation was not tissue specific, neutrophils were also recruited but not to lung while the mast cells were in great numbers in the peritoneum and basophils were not present. Pre-treatment of chitin with AMCase determined the loss of eosinophil and basophil recruitment, consistent with a role for intact chitin in these cellular events. This recruitment of eosinophils by chitin was independent of STAT6 and Rag (recombination activating gene), but dependent on BLT1, the high affinity receptor for leukotriene B4, a potent chemoattractant for eosinophils (Huang *et al.* 1998), suggesting an early innate response. These experiments suggest that administration of chitin in mice up regulated AMCase and consequently it attenuated inflammatory response. In fact mice over-expressing AMCase had no apparent abnormalities (Reese *et al.* 2007) and in addition it was demonstrated that chitin alternatively induced macrophage activation.

These results allow the conclusion that chitin determines infiltration by IL-4 producing immune cells and that AMCase abrogates this effect removing the stimulus for eosinophil and basophil recruitment. However other studies have suggested a role for chitin different from this work, sustaining that chitin is able to skew away from Th2 responses (Shibata *et al.* 2000). This last result is more consistent with our recent study in a rabbit experimental model. In fact we found that an ocular instillation of allosamidin, a chitinase inhibitor, in a rabbit model of LPS induced uveitis, drastically lowered the pathology manifestation reducing AMCase activity (Bucolo *et al.* 2008) confirming that the expression of AMCase correlates with the immune response. We cannot exclude that these differences may depend on the use of different animal models, chitin preparations, or other contaminants, thus more studies are necessary to clarify the role of chitin and chitinases in human pathologies, since up to now contrasting results are reported in the literature.

3. Chitinase and Innate Immunity

Innate and adaptative immunity combat pathogens by discriminating between self and non self. In fact mammalian innate immunity recognition, occurs e.g. via Toll-like receptors (TLR) (Medzhitov R, 2001) and intracellular nucleotide-binding oligomerization domain proteins (NODs) (Inohara N and Nunez G, 2003). In fact about these proteins and their triggering on immune cells, van Eijk *et al.* 2005 and 2007 published two interesting studies, which consider the role of Chit in the mechanism of innate immunity.

They induced PMNs with different stimuli to study the specific secretion, arriving to the conclusion that exposure to PAF+FMLP (platelet-activating factor + N-formyl-methionyl-leucyl-phenylalanine) determines release of Chit and that this secretion comes from lactoferrin-containing compartments (van Eijk *et al.* 2005). Moreover they demonstrated that the release of Chit in neutrophils (PMNs), from specific granules, probably operates via TLR, but not via NOD2 activation (van Eijk *et al.* 2007). The localization of chitotriosidase to neutrophil granules had been previously shown by Boussac and Garin (Boussac and Garin, 2000) using 2D-gel electrophoresis.

When van Eijk *et al.* 2007 used macrophages, they found that NOD2 activation, but not TLR stimulation, induced chitinase expression. The opposite of what they observed with neutrophils.

The phagocyte-specific regulation is important for efficient eradication of chitin-containing pathogens. In fact macrophages are, in general, considered to be important elements in natural resistance to pathogens and are strategically placed to protect the microenvironment, in which they are situated. It is known that the stimulation of macrophages with INF-gamma develops an increased cytocidal activity against intracellular micro-organism and tumour cells. Our study has provided direct evidence that monocyte-derived macrophages treated with INF-gamma, TNF-alfa, and LPS, but not with IL-10, express high level of Chit (Malaguarnera *et al.* 2005). The Chit mRNA levels in monocyte-derived macrophages was induced strongly 2h after treatment with INF-gamma and decreased gradually in time; a similar pattern was also observed after TNF-alfa stimulation. After 4 h of LPS stimulation Chit had the highest levels, while after 24 h it was undetectable (Figure 3). This observation could solve the conflict between our data and the data of van Eijk *et al* 2007.

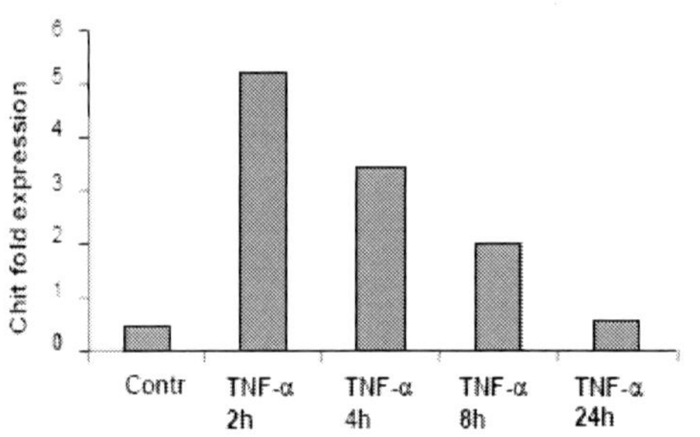

b

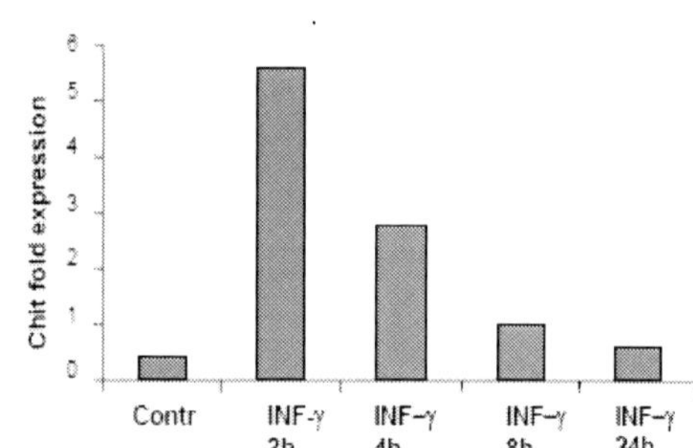

c

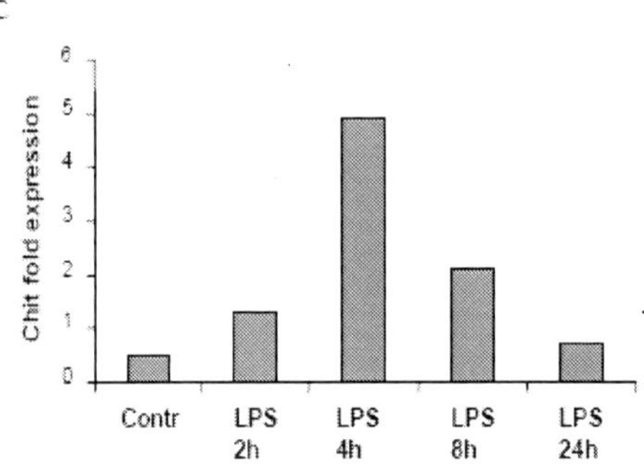

Figure 3. Detection of Chit expression by RT-PCR of RNA obtained from human macrophage untreated and treated at different times with TNF- alfa 100U/ml (a), INF-gamma 100U/ml (b) and LPS 50ng/ml (c). (Adapted from Malaguarnera *et al.* 2005).

In fact a tentative of explanation could be that LPS does not directly induce Chit expression in monocyte-derived macrophages (van Eijk *et al* 2007), but does certainly induce

TNF-α (more than hundredfold). TNF-alfa in turn induces Chit expression. This would also explain why the increase of Chit expression is maximal after 4 hours – just in time for some TNF-alfa to be produced and to act on the macrophages in an autocrine loop. These investigations are important since they suggest a key role of Chit in this cascade of events oriented to the protection against viral, bacterial and intracellular parasitic infections and put Chit in the cellular response elicited by regulatory cytokines.

Moreover Van Eijk *et al.* 2005 had also demonstrated that GM-CSF is a potent inducer of Chit expression in monocyte-derived macrophages, while INF-gamma and IL-4 prevented the expression of the enzyme. Probably this difference compared to our study is explained by the different culture methods and experimental procedures involved, suggesting that Chit shows prevalently a modulator mechanism in the immune response. This observation is important since an antifungal response is observed after GM-CSF treatment, probably via PMN and Mϕ [Jones, 1999; Roilides *et al.* 2001). In fact the Chit expression from PMNs by GM-CSF is important because this immune response has a strong role in the anti-fungal response inhibiting hyphal growth of chitin-containing pathogens (van Eijk *et al.* 2005). In vitro exposure to Chit of chitin containing fungi determines a strong growth inhibition, hyphal form and hyphal tip bursting. In "in vivo" Candidiasis and systemic Aspergillosis mouse models, the continuous intravenous administration of Chit determines a dose-dependent survival when compared with controls.

In another study we observed that the prolactin stimulation of macrophages induces a strong expression of Chit and O_2 production. A time course performed in human monocyte-derived macrophages cultured for 4 days and treated at different time with 25 ng/ml of prolactin, showed a strong Chit expression at 2 and 4 hours, while it decreased slowly after this time.

These results are consistent with notion that prolactin acts as cytokine and regulates immune response (Malaguarnera *et al.* 2004) (Figure 4).

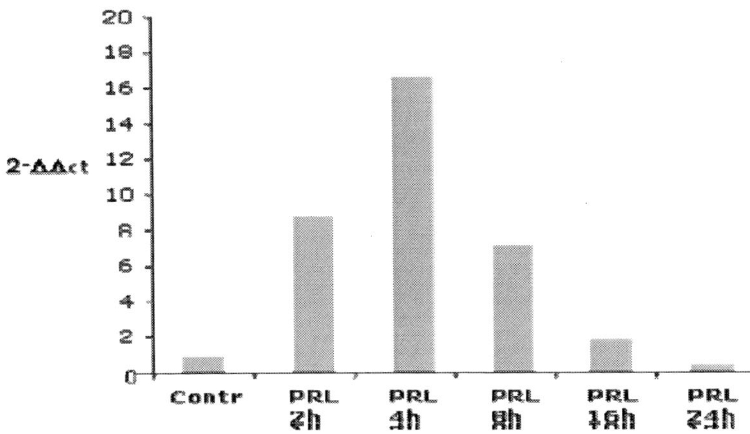

Figure 4. etection of Chit expression by quantitative real time PCR of RNA obtained from human monocyte-macrophages cultured for 4 days and after treated with prolactin (PRL) 25ng/ml in time. (Adapted from Malaguarnera *et al.* 2004).

Mostly this rapid macrophage activation may have biological relevance for the immune response in the early phase of infection where an elevated cytokine induction takes place, supporting the role of Chit in the first defence against pathogen.

The regulation of Chit expression by cytokines has been demonstrated also when in chimpanzee injection of IL-12 is associated with an enhanced Chit activity (Lauw *et al.* 1999). It seems relevant, at this moment, to consider that high levels of Chit was observed also in children with acute *Plasmodium falciparum* malaria where the IL-12 levels in plasma were considered protective against severe malaria (Musumeci *et al.* 2003), creating a network of immune functions among different proteins involved in the immune malaria response (Malaguarnera *et al.* 2002).

4. Chitinase-Like Proteins and Immunity

The regulation of immune response by chitinases was operated also by other chitinases not functionally active but able to bind chitin.

Chitinase-like protein as YKL-40 is also secreted by macrophages present in atherosclerotic plaques (Renkema *et al.* 1998; Boot *et al.* 1999), and in neutrophils (Volck *et al.* 1998). In addition, elevated levels were found in several tumors (Johansen *et al.* 2006). In meningitis and pneumonia it is secreted by neutrophils and macrophages (Kronborg *et al.* 2002; Nordenbaek *et al.* 1999) and by giant cells and macrophages in pulmonary sarcoidosis (Johansen *et al.* 2005) but its function it is not so clear as demonstrated for Chit and/or AMCase.

YKL-40 functions as auto-antigen in rheumatoid arthritis (Peltomaa *et al.* 2001), while YKL-39 was directly expressed by chondrocytes (Knorr *et al.* 2003). The auto antibodies to YKL-39 were found in early phases of osteoarthritis supporting that the autoimmune response starts when the degeneration of cartilage is in the initial phase (Du *et al.* 2005). YKL-40 also is mainly expressed in colonic epithelial cells (CEC) and in macrophages of inflamed colon of dextran sulfate sodium induced colitis. This chitinase can be up-regulated by proinflammatory cytokines and possesses a high ability to enhance the bacterial/CEC interaction (Mizoguchi *et al.* 2006; Kawada *et al.* 2007). Recently, YKL-40 has been involved as AMCase in asthma and it was correlated with exacerbation and progression of disease (Chupp *et al.* 2007).

Also stabilin-interacting chitinase-like protein (SI-CLP) was intracellularly accumulated in macrophages by the Th2 cytokine IL-4 and dexamethasone. Its possible use could be that of a marker for the response to glucocorticoid therapy (Kzhyshkowska *et al.* 2006), but it is not clear the direct relation with the immune system.

These last chitinases where an immunoregulatory function seems recently to acquire consistence, remain not clearly localized in the mammalian organism economy (Kzhyshkowska *et al.* 2006).

Conclusion

As we can note in the previous chapters of this book, all chitinases are phylogenetically related, showing their final point of different evolutive process. Considering to day the chitinases in the double face of instruments for the defence against chitin containing pathogens or of mediators in the inflammatory response, their archaic origin suggests not only a relevant role in innate immunity, but also in the organization of more elevated forms of life, through selective unknown mechanisms. The isolation of chitinases in mammals and humans who do not synthesize chitin, stimulate us to consider their conservation as a finishing line where the chitinases will show a new mostly important role in clinics.

References

Barone R, Bertrand G, Simporè J, Malaguarnera M, Musumeci S. Plasma chitotriosidase activity in beta-thalassemia major: a comparative study between Sicilian and Sardinian patients. *Clin Chim Acta.* 2001; 306:91-6.

Barone R, Simporé J, Malaguarnera L, Pignatelli S, Musumeci S. Plasma chitotriosidase activity in acute *Plasmodium falciparum* malaria. *Clin Chim Acta.* 2003; 331:79-85.

Barone R, Sotgiu S, Musumeci S. Plasma chitotriosidase in health and pathology.*Clin Lab.* 2007; 53:321-33.

Boot RG, van Achterberg TA, van Aken BE, Renkema GH, Jacobs MJ, Aerts JM, de Vries CJ. Strong induction of members of the chitinase family of proteins in atherosclerosis: chitotriosidase and human cartilage gp-39 expressed in lesion macrophages.*Arterioscler Thromb Vasc Biol.* 1999; 19:687-94.

Boussac M, Garin J. Calcium-dependent secretion in human neutrophils: a proteomic approach. *Electrophoresis*. 2000; 21:665-72.

Bucolo C, Musumeci M, Maltese A, Filippo Drago F, Musumeci S. Effect of chitinase inhibitors on endotoxin-induced uveitis (EIU) in rabbits. *Pharmacology Research* 2008; 57:247-52.

Castellani RJ, Siedlak SL, Fortino AE, Perry G, Ghetti B, Smith MA. Chitin-like polysaccharides in Alzheimer's disease brains.*Curr Alzheimer Res*. 2005; 2:419-23.

Choi EH, Zimmerman PA, Foster CB, Zhu S, Kumaraswami V, Nutman TB, Chanock SJ. Genetic polymorphisms in molecules of innate immunity and susceptibility to infection with Wuchereria bancrofti in South India.*Genes Immun.* 2001; 2:248-53.

Chung KF. Individual cytokines contributing to asthma pathophysiology: valid targets for asthma therapy? *Curr Opin Investig Drugs.* 2003; 4:1320-6.

Chupp GL, Lee CG, Jarjour N, Shim YM, Holm CT, He S, Dziura JD, Reed J, Coyle AJ, Kiener P, Cullen M, Grandsaigne M, Dombret MC, Aubier M, Pretolani M, Elias JA. A chitinase-like protein in the lung and circulation of patients with severe asthma. *N Engl J Med.* 2007; 357:2016-27.

Du H, Masuko-Hongo K, Nakamura H, Xiang Y, Bao CD, Wang XD, Chen SL, Nishioka K, Kato T. The prevalence of autoantibodies against cartilage intermediate layer protein, YKL-39, osteopontin, and cyclic citrullinated peptide in patients with early-stage knee osteoarthritis: evidence of a variety of autoimmune processes. *Rheumatol Int.* 2005; 26:35-41.

Di Rosa M, Dell'Ombra N, Zambito AM, Malaguarnera M, Nicoletti F, Malaguarnera L. Chitotriosidase and inflammatory mediator levels in Alzheimer's disease and cerebrovascular dementia. *Eur J Neurosci.* 2006; 23:2648-56.

Elias JA, Homer RJ, Hamid Q, Lee CG.Chitinases and chitinase-like proteins in T(H)2 inflammation and asthma. *J Allergy Clin Immunol.* 2005; 116:497-500.

Gianfrancesco F, Musumeci S.The evolutionary conservation of the human chitotriosidase gene in rodents and primates.*Cytogenet Genome Res.* 2004; 105:54-6.

Hollak CE, van Weely S, van Oers MH, Aerts JM.Marked elevation of plasma chitotriosidase activity. A novel hallmark of Gaucher disease. *J Clin Invest.* 1994; 93:1288-92.

Huang WW, Garcia-Zepeda EA, Sauty A, Oettgen HC, Rothenberg ME, Luster AD. Molecular and biological characterization of the murine leukotriene B4 receptor expressed on eosinophils. *J Exp Med.* 1998; 188: 1063-74.

Inohara N, Nunez G. NODs: intracellular proteins involved in inflammation and apoptosis *Nat Rev Immunol.* 2003; 3:371–382

Johansen JS, Milman N, Hansen M, Garbarsch C, Price PA, Graudal N. Increased serum YKL-40 in patients with pulmonary sarcoidosis--a potential marker of disease activity? *Respir Med.* 2005; 99:396-402.

Johansen JS, Jensen BV, Roslind A, Nielsen D, Price PA. Serum YKL-40, a new prognostic biomarker in cancer patients? *Cancer Epidemiol Biomarkers Prev.* 2006; 15:194-202.

Jones TC. Use of granulocyte-macrophage colony stimulating factor (GM-CSF) in prevention and treatment of fungal infections. *Eur J Cancer.* 1999;35 Suppl 3:S8-10.

Kawada M, Hachiya Y, Arihiro A, Mizoguchi E. Role of mammalian chitinases in inflammatory conditions. *Keio J Med.* 2007; 56:21-7.

Knorr T, Obermayr F, Bartnik E, Zien A, Aigner T. YKL-39 (chitinase 3-like protein 2), but not YKL-40 (chitinase 3-like protein 1), is up regulated in osteoarthritic chondrocytes. *Ann Rheum Dis.* 2003; 62:995-8

Kronborg G, Ostergaard C, Weis N, Nielsen H, Obel N, Pedersen SS, Price PA, Johansen JS. Serum level of YKL-40 is elevated in patients with Streptococcus pneumoniae bacteremia and is associated with the outcome of the disease. *Scand J Infect Dis.* 2002; 34:323-6.

Kurt I, Abasli D, Cihan M, Serdar MA, Olgun A, Saruhan E, Erbil MK. Chitotriosidase levels in healthy elderly subjects. *Ann N Y Acad Sci.* 2007; 1100:185-8.

Kzhyshkowska J, Mamidi S, Gratchev A, Kremmer E, Schmuttermaier C, Krusell L, Haus G, Utikal J, Schledzewski K, Scholtze J, Goerdt S. Novel stabilin-1 interacting chitinase-like protein (SI-CLP) is up-regulated in alternatively activated macrophages and secreted via lysosomal pathway. *Blood.* 2006;107:3221-8.

Labadaridis J, Dimitriou E, Costalos C, Aerts J, van Weely S, Donker-Koopman WE, Michelakakis H. Serial chitotriosidase activity estimations in neonatal systemic candidiasis.*Acta Paediatr.* 1998; 87:605.

Labadaridis I, Dimitriou E, Theodorakis M, Kafalidis G, Velegraki A, Michelakakis H. Chitotriosidase in neonates with fungal and bacterial infections.*Arch Dis Child Fetal Neonatal* 2005; 90: F531-2

Lauw FN, te Velde AA, Dekkers PE, Speelman P, Aerts JM, Hack CE, van Deventer SJ, van der Poll T. Activation of mononuclear cells by interleukin-12: an in vivo study in chimpanzees. *J Clin Immunol.* 1999; 19:231-8.

Maccarone C, Pizzarelli G, Barone R, Musumeci S. Plasma chitotriosidase activity is a marker of recovery in transplanted patients affected by beta-thalassemia major. *Acta Haematol.* 2001; 105:109-10.

Malaguarnera L, Musumeci S. The immune response to Plasmodium falciparum malaria. *Lancet Infect Dis*. 2002; 2:472-8.

Malaguarnera L, Simporè J, Prodi DA, Angius A, Sassu A, Persico I, Barone R, Musumeci S. A 24-bp duplication in exon 10 of human chitotriosidase gene from the sub-Saharan to the Mediterranean area: role of parasitic diseases and environmental conditions. *Genes Immun.* 2003; 4:570-4.

Malaguarnera L, Musumeci M, Licata F, Di Rosa M, Messina A, Musumeci S. Prolactin induces chitotriosidase gene expression in human monocyte-derived macrophages. *Immunol Lett.* 2004; 94:57-63.

Malaguarnera L, Musumeci M, Di Rosa M, Scuto A, Musumeci S. Interferon-gamma, tumor necrosis factor-alpha, and lipopolysaccharide promote chitotriosidase gene expression in human macrophages. *J Clin Lab Anal.* 2005; 19:128-32.

Malaguarnera L. Chitotriosidase: the yin and yang. *Cell Mol Life Sci.* 2006; 63:3018-29.

Medzhitov R. Toll-like receptors and innate immunity, *Nat Rev Immunol.* 2001; 1(2): 135–145.

Minagar A, Shapshak P, Fujimura R, Ownby R, Heyes M, Eisdorfer C. The role of macrophage/microglia and astrocytes in the pathogenesis of three neurologic disorders: HIV-associated dementia, Alzheimer disease, and multiple sclerosis. *J Neurol Sci.* 2002; 202:13-23.

Mizoguchi E. Chitinase 3-like-1 exacerbates intestinal inflammation by enhancing bacterial adhesion and invasion in colonic epithelial cells. *Gastroenterology.* 2006; 130:398-411.

Musumeci M, Malaguarnera L, Simporè J, Messina A, Musumeci S. Modulation of immune response in Plasmodium falciparum malaria: role of IL-12, IL-18 and TGF-beta. *Cytokine*. 2003; 21:172-8.

Musumeci M, Malaguarnera L, Simpore J, Barone R, Whalen M, Musumeci S. Chitotriosidase activity in colostrum from African and Caucasian women. *Clin Chem Lab Med.* 2005; 43:198-201.

Musumeci M, Simpore J, Barone R, Angius A, Musumeci S. Synchronic macrophage response and Plasmodium falciparum malaria. *J Vector Borne Dis.* 2006; 43:84-7.

Musumeci M, Bellin M, Maltese A, Aragona P, Bucolo C, Musumeci S. Chitinase levels in the tears of subjects with ocular allergies. *Cornea.* 2008; 27:168-73

Nair MG, Guild KG, Artis D. Novel effector molecules in type 2 inflammation: lesson drawn from helminth infection and allergy. *J Immunol.* 2006; 177, 1393-1399

Nordenbaek C, Johansen JS, Junker P, Borregaard N, Sørensen O, Price PA. YKL-40, a matrix protein of specific granules in neutrophils, is elevated in serum of patients with community-acquired pneumonia requiring hospitalization. *J Infect Dis.* 1999; 180:1722-6.

Palasik W, Fiszer U, Lechowicz W, Czartoryska B, Krzesiewicz M, Lugowska A. Assessment of relations between clinical outcome of ischemic stroke and activity of inflammatory processes in the acute phase based on examination of selected parameters. *Eur Neurol.* 2005; 53:188-93.

Peltomaa R, Paimela L, Harvey S, Helve T, Leirisalo-Repo M. Increased level of YKL-40 in sera from patients with early rheumatoid arthritis: a new marker for disease activity. *Rheumatol Int.* 2001; 20:192-6.

Ramanathan M Jr, Lee WK, Lane AP. Increased expression of acidic mammalian chitinase in chronic rhinosinusitis with nasal polyps. *Am J Rhinol.* 2006; 20:330-5.

Reese TA, Liang HE, Tager AM, Luster AD, Van Rooijen N, Voehringer D, Locksley RM. Chitin induces accumulation in tissue of innate immune cells associated with allergy. *Nature.* 2007; 447:92-6.

Relova AJ, Kampf C, Roomans GM. Effects of Th-2 type cytokines on human airway epithelial cells: interleukins-4, -5, and -13. *Cell Biol Int.* 2001; 25:563-6.

Renkema GH, Boot RG, Muijsers AO, Donker-Koopman WE, Aerts JM. Purification and characterization of human chitotriosidase, a novel member of the chitinase family of proteins. *J Biol Chem.* 1995; 270:2198-202.

Renkema GH, Boot RG, Au FL, Donker-Koopman WE, Strijland A, Muijsers AO, Hrebicek M, Aerts JM. Chitotriosidase, a chitinase, and the 39-kDa human cartilage glycoprotein, a chitin-binding lectin, are homologues of family 18 glycosyl hydrolases secreted by human macrophages. *Eur J Biochem.* 1998; 251:504-9

Roilides E, Farmaki E. Granulocyte colony-stimulating factor and other cytokines in antifungal therapy. *Clin Microbiol Infect.* 2001; 7 Suppl 2:62-7.

Shahabuddin M, Kaslow DC. Chitinase: a novel target for blocking parasite transmission? *Parasitol Today.* 1993; 9:252-5.

Shibata Y, Foster LA, Bradfield JF, Myrvik QN. Oral administration of chitin down-regulates serum IgE levels and lung eosinophilia in the allergic mouse. *J Immunol.* 2000;164:1314-21.

Sotgiu S, Barone R, Zanda B, Arru G, Fois ML, Arru A, Rosati G, Marchetti B, Musumeci S. Chitotriosidase in patients with acute ischemic stroke. *Eur Neurol.* 2005; 54:149-53

Sotgiu S, Barone R, Arru G, Fois ML, Pugliatti M, Sanna A, Rosati G, Musumeci S. Intrathecal chitotriosidase and the outcome of multiple sclerosis. *Mult Scler.* 2006; 12:551-7.

Sotgiu S, Piras MR, Barone R, Arru G, Fois ML, Rosati G, Musumeci S. Chitotriosidase and Alzheimer's disease. *Curr Alzheimer Res.* 2007; 4:295-6

Sotgiu S, Angius A, Embry A, Rosati G, Musumeci S. Hygiene hypothesis: Innate immunity, malaria and multiple sclerosis. *Med Hypotheses.* 2008; 70:819-25.

Taib M, Pinney JW, Westhead DR, McDowall KJ, Adams DJ. Differential expression and extent of fungal/plant and fungal/bacterial chitinases of Aspergillus fumigatus. *Arch Microbiol.* 2005; 184:78-81.

van Eijk M, van Roomen CP, Renkema GH, Bussink AP, Andrews L, Blommaart EF, Sugar A, Verhoeven AJ, Boot RG, Aerts JM. Characterization of human phagocyte-derived chitotriosidase, a component of innate immunity. *Int Immunol.* 2005; 17:1505-12.

van Eijk M, Scheij SS, van Roomen CP, Speijer D, Boot RG, Aerts JM. TLR- and NOD2-dependent regulation of human phagocyte-specific chitotriosidase. *FEBS Lett.* 2007; 581:5389-95.

Volck B, Price PA, Johansen JS, Sørensen O, Benfield TL, Nielsen HJ, Calafat J, Borregaard N. YKL-40, a mammalian member of the chitinase family, is a matrix protein of specific granules in human neutrophils. *Proc Assoc Am Physicians.* 1998; 110:351-60.

Zhu Z, Zheng T, Homer RJ, Kim YK, Chen NY, Cohn L, Hamid Q, Elias JA. Acidic mammalian chitinase in asthmatic Th2 inflammation and IL-13 pathway activation. *Science.* 2004; 304:1678-82.

In: Binomium Chitin-Chitinase: Recent Issues
Editor: Salvatore Musumeci and Maurizio G. Paoletti
ISBN 978-1-60692-339-9

Chapter XVIII

The Role of Chitin and Chitinases in Asthma

Max A. Seibold [1,2] ***and Esteban Gonzalez Burchard*** [1,2,3,4]
[1]Biopharmaceutical Sciences,
[2]Lung Biology Center,
[3]Department of Medicine and
[4]Institute for Human Genetics of the University of California
San Francisco (UCSF), USA

Abstract

Asthma is a disease characterized by chronic inflammation of the airway, thought to result from inappropriate activation of the Th2 immune response. A master regulator of Th2 inflammation is the cytokine IL-13, which stimulates the expression of many effectors responsible for the airway hyperresponsiveness and eosinophilic inflammation that is characteristic of asthma. Recent work has shown that both chitinase and chi-lectin proteins are strongly upregulated by IL-13 expression. Inhibition of the acidic mammalian chitinase (AMCase), blocked the inflammation and hyperresponsiveness observed in a mouse model of asthma. Furthermore, chitin, a widespread polymer of *N*-acetyl-β-D-glucosamine and substrate of chitinases, is found in many organisms including those for which exposure is linked to asthma such as, dust mites, fungi and cockroaches. Interestingly, in a mouse model of asthma, investigators have demonstrated that chitin can induce the recruitment of immune cells associated with allergic asthma to the lung. Moreover, they find that AMCase enzymatic activity negatively regulates this process. Herein, these seminal studies on the role of chitinases and chitin in Th2 inflammation are highlighted in the context of other human and murine data on chitinases.

1. Introduction

Chitin, a linear polymer of *N*-acetyl-β-D-glucosamine, is the second most abundant polysaccharide in nature and is used as an important structural component of many organisms including the cell wall of fungi, exoskeleton of arthropods such as insects and crustaceans, and the pharyngeal organs of parasitic nematodes (Neville *et al.* 1976; Fuhrman and Piessens 1985; Araujo *et al.* 1993; Debono and Gordee 1994; Shahabuddin and Kaslow 1994). Chitinases are the enzymes which hydrolytically cleave the chitin polymer into di- and tri-saccharide units (Chou *et al.* 2006). These enzymes are found in many organisms, where they fulfill roles ranging from structural remodeling of chitin components and nutrient scavenging, to host defense against chitin containing parasites and fungi (Debono and Gordee 1994; Techkarnjanaruk *et al.* 1997). Although mammals do not synthesize chitin, they do express an extensive repertoire of both chitin binding proteins (chi-lectins) such as YM-1, YM-2, OVGP1, and YKL-40, in addition to two chitin degrading enzymes (chitinases), acidic mammalian chitinase (AMCase) and chitotriosidase (CHIT1) (Bussink *et al.* 2007; Funkhouser and Aronson 2007). Both chi-lectins and chitinases are included in the Family 18 of glycosyl hydrolases, which all contain an N-terminal 8-stranded (β/α) barrel catalytic domain (Bussink *et al.* 2007). However, chi-lectin proteins have acquired active site mutations which prevent chitin degradation, but retain active site binding of chitin (Bussink *et al.* 2007). Chi-lectins also differ from chitinases in that they do not contain the C-terminal chitin-binding domain (Boot *et al.* 1995). Once thought to be an evolutionary remnant of our chitin-containing ancestors, recent studies have demonstrated that chitinases and chi-lectins are strongly induced in Th2 inflammation and asthma (Zhu *et al.* 2004; Musumeci *et al.* 2008). Furthermore, chitin itself has been implicated as a pattern recognition molecule capable of stimulating the accumulation of immune cells associated with allergic asthma (Reese *et al.* 2007). Herein, we review the role of chitinases and chi-lectins in mouse models of asthma, the genetics and expression of these proteins in humans, and the potential immunomodulatory properties of chitin itself.

2. Mouse Models of Asthma

The pathology of allergic asthma is characterized by persistent airway inflammation, which is driven in large part by a dysregulated Th2 immune response. Inappropriate activation of Th2 cells leads to the production of a panel of pro-allergic cytokines such as IL-4, IL-5, and IL-13. IL-13 appears to be foremost in this group of inflammatory cytokines, being both necessary and sufficient for the reproduction of allergic asthma in mouse models of asthma (Grunig *et al.* 1998; Wills-Karp *et al.* 1998). The pulmonary manifestations of IL-13 overexpression in mouse models include eosinophilic infiltration, excessive mucus secretion, and airway hyperresponsiveness to cholinergic stimuli (Grunig *et al.* 1998; Wills-Karp *et al.* 1998). Many of these inflammatory effects are mediated through signaling resulting in the activation of the inflammatory transcription factor STAT6 (Kuperman *et al.* 2002). In addition, both IL-13 and STAT6 have been observed to be overexpressed in the lungs of subjects with asthma (Humbert *et al.* 1997; Christodoulopoulos *et al.* 2001).

Therefore, genes regulated by (IL-13/STAT6)-dependent signaling have been explored as effectors of Th2 inflammation in asthma (Hoshino *et al.* 2004; Kuperman *et al.* 2005).

The acidic mammalian chitinase (AMCase) was investigated, as a potential mediator of Th2 airway inflammation and IL-13 signaling in the seminal work by Zhu *et al.* They demonstrated in an ovalbumin mouse model of asthma that sensitized and challenged mice elicited a dramatic increase in the expression of AMCase mRNA and protein in the lung and a corresponding increase in the level of chitinase activity in the bronchoalveolar lavage (BAL) (Zhu *et al.* 2004). However, no effect on chitotriosidase mRNA expression was observed. Using immunohistochemistry they revealed that the source of this expression was airway epithelial cells and macrophages. In experiments comparing mice reconstituted with either Th1 or Th2 polarized cells they found AMCase upregulation to be a result of Th2 mediated inflammation. Moreover, it was determined that AMCase upregulation was dependent on IL-13 rather than IL-4 signaling. Most interestingly, they observed that inhibition of AMCase enzymatic activity with either allosamidin (specific chitinase inhibitor) or anti-AMCase antibody resulted in a significant decrease in airway hyperresponsiveness, tissue eosinophilia, and inflammation. Importantly, they found this inhibition did not affect either IL-4 or IL-13 signaling nor did it inhibit activation of STAT6. Rather inhibition of AMCase diminished the induction of multiple IL-13 effector molecules including the chemokines eotaxin, eotaxin-2, monocyte chemotactic protein (MCP-1), MCP-2, macrophage inflammatory protein (MIP-1β), C10 and ENA-78 (Zhu *et al.* 2004). These experiments suggest within the Th2 inflammatory cascade that AMCase is downstream of IL-13 cytokine production. Additionally, AMCase expression and activity is required for airway hyperresponsiveness and inflammation, possibly through mediating expression of chemokines (See Figure 1).

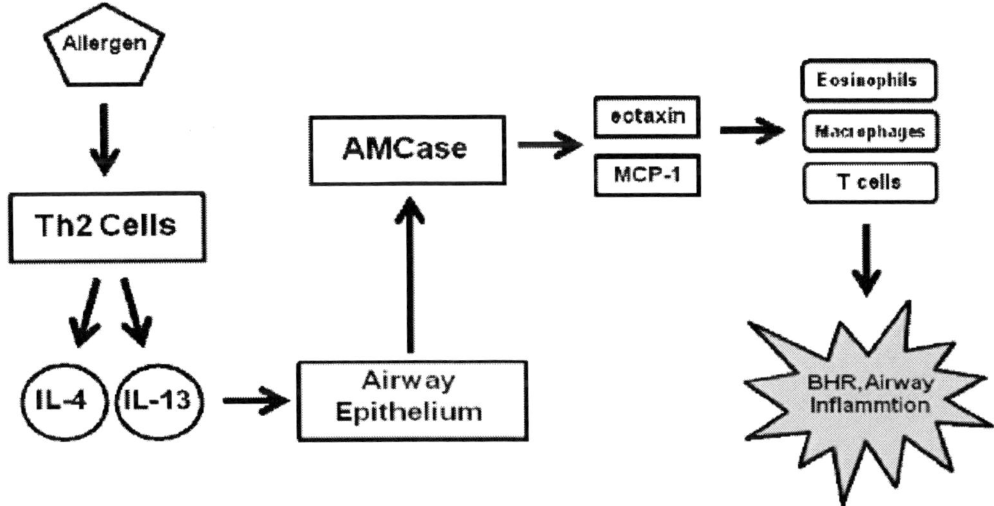

Figure 1. AMCase is a central mediator of the allergen induced Th2 inflammation in the airway of mice. (BHR – bronchial hyperresponsiveness)

Paradoxically, AMCase is also reported as a negative regulator of chitin induced recruitment of innate immune cells associated with allergy (discussed in Section 5) (Reese *et al.* 2007). Although the case may be made that AMCase is only protective in the presence of chitin induced inflammation, it is also true that no inflammatory effect of AMCase was observed in the experiments of Reese *et al.* Most puzzling is the fact that transgenic mice overexpressing lung AMCase were not predisposed to any exaggerated inflammation (Reese *et al.* 2007). Despite these inconsistencies it is clear that AMCase is strongly induced by Th2 inflammatory signaling and may well play multiple contextual roles in allergic inflammation. Further experiments elucidating the precise biology of chitinases and chi-lectins will be necessary to clarify the roles of AMCase in Th2 inflammation of the airway.

3. Active Chitinases in Human Asthma

The role of AMCase in mediating the Th2 inflammation and airway hyperresponsiveness in mouse models of asthma has led to examination of AMCase expression in human asthma. *In-situ* hybridization probing for AMCase mRNA in bronchial biopsies from both fatal asthmatics and subjects who have died of natural causes, revealed AMCase expression only among asthmatics (Zhu *et al.* 2004). AMCase expression was observed in both the epithelial and sub-epithelial cells of the fatal asthmatics (Zhu *et al.* 2004). These results are congruent with the overexpression of AMCase observed during Th2 inflammation in mice. However, a larger more recent study has examined airway epithelial cells from both asthmatics (N=30) and controls (N=28) failing to observe any AMCase expression in either group (Kuperman *et al.* 2005). This discrepancy with the finding of Zhu *et al.* may reflect the heterogeneous nature of asthma and in particular severe asthma, as the latter study only included mild to moderate asthmatics rather than severe asthmatics. Additionally, biopsies from the first study were taken from lung tissue isolated after a fatal asthma attack versus the second study which collected cells from stable asthmatics.

It is also important to note that AMCase mRNA expression cannot be equated with enzymatically active AMCase in the lung. The enzymatic activity of AMCase in humans is particularly relevant since the results of Zhu *et al.* revealed specific inhibition of AMCase enzymatic activity alleviated much of the bronchial hyperresponsiveness and inflammation in mouse models of asthma. In fact, an AMCase transcript (TSA-1902S) with the active site spliced out has been reported to be expressed in the human lung (Saito *et al.* 1999). Moreover, to date there has been no report of AMCase protein expression or specifically AMCase enzymatic activity in the lungs of human subjects. Any exploration into the role of AMCase activity in the human lung should take into account the presence of the other active chitinase in the human genome, chitotriosidase (CHIT1). Indeed, the presence of CHIT1 protein and chitinase activity has been observed in the lungs of healthy human controls (Bargagli *et al.* 2007; Bargagli *et al.* 2007). Moreover, these studies found CHIT1 expression and chitinase activity were both dramatically upregulated in the lungs of patients with interstitial lung diseases (ILD), such as sarcoidosis and pulmonary fibrosis. Due to the involvement of macrophages in ILDs the authors attributed this chitinase activity to CHIT1. However, the authors did not account for AMCase expression. Additional studies will be needed to determine the protein expression and activity of both active human chitinases in human asthma.

4. Chi-Lectins in Human Asthma

Chi-lectins, such as YM1 and YM2 are some of most highly expressed proteins during Th2 inflammation of the lung in mouse models of asthma (Webb *et al.* 2001; Song *et al.* 2008). This observation raises the potential that chi-lectins may also modulate human asthma phenotypes (Webb *et al.* 2001; Song *et al.* 2008). This hypothesis has been tested in examination of the human chi-lectin YKL-40 (also called CHI3L1) in asthma. Specifically, serum levels of YKL-40 were examined in two cohorts of asthmatic cases and healthy controls (Chupp *et al.* 2007). In this study YKL-40 levels were determined specifically by ELISA assay. In the Yale University cohort median serum levels of YKL-40 were found to be slightly higher in asthmatics (Median= 69.7 ng/ml, N=97) versus controls (Median= 58.3 ng/ml, N=38). These results were subsequently confirmed in the smaller replication cohort from Paris University, with asthmatics (Median=97.7 ng/ml, N=40) displaying over twice the median YKL-40 levels of control subjects (41.5 ng/ml, N=12). Using cross-sectional analysis they tested whether serum YKL-40 levels were correlated with asthma severity, as defined by the amount of medication required to achieve asthma control. The authors observed a significant positive correlation between YKL-40 levels and asthma severity in both the Paris and Yale cohorts, and an additional replication cohort of asthmatics (Wisconsin Cohort). Furthermore, the authors found YKL-40 serum levels were also negatively correlated with lung function, as measured by forced expiratory volume in one second (FEV_1) in all three groups of asthmatics (Chupp *et al.* 2007). Although these results suggest serum YKL-40 levels to be a marker of asthma and disease severity in general, it is important to note the considerable overlap of YKL-40 distributions observed between asthmatic cases and controls. Moreover, the correlations with asthma severity appear to be driven in large part by a few severe asthmatic cases with extremely high YKL-40 serum levels.

Extending this study to bronchial biopsy samples in the Paris cohort, immunohistochemistry revealed significantly larger numbers of YKL-40 expressing sub-epithelial cells and airway macrophages in asthmatics as compared to controls (Chupp *et al.* 2007). Positive correlation was also observed in the second cohort between asthma severity and the number of YKL-40 expressing cells in bronchial biopsies. Moreover, they found serum YKL-40 levels were associated with subepithelial basement membrane thickness, an important measure of airway remodeling (Ward *et al.* 2002; Chupp *et al.* 2007).

These results are supported by a recent report of significant increases in YKL-40 protein levels in the bronchoalveolar lavage (BAL) of subjects with allergic asthma 24 hours after segmental allergen challenge (Kuepper *et al.* 2008). Moreover, BAL eosinophil counts at 24 hours correlated positively with YKL-40 levels. These results extend the findings of Chupp *et al.* by revealing upregulation of YKL-40 levels during active inflammation in the human lung. However, the authors do not mention whether significant YKL-40 levels were observed in the asthmatics prior to allergen challenge. As this was a short letter to the editor, further comparison with the results of Chupp *et al.* is difficult as clinical characteristics and severity of these patients was not reported. Nonetheless, these are important observations and should drive the examination of other chi-lectin proteins in asthma such as OVGP1 and CHI3L2 (Funkhouser and Aronson 2007).

5. Chitinase Genetics in Human Asthma

The functional data in mouse models of asthma and differential expression of chitinases in human asthma has generated interest in the possible effects of chitinase genetic variation in susceptibility to asthma. Genetic variants within the AMCase gene have recently been tested for association with asthma in a German case-control cohort (Bierbaum *et al.* 2005). In this study, all exonic portions of the AMCase gene were resequenced in 30 German subjects to determine all coding gene variation. The AMCase gene exhibited considerable variation in this population with 12 polymorphisms being identified including 5 non-synonymous (amino acid changing) variants. All 6 exonic variants (rs12033184, rs2275253, rs2275254, rs2256721, F269S, and K17R) and one promoter polymorphism (rs3818822) were genotyped in the case-control group. Interestingly, the authors found significant associations, using both the adult and pediatric controls, between asthma and the promoter SNP rs3818822 (P=0.0003, P=0.017). Additionally, one of the novel coding SNPs, K17R, was also associated with asthma using both adult and pediatric controls (P=0.019, P=0.0003). Haplotype analysis of the seven SNPs was then performed, identifying 16 haplotype combinations with an allele frequency of at least 1%. Analysis of the distribution of haplotypes in cases versus adult controls resulted in a $\chi 2$ P-value of 0.000039. Similar haplotype analysis between cases and adult controls resulted in a $\chi 2$ P-value of less than 10^{-10}. Closer examination of the results finds four of the six common haplotypes (those with > 5% frequency) differed in haplotype frequency by greater than 4% between asthmatic cases and pediatric controls. However, haplotype tests were not performed on individual haplotypes, precluding the identification of a particular risk haplotype. These results strongly suggest a role for AMCase polymorphisms in the predisposition toward asthma. The associated polymorphisms included both a non-synonymous (amino acid changing) and a promoter SNP. Therefore, to aid in the elucidation of AMCase's role in human asthma, it will be important to determine how these variants may affect enzyme function and expression, respectively.

Based on the functional and expression overlap of AMCase and the chitotriosidase (CHIT1) enzymes, genetic variants in the CHIT1 gene were also tested for association with asthma disease status in a German case-control cohort (Bierbaum *et al.* 2006). Three coding variants of the CHIT1 gene were tested including SNPs Gly102Ser and Ala442Gly, and a 24 base pair duplication in exon 10. This duplication polymorphism has been fully characterized and results in a misspliced mRNA that is quickly degraded (Boot *et al.* 1998). Therefore, if loss of CHIT1 activity plays a role in predisposition to asthma this null variant should be strongly predictive. However, no association was observed between the 322 asthmatic cases and 270 adult controls for either SNPs or haplotypes. Despite the negative results it must be noted that resequencing analysis was not completed for the CHIT1 gene and judging by the importance of haplotypic structure across the AMCase gene in asthma risk, further analysis of sequence variation in the CHIT1 gene will be needed to confirm this result.

To date no studies have examined the relationship between genetic variation in the YKL-40 gene (also called CHI3L1) and risk for asthma. However, a 3 SNP haplotype in the YKL-40 promoter has recently been associated with increased risk of schizophrenia in two independent Chinese groups by both case-control and transmission disequilibrium (TDT) analysis (Zhao *et al.* 2007). The haplotype was determined to disrupt MYC/MAX regulated transcriptional activation leading to a significant decrease in YKL-40 transcription (Zhao *et al.* 2007). Based on strong correlations between YKL-40 expression and both asthma disease status and severity (discussed in Section 3), it will be important to investigate these variants in risk for asthma as well.

These studies have just begun to scratch the surface of the potential involvement of chitinase genetics in asthma traits. Future studies should explore the possibility of interactions between AMCase and CHIT1 genetic variants, since both genes are expressed in the human lung and overlap functionally. Furthermore, there is a possibility for effect modification based on exposure to environmental chitin, which has been shown to have inflammatory properties in the lung (discussed in Section 5). Therefore, it will be important to determine whether chitinase genetic results are consistent in different environmental settings (rural versus urban) which may act as a surrogate of environmental chitin exposure.

6. Immunomodulatory Properties of Chitin

Although the formerly discussed results reveal a role for AMCase in a mouse model of asthma, which is likely unrelated to chitin, recent work has implicated environmental chitin in polarizing the T-helper cell response in the lungs (Shibata *et al.* 2000; Reese *et al.* 2007). The interface of environmental factors (*i.e.* chitin) and effectors of asthma (*i.e.* AMCase) are thought to underlie the complex etiology of asthma. Exposure to chitin by inhalation is very common and it is an important component of aeroallergens. Furthermore, researchers have noted the similarities between the immune responses to chitin-containing parasites and those observed in allergic asthma, including Th2 inflammation driven by IL-4 and IL-13 expressing cells (eosinophils and basophils) (Holt 1996).

The most comprehensive study in this area was recently completed by Reese *et al.* In this study they used the chitin containing parasitic helminth, *Nippostrongylus brasiliensis,* to investigate the role of chitin in Th2 immunity (Reese *et al.* 2007). They found in mice infected with *Nippostrongylus brasiliensis* that STAT6-dependent expression of AMCase, YM-1, and YM-2 was induced in the lung and that the immune response was dependent on IL-4 and IL-13 expression. Since these worms molt in the lungs the authors hypothesized that the structural chitin remodeled during molting acts as a pattern recognition particle, triggering accumulation of IL-4 and IL-13 expressing innate immunity cells. This was tested by administering chitin to the lungs of genetically engineered 4get mice. These mice contain an IL-4 green fluorescent protein (GFP) knock-in, allowing detection of IL-4 expressing cells. Chitin challenge led to the accumulation of eosinophils and basophils in the lungs. To ensure that the immune cell recruitment was not due to contaminating lipopolysaccharide (LPS), these experiments were repeated in TLR4 and MyD88 deficient mice (mice unresponsive to LPS). Furthermore, pretreatment of chitin with enzymatically active mouse AMCase blocked this response. These results suggest that the observed effect was dependent on intact chitin. The chitin challenge experiments were repeated in AMCase overexpressing transgenic mice and immune cell recruitment was strongly reduced compared to wild type mice. Moreover, these effects were determined to be mediated by innate immunity cells, as chitin-induced eosinophil recruitment was functional in recombination activating gene (RAG) knockout mice (no adaptive immune system). Based on the presence of alternatively activated macrophages in allergic and parasitic immunity, the authors explored the role of these cells in mediating eosinophil and basophil recruitment. To test this hypothesis they used reporter mice (termed YARG) which co-express enhanced yellow fluorescent protein (eYFP) and arginase I, allowing for the detection of alternatively activated macrophages. Both *Nippostrongylus brasiliensis* infection and intranasal chitin challenge resulted in the accumulation of eYFP-positive macrophages in the lung. Through *in vitro* experiments they implicated the macrophage produced Leukotriene B_4 in eosinophil recruitement. Chitin challenge experiments in mice lacking the high-affinity receptor for leukotriene B4 (BLT1) and additional experiments in mice depleted of macrophages supported this hypothesis. When the chitin challenge was administered in YARG + AMCase over-expressing transgenic mice, the tissue accumulation of alternatively activated macrophages was greatly diminished. These experiments strongly implicate chitin as a pattern recognition molecule stimulating the tissue recruitment of innate immunity cells associated with allergic asthma, including eosinophils, basophils, and alternatively activated macrophages. In these experiments, rather than being pro-inflammatory (see Section 1) AMCase fulfills the role of a strong negative regulator of chitin induced allergic inflammation. These results are especially interesting considering the widespread presence of chitin in many aeroallergens. Interestingly, bacterial production of chitinases comprise the primary mechanism for chitin turnover in marine environments (Li and Roseman 2004). Therefore, reduction in bacterial populations through the pervasive use of anti-microbial cleaning products and antibiotics in developed countries may lead to diminished ambient chitinases and an increased load of environmental chitin. Strikingly, levels of chitin covered dust mite aeroallergens, measured in homes, were inversely correlated with poverty level (Kitch *et al.* 2000). Indeed, these ideas fit well within the hygiene hypothesis of asthma, which claims reduced exposure to bacterial products acting as pattern recognition molecules, such as endotoxin, are responsible for the recent increases in asthma prevalence (See Figure 2). Furthermore, these observations are supported by the high prevalence of occupational asthma among workers in shellfish processing plants, where chitin exposure is high (MMWR 1982). The risk of asthma in these processing plants has been linked to steps involving high-pressure meat removal from shells, resulting in particulate aerosols which may contain chitin (MMWR 1982).

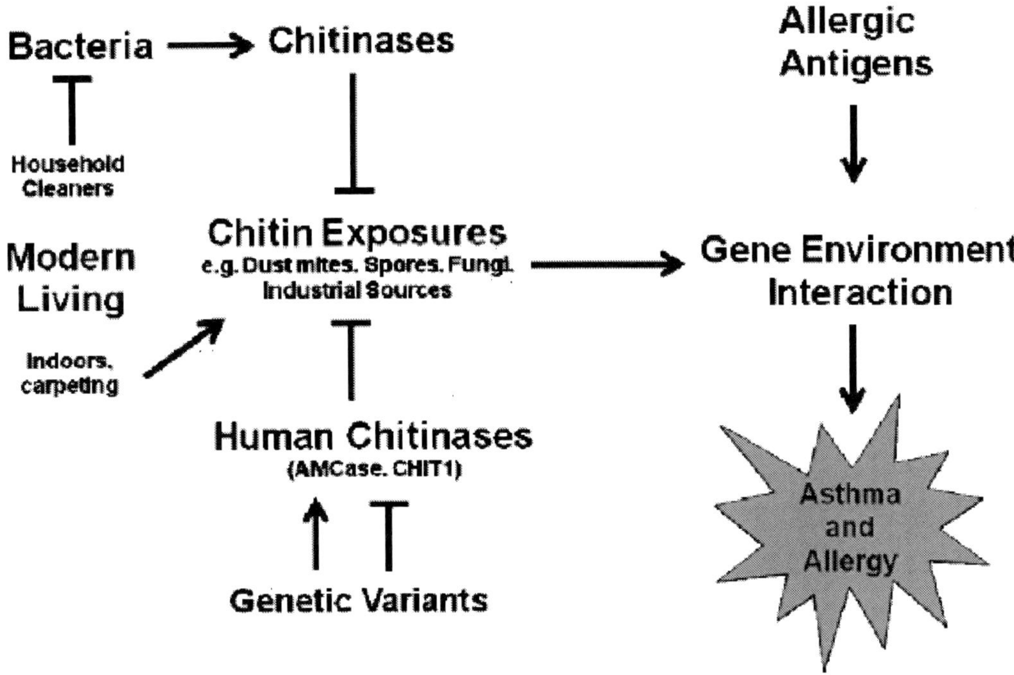

Figure 2. Model describing the potential interaction of pro-inflammatory environmental chitin exposures with allergic antigens resulting in the development of allergy and asthma. In line with the hygiene hypothesis we predict decreased bacterial populations due to the prevalence of anti-microbial products in developed societies, which may increase environmental chitin exposure. Moreover, genetic variants upregulating chitinase expression and activity would be protective against asthma according to the work of Reese *et al.*

Despite the evidence implicating chitin in Th2 inflammation, prior studies have shown the opposite effects of chitin on immune responses. Namely, Shibata *et al.* found chitin micro-particles (1-10 μM) are phagocytosed through mannose receptors and stimulate production of INF-gamma (Th1 cytokine) (Shibata *et al.* 1997). Additional experiments have demonstrated that the oral administration of chitin microparticles in a ragweed allergen mouse model of asthma reduces serum IgE levels and the recruitment of eosinophils to the lungs (Shibata *et al.* 2000). Chitin microparticles have even been proposed as a therapy for allergic diseases and for use as a Th1 adjuvant (Shibata *et al.* 2000; Shibata *et al.* 2001). Although these results appear to be contradictory, they likely are related to differences in the chitin used including the chitin particle size (phagocytosable micro-particles versus larger fragments) and the route of administration (intranasal and lung *vs.* oral). Additionally, commercial chitin may be contaminated with LPS, which despite purification procedures could account for some of the Th1 stimulatory effects seen in these studies. Reese *et al.* controlled for LPS contamination by using TLR4 (receptor for LPS) knockout mice, increasing confidence their results. Nonetheless, it is clear that chitin can modulate immune responses and that more carefully controlled experiments are needed to further elucidate the role of chitin in mediating and/or inhibiting the progression of Th2 inflammation in allergic asthma.

Conclusions and Outlook

In a short period of time chitin and chitinases have risen from environmental and genome obscurity, respectively, to be realized as key factors in asthma pathology and the possible target of new asthmatic therapies (Wills-Karp and Karp 2004; Dickey 2007). There is general agreement that the expression of chitinases and chi-lectins is induced during Th2 inflammation and is IL-13 dependent. Chitinases likely fulfill at least two contrasting roles in airway inflammation: 1) in the absence of chitin, chitinases and chi-lectins promote active airway inflammation possibly through stimulation of chemotactic factors and 2) chitinases can also negatively regulate chitin induced airway inflammation by enzymatic degradation of chitin. Both genetic and expression data in humans are limited and sometimes conflicting but generally seem to support a role for chitinases and chi-lectins in human asthma. Due to the complex inheritance of asthma involving genetic alterations of multiple pathways interfacing with environmental risk factors, no one approach will be completely successful in resolving these discrepancies. Cross-fertilization and interaction between many researchers examining diverse aspects of chitin-chitinase biology is needed including human studies (chitinase expression, genetics, and biochemistry) coupled with both epidemiological studies of chitin exposure and mouse studies of chitinase biology. Only through a comprehensive and multidimensional approach will we be able to clarify the roles of chitin and chitinases in human asthma.

References

Araujo AC, Souto-Padrón T, de Souza W. Cytochemical localization of carbohydrate residues in microfilariae of Wuchereria bancrofti and Brugia malayi. *J Histochem Cytochem.* 1993; 41: 571-8.

Bargagli E, Margollicci M, Luddi A, Nikiforakis N, Perari MG, Grosso S, Perrone A, Rottoli P. Chitotriosidase activity in patients with interstitial lung diseases. *Respir Med.* 2007; 101: 2176-81.

Bargagli E, Margollicci M, Perrone A, Luddi A, Perari MG, Bianchi N, Refini RM, Grosso S, Volterrani L, Rottoli P. Chitotriosidase analysis in bronchoalveolar lavage of patients with sarcoidosis. *Sarcoidosis Vasc Diffuse Lung Dis.* 2007; 24: 59-64.

Bierbaum S, Nickel R, Koch A, Lau S, Deichmann KA, Wahn U, Superti-Furga A, Heinzmann A. Polymorphisms and haplotypes of acid mammalian chitinase are associated with bronchial asthma. *Am J Respir Crit Care Med.* 2005; 172: 1505-9.

Bierbaum S, Superti-Furga A, Heinzmann A. Genetic polymorphisms of chitotriosidase in Caucasian children with bronchial asthma. *Int J Immunogenet.* 2006; 33: 201-4.

Boot RG, Renkema GH, Strijland A, van Zonneveld AJ, Aerts JM. Cloning of a cDNA encoding chitotriosidase, a human chitinase produced by macrophages. *J Biol Chem.* 1995; 270: 26252-6.

Boot RG, Renkema GH, Verhoek M, Strijland A, Bliek J, de Meulemeester TM, Mannens MM, Aerts JM. The human chitotriosidase gene. Nature of inherited enzyme deficiency. *J Biol Chem.* 1998; 273: 25680-5.

Bussink AP, Speijer D, Aerts JM, Boot RG. Evolution of mammalian chitinase(-like) members of family 18 glycosyl hydrolases. *Genetics* 2007; 177: 959-70.

Chou YT, Yao S, Czerwinski R, Fleming M, Krykbaev R, Xuan D, Zhou H, Brooks J, Fitz L, Strand J, Presman E, Lin L, Aulabaugh A, Huang X. Kinetic characterization of recombinant human acidic mammalian chitinase. *Biochemistry* 2006; 45: 4444-54.

Christodoulopoulos P, Cameron L, Nakamura Y, Lemière C, Muro S, Dugas M, Boulet LP, Laviolette M, Olivenstein R, Hamid Q. TH2 cytokine-associated transcription factors in atopic and nonatopic asthma: evidence for differential signal transducer and activator of transcription 6 expression. *J Allergy Clin Immunol.* 2001; 107: 586-91.

Chupp GL, Lee CG, Jarjour N, Shim YM, Holm CT, He S, Dziura JD, Reed J, Coyle AJ, Kiener P, Cullen M, Grandsaigne M, Dombret MC, Aubier M, Pretolani M, Elias JA. A chitinase-like protein in the lung and circulation of patients with severe asthma. *N Engl J Med.* 2007; 357: 2016-27.

Debono M, Gordee RS. Antibiotics that inhibit fungal cell wall development. *Annu Rev Microbiol.* 1994; 48: 471-97.

Dickey BF. Exoskeletons and exhalation. *N Engl J Med.* 2007; 357: 2082-4.

Fuhrman JA, Piessens WF. Chitin synthesis and sheath morphogenesis in Brugia malayi microfilariae. *Mol Biochem Parasitol.* 1985; 17: 93-104.

Funkhouser JD, Aronson NN Jr. (2007). Chitinase family GH18: evolutionary insights from the genomic history of a diverse protein family. *BMC Evol Biol.* 2007; 7: 96.

Grünig G, Warnock M, Wakil AE, Venkayya R, Brombacher F, Rennick DM, Sheppard D, Mohrs M, Donaldson DD, Locksley RM, Corry DB. Requirement for IL-13 independently of IL-4 in experimental asthma. *Science* 1998; 282(5397): 2261-3.

Holt PG. Infections and the development of allergy. *Toxicol Lett.* 1996; 86: 205-10.

Hoshino A, Tsuji T, Matsuzaki J, Jinushi T, Ashino S, Teramura T, Chamoto K, Tanaka Y, Asakura Y, Sakurai T, Mita Y, Takaoka A, Nakaike S, Takeshima T, Ikeda H, Nishimura T. STAT6-mediated signaling in Th2-dependent allergic asthma: critical role for the development of eosinophilia, airway hyper-responsiveness and mucus hypersecretion, distinct from its role in Th2 differentiation. *Int Immunol.* 2004; 16: 1497-505.

Humbert M, Durham SR, Kimmitt P, Powell N, Assoufi B, Pfister R, Menz G, Kay AB, Corrigan CJ. Elevated expression of messenger ribonucleic acid encoding IL-13 in the bronchial mucosa of atopic and nonatopic subjects with asthma. *J Allergy Clin Immunol.* 1997; 99: 657-65.

Kitch BT, Chew G, Burge HA, Muilenberg ML, Weiss ST, Platts-Mills TA, O'Connor G, Gold DR. Socioeconomic predictors of high allergen levels in homes in the greater Boston area. *Environ Health Perspect* 2000; 108: 301-7.

Kuepper M, Bratke K, Virchow JC. Chitinase-like protein and asthma. *N Engl J Med.* 2008; 358(10): 1073-5; author reply 1075.

Kuperman DA, Huang X, Koth LL, Chang GH, Dolganov GM, Zhu Z, Elias JA, Sheppard D, Erle DJ. Direct effects of interleukin-13 on epithelial cells cause airway hyperreactivity and mucus overproduction in asthma. *Nat Med.* 2002; 8: 885-9.

Kuperman DA, Lewis CC, Woodruff PG, Rodriguez MW, Yang YH, Dolganov GM, Fahy JV, Erle DJ. Dissecting asthma using focused transgenic modeling and functional genomics. *J Allergy Clin Immunol.* 2005; 116: 305-11.

Li X, Roseman S. The chitinolytic cascade in Vibrios is regulated by chitin oligosaccharides and a two-component chitin catabolic sensor/kinase. *Proc Natl Acad Sci U S A.* 2004; 101: 627-31.

MMWR. Epidemiologic Notes and Reports Asthma-Like Illness among Crab-Processing Workers –Alaska. *MMWR* 1982; 31: 95-96.

Musumeci M, Bellin M, Maltese A, Aragona P, Bucolo C, Musumeci S. Chitinase levels in the tears of subjects with ocular allergies. *Cornea* 2008; 27: 168-73.

Neville AC, Parry DA, Woodhead-Galloway J. The chitin crystallite in arthropod cuticle. *J. Cell Sci.* 1976; 21: 73-82.

Reese TA, Liang HE, Tager AM, Luster AD, Van Rooijen N, Voehringer D, Locksley RM. Chitin induces accumulation in tissue of innate immune cells associated with allergy. *Nature* 2007; 447(7140): 92-6.

Saito A, Ozaki K, Fujiwara T, Nakamura Y, Tanigami A. Isolation and mapping of a human lung-specific gene, TSA1902, encoding a novel chitinase family member. *Gene* 1999; 239: 325-31.

Shahabuddin M, Kaslow DC. Plasmodium: parasite chitinase and its role in malaria transmission. *Exp Parasitol.* 1994; 79: 85-8.

Shibata Y, Foster LA, Bradfield JF, Myrvik QN. Oral administration of chitin down-regulates serum IgE levels and lung eosinophilia in the allergic mouse. *J Immunol.* 2000; 164: 1314-21.

Shibata Y, Foster LA, Metzger WJ, Myrvik QN. Alveolar macrophage priming by intravenous administration of chitin particles, polymers of N-acetyl-D-glucosamine, in mice. *Infect Immun.* 1997; 65: 1734-41.

Shibata Y, Honda I, Justice JP, Van Scott MR, Nakamura RM, Myrvik QN. Th1 adjuvant N-acetyl-D-glucosamine polymer up-regulates Th1 immunity but down-regulates Th2 immunity against a mycobacterial protein (MPB-59) in interleukin-10-knockout and wild-type mice. *Infect Immun.* 2001; 69: 6123-30.

Song HM, Jang AS, Ahn MH, Takizawa H, Lee SH, Kwon JH, Lee YM, Rhim TY, Park CS. Ym1 and Ym2 expression in a mouse model exposed to diesel exhaust particles. *Environ Toxicol.* 2008; 23: 110-6.

Techkarnjanaruk S, Pongpattanakitshote S, Goodman AE. Use of a promoterless lacZ gene insertion to investigate chitinase gene expression in the marine bacterium Pseudoalteromonas sp. strain S9. *Appl Environ Microbiol.* 1997; 63: 2989-96.

Ward C, Pais M, Bish R, Reid D, Feltis B, Johns D, Walters EH. Airway inflammation, basement membrane thickening and bronchial hyperresponsiveness in asthma. *Thorax* 2002; 57: 309-16.

Webb DC, McKenzie AN, Foster PS. Expression of the Ym2 lectin-binding protein is dependent on interleukin (IL)-4 and IL-13 signal transduction: identification of a novel allergy-associated protein. *J Biol Chem.* 2001; 276: 41969-76.

Wills-Karp M, Karp CL. Chitin checking--novel insights into asthma. *N Engl J Med.* 2004; 351: 1455-7.

Wills-Karp M, Luyimbazi J, Xu X, Schofield B, Neben TY, Karp CL, Donaldson DD. Interleukin-13: central mediator of allergic asthma. *Science* 1998; 282(5397): 2258-61.

Zhao X, Tang R, Gao B, Shi Y, Zhou J, Guo S, Zhang J, Wang Y, Tang W, Meng J, Li S, Wang H, Ma G, Lin C, Xiao Y, Feng G, Lin Z, Zhu S, Xing Y, Sang H, St Clair D, He L. Functional variants in the promoter region of Chitinase 3-like 1 (CHI3L1) and susceptibility to schizophrenia. *Am J Hum Genet.* 2007; 80: 12-8.

Zhu Z, Zheng T, Homer RJ, Kim YK, Chen NY, Cohn L, Hamid Q, Elias JA. Acidic mammalian chitinase in asthmatic Th2 inflammation and IL-13 pathway activation. *Science* 2004; 304(5677): 1678-82.

In: Binomium Chitin-Chitinase: Recent Issues
Editor: Salvatore Musumeci and Maurizio G. Paoletti ISBN 978-1-60692-339-9

Chapter XIX

Chitinase-Like Lectins in Humans

Julia Kzhyshkowska[15]
From the Department of Dermatology and Allergology, University Medical Centre Mannheim, Ruprecht-Karls University of Heidelberg, Mannheim, Germany D-68167 and Isntitute of General Pathology and Pathophysiology, Russian Academy of Medical Sciences, Moscow, Russia

Abstract

Family of human Glyco_18-domain-containing proteins comprises catalytically active chitinases and chitinase-like proteins. Human chitinase-like proteins include YKL-39, YKL-40 and SI-CLP (stabilin-1 interacting chitinase-like protein). In addition, chitinase-like proteins YM1 and YM2 were identified in rodents, but their human homologues do not exist. In contrast to true chitinases, YKL-39, YKL-40 and SI-CLP are enzymatically inactive due to the lack of critical catalytic aminoacids in the enzymatic site within their Glyco_18 domain. While true chitinases bind chitin via C-terminal chitin-binding domain, YKL-39, YKL-40 and SI-CLP do not posses chitin-binding domain, and contain solely Glyco_18-domain. However the Glyco_18 domain of YKL-40 mediates binding to heparin, hyaluronan, and chitin. All three human chitinase-like proteins are secreted into the extracellular space. Elevated levels of YKL-40 are associated with several chronic inflammatory disorders and cancers. Biological activities of YKL-40 include regulation of cell proliferation, adhesion, migration and activation. YKL-40 promotes growth of human synovial cells, chondrocytes, skin and foetal lung fibroblasts. Two biological activities of YKL-39 are suggested to contribute to progression of osteoarthritis. One is the induction of autoimmune response, and second is participation in tissue remodeling. Biological activity of SI-CLP is currently under investigation in our laboratory. Both YKL-40 and SI-CLP are expressed by several cell types including tumour cells and macrophages. We found antagonistic regulation of

15 Address for correspondence: Dr. Julia Kzhyshkowska, Department of Dermatology, Venerology and Allergology, University Medical Centre Mannheim, University of Heidelberg, Theodor-Kutzer Ufer 1-3, 68167 Mannheim, Germany, Tel: +49 621 383 2440, Fax: +49 621 383 3815, E-Mail: julia.kzhyshkowska@haut.ma.uni-heidelberg.de.

expression of YKL-40 and SI-CLP in human macrophages. YKL-40 is strongly induced by IFNgamma, Th1 cytokine which initiates classical inflammation. Th2 cytokine IL-4 as well as glucocortiocid dexamethasone suppress YKL-40 expression. In contrast, IL-4 and dexamethasone synergistically activate SI-CLP, while IFNgamma abrogates this effect. YKL-39 was identified as an abundantly secreted protein in primary culture of human articular chondrocytes. YKL-39 is currently recognized as a biomarker for the activation of chondrocytes and the progress of the osteoarthritis in human. Recently we found that mRNA of YKL-39 is dramatically upregulated in human macrophages by IL-4 and TGF-beta, a crucial growth factor regulating tumour growth and atherosclerosis. Thus, all three mammalian chitinase-like proteins YKL-39, YKL-40 and SI-CLP are indicators of macrophage activation modes found in distinct pathological situations. Macrophages utilise several tightly regulated pathways for secretion of soluble mediators. The mechanism which regulates intracellular sorting for human chitinase-like proteins and their commitment for secretion was not elucidated. We investigated the mechanism of SI-CLP sorting and secretion in human alternatively activated macrophages. Using biochemical and cell biology approaches we showed, that SI-CLP directly interacts with multifunctional macrophage receptor stabilin-1. Stabilin-1 recognises newly synthesised SI-CLP in late Golgi compartment and targets it to the secretory lysosomes. However SI-CLP can be delivered into lysosomes also in the absence for stabilin-1, indicating that more than one intracellular receptor is involved in its sorting. We also showed, that glucocorticoid dexamethasone in therapeutic concentrations despite inducing expression of SI-CLP, blocks its secretion leading to the intracellular accumulation of high levels of SI-CLP. Significance of these findings for understanding of macrophage-mediated pathologies is discussed.

1. Introduction

The common structural feature of mammalian chitinases and chitinase-like proteins is the presence of highly conserved Glyco_18 domain. The Glyco_18 domain is characteristic for the evolutionary conservative chitinases which belong to the family 18 glycosyl hydrolases. Chitinases are distributed in a wide range of organisms, including bacteria, plants, fungi, insects, viruses, protozoan parasites, and more recently were identified in mammals. Chitinases catalyze the hydrolysis of chitin. An additional activity of these enzymes is trans-glycosylation resulting in formation of chitin oligomers longer than the initial substrate (Sasaki *et al.* 2002). After cellulose, chitin, a polymer of N-acetylglucosamine, is the second most abundant polysaccharide in nature. Chitin is a component of fungal cell walls, exoskeletons of insects and crustaceans, and microfilarial sheaths of parasitic nematodes. Evolutionary conserved function of chitinases in lower life forms is a host defence against chitin-containing organisms (Elias *et al.* 2005). There are 6 Glyco_18 domain containing proteins identified in human up to date (see Figure 1). In addition two closely related Glyco_18 domain containing proteins, YM1 and YM2, have been reported in rodents. However there are no human analogues for YM1/YM2 neither on genomic nor on protein level (reviewed in (Kzhyshkowska *et al.* 2007a).

Only two mammalian Glyco_18 containing proteins possess enzymatic activity: chitotriosidase (Renkema *et al.* 1995; Boot *et al.* 1995) and acidic mammalian chitinase AMCase (Boot *et al.* 2001). Chitotriosidase is able to hydrolise the artificial substrate 4-MU-chitotrioside as well as chitin (Renkema *et al.* 1995). In the excess of substrate, chitotriosidase exhibit transglycosylation activity toward substrate molecules (Aguilera *et al.* 2003). AMCase can cleave both natural chitin from fungal cell wall and crab shell, as well as artificial chitin-like substrates. In contrast to chitotriosidase, AMCase is acid-stable and shows second pH optimum around pH2 (Boot *et al.* 2001). The catalytic activity of Glyco_18 domain proteins is defined by the presence of a catalytic glutamic acid in the context of a DXDXE site located directly (with one aminoacid as a spacer) after the characteristic aminoacid triplet FDG (Fusetti *et al.* 2002). Both chitotriosidase and AMCase in addition to Glyco_18 domain, contain the functional chitin-binding domain on their C-terminus.

YKL-39, YKL-40, SI-CLP and YM1/YM2 contain only Glyco_18 domain but not a chitin-binding domain. Oviductin/MUC9 contains a Glyco_18 domain and a long fragment with numerous sites for O-glycosylation, characteristic of mucins. YKL-39, YKL-40, SI-CLP, YM1/YM2 and oviductin lack critical aminoacids within the catalytic site (Figure 1B) and do not exhibit enzymatic activity. YKL-40 and YKL-39 are the best investigated human chitinase-like proteins which are highly similar on the structural level and are associated with some similar pathologies in human. First discovered protein was YKL-40. The name YKL-40 derives from the on letter code for the tyrosine, lysine and leucine in its N-terminus and its apparent molecular weight (Johansen *et al.* 1992, Hakala *et al.* 1993). N-terminal sequencing of YKL-40 containing fraction of conditioned medium revealed the presence of a second closely related protein. In case of YKL-40 this sequence was YKLVCY**Y**T**S**WSQ**Y**R while in case of YKL39 the sequence was YKLVCY**F**T**N**WSQ**D**R. Since the new protein also had an YKL motif, it was called YKL-39 in accordance with its apparent molecular weight (MW). In contrast to YKL-40, YKL-39 is not a glycoprotein. Although the predicted number of aminoacids was slightly larger in YKL-39, the presence of carbohydrate in YKL-40, but not YKL-39 explained the differences in MW (Hu *et al.* 1996). This chapter will be focused on biochemical properties, intracellular sorting, biological activities and association with diseases of three human chitinase-like proteins YKL-39, YKL-40 and SI-CLP. When necessary the properties of rodent YM1/2 will be discussed in parallel.

2. Sugar Binding Specificity of Chitinase-Like Proteins

Glyco_18 domain of chitinase-like proteins despite the absence of the catalytic activity is responsible for the interaction with diverse carbohydrates. Biochemical and structural analysis clearly demonstrated that YKL-40 is a catalytically silent lectin. The exchange of glutamic residues in catalytic centre of chitotriosidase to leucine, which is present in YKL-40 in the same position, converts the catalytically active site into chitin binding site (Renkema *et al.* 1998). Accordingly, YKL-40 was established to be a chitin-specific lectin (Renkema *et al.* 1998). Binding of chitin fragments of different lengths identified nine sugar binding subsites in the groove in YKL-40. The specificity of chitin binding to YKL-40 was shown to depend on the length of oligosaccharides (Fusetti *et al.* 2003). Chitin disacharides were found to occupy distal subsites and longer chains preferentially occupied the central subsites in the groove.

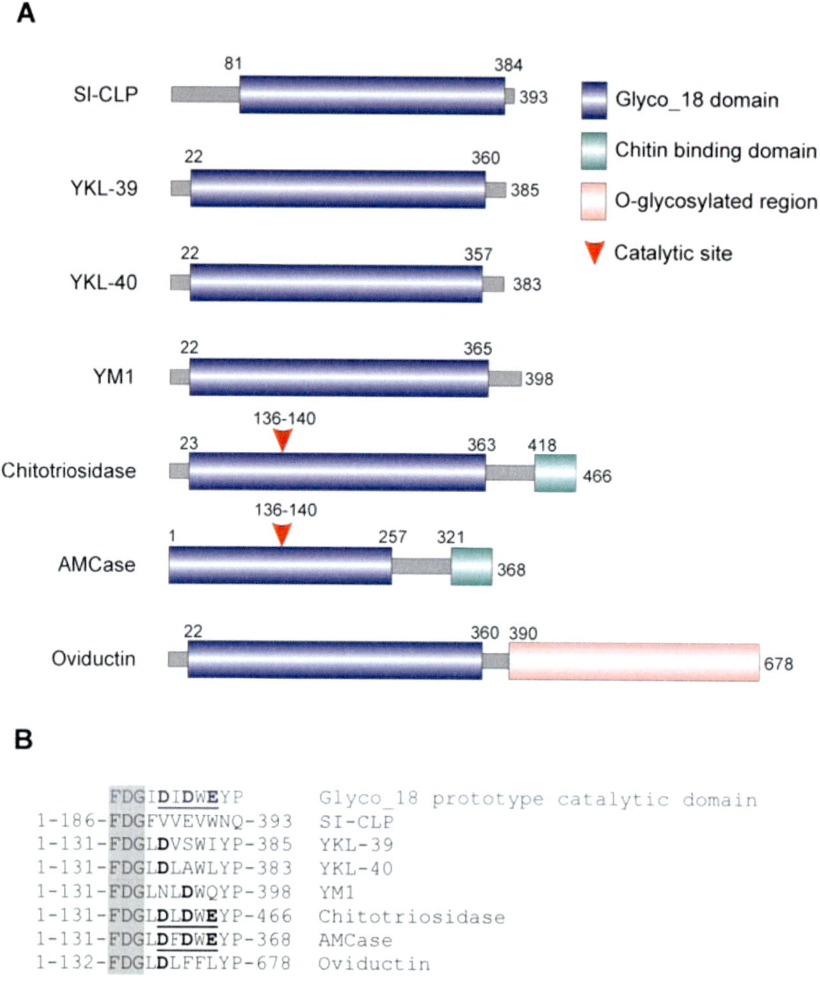

Figure 1. SI-CLP is a novel member of Glyco_18 domain containing human chitinases and chitinase-like proteins. (A) Schematic presentation of mammalian Glyco_18 domain containing proteins; (B) Critical aminoacid in the catalytic sites. The characteristic FDG sequence preceding catalytic motif is shown in shadowed box. Catalytic aminoacids are shown in bold. Complete active catalytic motifs are underlined. This research was originally published in Blood. Kzhyshkowska, J *et al.* Novel stabilin-1 interacting chitinase-like protein (SI-CLP) is up-regulated in alternatively activated macrophages and secreted via lysosomal pathway. Blood. 2006; 107:3221-3228. © the American Society of Hematology.

Crystal structures of native YKL-40 and its complex with $GLcNAc_8$ showed that YKL-40 binds to oligosaccharide ligand in a fashion identical to catalytically active family 18 chitinases. But in contrast to chitinases, in YKL-40 the oligosaccharide ligand induces large conformational change that might have regulatory significance during host response to fungi or nematodes (Houston *et al.* 2003).

Screen for oligosacharides interacting with YM-1 purified out of *T. spiralis* infected mice and immobilised on CM5 sensor chip was performed using surface plasmon resonance techniques (Sun *et al.* 2001). YM-1 was found to interact highly specifically with monosacharides GlcN and GalN as well as with GlcN oligosaccharides (Chang *et al.* 2001;

Sun *et al.* 2001). Purified YM-1 was also able to bind to the immobilised heparin. This interaction was suggested to contribute to biological effects of YM-1 during inflammation and tissue remodelling. Three-dimensional crystal structure of Ym1 at 2.5-A resolution revealed the presence of saccharide binding sites within TIM domain which is a part of Glyco_18 domain (Sun *et al.* 2001).

Our recent studies demonstrated that identified by us SI-CLP also has lectin-binding properties. We screened the ability of in vitro translated as well as recombinant purified SI-CLP to interact with various carbohydrates immobilised on beads and found, specific binding to heparin type I and II (Kzhyshkowska, Mamidi, unpublished data). Interaction of SI-CLP with heparin allows us to propose its function in various cell-physiological processes including cell proliferation, differentiation, adhesion and motility. Considering the high level of SI-CLP production by alternatively activated macrophages, novel molecular mechanism of macrophage-mediated control of tissue morphogenesis can be suggested.

In summary, enzymatically inactive Glyco_18 domain is indicative for lectin properties of the protein, and consequently the enzymatically inactive chitinase-like proteins constitute a novel class of secreted lectins which combine properties of cytokines and growth factors and mediate cellular cross-talk.

3. Cellular Specificity and Regulation of Expression of Chitinase-Like Proteins

The major cell types producing mammalian chitinases and chitinase-like proteins are macrophages, neutrophils, epithelial cells, chondrocytes and synovial cells, as well as cancer cells (for summary see Table 1). Expression of these proteins is regulated on the mRNA level by various cytokines and hormones.

YKL-40 (CHI3L1, HC gp-39) was the first human chitinase-like protein identified and is the best characterised from biological and clinical points of view. YKL-40 was first identified as one of the major secreted proteins by articular chondrocytes and synovial fibroblasts (Hakala *et al.* 1993), and its porcine homologue gp38k was found to be secreted by cultured porcine smooth muscle cells (Millis *et al.* 1986; Shackelton *et al.* 1995). YKL-40 was shown to be a differentiation marker for macrophages, where its expression is predominantly regulated by the transcription factor SP1 (Rehli *et al.* 1997; Rehli *et al.* 2003). YKL-40 is secreted in vitro by some human cancer cell lines of different origin, including glioblastoma, colon cancer, ovarian cancer, prostate cancer, osteosarcoma, malignant melanoma (reviewed in (Johansen *et al.* 2006)). In accordance to this, YKL-40 protein expression was detected in biopsies of glioblastomas, breast and colon cancer (Nutt *et al.* 2005; Pelloski *et al.* 2005; Johansen *et al.* 2006).

YKL-40 expression is associated with the intensive tissue remodelling in vivo. YKL-40 mRNA was detected in osteoarthritic cartilage (Hakala *et al.* 1993; Connor *et al.* 2000).

Table 1. Cell type specificity of expression of chitinases and chitinase-like proteins. The expression profile is described predominantly for human proteins. Expression in mouse cells is indicated in italic

	YKL-40 (CHI3L1)	YKL-39 (CHI3L2)	SI-CLP (CHID1)	*YM-1/2 (Chi3l3/l4)*
Macrophages (MΦ)	MΦ at late stages of differentiation; MΦ stimulated with IFNγ; tumour-associated MΦ; MΦ and giant cells in pulmonary sarcoid granuloma; subsets of MΦ in the atherosclerotic plaque; Microglia in AD	Microglia in AD; MΦ stimulated with IL4 and TGFβ	MΦ stimulated with IL-4/dex; MΦ in BAL	*Microglia in AD; murine model of chronic proliferative dermatitis; lung macrophages in IL-13-transgenic mice*
Neutrophils	Non-activated and activated neutrophils			*Neutrophils in murine model of CGD*
Mast cells	Mast cells in various tissues including spleen, bone marrow, thymus, thyroid, pancreas, adrenal gland, gastrointestinal organs.			*murine model of chronic proliferative dermatitis*
T-cells			Jurkat cells, CD3+ cells from peripheral blood	
B-cells			Raji cells	*B-cells in draining lymph node of mice implanted with B. malayi*
Epithelial cells	Epithelia in various tissues including liver, stomach, lung, hypothesis, testis, ovary, placenta Colonic epithelial cells			*muscus producing epithelial cells of distal airway (Homer et al 2006)*
synovial cells	fibroblast-like synovial cells	fibroblast-like synovial cells		

	YKL-40 (CHI3L1)	YKL-39 (CHI3L2)	SI-CLP (CHID1)	*YM-1/2 (Chi3l3/l4)*
Chondrocytes	RA and OA chondrocytes	articular cartilage chondrocytes		
Smooth-muscle cells	vascular smooth-muscle cells (VSMC)			
Malignant tumours and tumour cell lines	Glioblastoma (U87); osteosarcoma (MG-63); ovarian cancer (SW626); prostate cancer (DV-145); malignant melanoma (SK-MEL-28); cancer cells in biopsies of glioblastoma, breast and prostate cancer		HEK293, MCF7, HeLa, A-549	

Immunohistochemical analysis showed that YKL-40 protein is expressed in chondrocytes of osteoarthritic cartilage mainly in the superficial and middle zone of the cartilage rather than the deep zone (Volck *et al.* 1999). YKL-40 is expressed in human smooth muscle cells and upregulated in distinct subsets of macrophages in the atherosclerotic plaque. It is hypothesised, that chitotriosidase together with YKL-40 regulate cell migration and tissue remodelling during atherogenesis (Boot *et al.* 1999). Expression of YKL-40 mRNA is higher in macrophages in early atherosclerotic lesions and in macrophages which infiltrated deep in the lesion (Boot *et al.* 1999). YKL-40 expression is indicative for the differentiation of macrophages during formation of atherosclerotic plaque (Rathcke *et al.* 2006b). Proteomics study demonstrated elevated levels of YKL-40 in supernatants of macrophage cell line THP-1 treated with oxidised LDL (in vitro "foam cell" model) (Fach *et al.* 2004). In vitro, the same macrophages which secrete chitotriosidase, can secrete YKL-40 (Renkema *et al.* 1998), however expression of two proteins is independently regulated. Thus, YKL-40 mRNA was detected after 2 days of in vitro cultivation of macrophages and reached the maximum at day 14, while expression of chitotriosidase started after 7 days and dramatically increase with time. YKL40 is also expressed human neutrophils, which release YKL-40 in inflammatory conditions (Volck *et al.* 1998).

Production of YKL-40 is regulated by various hormones and cytokines. Studies in IL-6 knockout mice revealed that YKL-40 expression depends on IL-6 (Johansen *et al.* 2006). Real-time PCR analysis performed in our laboratory demonstrated that expression of YKL-40 mRNA in human peripheral-blood derived monocytes is strongly stimulated by Th-1 cytokine IFNγ and inhibited by Th2 cytokine IL-4 and dexamethasone (Figure 2), (Kzhyshkowska *et al.* 2006b). Pro-inflammatory hormones arg-vasopressin (AVP) and parathyroid hormone-related protein (PTHrP) differentially affect secretion of YKL-40 by cultured chondrocytes

(Petersson *et al.* 2006). Both PTHrP and AVP stimulated the secretion of YKL-40 in primary cultured chondrocytes derived from patients with rheumatoid arthritis. In contrast AVP decreased and PTHrP did not affect the secretion of YKL-40 in primary cultured chondrocytes from patients with osteoarthritis.

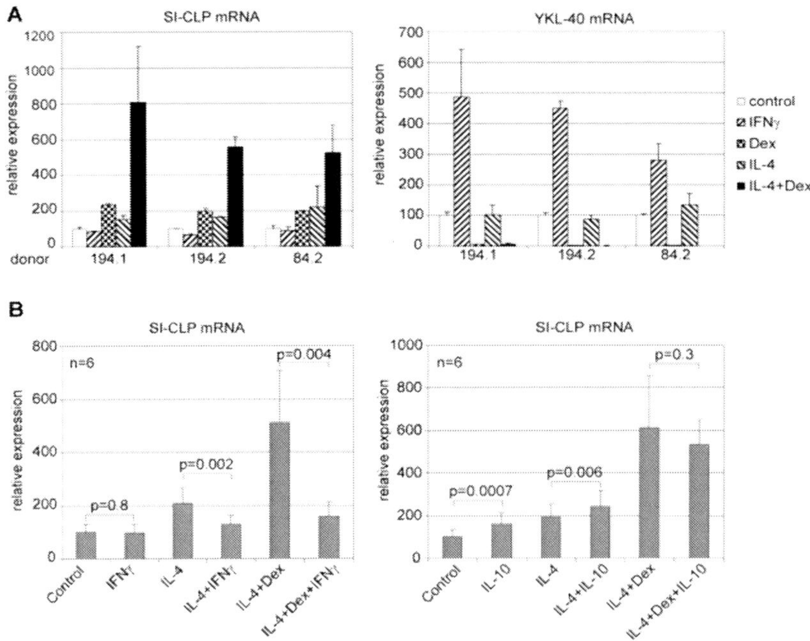

Figure 2. Real-time RT-PCR analysis of SI-CLP and YKL-40 expression in human macrophages. Peripheral blood derived monocytes non-stimulated (control) or stimulated with cytokines as indicated were propagated in culture for 6 days. Three representative donors with differential responses are presented. (A) IL-4, dexamethasone and combination of both induce SI-CLP mRNA upregulation in macrophage cultures, the lowest SI-CLP expression is detected in case of IFNγ stimulation; (B) IFNγ induces YKL-40 mRNA expression, whereas dexamethasone has strong inhibitory effect. This research was originally published in Blood. Kzhyshkowska, J *et al.* Novel stabilin-1 interacting chitinase-like protein (SI-CLP) is up-regulated in alternatively activated macrophages and secreted via lysosomal pathway Blood. 2006; 107:3221-3228. © the American Society of Hematology.

In chondrocytes derived from healthy subjects, PTHrP did not change the amounts of YKL-40, and AVP had an inhibitory effect. Thus disease-specific pre-programming of chondrocytes makes them susceptible for further hormone stimulation and inducible production of YKL-40. In cancer cells various types of physiological stress affect expression of YKL-40. Hypoxia, ionizing radiation, etoposide, ceramide, serum depletion and confluence stimulate YKL-40 expression, while FGF and TNFα repressed it (Junker *et al.* 2005). The authors hypothesised, that YKL-40 is a cellular survival factor. Whether the same stimulations have similar effects in primary, non-transformed cell remains to be identified. Especially interesting is the question, whether tumour-associated macrophages in hypoxic sites release elevated levels of YKL-40.

In contrast to broad range of YKL-40 expression, YKL-39 production was believed for a long time to be highly specific for chondrocytes and synoviocytes (Hu *et al.* 1996). First report suggesting that YKL-39 can be also produced by macrophages came from the investigation of patients with Alzheimer disease (Colton *et al.* 2006). Messenger RNAs for both YKL-39 and YKL-40 were strongly upregulated in the brain of patients with Alzheimer disease, and this fact was attributed to the alternative activation of microglial macrophage during the course of the disease (Colton *et al.* 2006). Comprehensive analysis of various stimulations for their ability to induce the expression of YKL-39 in macrophages was performed in our laboratory (Kzhyshkowska *et al.* 2006b; Gratchev *et al.* 2008). We used an in vitro model of human monocyte-derived macrophage differentiation. CD14+ MACS sorted human monocytes were cultured for 6 days under different stimulations including IFNγ, IL-4, dexamethasone and TGF-β. Expression of YKL-39 was examined by Real-time RT-PCR. We found that both IL-4 and TGF-β have weak stimulatory effect on YKL-39 expression if used separately. However the combination of IL-4 and TGF-β had strong stimulatory effect on the expression of YKL-39 in all individual macrophage cultures analyzed (Gratchev *et al.* 2008). INFγ did not show statistically significant effect on the YKL-39 mRNA expression. Presence of dexamethasone almost completely abolished the stimulatory effects of IL-4 and TGF-β (Figure 3). Thus, maturation of monocyte derived macrophages in the presence of Th2 cytokine IL-4 and TGF-β leads to the strong activation of YKL-39 expression. Consequently, the elevated levels of YKL-39 observed during chronic inflammations cannot be attributed solely to the activity of chondrocytes. In perspective, YKL-39 might serve as a useful biomarker to detect macrophage-specific response in pathologies like cancer, atherosclerosis and Alzheimer disease.

YM1 and YM2 were found only in rodents and have no human analogue neither on genomic nor on protein level (Chang *et al.* 2001;Raes *et al.* 2005). YM1 is predominantly produced by macrophages (Chang *et al.* 2001;Welch *et al.* 2002;Nair *et al.* 2003) . In murine models of Th2-type parasite infections, expression of both YM-1 and AMCase depends on the activity of IL-21 receptor, which has structural homology to IL-4Rα chain and responds to the Th2 cytokine IL-21 (Pesce *et al.* 2006).

Since both YM1 and AMCase are induced in Th2 environment, several researchers consider human AMCase to be a functional homolog of murine YM1. However, absence of chitin-binding domain and catalytic aminoacids within Glyco_18 domain of YM1 (see Figure 1) argues against this point of view (Boot *et al.* 2005). Moreover, in murine experimental model of asthma, YM1 and AMCase were shown to be useful markers for distinguishing between proximal and distal airway epithelium. AMCase is expressed by non-mucus-producing CCSP- expressing cells of the distal airway, and YM-1 is expressed by the mucus-producing cells of the proximal airways (Homer *et al.* 2006).

We have recently demonstrated, that human macrophages produce one more catalytically inactive Glyco_18 domain containing protein, SI-CLP (stabilin-1 interacting chitinase-like protein) (Kzhyshkowska *et al.* 2006b). The expression of SI-CLP is induced by Th2 cytokine IL-4 and glucocorticoid dexamethasone in human primary macrophages (Figure 3). Combined IL-4/dexamethasone stimulation has a synergistic effect and is impaired in the presence of prototype Th1 cytokine IFNγ. In vivo, high amounts of SI-CLP were detected in bronchoalveolar lavage cells of patients with chronic airway inflammation (Figure 4). SI-CLP protein expression is not restricted to primary macrophages and is found in Raji cells, Jurkat cells, various tumour cell lines (Kzhyshkowska *et al.* 2006b), as well as in CD3+ T-cells isolated from peripheral blood of healthy donors (Kzhyshkowska, unpublished data). We were able to show that human macrophages secrete SI-CLP (see below) however it is currently unknown whether other cells can secrete SI-CLP.

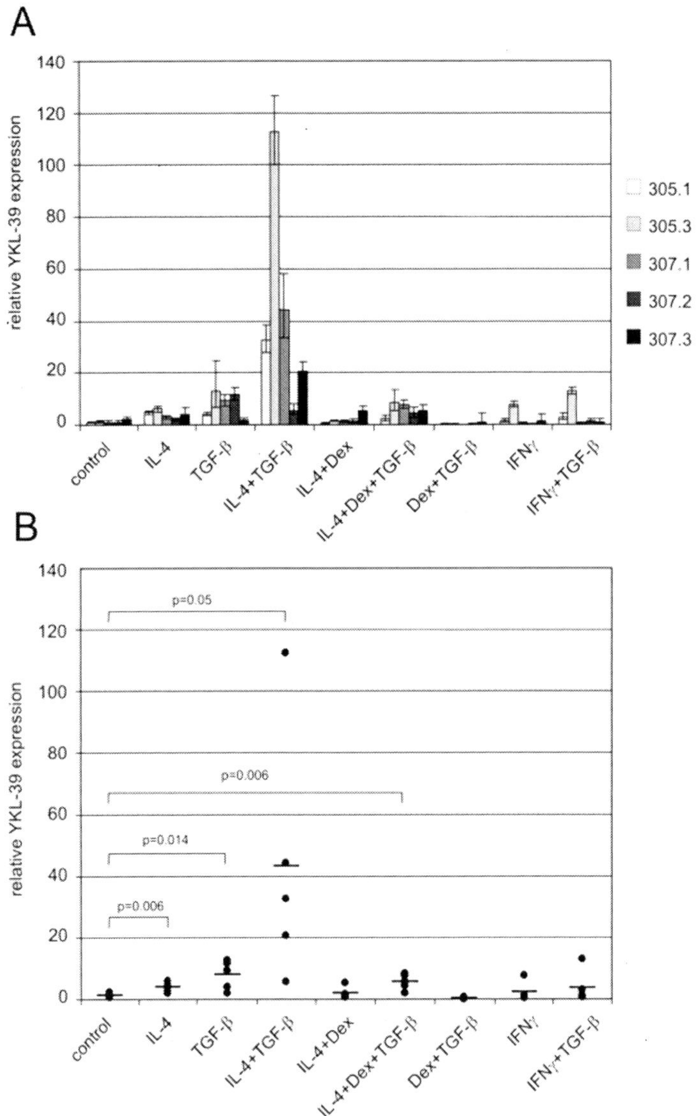

Figure 3. (A) Analysis of YKL-39 mRNA expression in primary human monocyte-derived macrophages. YKL-39 levels were normalized to the GAPDH mRNA expression. Five individual donors are presented. Normalized expression level in control sample of donor 305.1 was taken as a 1. Stimulations were performed for 6 days in X-vivo medium. (B) Statistical analysis of the effects of different stimuli on the expression of YKL-39 in primary human monocyte derived macrophages. Dots represent expression levels obtained for individual donors. Mean values are indicated as lines. P-values were obtained using ANOVA. This research was originally published in Bimarker Insights, Gratchev *et al.* 2008 (Gratchev *et al.* 2008)

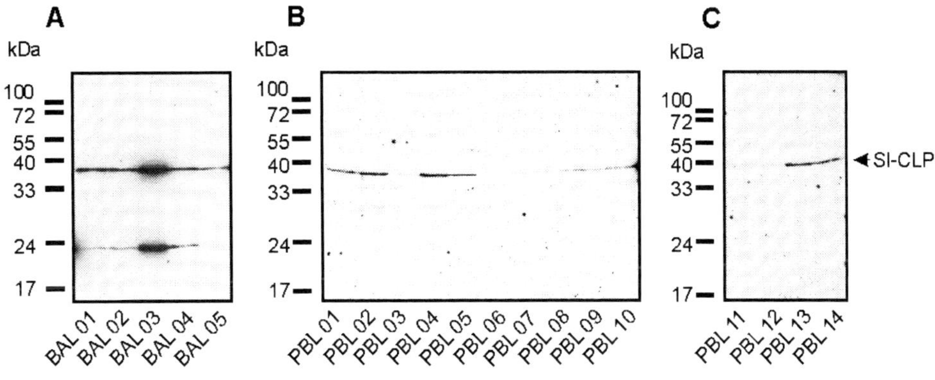

Figure 4. 1C11 monoclonal antibody detects SI-CLP in bronchoalveolar lavage samples and peripheral blood leukocytes. Western blot analysis was performed using 1C11 mAb. (A) Different pattern of SI-CLP expression was detected in cells isolated out of bronchoalveolar lavage (BAL). BAL 1, 2, 4, and 5 are obtained from patients with chronic bronchitis; BAL 3 was obtained from patient with sarcoidosis, undergoing corticoid therapy; (B) and (C) Peripheral blood leukocytes were isolated from fresh blood samples. PBL samples 01-05 correspond to BAL samples 01-05; PBL samples 06-14 are obtained from healthy donors. This research was originally published in Blood. Kzhyshkowska, J et al. Novel stabilin-1 interacting chitinase-like protein (SI-CLP) is up-regulated in alternatively activated macrophages and secreted via lysosomal pathway Blood. 2006; 107:3221-3228. © the American Society of Hematology.

4. Mechanism of Intracellular Sorting of Chitinase-Like Proteins in Human Macrophages

Two major cell types of innate immune system – macrophages and neutrophils use regulated secretory pathway to release chitinases and chitinase-like proteins. While neutrophils use specialised secretory granula, macrophages utilise lysosomal type of secretion. Mechanism of intracellular sorting was studies in details only for SI-CLP, but not for other chitinase-like proteins. Therefore, secretion pathways of true chitinase-chitotriosidase will be discussed below.

Intracellular localisation studies demonstrated that chitotriosidase and SI-CLP are primarily localised in lysosomes in macrophages (Renkema *et al.* 1997; Kzhyshkowska *et al.* 2006b). Besides the classical, constitutively operating ER/Golgi secretory pathway macrophages use non-classical (Nickel 2003) and lysosomal secretory pathways (Andrews 2000; Stinchcombe *et al.* 1999). The lysosomal secretion route in macrophages is regulated by specific sorting of newly synthesised products into secretory lysosomes (Logan *et al.* 2003). Two well-investigated mannose-6-phosphate receptors CI-MPR and CD-MPR are ubiquitously expressed and are responsible for the constitutive delivery of lysosomal enzymes to the lysosomes in numerous cell types (Ghosh *et al.* 2003; Kornfeld *et al.* 1989). However Glyco_18 domain containing proteins differ from other lysosomal proteins by the lack of N-glycosylation which is necessary for the recognition by MPRs (Renkema *et al.* 1997; Boot *et al.* 2001). Thus the receptor-mediated recognition and lysosomal routing of Glyco_18 containing proteins should differ from the classical MPR-mediated one.

We have recently demonstrated, that intracellular sorting of SI-CLP in alternatively activated macrophages is mediated by the multifunctional receptor stabilin-1. Stabilin-1 is specifically expressed on macrophages in healthy adult tissues, in placenta, on tumour-associated macrophages as well as on sinusoidal endothelial cells in liver, spleen and lymph nodes (reviewed (Kzhyshkowska *et al.* 2006a)). Both stabilin-1-positive macrophages and sinusoidal endothelial cells are characterised by the increased scavenging potential. First function which we have identified for stabilin-1 was endocytosis. We showed that stabilin-1 mediates internalization of acLDL (Kzhyshkowska *et al.* 2005), regulator of ECM-remodeling and cell adhesion SPARC (Kzhyshkowska *et al.* 2006c), and hormone placental lactogen (PL) (Kzhyshkowska *et al.* 2008). Both SPARC and acLDL are targeted via stabilin-1 for the degradation in lysosomes. In contrast, a portion of PL escapes degradation and is transported through the trans-Golgi network (TGN) to the novel storage vesicles. Stored PL can be secreted back to the extracellular space.

We also found that stabilin-1 shuttles between TGN and endosomes and interacts with sorting clathrin adaptors GGAs which mediate shuttling of MPRs between endosomes and TGN (Kzhyshkowska *et al.* 2004). Searching for an intracellular ligand which can be sorted by stabilin-1, we performed yeast two-hybrid screening and found novel Glyco_18 domain containing protein which we named SI-CLP for the stabilin-1-interacting chitinase-like protein (Kzhyshkowska *et al.* 2006b). Biochemical analysis revealed that SI-CLP interacts with fasciclin 7 domain (F7) of stabilin-1 which can be exposed to the extracellular or intra-vesicular space. Intracellular localisation studies performed in primary human macrophages revealed that SI-CLP can be recognised by stabilin-1 in the trans-Golgi network and delivered to the late endosomes and consequently into Lamp1-positive and secretion-committed CD63-positive lysosomes (Figure 5). Additional confirmation that stabilin-1 mediates intracellular sorting of SI-CLP was obtained in a model cellular system. H1299 cells which have no detectable level of endogenous SI-CLP protein and do not express stabilin-1 were stably transfected with SI-CLP-FLAG expressing construct. Overexpressed recombinant SI-CLP was artificially sorted to the nucleus and stored in the globular structures. Transient expression of stabilin-1 resulted in the complete re-localisation of SI-CLP into the cytoplasm (Figure 5). Final confirmation that stabilin-1 is involved in the sorting of SI-CLP was obtained by downregulation of stabilin-1 in primary human macrophages using siRNA. Macrophages differentiated the presence of IL-4 and dexamethasone and treated with satbilin-1 siRNA showed decreased ability to sort SI-CLP into lysosomes. Using long-term cultures of human primary macrophages we showed that SI-CLP can be released into conditioned medium, and dexamethasone is able to block this release (Figure 5) (Kzhyshkowska *et al.* 2006b).

Many questions concerning the role of stabilin-1 in the intracellular routing of chitinase-like protein remain to be resolved. It is currently unknown whether stabilin-1 can mediate lysosomal routing of other than SI-CLP chitinase-like proteins, and which other intracellular receptors can co-operate with stabilin-1 during this process. Another intriguing question is how endocytic and intracellular sorting functions of stbailin-1 communicate inside of macrophages, in other words, how stabilin-1 mediated endocytosis of acLDL or SPARC will affect delivery of newly synthesised SI-CLP from biosynthetic compartment to the secretory lysosomes. The complexity of stabilin-1 trafficking is schematically presented on Figure 6 (Kzhyshkowska *et al.* 2007b).

Synthesis, processing and intracellular sorting of human chitotriosidase was examined in details in primary monocyte derived macrophages, which were differentiated in culture in the absence of Th1 or Th2 cytokines (Renkema *et al.* 1997). Immunoelectron microscopy and subcellular fractionation demonstrated that chitotriosidase is sorted into lysosomes. Chitotriosidase was detected in lysosomal vesicles comparably to cathepsin D, and chitotriosidase activity was found in fractions positive for lysosomal enzyme beta-hexoaminidase. Chitotriosidase was shown to be synthesised as a 50 kDa protein and partially processed into a 39 kDA form. The 39-kDa isoform was predominantly found in dense core fractions of mature lysosomes, while the 50-kDa form was found in fractions of lower density, which were considered by the authors to contain pre-lysosomal vesicles. In the blood stream, the 50 kDa isoform was predominant, while the 39-kDa isoform was abundant in the tissues (Renkema *et al.* 1997). At this point it can be also hypothesised that the 50-kDa form was sorted into a specific subpopulation of secretory–committed lysosomes. It is currently unknown which intracellular receptors mediate delivery of distinct isoforms of chitotriosidase into distinct secretory compartments. Peripheral blood monocytes differentiated in culture into mature macrophages secrete both chitotriosidase and YKL-40 (Renkema *et al.* 1998). However no intracellular localisation studies were performed up to date to define whether YKL-40 and chitotriosidase are sorted into the same vesicles in macrophages.

In neutrophils, both chitotriosidase and YKL-40 are sorted into specific granules (Volck *et al.* 1998; Nordenbaek *et al.* 1999; van Eijk *et al.* 2005).

Differential induction of lysosomal or secretory granula secretion in polymorphonuclear neutrophils (PMN) demonstrated that chitotriosidase is not sorted into lysosome-like azurophilic granules. Using immunogold double labelling experiments, chitotriosidase was detected in specific lactoferrin-containing granules in PMN (van Eijk *et al.* 2005). Similarly, YKL-40 was found to colocalize and comobilize with the most abundant protein of specific granules - lactoferrin but not with gelatinase in subcellular fractionation studies on stimulated and unstimulated neutrophils (Volck *et al.* 1998). Double-labelling immunoelectron microscopy further confirmed the colocalization of YKL-40 and lactoferrin in specific granules of neutrophils. It is suggested that neutrophil-released YKL-40 acts as an autoantigen in rheumatoid arthritis (RA). Moreover release of YKL-40 from specific neutrophil granules was suggested to lead to the post transfusional complications and can be prevented by the prestorage leukocyte depletion by filtration of whole blood (Cintin *et al.* 2001). Further question which has to be addressed experimentally is whether chitotriosiadase and YKL-40 can be secreted simultaneously by the same PMNs, and what kind of biological effects can be regulated by the co-operative action of these two proteins.

The current picture of the intracellular localisation and regulation of chitinase and chitinase-like proteins is far from being complete. Especially important are questions, whether these proteins are sorted into specific subclasses of lysosomes in macrophages, and what are the specific intracellular receptors which define specificity of intracellular routing of newly synthesised chitinase-like proteins to the secretory pathways. The mechanisms of the chitinase-like protein secretion in the cells of non-myeloid origin, for example fibroblasts epithelial and cancer cells have to be investigated in the future.

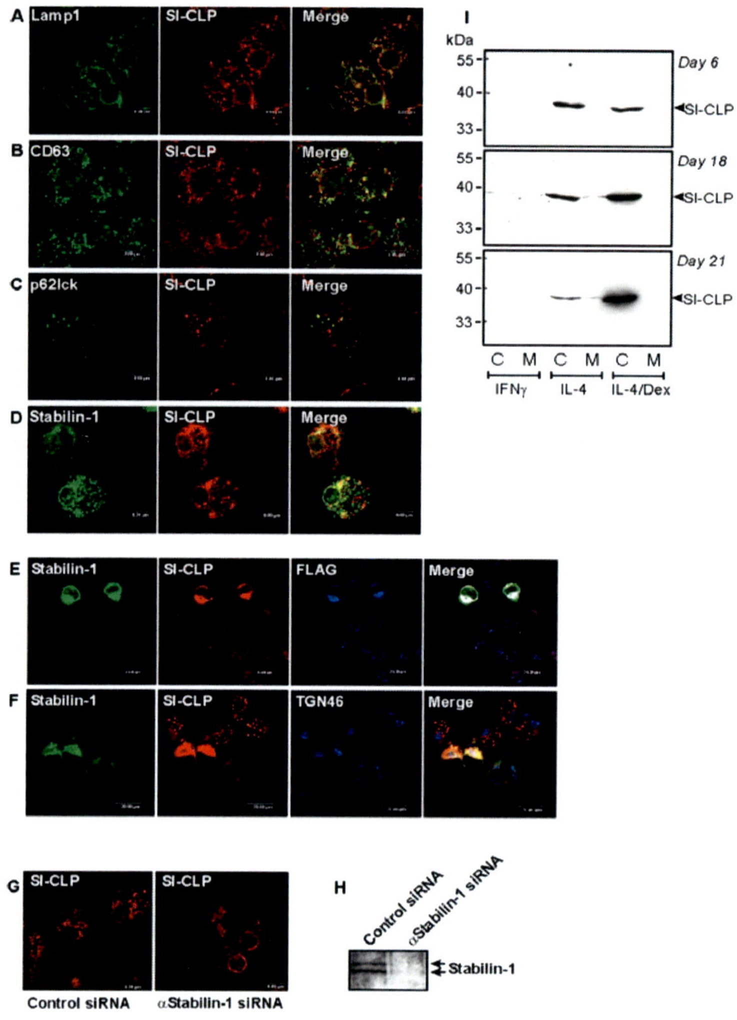

Figure 5. Intracellular sorting of SI-CLP in human macrophages. Endogenous SI-CLP was detected with rat mAb 1C11 both in immunofluorescent/confocal microscopy and Western blot analysis. Endogenous stabilin-1 was detected with rabbit polcyclonal F4 antibody. (A) SI-CLP is strongly co-localises with Lamp1; (B) SI-CLP strongly co-localises with CD63, marker for secretory lysosomes; (C) SI-CLP is occasionally found in p62lck positive late endosomes; (D) SI-CLP partially co-localises with stabilin-1 in TGN, but is absent from stabilin-1 positive early endosomes;.(E) and (F) Recombinant FLAG-tagged SI-CLP is artificially sorted in globular nuclear structures in stably transfected H1299 cells. Transient overexpression of stabilin-1 results in re-localisation of SI-CLP into the cytoplasm, where stabilin-1 and SI-CLP partially co-localise with TGN46. (G) $M_{IL-4/Dex}$ were transfected with control siRNA and stabilin-1 siRNA. The decreased sorting into lysosomes and abnormal concentration in nuclear rim structures was detected in part of cells transfected with stabilin-1 siRNA; (H) Stabilin-1 protein expression in $M_{IL-4/Dex}$ is efficiently suppressed by stabilin-1 siRNA; (G) Western blot analysis of SI-CLP secretion in long-term macrophage cultures. SI-CLP is detected in conditioned medium of IL-4–stimulated macrophages after 18 and 21 days of stimulation. Co-stimulation with dexamethasone results in intracellular accumulation of SI-CLP and blocks its secretion. This research was originally published in Kzhyshkowska, J *et al.* Novel stabilin-1 interacting chitinase-like protein (SI-CLP) is up-regulated in alternatively activated macrophages and secreted via lysosomal pathway. Blood. 2006; 107:3221-3228. © the American Society of Hematology.

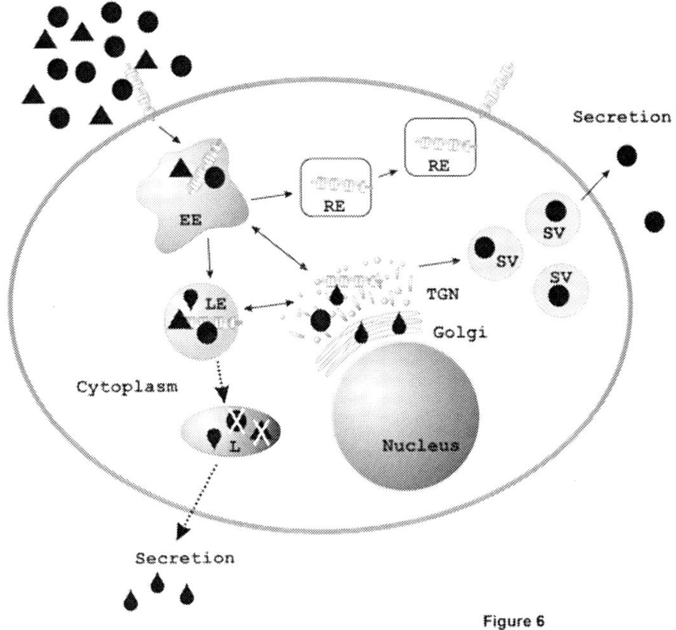

Figure 6. Complex trafficking pathways of stabilin-1 and its ligands. Stabilin-1 recognises extracellular endocytic ligands SPARC (shown as a filled triangles) and placental lactogen (show as a filled circles), as well as newly synthesised intracellular sorting ligand SI-CLP (filled drops). Upon binding to the surface-expressed stabilin-1, endocytic ligands are internalized and delivered to early/sorting endosomes (EE). A portion of the ligand-free receptor can recycle back to the cell surface via recycling endosomes (RE). Stabilin-1 targets both SPARC and placental lactogen to late endosomes (LE), and consequently for the degradation in lysosomes (L). Part of placental lactogen escapes degradation and is delivered by stabilin-1 to the trans-Golgi network (TGN). In TGN placental lactogen dissociates from stabilin-1, is further transported to the new type of storage vesicles (SV), and can be secreted to the extracellular space. Stabilin-1 is also involved in intracellular sorting process; it shuttles between endosomes and TGN. We propose a model whereby newly synthesized SI-CLP is recognized by stabilin-1 in the late Golgi compartment and delivered late endosomes. In LE SI-CLP dissociates from stabilin-1, is transported to the lysosomes (L). Lysosme, inturn, can undergo stimuli-dependent secretion (from (Kzhyshkowska *et al.* 2007b))

5. Biological Activity of YKL-39, YKL-40 and YM1/2

Biological activities of two human chitinase-like proteins, YKL-40 and YKL-39, were demonstrated experimentally and will be discussed here. Despite the absence in human, murine YM1/2 proteins have remarkable biological effects. Similarity in domain organisation, expression profile and biochemical parameters suggest that human SI-CLP can be a functional homologue of rodent YM1 protein. Thus, data about biological effects of YM1/2 are also included in this section.

5.1. YkL-40

Biological activities of YKL-40 include regulation of cell proliferation, adhesion, migration and activation of various cell types. YKL-40 at nanomolar concentrations promotes the growth of human synovial cells, chondrocytes, skin and foetal lung fibroblasts. The proliferative effect of YKL-40 synergizes with the effect of insulin-like growth factor-1 (Recklies *et al.* 2002). Mitogenic response to YKL-40 depends on the activity of both mitogen-activated (MAP) kinase and protein kinase B/Akt signalling. At the same time YKL-40 is able to suppress the TNFα and IL-1-indused secretion of matrix metalloproteases and IL-8 in both human skin fibroblasts and articular chondrocytes. Thus, YKL-40 is able to promote the proliferation and antagonizes catabolic or degradation processes during the inflammatory response of connective tissues (Ling *et al.* 2004). *Cg*-Clp1, the conserved molluscan homologue of YKL-40, stimulates proliferation and regulates synthesis of extracellular matrix components in rabbit articular chondrocytes (Badariotti *et al.* 2006). Porcine homologue of YKL-40, gp38k (CHI3L1), was shown to have cell-type specific effect on cell migration. It induced the migration of vascular smooth muscle cells (VSMC), but not fibroblasts. Moreover, gp38k promoted the attachment and spreading of VSMC (Nishikawa *et al.* 2003). Recently, the existence of three isoforms YKL-40 was reported: major and minor forms were found in resorbing cartilage and a third isoform was detected in chondrocytes (Bigg *et al.* 2006). Affinity chromatography experiments with purified YKL-40 demonstrated specific binding of all three isoforms to collagens types I, II, and III. The chondrocyte-derived YKL-40 isoform was able to prevent collagenolytic cleavage of type I collagen and to stimulate the rate of type I collagen fibril formation. By contrast, the cartilage major form had an inhibitory effect on type I collagen fibrillogenesis. Differential biological activities of various YKL-40 isoforms indicate that expression of specific isoforms has to be taken into consideration in the epidemiological and functional studies.

The ability of YKL-40 to regulate cell proliferation, adhesion, migration, and activation, as well as to regulate extracellular matrix assembly, correlates well with elevated levels of YKL-40 in the sites of chronic inflammation and active connective tissue turnover. Interesting question is, whether differentially activated macrophages can express various isoforms of YKL-40, and how YKL-40 secreted by different cell types contributes to the pathological situations in human body. For more information about YKL-40 see Chapter XIII in this book.

5.2. YKL-39

Two biological activities of YKL-39 are suggested to contribute to the disease progression. One is the induction of autoimmune response (Sekine *et al.* 2001; Tsuruha *et al.* 2002; Du *et al.* 2005), and second is the participation in tissue remodelling. Immunisation with purified YKL-39 induced arthritis in different strains of mice (Sakata *et al.* 2002). Histological examination revealed synovial proliferation and irregularity of the cartilage surface in BALB/c mice. After injection of purified YKL-39, not only anti-YKL-39 antibody was detected, but also the antibody against type II collagen, suggesting the spreading of autoimmune reactions (Sakata *et al.* 2002). While role of YKL-39 as an inducer of autoimmunity is supported by the experimental data in animal models, its function in tissue remodelling was suggested on the basis of its high level in chondrocyte cultures and close homology to YKL-40 that was shown to induce cell proliferation and migration. Thus biological activity of YKL-39 needs further investigation in vivo and in vitro.

5.3 YM1 (ECF-L) and YM2

Expression of YM1, similarly to AMCase and SI-CLP, is upregulated during Th2-driven immunological reactions. YM2 is a very close homologue of YM1, and its expression depends on interleukin (IL)-4 and IL-13 signal transduction (Webb *et al.* 2001). YM1/2 are mainly associated with two pathological situations: parasite infections and lung disorders. It was shown in the experimental model of murine trypanosomosis, that YM1 is expressed at the late stage of the infection characterised by the conversion of classical macrophage activation to the alternative one (Raes *et al.* 2002). Expression of YM1 is a generalised feature of nematode infection, and YM1 together with murine AMCase are highly upregulated in lung of mice infected with *N. brasiliensis* (Nair *et al.* 2005). During the course of allergic peritonitis, macrophages secrete YM1 in IL-4 and STAT6-dependent manner (Welch *et al.* 2002). YM2 is strongly upregulated in the lung of BALB/c mice during OVA-induced allergic airways inflammation (animal model of asthma), where overproduction of YM2 was dependent on CD4+ T-cells and on signalling of IL-4 and IL-13 through the IL-4Rα subunit (Webb *et al.* 2001). Further, polymorphism in IL-4Rα in mice was shown to correlate with YM2 protein expression as well as with airways hypersensitivity and eosinophilia (Webb *et al.* 2001).

Both YM1 and YM2 were found to be strongly upregulated in a murine model of proliferative dermatitis, which is characterised by accumulation of eosinophils (Hogenesch *et al.* 2006). YM1 is crystallised in lung of motheaten mice characterised by hyperreactivity of alveolar macrophages and consistent formation of intrapulmonary eosinophilic crystals (Guo *et al.* 2000). Crystals of YM1 were also found within the aged lung at sites of chronic inflammation in the murine model of chronic granulomatosis disease (CGD) (Harbord *et al.* 2002) and in bronchoalveolar lavages of Abcg1(-/-) mice (knock-out mice which luck ABCG1, a member of the ATP-binding cassette transporter superfamily) characterised by progressive and chronic inflammation in lung accompanied by the lipidosis and elevated numbers of foamy macrophages and leukocytes in bronchoalveolar lavages (Baldan *et al.* 2008). These data suggested that cholesterol loading of macrophages induces YM1/2 production.

YM1, also named ECF-L for the Eosinophil chemotactic factor-L, purified out of supernatant of splenocytes of C57BL6 mice was demonstrated to possess chemotactic activity towards eosinophils, T lymphocytes and polymorphonuclear leukocytes in vitro, and to cause selective extravasation of eosinophils in vivo (Owhashi *et al.* 2000). Induction of YM-1 (ECF-L) expression in alveolar macrophages was observed soon after allergen exposure but before the onset of airway inflammation in the murine model of airway hyperresponsiveness (AHR) (Iwashita *et al.* 2006). The fact that intratracheal administration of an adenoviral vector expressing antisense ECF-L RNA resulted in the suppression of eosinophil infiltration and AHR indicated that biological activity of ECF-L is critical for the pathogenesis of allergic inflammation and bronchial asthma.

In addition to its chemotactic activity toward immune cells, YM-1 (ECF-L) acts as an osteoclast (OCL) stimulating factor (Oba *et al.* 2003). It was shown to be a potent mediator of OCL formation which enhances the effects of RANKL by stimulation of expression of ICAM-1 and LFA-1 (Garcia-Palacios *et al.* 2007; Garcia-Palacios *et al.* 2006). In conclusion, YM1 protein participates in Th2-drived allergic processes and functions as a cellular differentiation factor. Role of SI-CLP, which we presume is a functional analogue of YM1, in chemotaxis of various immune cells and in cellular differentiation is in focus of our current research.

6. Chitinase-Like Proteins as Biomarkers of Inflammation and Cancer

Mammalian chitinases and chitinase-like proteins are secreted into the extracellular space and can be detected in the tissues and in blood circulation. They possess numerous of cell physiological and immunomodulatory activities. Accumulating data point toward connection of their biological activities with various human diseases.

6.1. YKL-40

YKL-40, also called human cartilage glycoprotein-39 (HC gp-39) and Chitinase-3-like-1 (CHI3L1), is the best investigated human chitinase-like protein regarding its association with various disorders. Several comprehensive reviews have been published recently which address in details association of YKL-40 with human disorders, methods of YKL-40 detection and evaluation of YKL-40 as a diagnostic and prognostic factor (Johansen *et al.* 2006; Rathcke *et al.* 2006b; Johansen *et al.* 2007). Here we summarise major pathologies associated with elevated levels of YKL-40. For more information about YKL-40 see Chapter XIII in this book.

Circulating YKL-40 can be detected by non-invasive methods in human serum or plasma using in-house RIA (Johansen *et al.* 1993; Johansen *et al.* 2006) as well as by commercially available ELISA (Quidel, Santa Clara, CA) (Harvey *et al.* 1998). Increased concentrations of YKL-40 were detected not only in sites of inflammation, but also in serum of patients with rheumatoid arthritis (RA). Several independent groups demonstrated that the elevated levels of YKL-40 in serum reflect the degree of the synovial inflammation and joint destruction in patients with RA and OA (Johansen *et al.* 1999; Johansen *et al.* 2001; Matsumoto *et al.* 2001; Peltomaa *et al.* 2001; Conrozier *et al.* 2000). Elevated level of YKL-40 is a marker for joint involvement in inflammatory bowel disease (IBD) where rheumatic symptoms are also common (Bernardi *et al.* 2003; Vind *et al.* 2003). Plasma levels of both YKL-40 and high sensitive C-reactive protein were found to be related to insulin resistance in patients with type 2 diabetes mellitus in an independent way (Rathcke *et al.* 2006a). YKL-40 was suggested to be an independent biomarker for the inflammatory/atherosclerotic processes in patients with type 2 diabetes mellitus (Rathcke *et al.* 2006b; Rathcke *et al.* 2006a).

YKL-40 was also found to be associated with the lung pathologies - pulmonary sarcoidosis (Johansen *et al.* 2005b) and asthma (Chupp *et al.* 2007). Both macrophages and giant cells in pulmonary sarcoid granuloma express YKL-40, and serum levels of YKL-40 are indicative for sarcoid disease activity and ongoing fibrosis (Johansen *et al.* 2005b). Serum levels of YKL-40 were found to correlate positively with severity of asthma, in particular with frequency of rescue-inhaler use, oral corticoid use, and rate of hospitalisation (Chupp *et al.* 2007). Increased lung levels of YKL-40 correlated with its levels in the circulation. In bronchial biopsy samples obtained from asthmatic patients, YKL-40 was elevated in subepitelial cells, and YKL-40 was detected in macrophages and neutrophils in broncoalveilar-lavage samples (Chupp *et al.* 2007). Genomewide association study of serum YKL-40 levels, implicated SNP and asthma revealed that CHI3L1 (YKL-40 coding gene) is a susceptibility gene for asthma, bronchial hyperresponsiveness, and reduced lung function (Ober *et al.* 2008). Serum and spinal fluid levels of YKL-40 are elevated in patients with pathogen-induced inflammation, including purulent meningitis, pneumonia and endotoxemia caused by endotoxin of *E.coli* (Ostergaard *et al.* 2002; Nordenbaek *et al.* 1999; Kronborg *et al.* 2002; Johansen *et al.* 2005a). In both meningitis and pneumonia, YKL-40 is secreted by locally activated macrophages (Ostergaard *et al.* 2002) and neutrophils (Nordenbaek *et al.* 1999), and was proposed as a specific supplementary serological marker for the activation of granulocytes and macrophages in inflamed tissues (Rathcke *et al.* 2006b).

Increased production of YKL-40 is indicative for liver pathology. YKL-40 is differentially upregulated in cirrhotic liver on the end-stage of hepatitis C virus (HCV) induced liver cirrhosis (Shackel *et al.* 2003). Serum levels of YKL-40 correlate with YKL-40 mRNA expression in liver (Kamal *et al.* 2006), therefore YKL-40 was suggested to be an useful non-invasive marker for evaluation of the degree of fibrosis as well as for the efficiency of therapy in patients with HCV-associated liver disorders (Saitou *et al.* 2005). Increased plasma levels of YKL-40 were also suggested to reflect the progression of liver fibrosis in alcoholics (Tran *et al.* 2000).

Elevated level of YKL-40 in the circulation was found in number of solid tumors including breast cancer, colorectal cancer, ovarian cancer, glioblastoma, metastatic renal and prostate cancer and malignant melanoma (reviewed in (Johansen *et al.* 2006)). Serum levels of YKL-40 are indicative for the poor prognosis and efficiency of metastatic process. For example, increased plasma concentration of YKL-40 is related to poor prognosis and shorter survival in patients with ovarian cancer (Hogdall *et al.* 2003), colorectal carcinoma (Cintin *et al.* 2002), metastatic prostate carcinoma (Brasso *et al.* 2006), and melanoma (Schmidt *et al.* 2006a; Schmidt *et al.* 2006b). Expression of YKL-40 is characteristic for glioblastoma – highly malignant glioma characterised by invasive phenotype (Colin *et al.* 2006). Study aimed to assess the efficiency of vascular endothelial growth factor (VEGF) inhibitors as anti-tumour treatment revealed, that human glioma cells transfected with short-interfering RNAs against VEGF-A and implanted on the chick chorio-allantoic membrane expressed both YKL-40 (CHI3L1) and YKL-39 (CHI3L2). The authors concluded that YKL-40 may be useful as new prognostic markers and new therapeutic target (Saidi *et al.* 2008). YKL-40 was recently proposed as a novel marker for the detection of endometrial cancer (Diefenbach *et al.* 2006). However, in most solid tumours, the serum concentrations of YKL-40 do not show high sensitivity for identification of primary cancer, and determination of YKL-40 cannot be used as a single screening marker for diagnosis of cancer (Johansen *et al.* 2006; Kzhyshkowska *et al.* 2007a). In addition there is no direct correlation between the serum YKL-40 and various prostate specific antigens (Brasso *et al.* 2006). Out of six necessary criteria of "tumor marker utility system" (Hayes *et al.* 1996; Hayes 1998), YKL-40 is positive only for three, and therefore is currently considered to be investigational (Johansen *et al.* 2006). It is suggested, that differential levels of YKL-40 reflect difference in biology of cancer cells which produce YKL-40 or do not produce YKL-40. At the same time, peritumoral macrophages in human small lung cancer (SCLC) were shown to be the predominant source of YKL-40 in patients' serum (Junker *et al.* 2005). It has to be taken into consideration that many solid tumours consist not only out of tumour cells, but also out of stromal and immune cells, where tumour-associated macrophages can be predominant. Since macrophages are one of the major sources for YKL-40, differential levels of YKL-40 in the circulation of cancer patients may reflect also level of activity and polarisation vector of tumour-associated macrophages.

6.2. YKL-39

YKL-39 was identified as an abundantly secreted protein in primary culture of human articular chondrocytes (Hu *et al.* 1996). YKL-39 accounted for 4% and YKL-40 for 33% of the secreted protein in chondrocyte-conditioned medium. Despite the high homology on the protein level (more than 50%), the radioimmunoassay developed for the detection of YKL-40 in serum was shown to be very specific and does not detect YKL-39 (Hu *et al.* 1996). YKL-39 is currently recognised as a biochemical marker more accurate than YKL-40 for the activation of chondrocytes and osteoarthritis progression in human (Knorr *et al.* 2003).

Comparison of the expression of YKL-39 and YKL-40 in osteoarthritic cartilage revealed, that YKL-39 mRNA is significantly upregulated in cartilage of patients with osteoarthritis versus normal subjects. Increased expression of YKL-39 mRNA correlated with the upregulation of collagen 2, while YKL-40 mRNA showed no significant upregulation in OA cartilage (Steck *et al.* 2002). Another study showed that normal human chondrocytes express mRNA for both YKL-39 and YKL-40. While the expression of YKL-39 was upregulated both in early degenerative and late stage osteoarthritis, the expression of YKL-40 was downregulated during the progression of osteoarthritis (Knorr *et al.* 2003). Recent proteomic analysis identified YKL-39, but not YKL-40 to be secreted by human osteoarthritic cartilage in culture (De Ceuninck *et al.* 2005), however the earlier study of Johansen et al demonstrated that isolated chondrocytes spontaneously release newly synthesised YKL-40. Moreover, both IL-1beta and TGFbeta were able to downregulate production is YKL-40 by chondorocytes. The spontaneous release of YKL-40 by chondorocytes in culture was suggested to reflect their response to changes in exracellular invoronment (Johansen J *et al.* 2001). Further in vivo evidence for the expression of YKL-40 in chondrocytes was obtained by Volck et al. This immunohistological study showed that in synovium YKL-40 can be detected in lining cells and stromal cells (macrophages), while in arthritic cartilage, YKL-40 was located to chondrocytes (Volck *et al.* 2001). Thus elevation of YKL-40 in serum and synovial fluid of patients with arthritis might result from its overproduction by synovial cell, by macrophages and/or by chondrocytes. Our recent studies revealed that YKL-39 can be also secreted by human peripheral blood derived macrophages under combined stimulation with Th2 cytokine IL-4 and multifunctional regulator of cell differentiation and tissue remodelling TGFβ (Gratchev *et al.* 2008). Thus, circulating levels of YKL-39 would reflect not only activity of chondrocytes, but also polarisation of macrophages highly abundant in sites of inflammation.

Functional analysis in animal models suggested the role of YKL-39 as an inducer of autoimmune processes related to arthritis (Sakata *et al.* 2002). Accordingly, antibodies to YKL-39 can be detected in human serum with ELISA and Western blotting, and were found in patients with rheumatoid arthritis (RA) and osteoarthritis (OA) (Sekine *et al.* 2001; Du *et al.* 2005;Tsuruha *et al.* 2002). Autoantibodies to YKL-39 were detected in 8-11,8% (Sekine *et al.* 2001;Tsuruha *et al.* 2002) of patients with RA, and in 11.1% (Tsuruha *et al.* 2002) of patients with OA, while only 1% of patients with RA had autoantibodies to YKL-40 (Sekine *et al.* 2001). The immune response to YKL-39 was independent of that to YKL-40. In patients with OA, the prevalence of autoantibodies to YKL-39 and other autoantigens on early stages of disease suggested, that the autoimmune response occurs during the initial phase of cartilage degeneration (Du *et al.* 2005).

6.3. SI-CLP

SI-CLP (stabilin-interacting chitinase-like proteins) is the most recent identified human Glyco_18 domain containing protein (Kzhyshkowska *et al.* 2006b). It was found as an interacting partner and sorting ligand for the multifunctional receptor stabilin-1, which is specifically expressed on subpopulations of tissue macrophages and sinusoidal endothelial cells (Kzhyshkowska *et al.* 2006a; Martens *et al.* 2006). In parallel with stabilin-1, expression of SI-CLP mRNA was strongly upregulated in macrophages by the Th2 cytokine IL-4 and by dexamethasone. We developed rat monoclonal antibody 1C11 recognising N-terminal epitope in SI-CLP. This epitope is located upstream of conservative Glyco_18 domain and has no similarity with sequences of other human Glyco_18 containing proteins. Using the 1C11 antibody we demonstrated that IL-4 and dexamethasone in combination increase SI-CLP protein levels in macrophages, the extent of which varied between donors. The 1C11 mAb recognised SI-CLP in the cellular fraction of bronchoalveolar lavage specimens obtained from patients with chronic inflammatory disorders of the respiratory tract and in PBLs from these patients as well as from healthy donors (Figure 4). Thus 1C11 mAb can be applied for the examination of association of SI-CLP with human disorders.

SI-CLP is the only chitinase-like protein which is upregulated by glucocorticoids (Kzhyshkowska *et al.* 2006b). Remarkably, the highest expression level of SI-CLP was found by us in a patient with sarcoidosis undergoing corticoid therapy. We observed strong differences in stimulatory effect of dexamethasone on SI-CLP production in macrophages obtained from different healthy individuals. These facts indicate that SI-CLP is a promising marker for the individual response to glucocorticoids and prediction of side effects of corticoid treatment.

Conclusions and Perspectives

Human chitinase-like lectins, YKL-40, YKL-39 and SI-CLP constitute novel class of extracellular mediators which combine properties of cytokines and growth factors. Many open questions about chitinase-like lectins remain to be addressed experimentally: what are the intracellular sorting machineries responsible for their selective delivery to the secretory pathway, can these proteins co-operate in terms of their biological activities or will they act antagonistically; what are the biological effects of chitinase-like lectins on various cell types, including neurons, and cellular components of innate and adaptive immunity. To answer the last question, primary experimental task is the identification of specific receptors which recognise chitinase-like lectins. Due to the high value of human chitinase-like proteins as biomarkers of various human pathologies, answering these basic questions will bring a great advantage for the understanding of molecular and cellular mechanisms for such life threatening human diseases like cancer, atherosclerosis and asthma. Chitinase-like lectins can be expressed by different cells types, however one of the major cellular types which release chitinase-like lectins in pathologies are macrophages and neutrophils. Thus elevated levels of chitinase-like proteins in tissues and circulation provide us with new highly valuable diagnostic tools for monitoring of various types of activation of innate immune system in human pathologies.

Acknowledgments

This work was supported by Margarete von Wrangell Habilitationprogram, Ministry of Science, Research and Art of Baden-Württemberg

References

Aguilera B, Ghauharali-van der Vlugt K, Helmond MT, Out JM, Donker-Koopman WE, Groener JE, Boot RG, Renkema GH, van der Marel GA, van Boom JH, Overkleeft HS, Aerts JM. Transglycosidase activity of chitotriosidase: improved enzymatic assay for the human macrophage chitinase. *J Biol Chem*. 2003; 278: 40911-6.

Andrews NW. Regulated secretion of conventional lysosomes. *Trends Cell Biol*. 2000; 10: 316-21.

Badariotti F, Kypriotou M, Lelong C, Dubos MP, Renard E, Galera P, Favrel P. The Phylogenetically Conserved Molluscan Chitinase-like Protein 1 (Cg-Clp1), Homologue of Human HC-gp39, Stimulates Proliferation and Regulates Synthesis of Extracellular Matrix Components of Mammalian Chondrocytes. *J Biol Chem*. 2006; 281: 29583-96.

Baldan A, Gomes AV, Ping P, Edwards PA. Loss of ABCG1 Results in Chronic Pulmonary Inflammation. *J Immunol.* 2008; 180: 3560-8.

Bernardi D, Podswiadek M, Zaninotto M, Punzi L, Plebani M. YKL-40 as a marker of joint involvement in inflammatory bowel disease. *Clin Chem.* 2003; 49: 1685-8.

Bigg HF, Wait R, Rowan AD, Cawston TE. The mammalian chitinase-like lectin, YKL-40, binds specifically to type I collagen and modulates the rate of type I collagen fibril formation. *J Biol Chem.* 2006; 281: 21082-95.

Boot RG, Blommaart EF, Swart E, Ghauharali-van der Vlugt K, Bijl N, Moe C, Place A, Aerts JM. Identification of a novel acidic mammalian chitinase distinct from chitotriosidase. *J Biol Chem.* 2001; 276: 6770-8.

Boot RG, Bussink AP, Aerts JM. Human acidic mammalian chitinase erroneously known as eosinophil chemotactic cytokine is not the ortholog of mouse YM1. *J Immunol.* 2005; 175: 2041-2.

Boot RG, Renkema GH, Strijland A, van Zonneveld AJ, Aerts JM. Cloning of a cDNA encoding chitotriosidase, a human chitinase produced by macrophages. *J Biol Chem.* 1995; 270: 26252-6.

Boot RG, van Achterberg TA, van Aken BE, Renkema GH, Jacobs MJ, Aerts JM, de Vries CJ. Strong induction of members of the chitinase family of proteins in atherosclerosis: chitotriosidase and human cartilage gp-39 expressed in lesion macrophages. *Arterioscler Thromb Vasc Biol.* 1999; 19: 687-94.

Brasso K, Christensen IJ, Johansen JS, Teisner B, Garnero P, Price PA, Iversen P. Prognostic value of PINP, bone alkaline phosphatase, CTX-I, and YKL-40 in patients with metastatic prostate carcinoma. *Prostate* 2006; 66: 503-13.

Chang NC, Hung SI, Hwa KY, Kato I, Chen JE, Liu CH, Chang AC. A macrophage protein, Ym1, transiently expressed during inflammation is a novel mammalian lectin. *J Biol Chem.* 2001; 276: 17497-506.

Chupp GL, Lee CG, Jarjour N, Shim YM, Holm CT, He S, Dziura JD, Reed J, Coyle AJ, Kiener P, Cullen M, Grandsaigne M, Dombret MC, Aubier M, Pretolani M, Elias JA. A chitinase-like protein in the lung and circulation of patients with severe asthma. *N Engl J Med.* 2007; 357: 2016-27.

Cintin C, Johansen JS, Christensen IJ, Price PA, Sorensen S, Nielsen HJ. High serum YKL-40 level after surgery for colorectal carcinoma is related to short survival. *Cancer* 2002; 95: 267-74.

Cintin C, Johansen JS, Skov F, Price PA, Nielsen HJ. Accumulation of the neutrophil-derived protein YKL-40 during storage of various blood components. *Inflamm Res.* 2001; 50: 107-11.

Colin C, Baeza N, Bartoli C, Fina F, Eudes N, Nanni I, Martin PM, Ouafik L, Figarella-Branger D. Identification of genes differentially expressed in glioblastoma versus pilocytic astrocytoma using Suppression Subtractive Hybridization. *Oncogene.* 2006; 25: 2818-26.

Colton CA, Mott RT, Sharpe H, Xu Q, Van Nostrand WE, Vitek MP. Expression profiles for macrophage alternative activation genes in AD and in mouse models of AD. *J Neuroinflammation.* 2006; 3: 27.

Connor JR, Dodds RA, Emery JG, Kirkpatrick RB, Rosenberg M, Gowen M. Human cartilage glycoprotein 39 (HC gp-39) mRNA expression in adult and fetal chondrocytes, osteoblasts and osteocytes by in-situ hybridization. *Osteoarthritis Cartilage.* 2000; 8: 87-95.

Conrozier T, Carlier MC, Mathieu P, Colson F, Debard AL, Richard S, Favret H, Bienvenu J, Vignon E. Serum levels of YKL-40 and C reactive protein in patients with hip osteoarthritis and healthy subjects: a cross sectional study. *Ann Rheum Dis.* 2000; 59: 828-31.

De Ceuninck F, Marcheteau E, Berger S, Caliez A, Dumont V, Raes M, Anract P, Leclerc G, Boutin JA, Ferry G. Assessment of some tools for the characterization of the human osteoarthritic cartilage proteome. *J Biomol Tech.* 2005; 16: 256-65.

Diefenbach CS, Shah Z, Iasonos A, Barakat RR, Levine DA, Aghajanian C, Sabbatini P, Hensley ML, Konner J, Tew W, Spriggs D, Fleisher M, Thaler H, Dupont J. Preoperative serum YKL-40 is a marker for detection and prognosis of endometrial cancer. *Gynecol Oncol.* 2006.

Du H, Masuko-Hongo K, Nakamura H, Xiang Y, Bao CD, Wang XD, Chen SL, Nishioka K, Kato T. The prevalence of autoantibodies against cartilage intermediate layer protein, YKL-39, osteopontin, and cyclic citrullinated peptide in patients with early-stage knee osteoarthritis: evidence of a variety of autoimmune processes. *Rheumatol Int.* 2005; 26: 35-41.

Elias JA, Homer RJ, Hamid Q, Lee CG. Chitinases and chitinase-like proteins in T(H)2 inflammation and asthma. *J. Allergy Clin Immunol. 2005*; 116: 497-500.

Fach EM, Garulacan LA, Gao J, Xiao Q, Storm SM, Dubaquie YP, Hefta SA, Opiteck GJ. In vitro biomarker discovery for atherosclerosis by proteomics. *Mol Cell Proteomics.* 2004; 3: 1200-10.

Fusetti F, Pijning T, Kalk KH, Bos E, Dijkstra BW. Crystal structure and carbohydrate-binding properties of the human cartilage glycoprotein-39. *J Biol Chem.* 2003; 278: 37753-60.

Fusetti F, von Moeller H, Houston D, Rozeboom HJ, Dijkstra BW, Boot RG, Aerts JM, van Aalten DM. Structure of human chitotriosidase. Implications for specific inhibitor design and function of mammalian chitinase-like lectins. *J Biol Chem.* 2002; 277: 25537-44.

Garcia-Palacios V, Chung HY, Choi SJ, Kurihara N, Lee JW, Ehrlich LA, Collins R, Roodman GD. Eosinophil chemotactic factor-L (ECF-L) enhances osteoclast formation by increasing ICAM-1 expression. *Ann NY Acad Sci.* 2006; 1068: 240-3.

Garcia-Palacios V, Chung HY, Choi SJ, Sarmasik A, Kurihara N, Lee JW, Galson DL, Collins R, Roodman GD. Eosinophil chemotactic factor-L (ECF-L) enhances osteoclast formation by increasing in osteoclast precursors expression of LFA-1 and ICAM-1. *Bone.* 2007; 40: 316-22.

Ghosh P, Dahms NM, Kornfeld S. Mannose 6-phosphate receptors: new twists in the tale. *Nat Rev Mol Cell Biol.* 2003; 4: 202-12.

Gratchev A, Schmuttermaier C, Mamidi S, Gooi L, Goerdt S, Kzhyshkowska J. Expression of Osteoarthritis Marker YKL-39 is Stimulated by Transforming Growth Factor Beta (TGF-beta) and IL-4 in Differentiating Macrophages. *Biomarker Insights.* 2008; 39-44.

Guo L, Johnson RS, Schuh JC. Biochemical characterization of endogenously formed eosinophilic crystals in the lungs of mice. *J Biol Chem* .2000; 275: 8032-7.

Hakala BE, White C, Recklies AD. Human cartilage gp-39, a major secretory product of articular chondrocytes and synovial cells, is a mammalian member of a chitinase protein family. *J Biol Chem.* 1993; 268: 25803-10.

Harbord M, Novelli M, Canas B, Power D, Davis C, Godovac-Zimmermann J, Roes J, Segal AW. Ym1 is a neutrophil granule protein that crystallizes in p47phox-deficient mice. *J Biol Chem.* 2002; 277: 5468-75.

Harvey S, Weisman M, O'Dell J, Scott T, Krusemeier M, Visor J, Swindlehurst C. Chondrex: new marker of joint disease. *Clin Chem.* 1998; 44: 509-16.

Hayes DF. Determination of clinical utility of tumor markers: a tumor marker utility grading system. *Recent Results Cancer Res.* 1998; 152: 71-85.

Hayes DF, Bast RC, Desch CE, Fritsche H, Jr., Kemeny NE, Jessup JM, Locker GY, Macdonald JS, Mennel RG, Norton L, Ravdin P, Taube S, Winn RJ. Tumor marker utility grading system: a framework to evaluate clinical utility of tumor markers. *J Natl Cancer Inst.* 1996; 88: 1456-66.

Hogdall EV, Johansen JS, Kjaer SK, Price PA, Christensen L, Blaakaer J, Bock JE, Glud E, Hogdall CK. High plasma YKL-40 level in patients with ovarian cancer stage III is related to shorter survival. *Oncol Rep.* 2003; 10: 1535-8.

Hogenesch H, Dunham A, Seymour R, Renninger M, Sundberg JP. Expression of chitinase-like proteins in the skin of chronic proliferative dermatitis (cpdm/cpdm) mice. *Exp Dermatol.* 2006; 15: 808-14.

Homer RJ, Zhu Z, Cohn L, Lee CG, White WI, Chen S, Elias JA. Differential expression of chitinases identify subsets of murine airway epithelial cells in allergic inflammation. *Am. J Physiol Lung Cell Mol Physiol. 2006*; 291: L502-L511.

Houston DR, Recklies AD, Krupa JC, van Aalten DM. Structure and ligand-induced conformational change of the 39-kDa glycoprotein from human articular chondrocytes. *J. Biol Chem. 2003;* 278: 30206-12.

Hu B, Trinh K, Figueira WF, Price PA. Isolation and sequence of a novel human chondrocyte protein related to mammalian members of the chitinase protein family. *J Biol Chem.* 1996; 271: 19415-20.

Iwashita H, Morita S, Sagiya Y, Nakanishi A. Role of eosinophil chemotactic factor by T lymphocytes on airway hyperresponsiveness in a murine model of allergic asthma. *Am J Respir Cell Mol Biol.* 2006; 35: 103-9.

Johansen JS, Jensen BV, Roslind A, Nielsen D, Price PA. Serum YKL-40, a new prognostic biomarker in cancer patients? *Cancer Epidemiol Biomarkers Prev.* 2006; 15: 194-202.

Johansen JS, Jensen BV, Roslind A, Price PA. Is YKL-40 a new therapeutic target in cancer? *Expert Opin Ther Targets* 2007; 11: 219-34.

Johansen JS, Jensen HS, Price PA. A new biochemical marker for joint injury. Analysis of YKL-40 in serum and synovial fluid. *Br J Rheumatol.* 1993; 32: 949-55.

Johansen JS, Kirwan JR, Price PA, Sharif M. Serum YKL-40 concentrations in patients with early rheumatoid arthritis: relation to joint destruction. *Scand J Rheumatol.* 2001; 30: 297-304.

Johansen JS, Krabbe KS, Moller K, Pedersen BK. Circulating YKL-40 levels during human endotoxaemia. *Clin Exp Immunol.* 2005a; 140: 343-8.

Johansen JS, Milman N, Hansen M, Garbarsch C, Price PA, Graudal N. Increased serum YKL-40 in patients with pulmonary sarcoidosis--a potential marker of disease activity? *Respir Med.* 2005b; 99: 396-402.

Johansen JS, Olee T, Price PA, Hashimoto S, Ochs RL, Lotz M. Regulation of YKL-40 production by human articular chondrocytes. *Arthritis Rheum.* 2001 Apr;44(4):826-37.

Johansen JS, Stoltenberg M, Hansen M, Florescu A, Horslev-Petersen K, Lorenzen I, Price PA. Serum YKL-40 concentrations in patients with rheumatoid arthritis: relation to disease activity. *Rheumatology* (Oxford). 1999; 38: 618-26.

Junker N, Johansen JS, Andersen CB, Kristjansen PE. Expression of YKL-40 by peritumoral macrophages in human small cell lung cancer. *Lung Cancer.* 2005; 48: 223-31.

Kamal SM, Turner B, He Q, Rasenack J, Bianchi L, Al Tawil A, Nooman A, Massoud M, Koziel MJ, Afdhal NH. Progression of fibrosis in hepatitis C with and without schistosomiasis: correlation with serum markers of fibrosis. *Hepatology.* 2006; 43: 771-9.

Knorr T, Obermayr F, Bartnik E, Zien A, Aigner T. YKL-39 (chitinase 3-like protein 2), but not YKL-40 (chitinase 3-like protein 1), is up regulated in osteoarthritic chondrocytes. *Ann Rheum Dis.* 2003; 62: 995-8.

Kornfeld S, Mellman I. The biogenesis of lysosomes. *Annu Rev Cell Biol.* 1989; 5: 483-525.

Kronborg G, Ostergaard C, Weis N, Nielsen H, Obel N, Pedersen SS, Price PA, Johansen JS. Serum level of YKL-40 is elevated in patients with Streptococcus pneumoniae bacteremia and is associated with the outcome of the disease. *Scand J Infect Dis.* 2002; 34: 323-6.

Kzhyshkowska J, Gratchev A, Brundiers H, Mamidi S, Krusell L, Goerdt S. Phosphatidylinositide 3-kinase activity is required for stabilin-1-mediated endosomal transport of acLDL. *Immunobiology.* 2005; 210: 161-73.

Kzhyshkowska J, Gratchev A, Goerdt S. Stabilin-1, a homeostatic scavenger receptor with multiple functions. *J Cell Mol Med.* 2006a; 10: 635-49.

Kzhyshkowska J, Gratchev A, Goerdt S. Human Chitinases and Chitinase-Like Proteins as Indicators for Inflammation and Cancer. *Biomarker Insights.* 2007a; 128-46.

Kzhyshkowska J, Gratchev A, Martens JH, Pervushina O, Mamidi S, Johansson S, Schledzewski K, Hansen B, He X, Tang J, Nakayama K, Goerdt S. Stabilin-1 localizes to endosomes and the trans-Golgi network in human macrophages and interacts with GGA adaptors. *J Leukoc Biol.* 2004; 76: 1151-61.

Kzhyshkowska J, Gratchev A, Schmuttermaier C, Brundiers H, Krusell L, Mamidi S, Zhang J, Workman G, Sage EH, Anderle C, Sedlmayr P, Goerdt S. Alternatively Activated Macrophages Regulate Extracellular Levels of the Hormone Placental Lactogen via Receptor-Mediated Uptake and Transcytosis. *J Immunol.* 2008; 180: 3028-37.

Kzhyshkowska J, Mamidi S, Gratchev A, Kremmer E, Schmuttermaier C, Krusell L, Haus G, Utikal J, Schledzewski K, Scholtze J, Goerdt S. Novel stabilin-1 interacting chitinase-like protein (SI-CLP) is up-regulated in alternatively activated macrophages and secreted via lysosomal pathway. *Blood.* 2006b; 107: 3221-8.

Kzhyshkowska J, Marciniak-Czochra A, Gratchev A. Perspectives of mathematical modelling for understanding of intracellular signalling and vesicular trafficking in macrophages. *Immunobiology* 2007b; 212: 813-25.

Kzhyshkowska J, Workman G, Cardo-Vila M, Arap W, Pasqualini R, Gratchev A, Krusell L, Goerdt S, Sage EH. Novel function of alternatively activated macrophages: Stabilin-1-mediated clearance of SPARC. *J Immunol.* 2006c; 176: 5825-32.

Ling H, Recklies AD. The chitinase 3-like protein human cartilage glycoprotein 39 inhibits cellular responses to the inflammatory cytokines interleukin-1 and tumour necrosis factor-alpha. *Biochem J.* 2004; 380: 651-9.

Logan MR, Odemuyiwa SO, Moqbel R. Understanding exocytosis in immune and inflammatory cells: the molecular basis of mediator secretion. *J Allergy Clin Immunol.* 2003; 111: 923-32.

Martens JH, Kzhyshkowska J, Falkowski-Hansen M, Schledzewski K, Gratchev A, Mansmann U, Schmuttermaier C, Dippel E, Koenen W, Riedel F, Sankala M, Tryggvason K, Kobzik L, Moldenhauer G, Arnold B, Goerdt S. Differential expression of a gene signature for scavenger/lectin receptors by endothelial cells and macrophages in human lymph node sinuses, the primary sites of regional metastasis. *J Pathol.* 2006; 208: 574-89.

Matsumoto T, Tsurumoto T. Serum YKL-40 levels in rheumatoid arthritis: correlations between clinical and laborarory parameters. *Clin Exp Rheumatol.* 2001; 19: 655-60.

Millis AJ, Hoyle M, Kent L. In vitro expression of a 38,000 dalton heparin-binding glycoprotein by morphologically differentiated smooth muscle cells. *J Cell Physiol.* 1986; 127: 366-72.

Nair MG, Cochrane DW, Allen JE. Macrophages in chronic type 2 inflammation have a novel phenotype characterized by the abundant expression of Ym1 and Fizz1 that can be partly replicated in vitro. *Immunol Lett.* 2003; 85: 173-80.

Nair MG, Gallagher IJ, Taylor MD, Loke P, Coulson PS, Wilson RA, Maizels RM, Allen JE. Chitinase and Fizz family members are a generalized feature of nematode infection with selective upregulation of Ym1 and Fizz1 by antigen-presenting cells. *Infect Immun.* 2005; 73: 385-94.

Nickel W. The mystery of nonclassical protein secretion. A current view on cargo proteins and potential export routes. *Eur J Biochem.* 2003; 270: 2109-19.

Nishikawa KC, Millis AJ. gp38k (CHI3L1) is a novel adhesion and migration factor for vascular cells. *Exp Cell Res.* 2003; 287: 79-87.

Nordenbaek C, Johansen JS, Junker P, Borregaard N, Sorensen O, Price PA. YKL-40, a matrix protein of specific granules in neutrophils, is elevated in serum of patients with community-acquired pneumonia requiring hospitalization. *J Infect Dis.* 1999; 180: 1722-6.

Nutt CL, Betensky RA, Brower MA, Batchelor TT, Louis DN, Stemmer-Rachamimov AO. YKL-40 is a differential diagnostic marker for histologic subtypes of high-grade gliomas. *Clin Cancer Res.* 2005; 11: 2258-64.

Oba Y, Chung HY, Choi SJ, Roodman GD. Eosinophil chemotactic factor-L (ECF-L): a novel osteoclast stimulating factor. *J Bone Miner Res.* 2003; 18: 1332-41.

Ober C, Tan Z, Sun Y, Possick JD, Pan L, Nicolae R, Radford S, Parry RR, Heinzmann A, Deichmann KA, Lester LA, Gern JE, Lemanske RF, Jr., Nicolae DL, Elias JA, Chupp GL. Effect of Variation in CHI3L1 on Serum YKL-40 Level, Risk of Asthma, and Lung Function. *N Engl J Med.* 2008.

Ostergaard C, Johansen JS, Benfield T, Price PA, Lundgren JD. YKL-40 is elevated in cerebrospinal fluid from patients with purulent meningitis. *Clin Diagn Lab Immunol.* 2002; 9: 598-604.

Owhashi M, Arita H, Hayai N. Identification of a novel eosinophil chemotactic cytokine (ECF-L) as a chitinase family protein. *J Biol Chem.* 2000; 275: 1279-86.

Pelloski CE, Mahajan A, Maor M, Chang EL, Woo S, Gilbert M, Colman H, Yang H, Ledoux A, Blair H, Passe S, Jenkins RB, Aldape KD. YKL-40 expression is associated with poorer response to radiation and shorter overall survival in glioblastoma. *Clin Cancer Res.* 2005; 11: 3326-34.

Peltomaa R, Paimela L, Harvey S, Helve T, Leirisalo-Repo M. Increased level of YKL-40 in sera from patients with early rheumatoid arthritis: a new marker for disease activity. *Rheumatol Int.* 2001; 20: 192-6.

Pesce J, Kaviratne M, Ramalingam TR, Thompson RW, Urban JF, Jr., Cheever AW, Young DA, Collins M, Grusby MJ, Wynn TA. The IL-21 receptor augments Th2 effector function and alternative macrophage activation. *J Clin Invest.* 2006; 116: 2044-55.

Petersson M, Bucht E, Granberg B, Stark A. Effects of arginine-vasopressin and parathyroid hormone-related protein (1-34) on cell proliferation and production of YKL-40 in cultured chondrocytes from patients with rheumatoid arthritis and osteoarthritis. *Osteoarthritis Cartilage.* 2006; 14: 652-9.

Raes G, de Baetselier P, Noel W, Beschin A, Brombacher F, Hassanzadeh GG. Differential expression of FIZZ1 and Ym1 in alternatively versus classically activated macrophages. *J Leukoc Biol.* 2002; 71: 597-602.

Raes G, Van den BR, de Baetselier P, Ghassabeh GH, Scotton C, Locati M, Mantovani A, Sozzani S. Arginase-1 and Ym1 are markers for murine, but not human, alternatively activated myeloid cells. *J Immunol.* 2005; 174: 6561-2.

Rathcke CN, Johansen JS, Vestergaard H. YKL-40, a biomarker of inflammation, is elevated in patients with type 2 diabetes and is related to insulin resistance. *Inflamm Res.* 2006a; 55: 53-9.

Rathcke CN, Vestergaard H. YKL-40, a new inflammatory marker with relation to insulin resistance and with a role in endothelial dysfunction and atherosclerosis. *Inflamm Res.* 2006b; 55: 221-7.

Recklies AD, White C, Ling H. The chitinase 3-like protein human cartilage glycoprotein 39 (HC-gp39) stimulates proliferation of human connective-tissue cells and activates both extracellular signal-regulated kinase- and protein kinase B-mediated signalling pathways. *Biochem J.* 2002; 365: 119-26.

Rehli M, Krause SW, Andreesen R. Molecular characterization of the gene for human cartilage gp-39 (CHI3L1), a member of the chitinase protein family and marker for late stages of macrophage differentiation. *Genomics.* 1997; 43: 221-5.

Rehli M, Niller HH, Ammon C, Langmann S, Schwarzfischer L, Andreesen R, Krause SW. Transcriptional regulation of CHI3L1, a marker gene for late stages of macrophage differentiation. *J Biol Chem.* 2003; 278: 44058-67.

Renkema GH, Boot RG, Au FL, Donker-Koopman WE, Strijland A, Muijsers AO, Hrebicek M, Aerts JM. Chitotriosidase, a chitinase, and the 39-kDa human cartilage glycoprotein, a chitin-binding lectin, are homologues of family 18 glycosyl hydrolases secreted by human macrophages. *Eur J Biochem.* 1998; 251: 504-9.

Renkema GH, Boot RG, Muijsers AO, Donker-Koopman WE, Aerts JM. Purification and characterization of human chitotriosidase, a novel member of the chitinase family of proteins. *J Biol Chem.* 1995; 270: 2198-202.

Renkema GH, Boot RG, Strijland A, Donker-Koopman WE, van den BM, Muijsers AO, Aerts JM. Synthesis, sorting, and processing into distinct isoforms of human macrophage chitotriosidase. *Eur J Biochem.* 1997; 244: 279-85.

Saidi A, Javerzat S, Bellahcene A, De Vos J, Bello L, Castronovo V, Deprez M, Loiseau H, Bikfalvi A, Hagedorn M. Experimental anti-angiogenesis causes upregulation of genes associated with poor survival in glioblastoma. *Int J Cancer.* 2008; 122: 2187-98.

Saitou Y, Shiraki K, Yamanaka Y, Yamaguchi Y, Kawakita T, Yamamoto N, Sugimoto K, Murata K, Nakano T. Noninvasive estimation of liver fibrosis and response to interferon therapy by a serum fibrogenesis marker, YKL-40, in patients with HCV-associated liver disease. *World J Gastroenterol.* 2005; 11: 476-81.

Sakata M, Masuko-Hongo K, Tsuruha J, Sekine T, Nakamura H, Takigawa M, Nishioka K, Kato T. YKL-39, a human cartilage-related protein, induces arthritis in mice. *Clin Exp Rheumatol.* 2002; 20: 343-50.

Sasaki C, Yokoyama A, Itoh Y, Hashimoto M, Watanabe T, Fukamizo T. Comparative study of the reaction mechanism of family 18 chitinases from plants and microbes. *J Biochem* (Tokyo). 2002; 131: 557-64.

Schmidt H, Johansen JS, Gehl J, Geertsen PF, Fode K, von der MH. Elevated serum level of YKL-40 is an independent prognostic factor for poor survival in patients with metastatic melanoma. *Cancer.* 2006a; 106: 1130-9.

Schmidt H, Johansen JS, Sjoegren P, Christensen IJ, Sorensen BS, Fode K, Larsen J, von der MH. Serum YKL-40 predicts relapse-free and overall survival in patients with American Joint Committee on Cancer stage I and II melanoma. *J Clin Oncol.* 2006b; 24: 798-804.

Sekine T, Masuko-Hongo K, Matsui T, Asahara H, Takigawa M, Nishioka K, Kato T. Recognition of YKL-39, a human cartilage related protein, as a target antigen in patients with rheumatoid arthritis. *Ann Rheum Dis.* 2001; 60: 49-54.

Shackel NA, McGuinness PH, Abbott CA, Gorrell MD, McCaughan GW. Novel differential gene expression in human cirrhosis detected by suppression subtractive hybridization. *Hepatology.* 2003; 38: 577-88.

Shackelton LM, Mann DM, Millis AJ. Identification of a 38-kDa heparin-binding glycoprotein (gp38k) in differentiating vascular smooth muscle cells as a member of a group of proteins associated with tissue remodeling. *J Biol Chem.* 1995; 270: 13076-83.

Steck E, Breit S, Breusch SJ, Axt M, Richter W. Enhanced expression of the human chitinase 3-like 2 gene (YKL-39) but not chitinase 3-like 1 gene (YKL-40) in osteoarthritic cartilage. *Biochem Biophys Res Commun.* 2002; 299: 109-15.

Stinchcombe JC, Griffiths GM. Regulated secretion from hemopoietic cells. *J Cell Biol.* 1999; 147: 1-6.

Sun YJ, Chang NC, Hung SI, Chang AC, Chou CC, Hsiao CD. The crystal structure of a novel mammalian lectin, Ym1, suggests a saccharide binding site. *J Biol Chem.* 2001; 276: 17507-14.

Tran A, Benzaken S, Saint-Paul MC, Guzman-Granier E, Hastier P, Pradier C, Barjoan EM, Demuth N, Longo F, Rampal P. Chondrex (YKL-40), a potential new serum fibrosis marker in patients with alcoholic liver disease. *Eur J Gastroenterol Hepatol.* 2000; 12: 989-93.

Tsuruha J, Masuko-Hongo K, Kato T, Sakata M, Nakamura H, Sekine T, Takigawa M, Nishioka K. Autoimmunity against YKL-39, a human cartilage derived protein, in patients with osteoarthritis. *J Rheumatol.* 2002; 29: 1459-66.

van Eijk M, van Roomen CP, Renkema GH, Bussink AP, Andrews L, Blommaart EF, Sugar A, Verhoeven AJ, Boot RG, Aerts JM. Characterization of human phagocyte-derived chitotriosidase, a component of innate immunity. *Int Immunol.* 2005; 17: 1505-12.

Vind I, Johansen JS, Price PA, Munkholm P. Serum YKL-40, a potential new marker of disease activity in patients with inflammatory bowel disease. *Scand J Gastroenterol.* 2003; 38: 599-605.

Volck B, Johansen JS, Stoltenberg M, Garbarsch C, Price PA, Ostergaard M, Ostergaard K, Løvgreen-Nielsen P, Sonne-Holm S, Lorenzen I. Studies on YKL-40 in knee joints of patients with rheumatoid arthritis and osteoarthritis. Involvement of YKL-40 in the joint pathology. *Osteoarthritis Cartilage.* 2001 Apr;9(3):203-14

Volck B, Ostergaard K, Johansen JS, Garbarsch C, Price PA. The distribution of YKL-40 in osteoarthritic and normal human articular cartilage. *Scand J Rheumatol.* 1999; 28: 171-9.

Volck B, Price PA, Johansen JS, Sorensen O, Benfield TL, Nielsen HJ, Calafat J, Borregaard N. YKL-40, a mammalian member of the chitinase family, is a matrix protein of specific granules in human neutrophils. *Proc Assoc Am Physicians.* 1998; 110: 351-60.

Webb DC, McKenzie AN, Foster PS. Expression of the Ym2 lectin-binding protein is dependent on interleukin (IL)-4 and IL-13 signal transduction: identification of a novel allergy-associated protein. *J Biol Chem.* 2001; 276: 41969-76.

Welch JS, Escoubet-Lozach L, Sykes DB, Liddiard K, Greaves DR, Glass CK. TH2 cytokines and allergic challenge induce Ym1 expression in macrophages by a STAT6-dependent mechanism. *J Biol Chem.* 2002; 277: 42821-9.

In: Binomium Chitin-Chitinase: Recent Issues
Editor: Salvatore Musumeci and Maurizio G. Paoletti
ISBN 978-1-60692-339-9

Chapter XX

Role of Chitinases in Human Stomach for Chitin Digestion: AMCase in the Gastric Digestion of Chitin and Chit in Gastric Pathologies

***Maurizio G. Paoletti*[1], *Lorenzo Norberto*[2], *Elisa Cozzarini*[1] and *Salvatore Musumeci*[3]**

[1]Department of Biology, Laboratory of Agroecology and Ethnobiology, University of Padova, Padova, Italy

[2]Department of Surgical and Gastroenterological Sciences, Section 1st of General Surgery, Surgical Endoscopy Unit, University of Padova, Padova, Italy

[3]Department of Neurosciences and Mother and Child Sciences, University of Sassari and Institute of Biomolecular Chemistry, CNR, Li Punti (SS), Italy

Abstract

Chitin-containing food is an interesting but underestimated source of locally available, in most cases sustainable, food although chitin digestion by humans has generally been questioned or denied. Only in recent times chitinases have been found in several human tissues and their role has been associated with defence against parasite infections as well as with some allergic conditions. We reflected that crustaceans, and to some extent molluscs, mushrooms and most arthropods containing chitin, are sometime a consistent part of food regimes for local communities. Finally, we demonstrated that AMCase is present in gastric juices and it is associated with chitin digestion. In most tropical and some temperate countries, such as Japan and Korea, a significant number of adult insects and larvae are consumed raw, or cooked along with diverse local specialities. At present, up to 2,000 species of insects and other terrestrial arthropods have been listed as edible in Africa, Asia, Central and South America, Australia and Europe. Both insects and crustaceans are covered by chitin teguments and mushrooms contain some chitin. In most cases, the hard covering of polysaccharide chitin on insects accounts for 5-20% of their dry weight. In general, chitinases can digest chitin and reduce it to simple compounds such as *N*-acetyl-glucosamine. Western society does not consider insects an important food, however: crustaceans, such as lobsters and crabs, are commonly eaten after discarding the hardened chitin-rich tegument, with the exception of small shrimps, which are generally eaten fried. Therefore, Western nutrition does not seem to depend on chitinases. These and other considerations, including the absence of chitin as a human body component, have led us to ask whether humans are capable of chitin digestion.

To assess chitinases' function as tools to digest chitin, we have examined 48 patient's gastric juices, obtained during gastroscopy, at Padova University Hospital. We found that 14.6% of total samples studied showed AMCase activity from 36.270 to 3.540 nmol/ml/h. The majority of involved subjects (75%) had lower values, from 2.800 to 0.178 nmol/ml/h; while in 10.4% of subjects the chitinolitic activity varied from 0.086 to 0.013 nmol/ml/h, and could be considered absent. We reported superficial digestion of fly forewings, utilizing gastric juice of a patient with an AMCase activity of 19.410 nmol/ml/h.

If AMCase enzyme, present in gastric juice, is truly involved in chitin digestion, we should expect a higher presence of expressed AMCase in populations currently accustomed to eating mushrooms and/or invertebrates bearing chitin.

We also found a positive relationship between *CHIT* expression level in antral gastric mucosa and both flogosis and *Helicobacter pylori* infection.

1. Introduction

Food containing some amount of chitin, such as crustaceans, molluscs, mushrooms and insects, is part of the human diet, especially for local communities in tropical and subtropical countries such as in Africa (Malaisse, 1997; Paoletti, 2005). This overlooked resource represents an important base for everyday survival, as well as a sustainable local resource available without the use of sophisticated traps or weapons even to the less empowered community members, like women and children (Dufour, 1987; Paoletti *et al.* 2000; Paoletti, 2005a) (Figure 1).

In most tropical and some temperate countries, such as Japan and Korea, a significant number of adult insects and larvae are consumed raw or cooked along with diversified local specialities. At present up to 2000 species of insects and other terrestrial arthropods have been listed as edible in Africa, Asia, Central and South America, Australia and Europe (Paoletti and Bukkens, 1997; DeFoliart, 2002; Paoletti, 2005).

Both insects and crustaceans are covered by chitin teguments. In most cases, the hard covering of polysaccharide chitin on insects accounts for 5-20% of the dry weight.

In general, chitinases can digest chitin and, usually in combination with other enzymatic activities, convert it into more readily absorbable components such as *N*-acetyl-glucosamine (Talent and Gracy, 1996; Jollès and Muzzarelli, 1999; Gardiner, 2000). Two kinds of chitinase have been reported with an exochitinase and endochitinase activity. The first releases *N*-acetyl-glucosamine from chitin on each occasion, while the second releases not *N*-acetylglucosamine, but instead a mixture of chitobiose and larger oligomers (Cohen-Kupiec and Chet, 1998; Fusetti *et al.* 2002; Tikhonov *et al.* 2004).

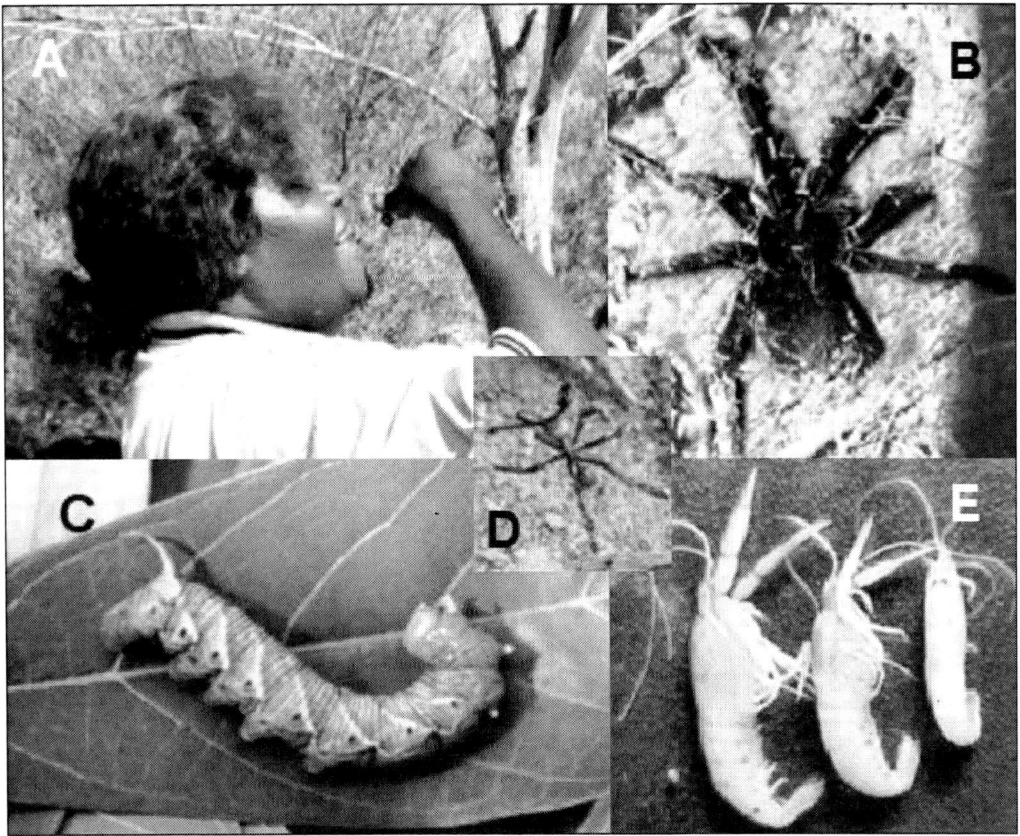

Figure 1. Edible invertebrates bearing chitin. A) Australian eating sweet ant (*Camponotus* sp.); B) Terrestrial spieders (*Theraphosa* sp.) eaten in Alto Orinoco (Venezuela); C) Caterpillars (*Erinnys ello*) are among the most appreciated insects in the Amazon; D) Many ants and termits are eaten in the tropics. Here the army ant *Eciton* sp. from Alto Orinoco; E) Small freshwater shrimpf raw or cooked are eaten in most areas included Amazon and Alto Orinoco.

Western society, interestingly, does not consider insects an important food (DeFoliart, 1999; DeFoliart, 2002), although crustaceans, such as lobsters and shrimps, are commonly eaten, albeit mostly after discarding the hardened chitin-rich tegument; although small shrimps are eaten with their teguments. Therefore, Western nutrition does not seem to depend on chitinases (DeFoliart, 1992). This and other considerations, including the absence of chitin as a human body component, have led some authors to question whether humans are capable of digesting chitin (Bukkens, 1997; Boot *et al.* 2005; Bukkens, 2005), and others still to suggest that its role is merely that of a dietetic fibre (Muzzarelli, 2001). Most nutritionists have argued the impossibility of humans feeding on and digesting chitin-containing food.

Evidence of chitinases in the human body is relatively recent: since 1994 for Chitotriosidase (Chit) and 2001 for Acidic Mammalian Chitinase (AMCase), (Hollak *et al.* 1994; Boot *et al.* 2001).

The function of human chitinases in diseases is still largely unknown. The only evidence available is that it occurs at high levels in certain disease states. For instance, Renkema *et al.* (1997, 1998) and Aerts (2009) in this book have described the occurrence in the plasma of patients affected by Gaucher disease of elevated levels of Chitotriosidase, a hydrolytic enzyme produced by macrophage cells, which exhibits optimum activity at pH 6. More recently, it has been reported that Chit may also be involved in innate immune responses (Van Eijk *et al.* 2005). Moreover, Chit levels in plasma have been shown to be high in the case of diseases such as acute malaria (Barone *et al.* 2003), beta-thalassemia (Barone *et al.* 1998; Barone *et al.* 2001), and other hemoglobinopathies (see other chapters in this book) indicating that macrophage activation is responsible for Chit expression (Bouzas *et al.* 2003; Musumeci *et al.* 2005).

Another chitinase, Acidic Mammalian Chitinase (AMCase), produced in different tissues (Boot *et al.* 2001), exhibits optimum activity in the acid pH range and recently has been implicated in allergic bronchial asthma (Zhu *at al.*, 2004). To date, however, its function is obscure (Boot *et al.* 2005). AMCase has also been found in mouse and rat stomachs, where it has been shown at the cellular level (immunohistochemically) and at the level of RNA expression (Suzuki *et al.* 2002; Goto *et al.* 2003; Boot *et al.* 2005;).

Considering that chitinases are highly conserved in mammals (from humans to rats) (Gianfrancesco and Musumeci, 2004), it is to be expected that also humans could produce AMCase in the gastric epithelium, where it would digest chitin from parasites and food, including arthropods (insects and crustaceans) and mushrooms containing some chitin.

Boot *et al.* showed in 2005 that AMCase is expressed in the human stomach and, to a lesser extent, in the lung; but its presence and chitinolytic activity in human gastric juice have been confirmed only recently (Paoletti *et al.* 2007).

This chapter aims: a) to analyse AMCase activity in human gastric juice; b) to assess whether humans are able to digest the cover of chitinous arthropods (including insects) by analysing digestion of chitinous insect wings by gastric juices; c) to show the results of expression of *CHIT* and *AMCase* in human gastric mucosa in relation to stomach pathologies such as phlogosis and *Helicobacter pylori* infection (Figure 2).

2. Sampling Chitinases in Gastric Juices

At Padova University Polyclinic, 48 Italian subjects (26 males and 22 females), aged 23-76 years were submitted to gastroscopy because of symptoms of dyspepsia and postprandial pains (Table 1). All subjects lived on the outskirts of Padova. The study was conducted between April 2006 and July 2007. The patients consumed no food for at least 14 hours prior to the gastroscopy.

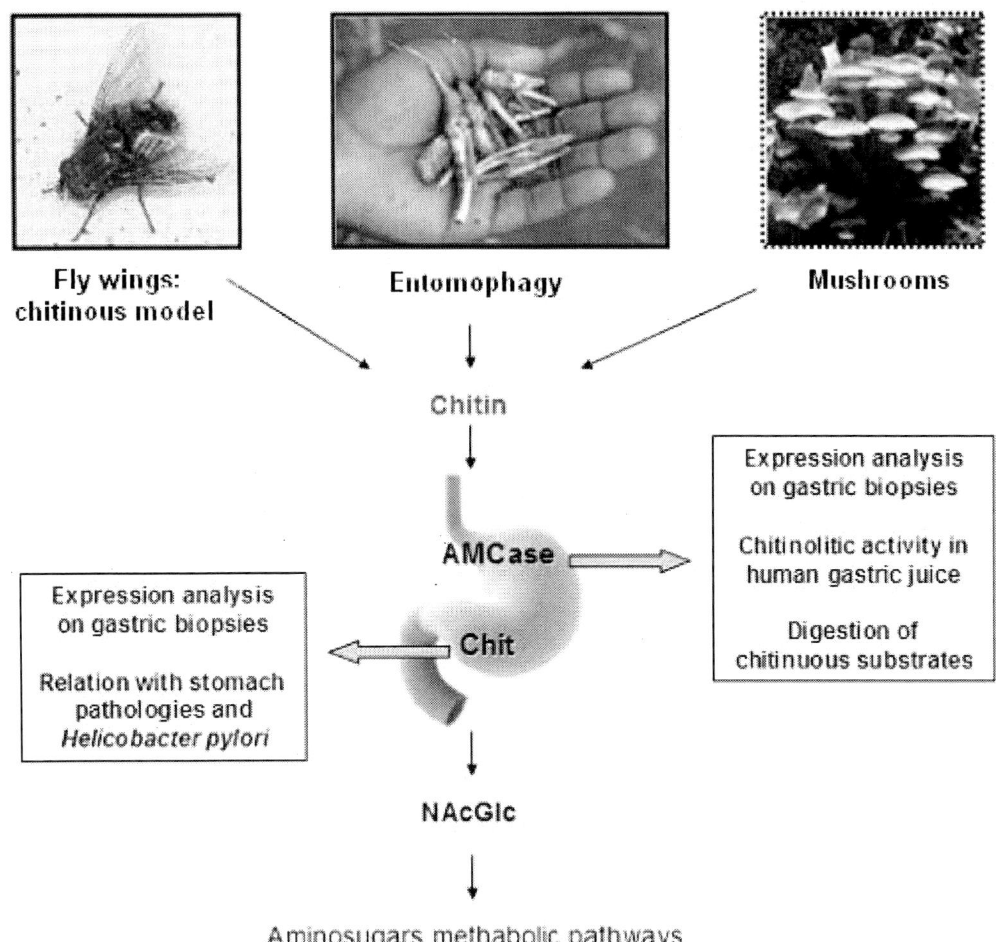

Figure 2. Schematic description of the aims of this chapter.

After the optical examination of the gastric mucosa a modified Giemsa test for detection of *Helicobacter pylori* (detected as a Giemsa positive bacillus) was also performed on all the patients. According to the Sydney Sistem (Dixon *et al.* 1996), the gastritis score relating to each biopsy site was then evaluated using a visual analog scale (+ mild; ++ moderate; +++ severe).

A fragment of gastric mucosa was collected for immunohystochemical analysis and to measure mRNA for *CHIT* and *AMCase* by Quantitative Real Time PCR. No drugs that could potentially interfere with the gastric juice secretion were administered before the gastroscopy. Before removing the gastroscope, ~ 10 ml of gastric juice were collected from each subject. Samples were held in ice until they were transferred to the Department of Biology at the University of Padova, where they were rapidly frozen at -80°C. Once the collection of gastric juice was completed, the frozen samples were transported in dry ice to the Department of Pediatrics at the University of Catania, where they remained frozen at -80°C until the determination of chitinase activities. The study was approved by the Ethics Committees of the University of Padova.

Table 1. Patient's personal data (sex and age), AMCase activity in gastric juice measured with the spectrofluorimetric method developed by Boot et al. expressed as nmol/ml/h (Boot, 2001), gastroscopic diagnosis and results of *Helicobacter pylori* test of each patient envolved in our studies (Paoletti *et al.* 2007; Cozzarini, 2007)

Subject No.	Sex M - F	Age (years)	AMCase activity (nmol/ml/hr)	Gastroscopic diagnosis
1	M	73	0.400	No affection
2	F	53	3.540	No affection
3	F	74	6.800	Chronic antral gastritis and gastric micropolyposis
4	F	58	36.270	No affection
5	F	59	0.260	Chronic antral gastritis, suspected Short Barrett esophagitis
6	M	64	0.086	Reflux disease
7	M	43	0.025	Chronic antral gastritis and duodenitis
8	M	45	0.065	Chronic gastritis
9	M	40	0.013	Antral gastritis
10	F	50	0.046	No affection
11	F	65	4.520	Antral gastritis, *H. pylori* test positive
12	F	39	0.178	Antral gastritis and reflux disease
13	M	74	0.256	Antral gastritis, reflux disease, *H. pylori* test positive
14	F	60	0.446	Slight antral gastritis
15	F	52	0.330	Antral gastritis and leiomioma, hiatal hernia
16	F	53	0.303	Antral gastritis, reflux disease, *H. pylori* test positive
17	M	50	0.299	Antral gastritis, reflux disease, *H. pylori* test positive
18	M	51	0.558	Antral gastritis, reflux disease, *H. pylori* test positive
19	M	62	19.410	Diffuse gastritis and reflux disease
20	M	48	0.249	Antral gastritis, reflux disease, *H. pylori* test positive
21	M	45	0.480	Antral gastritis, reflux disease, *H. pylori* test positive
22	F	54	0.549	Antral gastritis, reflux disease, gastric micropolyposis
23	M	46	0.181	Antral gastritis and reflux disease
24	M	40	0.230	Antral gastritis, reflux disease, *H. pylori* test positive

Subject No.	Sex M - F	Age (years)	AMCase activity (nmol/ml/hr)	Gastroscopic diagnosis
25	M	60	0.210	Diffuse gastritis, reflux disease, *H. pylori* test positive
26	F	62	0.252	Diffuse gastritis, reflux disease, *H. pylori* test positive
27	F	25	0.231	Reflux disease and suspected Short Barrett esophagitis
28	M	71	0.252	Chronic antral gastritis, reflux disease and duodenitis
29	F	44	0.249	Slight reflux disease
30	M	73	27.990	Slight reflux disease
31	F	76	2.438	Reflux disease, gastric micropolyposis, *H. pylori* test pos.
32	F	23	0.267	Chronic antral gastritis, reflux disease and metaplasia
33	M	35	0.350	Reflux disease, duodenitis, *H. pylori* test positive
34	M	69	0.370	Chronic antral gastritis and reflux disease
35	M	60	0.350	Gastric micropolyposis
36	M	66	0.510	Slight gastritis and duodenitis
37	M	73	14.500	Chronic antral gastritis, reflux disease, gastric micropolyposis
38	M	60	0.300	Chronic antral gastritis
39	F	40	0.230	Diffuse gastritis, reflux disease and esophageal candidiasis
40	F	71	0.270	Diffuse gastritis and reflux disease
41	M	35	0.250	Antral gastritis and reflux disease
42	F	55	0.370	Reflux disease and slight congestive gastropathy
43	F	74	0.250	Chronic antral gastritis and gastic micropolyposis
44	F	58	0.250	Slight gastritis, reflux disease and esophageal candidiasis
45	M	50	2.800	Chronic antral gastritis and reflux disease
46	M	46	0.300	Slight gastritis and reflux disease
47	M	75	0.400	Antral gastritis and slight reflux disease
48	F	25	0.200	Chronic antral gastritis

2.1. Preparation of Gastric Juice for AMCase Activity Determination

The tubes containing the gastric juice were left to defrost at 4°C and centrifuged for 30 minutes at 15,000 *g* at 4°C. The supernatant was aspired and fractioned into 1 ml Eppendorf tubes and stored at -80°C until examination.

2.2. AMCase Activity

We used the spectrofluorimetric methodologies developed by Boot et al. (Boot *et al.* 2001) and Tikhonov et al. (Tikhonov *et al.* 2004). Since these two methods are comparable, here we report only the results related to the first one (the Boot *et al.* method).

After the gastric samples were defrosted the pH was measured and corrected to pH 2 with 0.1 M HCL The value of pH 2 was chosen because it is the optimum pH for the dosage of AMCase (Talent and Gracy, 1996). Fifty µl of gastric juice were incubated with 0.1 ml of a solution containing 22 mmol/l of the artificial substrate 4-methylumbelliferyl-β-D-N,N'-diacetylchitobiose (Sigma Chemical Co catalogue M 9763) in 0.5 M citrate-phosphate buffer pH 4.5 for 30 minutes at 37°C. The reaction was stopped by using 2 ml of 0.5 mol/L Na_2CO_3-$NaHCO_3$ buffer, pH 10.7. The fluorescence was read by a spettrofluorimeter Hitachi 2500 (Hitachi, Europe Ltd, Herts, UK), on 365 nm excitation and 450 nm emission. AMCase activity was expressed as nanomoles of substrate hydrolyzed per ml per hour (nmol/ml/h). A blank for control, composed of reagents (0.5 M citrate-phosphate buffer + 4-methylumbelliferyl-β-D-N,N'-diacetylchitobiose) was used in each measurement and two standard samples at different chitinase activity (10 and 100 nmol/ml/h) were also added. The interassay variation coefficient (reproducibility) was <5%. The reported results are the mean of three determinations and the graphic representations are obtained by using the Prisma software. Median and range were calculated for each group.

2.3. Inibition Test with Allosamidin

To confirm that the hydrolytic activity was due to AMCase the dosage was repeated after neutralisation of gastric juice chitinase activity with 9 µM of allosamidin (kindly provided by Dr S. Sakuda, Sakuda and Sakurada, 1998), for 90 minutes at 37°C in a shaking water bath (160 rpm). This concentration is likely to completely inhibit chitinase activity (Jollès and Muzzarelli, 1999). The neutralisation reaction was stopped by adding 180 µl of sodium dodecylsulfate (DS) 10% wt/vol.

2.4. Chitinolytic Activity at Different pH Values

Samples of gastric juice initially at pH value > 6 were adjusted at different values decreasing of one unit with 0.1 M HCL and the chitinolytic activity was measured in each sample with the Boot *et al* 2001 method.

The activity of each sample measured by fluorescent emission was stable after repeated measurements. The operator's great experience and the simultaneity of determinations excluded processing artefacts.

3. Results

Among 48 studied patients, 4 (patients 1, 2, 4, 10), who at physical and clinical examination were otherwise healthy resulted negative for gastritis and they represented the healthy controls. The majority of patients (38/48) showed acute or chronic gastritis and gastric micropoliposis, while 32/48 patients showed signs of gastroesophageal reflux disease (GERD) and more in general dyspeptic symptoms and postprandial pain. Only 12/48 patients tested resulted positive for *Helicobacter pylori* (Table 1).

Table 1 reports the chitinase activity of each gastric sample, reported in nmol/ml/h of 4-methylumbelliferyl-beta-D-N,N′-diacetylchitobiose hydrolised.

In Table 2, the patients were assigned to one of three groups according to the values of AMCase activity. 14.6% of total studied samples showed AMCase activity from 36.270 to 3.540 nmol/ml/h, the majority of involved subjects (75%) had lower values from 2.800 to 0.178 nmol/ml/h, while in 10.4% of subjects the chitinolitic activity varied between 0.086 and 0.013 nmol/ml/h and could be considered absent. The prevalence of *Helicobacter pylori* was distributed among the three group examined. No statistically significant correlation was found between the AMCase activity and the sex or age of patients.

In the two control samples, where no gastric juice was added, the level of activity was < 0.1 nmol/ml/h. The standard samples added in each determination confirmed the reproducibility of the method with a variation coefficent of < 5% .

In 11 separate samples, after neutralisation of chitinase activity with allosamidine the AMCase activity disappeared, confirming that the results are due specifically to the chitinolytic activity of gastric juice (Figure 3).

The curve of chitinase activity in function of the pH demonstrated that the chitinolytic activity was high at pH 2, was stable up to pH 5-6 and decreased slowly to a value of pH > 6, confirming that chitinolytic activities obtained at different pH always belong to AMCase (Figure 4), similarly to mouse AMCase.

Table 2. Chitinolytic activity (AMCase) at pH 2 expressed in nmol/ml/h of hydrolyzed substracts (Paoletti *et al.* 2007)

Patients	%	AMCase activity (nmol/ml/h)	Age	Sex M/F
7	14.6	16.147 (36.270 - 3.540)	65 (53 - 74)	3/4
36	75	0.439 (2.800 - 0.178)	53 (23 - 76)	19/17
5	10.4	0.047 (0.086 - 0.013)	48 (40 - 64)	4/1

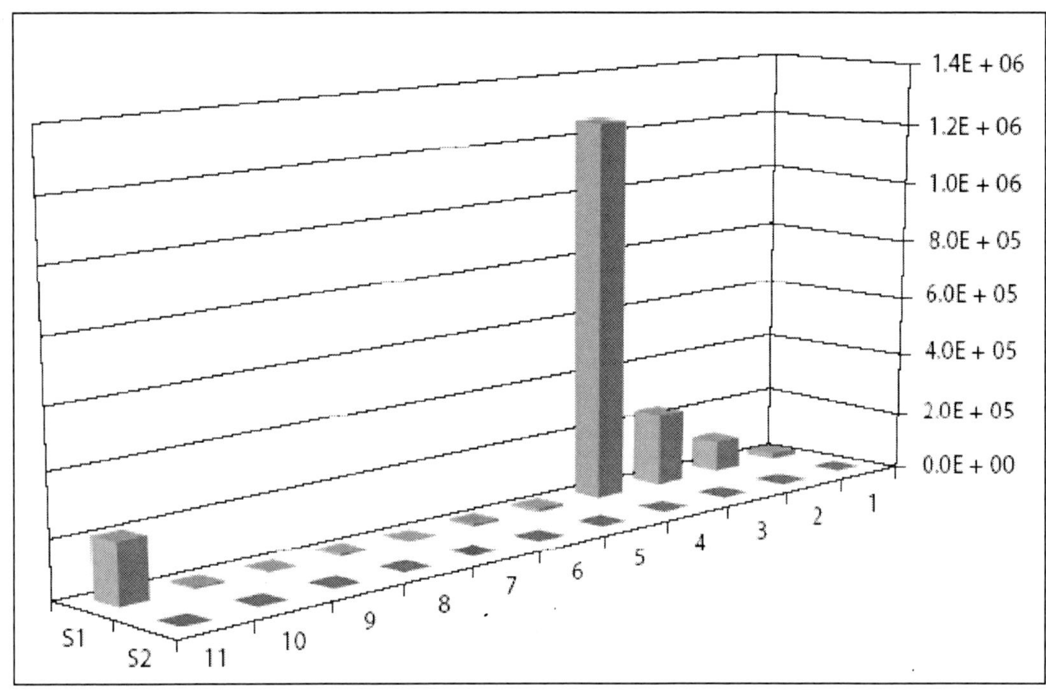

Figure 3. FITC-chitin activity expressed in fluorescence emission (CPS) before (S1) and after (S2) inhibition with allosamidin (9 μM), in the first 11 subjects (Paoletti *et al.* 2007).

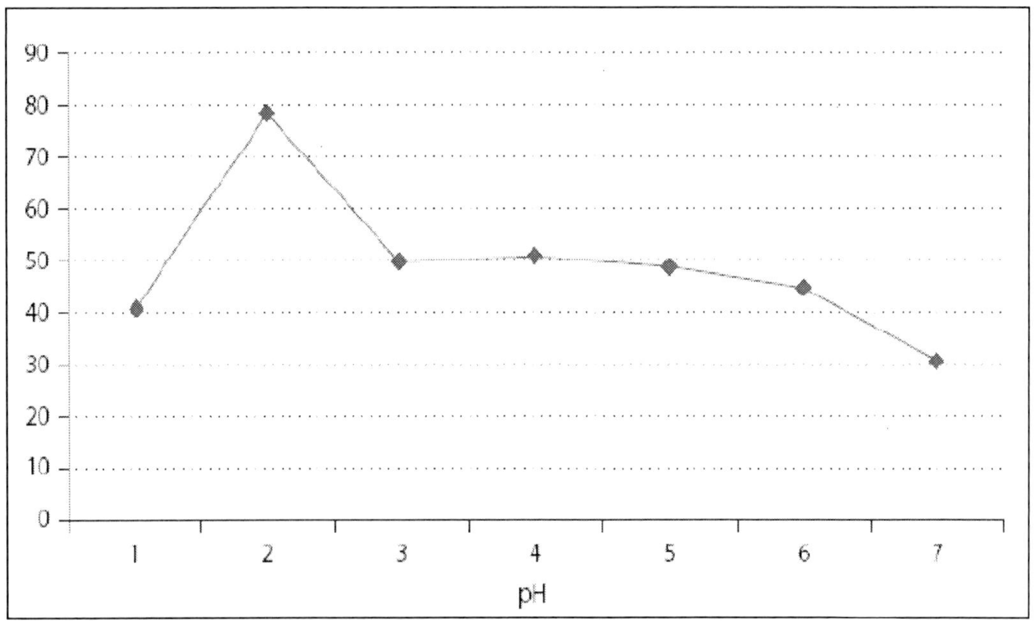

Figure 4. AMCase activity (nmol/ml/h) of gastric juice of subject 4 at different pH values (Paoletti *et al.* 2007) .

4. Is Chitin Digestion Really Affected by Chitinase Presence in Gastric Juice?

We performed microscopic observations of chitin degradation, utilising fly wings as chitinous substrates (Karelina, 2007) and the gastric juice of a patient having an AMCase activity of 19.41 nmol/ml/h. After immersion at 37°C for 4 hours of fly forewings (obtained from *Calliphora vomitoria* L. and *C. vicina* R.-D.) in human gastric juice, we assessed morphological changes using a straight microscope (Leica DMR, software IM500).

Figure 5 shows the damage caused by the gastric juice treatment after 2, 4 and 6 hours (Figures 5b-d), compared with a control wing processed with HCl 1 M at pH 2, in which the surface morphology appears intact (Figure 5a).

Figure 6 shows: microscopic observations (stereomicroscope Leica MZ16, software IM500) of chitin degradation after 8 hour treatment with human gastric juice, caused by the chitinase activity, on the border of two whole wings (Figures 6a and 6c) and a particular of these wings (Figures 6b and 6d); the effect of gastric juice after 8 hours treatment on wing's surface (Figure 6e) at ESEM microscope (ESEM XL30, software XL30); the particular of a wing (Figure 6f) treated only with HCl 1 M at SEM microscope (CAMBRIDGE STEREOSCAN 260, software Matrox Intellicam v. 2.0).

The chitinase effects on wing integrity, as judged by morphological examination, appear rather superficial, and could be considered as evidence of digestion due to AMCase presence in gastric juice. Since no modification was observed incubating the wing with HCl 1 M, the only possible conclusion is that modification of fly wings could be the effect of gastric juice. However, when we measured the amount of chitin metabolite in the gastric juice, where the fly wings were incubated, the *N*-acetylglucosamine (GlcNac) content, measured in solution, was very low. This result was not surprising, because the main product of chitin breakdown after incubation with endochitinase is a mixture of chito-oligomers $GlcNAc_n$ with $n > 2$ and chitobioside ($n=2$). In our previous article (Paoletti *et al.* 2007) we demonstrated that in humans the gastric chitinase functions differently as an endochitinase from the other chitinases present in nature. In fact, when we varied the incubation time from 30 to 120 minutes the fluorescence intensity of FITC-chitin increased rapidly, reaching values from 200,000 to 1,400,000 CPS (data not shown). This suggests that hydrolysis starts randomly in the middle of the FITC-chitin polysaccharide chain, generating fluorescent fragments. On the contrary exochitinase activity, typical of bacterial chitinases (Fusetti *et al.* 2002), is characterised by a slow increment of fluorescence, since the hydrolysis takes place at the terminal of chitin molecules.

We exclude the possibility that gastric chitinase activity in our patients could be associated with food residues, since the patients did not consume food for at least 14 hours prior to the gastroscopy; and we are certain that the chitinases present are not produced by gastric flora, since the several species present in gastric juice secrete exochitinases and not endochitinase (Fusetti *et al.* 2002). This is also confirmed by extremely low *N*-acetyl-glucosamine content after incubation with fly wings.

As negative control we treated gastric juices with allosamidin in order to inhibite chitinase activity (Sakuda and Sakurada, 1998).

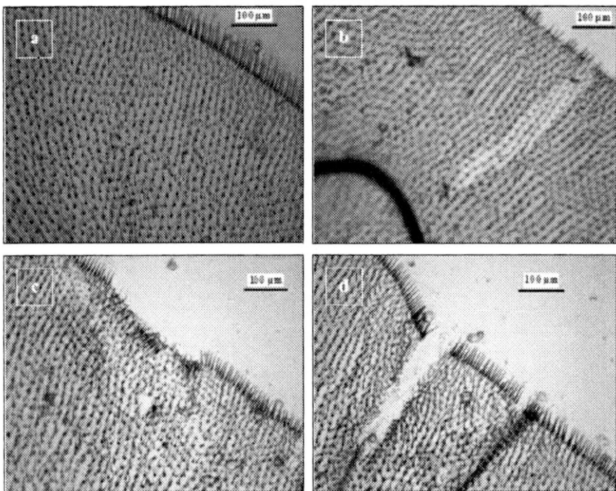

Figure 5. Microscopic observations with the straight microscope (Leica DMR, software IM500) of chitin degradation after 2 (b), 4 (c) and 6 (d) hour treatment with gastric juice of patient n°19 (19.410 nmol/ml/h), utilizing fly wings (C. vomitoria) as chitinous substrate. Damage probably caused by the chitinase activity (b, c, d) compared with blank wing processed with HCl 1 M (a). (Images from Kira Karelina, Thesis, University of Padova, 2007).

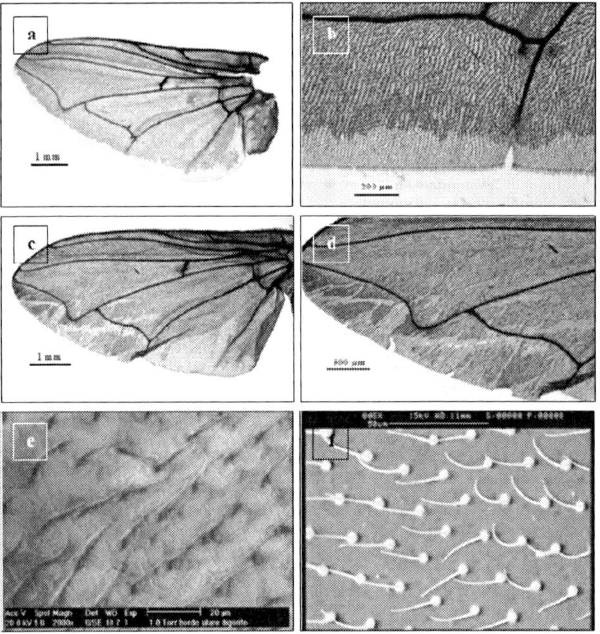

Figure 6. Microscopic observations of chitin degradation after 8 hour treatment with human gastric juice utilizing fly wings (*C. vomitoria*) as chitinous substrate. Vision at the stereomicroscope (Leica MZ16, software IM500) of damages, probably caused by the chitinase activity, on the border of two whole wings (a and c) treated with gastric juice and a particular of these wings (b and d). Effect of gastric juice after 8 hours treatment on wing's surface (e) at ESEM microscope (ESEM XL30, software XL30). Particular of a wing (f) treated only with HCl 1 M at SEM microscope (CAMBRIDGE STEREOSCAN 260, software Matrox Intellicam v. 2.0). (Images from Andrea Alfieri, Thesis, University of Padova, 2008).

In fact digestion blockage has been demonstrated because we observed no production of *N*-acetyl-glucosamine. We infer that chitinase and not other potential enzymes are involved in wings digestion.

Since the major visible effect is the softening and corrosion of the margins of wings, we think, based on these observations, that AMCase function in gastric juice may be to provide seasoning rather than extended fragmentation of the chitinous cover, permitting, however, a better digestion of the internal contents of chitin-covered organisms like caterpillars, insect larvae or adults.

5. Analysis of CHIT and AMCase Gene Expression with Qrt-PCR in Human Gastric Mucosa Biopsy

Stomach biopsies (antral mucosa specimens collected during gastroscopy on Italian patients at Padova University Hospital) from 27 patients have been studied for *CHIT* (Figure 7) and *AMCase* (Figure 8) gene expression with QRT-PCR in human gastric antral mucosa (Cozzarini, 2007; Cozzarini *et al.* 2008).

5.1. Total RNA Extraction and Quantification

Tissue samples were homogenised in 1 ml of Trizol® reagent (Invitrogen) using a power homogenizer (Ultra-turrax-T8; IKA® WERKE). After incubation of the homogenised samples for 5 min at room temperature, 200 µl of chloroform was added. Samples were mixed vigorously and held in ice for 15 min, centrifuged for 15 min at 16,100 *g* at 4°C. The RNA was precipitated from the aqueous phase by adding an equal volume of isopropanol. Then RNA was dissolved in RNase-free water.

For each sample, total RNA concentration was determined with spectrophotometrical analysis with NanoDrop® ND-1000 Spectrophotometer (NanoDrop Technologies, Wilmington, USA).

The quality of each sample total RNA was measured with the Agilent 2100 BioAnalyzer System (Agilent Technologies). Only good quality RNA was used for further experiments.

5.2. Quantitative Polymerase Chain Reaction

cDNA was prepared from 1 µg of single patient total RNA using Superscript II (Invitrogen) and oligo (dT), following manufacturer's instruction.

Quantitative real-time PCR (QRT-PCR) based on the SYBR™ Green chemistry (Applied Biosystems, Foster City, CA) was carried out to test the expression level of *CHIT* and *AMCase*. 50 ng of cDNA reverse-transcribed from patient's total RNA was amplified using DyNAmo HS SYBR Green qPCR Kit (Finnzymes).

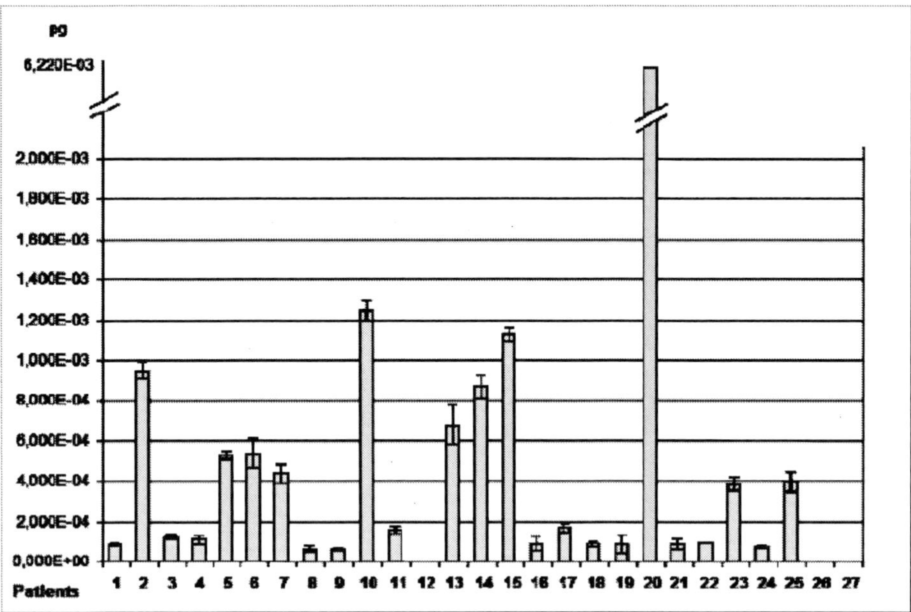

Figure 7. Quantitative real-time PCR analysis of Chitotriosidase (CHIT) mRNA in stomach biopsies from 27 patients. Patients 12, 26 and 27 showed no CHIT expression, while the CHIT-specific mRNA mean quantity was 5.399*10-4 ± 4.887*10-5 pg. Absolute expression levels are shown in mRNA pg, based on a standard curve. Standard Deviations (SD) calculated on the experimental replicates are also shown. (Figure from Elisa Cozzarini Thesis, University of Padova, 2007).

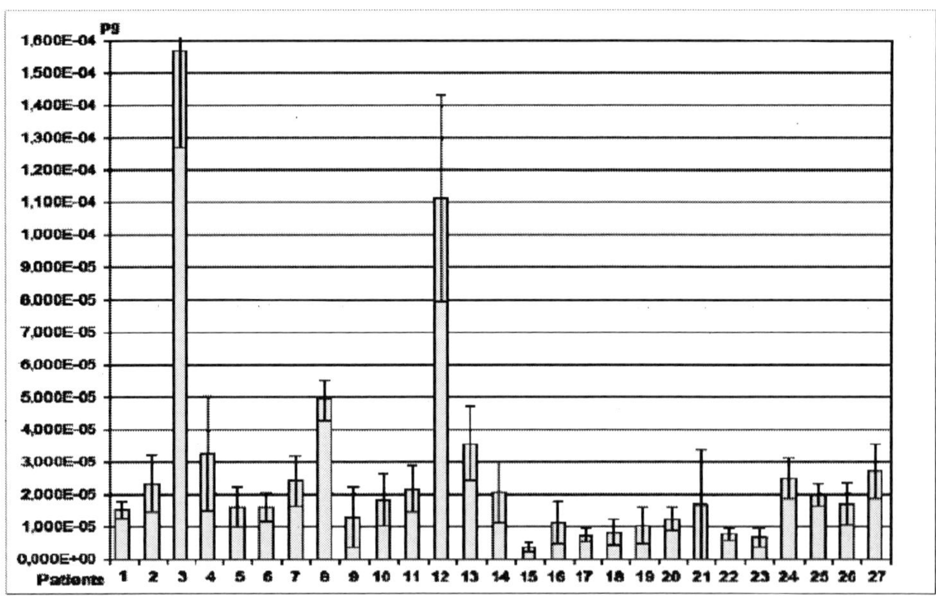

Figure 8. Quantitative real-time PCR of Acidic Mammalian Chitinase (AMCase) in stomach biopsies from 27 patients. Absolute expression levels are shown in mRNA pg, based on a standard curve. Standard Deviations (SD) are also shown. (Figure from Elisa Cozzarini Thesis, University of Padova, 2007).

PCR reactions were performed in a GeneAmp 9600 thermocycler, coupled with a GeneAmp 5700 Sequence Detection System (Applied Biosystems, Foster City, CA). Gene-specific oligonucleotides were designed using Primer 3 software (http://www-genome.wi.mit.edu/cgi-bin/primer/primer3.cgi).

CHIT primer sequences were designed according to the deposited mRNA ref-sequence (GenBank Accession No. NM_003465). *CHIT* primer sequences were 5'-GGGATGCTGGCCTACTATGA-3' (forward) and 5'-TAGGGCACCTTCTGATCCTG-3' (reverse). Primer sequences for *AMCase* were designed on the basis of the two deposited transcript variants (GenBank Accession NM_021797 and NM_201653). *AMCase* primer sequences were 5'-CTACGACCTCCATGGCTCCT-3' (forward) and 5'-TGCTCCATTGTCCTTCCAGT-3' (reverse) and were chosen to specifically identify both *AMCase* transcript variants.

In order to quantify *CHIT* and *AMCase* expression, we used the absolute quantification method, comparing patient samples to a standard curve (Bustin, 2000). The standard curves for each gene were constructed with five serial ten-fold dilutions of a purified and quantified PCR product (obtained with the same primers used for the qRT-PCR), with a sensitivity range of $7.4*10^{-2}$ pg – $7.4*10^{-5}$ pg (*CHIT*) and $3.8*10^{-2}$ pg – $3.8*10^{-6}$ pg (*AMCase*). Threshold cycles (Cts) obtained with patient's samples were compared to Cts generated from the specific standard curve. Results were expressed in quantity (pg) of specific amplified nucleic acid.

The equation used to calculate the pg quantity of template (mRNA) that was present at the beginning of the qRT-PCR reaction was: $pg = 10^{(Ct - b)/a}$, where a: the angular coefficient (slope) and b: the intercept.

Since the two genes we studied (*CHIT* and *AMCase*) were expressed at a very low level in the gastric mucosa, a classic qRT-PCR reaction (40 cycles) was not able to quantify the initial mRNA quantity. We then decided to stop the qRT-PCR reaction during the linear phase (i.e. after 10 amplification cycles), to add new mix (master mix and half concentration of primers) and to start with a second classic qRT-PCR reaction (40 cycles). The linearity of the reaction was preserved and the sample Cts fell in the standard curve sensitivity range.

Three replicates for each patient cDNA were performed, and mean and SD were calculated.

The correlation of the *CHIT* and *AMCase*'s expression level in human gastric mucosa with the severity of stomach inflammation and the *H. pylori* infection was analysed by the χ^2 test and significant P value < 0.05 was accepted.

6. Results

CHIT was expressed at low levels in the stomach of all the studied individuals except three patients who showed no expression (Figure 7). In particular, *CHIT* mRNA quantity varied between $6.315*10^{-5}$ pg to $6.220*10^{-3}$ pg, with a mean quantity of $5.427*10^{-4} \pm 4.703*10^{-5}$ pg.

Quantitative real-time PCR confirmed the presence of *AMCase* mRNA in human gastric mucosa, even though this gene was expressed at a very low level (Figure 8). In particular the majority of our patients examined (19/27) showed *AMCase* mRNA quantity between $1.014*10^{-5} \pm 2.432*10^{-6}$ pg and $3.529*10^{-5} \pm 7.815*10^{-6}$ pg. In 5 patients *AMCase* was weakly expressed (about 10^{-6} pg). Three patients had the highest *AMCase* mRNA expression corresponding to a mean value of $1.056*10^{-4} \pm 7.592*10^{-5}$ pg.

This research indicated an interesting positive correlation between *CHIT* expression and gastric inflammation (P value= 0.026) and *Helicobacter pylori* infection (P value= 0.016). We did not observe any positive correlation between AMCase activity in gastric juice and mRNA expression in gastric mucosa. This could be due to the complexity of the human stomach tissue, which has different biological attitudes in fundus, corpus and antrum portions.

Our study has demonstrated for the first time that *CHIT* mRNA is also present in gastric mucosa and this result could represent further evidence of the involvement of Chitotriosidase enzyme in human immune response (Cozzarini *et al.* 2008).

7. Discussion

In this study, we demonstrated AMCase activity in human stomach associated with chitin digestion and possibly with digestion of chitin-containing foods such as invertebrates and mushrooms. The function of protection from gastric parasites is another possibility. The fraction of the population (14.6% in Caucasian, Padova-living peoples studied) having high activity of AMCase in the stomach could have more potential for chitinase digestion. No one has, to our knowledge, looked at the link between AMCase genotypes and intestinal parasites, but if we want to put forward the hypothesis that gastric AMCase can protect against parasitic helmints, we need to consider that many parasites only have a very short gastric phase, but have a longer developmental phase in the lung (*Ascaris lumbricoides*), where AMCase could also play a role. On the contrary AMCase or macrophagic Chit present in the stomach may possibly better protect from parasites such as nematodes like *Ascaris* and flatworms like *Taenia*, because the eggs of both have some chitin cover (Wimmer *et al.* 1998; Harter *et al.* 2003). Chitin is definitely present in the egg-shells of *Ascaris lumbricoides* (Sromová and Lýsek, 1990), but it is doubtful whether external chitinases can access it at all. Since the eggs are produced in the intestine where the adult worms reside, so they are not in contact with gastric chitinases. Only for ingested Ascaris or *Taenia* eggs, therefore, could this protective mechanism be operative.

It is probable that the macrophage cells infiltrating gastric mucosa of patients who had gastroscopy for clinical symptoms of gastritis produced Chit in response to the inflammation, with an optimum at pH 5.2 or higher. In fact, we found a high correlation between *CHIT* mRNA expression and *Helicobacter pylori* infection (P value = 0.016). Phlogosis and *CHIT* mRNA have also shown consistent correlation (P value= 0.026).

Chitotriosidase apparently has similar effects on chitin, but the *CHIT* expression in gastric mucosa respond quickly to inflammation and *Helicobacter pylori* infection as a consequence of macrophages activation.

Even if the potential to digest chitin has been documented in 43/48 (i.e. 89.6%) of our Italian patients, we do not know the minimum quantity of chitinases needed to effectively digest the chitin associated with insects and crustaceans.

Food bearing chitin can, possibly, be better digested by peoples having higher rate of chitinase expression. Our current experimental evidence on fly forewings suggest chitinases have a role in the seasoning and shallow digestion of this wing substrate. We infer that chitinases may act as complements in breaking down chitinous cuticles and allowing other enzymes to better utilise invertebrates' or mushrooms' content. However, more work is needed to elucidate these steps in greater detail, including a more general view of chitin digestion in animals not limited to humans.

Do people whose diet relies on chitinous food have a different gut flora from people who do not? in other words, does the gut flora of entomophagous people contain bacterial populations where chitinolytic bacteria are more prevalent? An example could be the relation between termites and cellulosa or bacterial chitinases found in the stomach of fish, or the occurrence of several chitinolytic bacterial strains in the gut of entomophagous bats (Alwin Prem Anand and Sripathi, 2004).

Population-epidemiologic studies are needed to assess different populations in adopting different food resources containing and not containing chitin.

Chitin has a very long history in animal and human evolution, and chitinases have coevolved within different lifestyles as evolutionary consideration in several parts of this book seem to demonstrate.

The higher chitinase activity in tropical human populations with higher rates of entomophagy could represent an adactative response to alimentary habits, conferring increased resistance against parasitic infection in these areas and faciliting the digestion of chitin through bacterial chitinases. This view, however, needs to be more carefully substantiated.

Acknowledgments

We are greately indebted to Franco Falcone for discussion of several points of our work. Andrew Baldwin has provided assistance with manuscript review. Many thanks to Kira Karelina and Andrea Alfieri for the fly forewing's photos.

References

Alwin Prem Anand A, Sripathi K. Digestion of cellulose and xylan by symbiotic bacteria in the instestine of the Indian flying fox (*Pteropus giganteus*). *Comp Biochem Phisiol.* 2004; 139: 65-69.

Barone R, Di Gregorio F, Romeo MA, Schilirò G, Pavone L. Plasma chitotriosidase activity in patients with β–thalassemia. *Blood Cells Mol Dis.* 1998; 25: 1-8.

Barone R, Bertrand G, Simpore J, Malaguarnera M, Musumeci S. Plasma chitotriosidase activity in beta-thalassemia major: a comparative study between Sicilian and Sardinian patients. *Clin Chim Acta.* 2001; 306: 91-96.

Boot RG, Blommaart EFC, Swart E, Ghauharali-Van der Vlugt K, Bijl N, Moe C, Place A, Aerts JMFG. Identification of a novel acidic mammalian chitinase distinct from chitotriosidase. *J Biol Chem.* 2001; 276: 6770-6778.

Barone R, Simporé J, Malaguarnera L, Pignatelli S, Musumeci S. Plasma chitotriosidase activity in acute *Plasmodium falciparum* malaria. *Clin Chim Acta.* 2003; 331: 79-85.

Boot RG, Bussink AP, Verhoek M, de Boer PA, Moorman AF, Aerts JM. Marked differences in tissue-specific expression of chitinases in mouse and man. *J Histochem Cytochem.* 2005; 53: 1283-1292.

Bouzas L, Carlos Guinarte J, Carlos Tutor J. Chitotriosidase activity in plasma and mononuclear and polymorphonuclear leukocyte populations. *J Clin Lab Anal.* 2003; 17: 271-275.

Bukkens SGF. The Nutritional Value of edible Insects. *Ecol Food Nutr.* 1997; 36: 287-319.

Bukkens SGF. Insects in the Human Diet: Nutritional Aspects. In: Paoletti M.G. Ecologic Implications of Minilivestock. *Science Publishers Inc.* Enfield N.H., 2005; pp. 545-578.

Bustin S.A. Absolute quantification of mRNA using real-time reverse transcription polymerase chain reaction assays. *J Mol Endocrinol.* 2000; 25: 169-193.

Cataldo F, Simporè J, Greco P, Ilboudo D, Musumeci S. *Helicobacter pylori* infection in Burkina Faso: an enigma within an enigma. *Dig Liv Dis*, 2004; 36: 589-593.

Cohen-Kupiec R, Chet I. The molecular biology of chitin digestion. *Curr Opin Biotechnol.* 1998; 9: 270-277.

Cozzarini E. Analisi dell'espressione del gene *Acidic Mammalian Chitinase* in mucosa gastrica umana. *Thesis on Biology,* Padova University Dept Biology, Padova, Italy, 2007.

Cozzarini E, Bellin M, Norberto L, Polese L, Musumeci S, Lanfranchi G, Paoletti MG. *CHIT* and *AMCase* expression in human gastric mucosa correlates with inflammation and *Helicobacter pylori* infection? *European Journal of Gastroenterology & Hepatology* (in press) 2008 .

Damini R. La digeribilita' della chitina per l'uomo e la sua importanza per le popolazioni che si nutrono anche di insetti. *Thesis on Biology,* Padova University Dept Biology. Padova, Italy, 2006.

DeFoliart GR. Insects as human food. *Crop Protection*, 1992; 11: 395-399.

DeFoliart GR. Insects as food : why the western attitude is important. *Ann Rev Entomol.* 1999; 44: 21-50.

DeFoliart GR. The human use of insects as a food resource: a bibliographic account in progress. 2002. In: http://www.food-insects.com/

Dixon MF, Genta RM, Yardley JH. Classification and granding of gastritis. The updates Sydney System. *Am J Surg Pathol.* 1996; 20: 1161-1181.

Dufour DL. Insects as food: a case study from the northwest Amazon. *Amer Anthropol.* 1987; 89: 383-396.

Fusetti F, von Moeller H, Houston D, Rozeboom HJ, Dijkstra BW, Boot RG, Aerts JM, van Aalten DM. Structure of human chitotriosidase. Implications for specific inhibitor design and function of mammalian chitinase-like lectins. *J Biol Chem.* 2002; 277: 25537-25544.

Gardiner T. Dietary N-acetylglucosamine (GlcNac): absorption, distribution, metabolism, excretion (ADME), and biological activity. *Glycoscience and Nutrition.* 2000; 1: 1-3.

Gianfrancesco F, Musumeci S. The evolutionary conservation of the human chitotriosidase gene in rodents and primates. *Cytogenet Genome Res.* 2004; 105: 54-56.

Goto M, Fujimoto W, Nio J, Iwanaga T, Kawasaki T. Immunohistochemical demonstration of acidic mammalian chitinase in the mouse salivary gland and gastric mucosa. *Arch Oral Biol.* 2003; 48: 701-707.

Harter S, Le Bailly M, Janot F, Bouchet F. First Paleoparasitological Study of an Embalming Rejects Jar Found in Saqqara, Egypt. *Mem Inst Oswaldo Cruz.* 2003; 98: 119-121.

Hollak CE, van Weely S, van Oers MH, Aerts JM. Marked elevation of plasma chitotriosidase activity. A novel hallmark of Gaucher disease. *J Clin Invest.* 1994; 93: 1288-1292.

Jollès P, Muzzarelli RAA ed. Chitin and chitinases, EXS 87, Birkhauser Verlag, Basel-Switzerland, 1999.

Karelina K. L'enzima *Acidic Mammalian Chitinase* presente nei succhi gastrici umani digerisce la chitina dell'esoscheletro degli insetti. *Thesis on Biology.* Padova University Dept Biology, Padova, Italy, 2007.

Lee P, Waalen J, Crain K, Smargon A, Beutler E. Human chitotriosidase polymorphisms G354R and A442V associated with reduced enzyme activity. *Blood Cells Mol Dis.* 2007; 39: 353-360.

Malaisse F. Se nourrir en foret claire africaine. Approche ecologique et nutritionelle. Presses Agronomiques de Gembloux, Gembloux, Belgium, 1997.

Musumeci M, Simpore J, Barone R, Angius A, Malaguarnera L, Musumeci S. Synchronic macrophage response and *Plasmodium falciparum* malaria. *Pak J Biol Sci.* 2005; 8: 954-958.

Muzzarelli RAA ed. Chitin Enzymology, Atec, 2001.

Paoletti MG, Bukkens SGF, eds. Minilivestock, special issue. *Ecol Food Nutr.* 1997; 36: 95-346.

Paoletti MG, Dufour DL, Cerda H, Torres F, Pizzoferrato L, Pimentel D. The importance of leaf- and litter-feeding invertebrates as sources of animal protein for the Amazonian Amerindians. *Proc R Soc Lond B Biol Sci.* 2000; 267: 2247-2252.

Paoletti MG ed. Ecological implication of Minilivestock (Role of rodents, frogs, snails, and insects for sustainable development), *Science Publishers Inc.* Enfield NH,USA, 2005

Paoletti MG. State of Amazon. Biodiversity management and loss of traditional knowledge in the largest forest. In: Morgan De Dapper Tropical Forests in a Changing Global Context. The Royal Academy of Overseas Sciences, Bruxelles, 2005a; pp.: 93-111.

Paoletti MG, Norberto L, Damini R, Musumeci S. Human gastric juice contains chitinase that can degrade chitin. *Ann Nutr Metab.* 2007; 51: 244-251.

Renkema GH, Boot RG, Strijland A, Donker-Koopman WE, Van den Berg M, Muijsers AO, Aerts JMFG. Synthesis, sorting, and processing into distinct isoforms of human macrophage chitotriosidase. *Eur J Biochem.* 1997; 244: 279-285.

Renkema GH, Boot RG, Au FL, Donker-Koopman WE, Strijland A, Muijsers AO, Hrebicek M, Aerts JMFG. Chitotriosidase, a chitinase, and the 39-kDa human cartilage glycoprotein, a chitin-binding lectin, are homologues of family 18 glycosyl hydrolases secreted by human macrophages. *Eur J Biochem*. 1998; 251: 504-509.

Sakuda S, Sakurada M. Preparation of biotinylated allosamidins with strong chitinase inhibitory activities. *Bioorg Med Chem Lett*. 1998; 8: 2987-2990.

Sromová D, Lýsek H. Visualization of chitin-protein layer formation in *Ascaris lumbricoides* egg-shells. *Folia Parasitol*. 1990; 37: 77-80.

Suzuki M, Fujimoto W, Goto M, Morimatsu M, Syuto B, Iwanaga T. Cellular expression of gut chitinase mRNA in the gastrointestinal tract of mice and chickens. *J Histochem Cytochem*. 2002; 50: 1081-1089.

Talent JM, Gracy RW. Pilot study of oral polymeric N-acetyl-D-glucosamine as a potential treatment for patients with osteoarthritis. *Clin Ther*. 1996; 910: 1184-1190.

Tikhonov VE, Lopez-Llorka LV, Salinas J, Monfort E. Endochitinase activity determination using N-fluorescein-labeled chitin. *J Biochem Biophys Methods*. 2004; 60: 29-38.

Van Eijk M, Van Roomen C, Renkema H, Bussink A, Andrews L, Blommaart F.C, Sugar A, Verhoeven A, Boot R, Aerts JMFG. Characterization of human phagocyte-derived chitotriosidase, a component of innate immunity. *Int Immunol*. 2005; 17: 1505-1512.

Whary MT, Sundina N, Bravo LE, Correa P, Quinones F, Caro F, Fox JG. Intestinal Helminthiasis in Colombian Children Promotes a Th2 Response to *Helicobacter pylori*: Possible Implications for Gastric Carcinogenesis. *Cancer Epidemiol Biomarkers Prev*. 2005; 14: 1464-1469.

Wimmer M, Schmid B, Tag C, Hofer HW. *Ascaris suum*: Protein Phosphotyrosine Phosphatases in Oocytes and Developing Stages. *Exp parasitol*. 1998; 88: 139-145.

Zhu Z, Zheng T, Homer RJ, Kim YK, Chen NY, Cohn L, Hamid Q, Elias JA. Acidic mammalian chitinase in asthmatic Th2 inflammation and IL-13 pathway activation. *Science*. 2004; 304: 1678-1682.

In: Binomium Chitin-Chitinase: Recent Issues
Editor: Salvatore Musumeci and Maurizio G. Paoletti
ISBN 978-1-60692-339-9

Chapter XXI

Role of Chitinase in Gastroenterology

Emiko Mizoguchi[16] and Mayumi Kawada
Department of Medicine and Center for the Study of Inflammatory Bowel Disease, Gastrointestinal Unit, Massachusetts General Hospital and Harvard Medical School, Boston, MA 02114, U.S.A.

Abstract

Inflammatory bowel diseases (IBD), including Crohn's disease (CD) and ulcerative colitis (UC), are a group of chronic inflammatory disorders that affect individuals throughout life. The etiology and pathogenesis of these two major forms of IBD is largely unknown. Several studies have indicated that dysregulated host/enteric microbial interactions are required for the development of IBD. Both the colonic epithelial cells (CECs) that form a barrier between the luminal contents (including microorganisms and other antigens) and the underlying immune cells play important roles in maintaining adequate host/microbial interactions. In fact, CECs actively participate in the induction of both innate and adaptive immune responses to the luminal contents by inducing several specific molecules. By utilizing DNA microarray screening technology, our group has unexpectedly identified a novel intestinal inflammation-associated molecule, Chitinase 3-like-1 (CHI3L1, YKL-40 or HC-gp39), which is produced mainly by CECs and macrophages only under inflammatory conditions. We have also provided novel insight into the pathophysiological role of CHI3L1 for enhancing bacterial adhesion and invasion on/into CECs. CHI3L1 is characterized by a strong binding affinity to chitin without enzymatic activity. The ability of a host to produce chitinases, which have enzymatic activity, could be a critical factor in the regulation of the initial immune response against pathogen (e.g., fungi, parasites)-derived chitin. In contrast, exaggerated production of mammalian chitinases may cause harmful and pathogenic effects in mucosal regions. Although bacteria do not possess chitin as a structural component, some strains of bacteria can express chitin-binding proteins (CBPs) upon exposure to chitin. In

16 Corresponding Author: Emiko Mizoguchi, M.D., Ph.D. Gastrointestinal Unit, GRJ-702, Massachusetts General Hospital, Boston, MA 02114, USA, Phone: (617) 726-7892, Fax: (617) 726-3673, E-mail: emizoguchi@partners.org.

fact, our recent experimental results suggest that over-expression of CHI3L1 on CECs and CBP on bacteria can significantly enhance the bacterial adhesion on CECs. Therefore, bacterial CBPs may form an important bridge directly or indirectly in facilitating the binding of luminal bacteria to CHI3L1 on the colonic epithelial surface. Interestingly, many pathogenic and potentially pathogenic bacteria are presumably able to express CBPs. In this chapter, we will discuss about the physiological function of mammalian chitinases and bacterial CBPs in the intestine and their potentially pathogenic role during the development of human IBD.

Abbreviations

Ab: antibody;
AIEC: adherent invasive *Escherichia coli*;
AMCase: acidic mammalian chitinase;
CBM: carbohydrate binding module;
CBP: chitin binding protein;
CD: Crohn's disease;
CECs: colonic epithelial cells;
CFU: colony-forming units;
CHI3L1: chitinase 3-like-1;
CLP: chitinase-like protein;
ECM: extracellular matrix;
GbpA: GlcNAc binding protein A;
GlcNAc; N-acetylglucosamine;
HA: hyaluronic acid;
IBD: inflammatory bowel disease;
KO: knockout;
LP: lamina propria;
PBMC: peripheral blood mononuclear cells;
RA: rheumatoid arthritis;
3 Mu: mutations of Thy-54,
Glu-55 and Glu-60;
siRNA: short interfering RNA,
UC: ulcerative colitis;
VSMC: vascular smooth muscle cell;
WT: wild type;

1. Introduction

Chitin, the linear polymer of β 1,4-linked N-acetylglucosamine (GlcNAc), is the second most abundant glycopolymer found in nature next to cellulose, and chitin is the key component of many species including exoskeletons of insects, shells of crustaceans, and cell walls of fungi since chitin effectively protects these organisms from potentially harmful aspects of their environment (Gooday 1999; Suzuki *et al.* 2002b). Chitin accumulation is regulated by special degradation enzyme chitinases (Shahabuddin *et al.* 1993), which are highly conserved between species during the course of evolution (Gianfrancesco *et al.* 2004), and chitinase production is a common feature of anti-parasite responses of lower life forms against chitin-containing organisms (Herrera-Estrella *et al.* 1999; Palli *et al.* 1999). For a long time, it was thought that endo-glucosaminidases that fragment chitin was not produced by human because chitin does not exist in mammals. However, human chitinase family members such as chitotriosidase (chitinase-1) and acidic mammalian chitinases (AMCase) have been recently identified (Hollak *et al.* 1994; Renkema *et al.* 1995; Boot *et al.* 1995; Reese *et al.* 2007). Zhu et al. elegantly demonstrated that AMCase is mainly induced on bronchial epithelial cells and macrophages via a T helper-2 (Th2)-specific, interleukin-13 (IL-13)-mediated pathway in a murine asthma model (Zhu *et al.* 2004). Paradoxically, recent study also suggested that chitin induces the alternative macrophage activation *in vivo* and subsequently recruits the IL-4-expressing eosinophils and basophils that are associated with allergic immune responses, and AMCase efficiently blocks the chitin-mediated tissue infiltration (Boot *et al.* 2001). At the same timing, our group identified that another type of mammalian chitinase, chitinase 3-like-1 (CHI3L1/YKL-40/HC-gp39), is over-expressing on CECs and macrophages in animal models of colitis and active but not quiescent IBD (Mizoguchi, 2006). CHI3L1 belongs to the same family 18 glycohydrolases, which includes a broad range of prokaryotic and eukaryotic chitinases, but has no apparent enzymatic activity and grouped in the family of chi-lectins (chitinase-like-lectins) (Hakala *et al.* 1993). Interestingly, chi-lectins possess the substrate (e.g., chitin)-binding cleft of chitinases although they lack of apparent enzymatic activity (van Aalten *et al.* 2001). Three amino acid residues Asp, Glu, and Asp in chitinases play an essential role for the enzymatic activity. However, the corresponding residues in human CHI3L1 are Asp^{115}, Leu^{119}, and Asp^{186}, and the mutation of Glu to Leu deprives CHI3L1of glycohydrolase activity (Watanabe *et al.* 1993; Recklies *et al.* 2002).

CHI3L1 seems to be involved in inflammation and tissue remodeling since elevated CHI3L1 levels are present in patients with meningitis, rheumatoid arthritis, hepatic fibrosis, bronchial asthma, IBD, and breast or lung cancer (Kelleher *et al.* 2005; Ostergaard *et al.* 2002; Nordenbaek *et al.* 1999; Vind *et al.* 2003; Johansen, 2006b; Johansen *et al.* 2004; Chupp *et al.* 2007, Johansen *et al.* 1999), also described in details in Chapter 13 in this book. Elevated circulating CHI3L1 levels are positively associated with the bad prognosis of the disease, suggesting CHI3L1 may participate in the pathogenesis of these inflammatory disorders. In this chapter we will summarize the latest information of mammalian chitinases, mainly CHI3L1 in the pathogenesis of IBD.

2. Potential Role of CHI3L1 in Chronic Inflammatory Disorders

CHI3L1 is a 40 kDa mammalian glycoprotein and the gene is located on chromosome 1q32.1 consisting of 10 exons (Zhu *et al.* 2004; Johansen *et al.* 1992; Rehli *et al.* 1997). CHI3L1 is mainly produced by human synovial cells (Nyirkos *et al.* 1990), osteosarcoma cells (Johansen *et al.* 1992), chondrocytes (Zhu *et al.* 2004), smooth muscle cells (Shackelton *et al.* 1995), macrophages (Renkema *et al.* 1998), neutrophils (Volck *et al.* 1998) and CECs (Mizoguchi, 2006). CHI3L1 could be used as a sensitive clinical marker for early detection of disease activity, prognosis of disease, and for testing the effectiveness of therapy (Peltomaa *et al.* 2001, Johansen 2006a). The precise physiological function of CHI3L1 is not completely revealed yet, but recent report suggests that it stimulates cell proliferation of connective tissue acting synergistically with the insulin-like growth factor utilizing the phosphoinositide 3-kinase (PI3K) signaling pathways (Recklies, 2002). Interestingly, Ling et al. demonstrated CHI3L1 down-regulates TNFα or IL-1 induced matrix metalloproteases and IL-8 production by human skin fibroblasts or articular chondrocytes, suggesting that this protein plays a pivotal role in regulating the inflammatory responses of connective tissues (Ling *et al.* 2004).

In patients with rheumatoid arthritis (RA), a chronic and systemic inflammation characterized by joint pain, swelling and stiffness, increased level of CHI3L1 has been detected in the blood (Vos *et al.* 2000b), inflamed tissues (Kirkpatrick *et al.* 1997; Baeten *et al.* 2000). CHI3L1 seems to be a naturally occurring autoantigen in RA and its class II mediated presentation by dendritic cells and macrophages to CD4 positive T cells *in vivo* has a critical role in the pathogenesis of human RA (Tsark *et al.* 2002, Verheijden *et al.* 1997). Interestingly, healthy donor derived peripheral blood mononuclear cells (PBMC) stimulated with CHI3L1 produce high amount of immunoregulatory cytokine IL-10. In contrast, 50% of patients with RA exhibit polarized Th1 phenotype by producing IFNγ after stimulating with CHI3L1 (van Bilsen *et al.* 2004). Therefore, CHI3L1 seems to shift from an anti-inflammatory toward a pro-inflammatory phenotype in RA patients and presumably in patients with other chronic type of inflammation including diabetes and multiple sclerosis since the autoreactive T cell response is highly polarized toward a proinflammatory Th1 phenotype, whereas a regulatory response is observed in health (Arif *et al.* 2004; Viglietta *et al.* 2004). These results suggest that the loss of CD4+ CD25+ regulatory T cells function that normally protects against autoimmunity as well as the emergence of pathogenic T cells (Th1, Th2, and/or Th17) must be important in the induction of autoimmune disease and it may be regulated by the presence of CHI3L1 in the locally inflamed tissues.

3. Role of CHI3L1 in Bacterial Adhesion and Invasion *In Vitro*

Recent clinical and experimental studies have shown that impaired intestinal barrier function permits penetration of harmful molecules into the underlining colonic lamina propria (LP) and initiates and/or perpetuates the intestinal inflammation (Sartor, 2001). It has been well documented that commensal bacteria are important environment-regulators in gastrointestinal tracts (Sartor, 2004). In fact, some potentially pathogenic commensal bacteria including adherent invasive *Escherichia coli* (AIEC) (Boudeau *et al.* 1999) and *Bacteroides* species (Sartor, 2004) are likely to be associated with the development of IBD. In patients with IBD, especially CD patients, a thick bacterial film (so called bio-film) was observed tightly attached to the mucosal surface, but not in control individuals who have self-limiting colitis (Swidsinski *et al.* 2002). However, the factors that are actually involved in the host/microbial interaction have remained to be unidentified. Characterization and identification of key molecule(s), which will regulate commensal bacterial adhesion and invasion on/into CECs are crucial to understand the initiation, exacerbation and perpetuation of IBD. We previously demonstrated that the expression of CHI3L1 molecule is highly induced in the CECs and LP cells (mainly macrophages) with several intestinal inflammation observed in T cell receptor α- and IL-10-knockout (KO) mice as well as dextran sulfate sodium (DSS)-induced colitis (Mizoguchi, 2006). In contrast, the expression of CHI3L1 is undetectable in the normal colon of not only WT mice but also of these IBD models without colitis. Enhanced CHI3L1 mRNA expression is also detectable in patients with active but not quiescent IBD (both UC and CD) or normal individual (Mizoguchi, 2006). Therefore, it could be predicted that CHI3L1 is upregulated specifically in inflammatory conditions, including both acute and chronic colitis. Studies were performed to investigate whether CHI3L1 possesses the ability to interact with pathogenic bacteria and subsequently digest them. As a method of bacterial invasion and adhesion assays, CEC lines (CMT93, HT29, Caco-2 and SW480) were cultured on 24-well tissue culture plates. After 8 hours, these cells were transfected with CHI3L1 expression vector or empty vector (mock). Twenty-four hours after transfection the cells were infected with *Salmonella typhimurium* or *E. coli* strains a multiplicity of infection of 20, and incubated for 1-2 hours. After the incubation, monolayers were washed with PBS twice, and cultured for 1 hour in the presence of 100 μg/ml gentamicin in the complete culture medium (penicillin/streptomycin free) to kill any extracellular bacteria. Cells were washed once with PBS and resolved with 500 μl of 1% Triton-X100 for 30 minutes. The cell lysates were plated on a Luria-Bertani agar plate (antibiotic free) for culture of intracellular invaded bacteria. Plates were cultured at 37°C overnight and colonies were counted the following day. To determine the numbers of colony-forming units (CFUs), the total numbers of bacterial colonies were counted and the ratio was calculated compared with the control mock-transfection. Total number of cell-associated bacteria (adherent and intracellular) was examined by washing with PBS 3 times after the bacterial infection, lysed with 1% Triton-X100.

To investigate the potential role of CHI3L1 in host/microbial interaction, CHI3L1-overexpressing CECs were infected with *S. typhimurium* and AIEC. An AIEC LF82 strain has been characterized as potentially pathogenic bacteria in the colons of CD patients by the strong and efficient ability in adhesion to and invasion into intestinal epithelial cells and macrophages (Boudeau *et al.* 1999). Surprisingly, significantly enhanced adhesion and invasion rates (4- to 6- fold) to CECs were found in *S. typhimurium* and AIEC on CHI3L1-overexpressing cells compared to the control mock-transfection. In contrast, non-pathogenic *E. coli* such as DH10B and DH5α strains did not adhere to or invade into CHI3L1 over-expressing CECs compared with mock control (Mizoguchi, 2006). Therefore, it is predicted that CHI3L1 on CECs may specifically enhance the adhesion and invasion of pathogenic (*S. typhimurium*) and potentially pathogenic (AIEC) but not non-pathogenic (DH10B, DH5α) bacteria.

To confirm that the enhanced bacterial adhesion and invasion into CECs is mediated by CHI3L1, studies to inhibit CHI3L1 activity were performed. For the inhibition of CHI3L1 activity, 100 μg/ml of anti-CHI3L1 antibody (Ab) was added to the cell culture medium before infection and was present through the bacterial infection period. CHI3L1-mediated bacterial adhesion and invasion were significantly inhibited by pretreatment with anti-CHI3L1 specific Ab. In contrast, control rabbit IgG did not affect the bacterial adhesion/invasion. In addition, a dose-dependent effect (from 50 to 250 μg/ml) of the anti-CHI3L1 Ab treatment was observed in bacterial adhesion and invasion assays (Mizoguchi, 2006). RNA interference (siRNA) is an effective method to knockdown specific gene expression (McManus *et al.* 2004). By utilizing the siRNA technology, the functional role of CHI3L1 in *S. typhimurium* adhesion and invasion into CECs was examined. Four different short interfering RNA (siRNA) duplexes specific for mouse CHI3L1 were custom-designed and synthesized. The siRNA duplexes were designed using the HiPerformance design algorithm integrated with a stringent homology analysis tool. The 4 highest-ranking siRNA duplexes generated by the algorithm were chosen, and combinations of 2, 3, or 4 siRNA duplexes were transfected into CMT93 cells. Combination of 4 siRNA duplexes significantly (256-fold) suppressed the both message and protein levels of CHI3L1, whereas combinations of 2 siRNA duplexes showed much less knockdown efficiency (2-to 16- fold). After inoculation of *S. typhimurium*, the highly efficient CHI3L1 knockdown group showed a resistance against the adhesion/invasion as indicated by the significantly fewer numbers of attached (1.7-fold) and intracellular bacteria (1.4- to 1.7-fold) in CMT93 cells transfected with a combination of 4 siRNA compared with the control siRNA-transfected cells. In contrast, there was no significant difference in the adhesion/invasion between less-efficient CHI3L1 knockdown groups (transfectants with combinations of 2 siRNA) and the control siRNA-transfected group (Mizoguchi, 2006).

Taken together, these results strongly suggest that CHI3L1 is induced on CECs specifically during the course of intestinal inflammation and possesses the ability to enhance the adhesion and invasion of intracellular bacteria. Therefore, it could be predicted that CEC-derived CHI3L1 more actively contributes to induction and /or exacerbation of colitis presumably by enhancing the adhesion and invasion of enteric bacteria to CECs during the initial phase of intestinal inflammation and acts as a pathogenic mediator of colitis.

4. Affinity of CHI3L1 and Chitin

Chitin, a polymer of GlcNAc, is the second most abundant polysaccharide in nature next to cellulose and found richly in fungi, crustaceans, insects, amphibians, nematodes, and house dust mites, but not in mammals (Shahabuddin *et al.* 1993; Debono *et al.* 1994). Chitin is degraded by chitinases that belong to members of the glycohydrolase family 18. This family that is characterized by an eight-fold α/β barrel structure includes bacterial as well as plant chitinases (Henrissat *et al.* 1997). In addition to being important for energy exchange, chitin degradation is essential in a variety of biological processes. For example, chitin-containing organisms produce chitinases to remodel their structure (Herrera-Estrella *et al.* 1999). Plant chitinases have been shown previously to possess antifungal and antivirus effects through binding to chitin (Shahabuddin *et al.* 1993). The chitin-binding domain of plant chitinases efficiently interacts with as well as digests chitin-containing organisms (Collinge *et al.* 1993).

Interestingly, the chitin-binding domain interaction with chitin-containing organisms also is preserved within mammalian chitinases. The physiological purpose for mammalian chitinases has been unrevealed yet, because chitin does not exist in mammals. Mammalian chitinases that possess enzymatic activity have a chitin-binding domain that contains six cysteine residues responsible for their binding to chitin (Tjoelker *et al.* 2000). In contrast, chitinase-like protein (CLP) does not contain such typical domains (Kzhyshkowska *et al.* 2006). In general, mammalian chitinases possess a conserved sequence motif (DXXDXDXE) on strand β4, and catalytic activity is mediated by the glutamic acid (E), which protonates the glycosidic bond with chitin. The neighboring aspartic acid (D) plays a key role in orienting the N-acetyl group of the -1 sugar for nucleophilic attack on the anomeric carbon, and stabilizes the subsequently formed oxazolium ion intermediate (van Aalten, 2001). In contrast, due to the substitution of an essential glutamic acid residue to leucine, CHI3L1 has no chitinase activity (Hakala *et al.* 1993), but still can bind to chitin and chito-oligosaccharides with high affinity through a preserved hydrophobic substrate binding cleft. This binding ability of CHI3L1 occurs via van der Waals interactions with the side chains of aromatic amino acid residues and several hydrogen-bonding interactions involving the sugar-hydroxyl groups (Renkema *et al.* 1998; Fusetti *et al.* 2003). CHI3L1 consists of multiple domains including signal sequence, catalytic domain and chitin-binding domain. The putative catalytic domain, predicted on the basis of its extensive homology with the catalytic domains on the other chitinases, is contained within the N-terminal 75% of the proteins. A chitin-binding cleft is present at C-terminal side of the β-strands in the eight-fold α/β barrel structure. The absence of the typical chitin-binding domain does not affect the ability of the enzyme to hydrolyze the soluble substrate, but abolishes hydrolysis of insoluble chitin (Tjoelker *et al.* 2000). Despite the absence of enzymatic activity, long chitin fragments are distorted on binding, with the GlcNAc at subsite -1 in a boat conformation, similar to what has been observed in other chitinases. Interestingly, whereas chitin fragments of 4 or more GlcNAc residues tend to occupy the central part of the groove, shorter oligosaccharides bind preferentially at the more distant subsites on the protein surface (Fusetti *et al.* 2003). The existence of two distinct binding sites with selective affinity for long and short oligosaccharides has never been documented for chitinases and could be unique for CHI3L1.

CHI3L1 also interacts with glycosaminoglicans such as heparin and hyaluronan (Shackelton *et al.* 1995; Houston *et al.* 2003). Furthermore, Bigg *et al.* have recently reported an ability of CHI3L1 to bind to collagen type 1, 2 and 3 (Bigg *et al.* 2006). The putative heparin-binding site is GRRDKQH (residues 143-149) located in a surface loop, however, heparan sulfate, not heparin, is more likely the physiological ligand of CHI3L1 by using crystallization methods, and unsulfated fragments of heparan sulfate can be accommodated in the binding groove of CHI3L1 (Fusetti *et al.* 2003). CHI3L1 may interact with heparin-like molecules in the extracellular matrix (ECM) or on the cell surface. It has been found in vertebrates that short chito-oligosaccharides are used as primers for the synthesis of hyaluronic acid (HA) (Varki *et al.* 1996). HA is a linear polysaccharide composed of repeating disaccharide units of GlcNAc and D-glucuronic acid linked together by alternating β (1, 4) and β (1, 3) glycosidic bonds. HA is located in the ECM of many tissues and plays an important biological role in embryogenesis, cell proliferation, tissue remodeling and in acute and chronic inflammatory processes (Lee *et al.* 2000). Thus the function of CHI3L1 may be linked to the functions of HA. Because of its chitin-binding ability CHI3L1 could participate in specific signaling processes by perceiving the presence of newly synthesized HA chains that still contain the chito-oligosaccharide, and consequently influence the extent of cell adhesion and migration during the tissue remodeling processes that take place during inflammation, fibrosis, atherogenesis and metastasis. CHI3L1 has been shown to stimulate the proliferation of connective tissue cells through activation of mitogen-activated protein kinase (MAPK) and protein kinase B-mediated signaling pathways (Recklies *et al.* 2002). The cellular receptors responsible for mediating these effects have not yet been identified. A clearer understanding of the exact physiological ligands with which CHI3L1 interacts is therefore critical to uncovering its precise role and function *in vivo*. CHI3L1 has these characteristics of ability to interact with oligomer or monomer of GlcNAc and absence of glycohydrolase activity and, therefore, CHI3L1 has been defined as a CLP or chitinase-like lectin (Chi-lectin) (Houston *et al.* 2003).

5. Bacterial Chitinases and Chitin Binding Proteins (CBP)

Although bacteria do not contain chitin as their structural component, bacteria are capable of degrading chitin usually produce a battery of chitinases (Watanabe *et al.* 1997), consisting of a catalytic domain and often, one or more smaller non-catalytic domains involved in substrate binding (Perrakis *et al.* 1994; Uchiyama *et al.* 2001). While several individual chitinases have been characterized in detail (Aronson *et al.* 2003; Suzuki *et al.* 2002a; Brurberg *et al.* 1996; Fukamizo *et al.* 2001), little is known about important issues such as exo- vs. endo-action and processivity. Generally, enzymatic degradation of polysaccharides occurs from one of the chain ends (exo-mechanism) or from a random point along the polymer chain (endo-mechanism). Each of these two mechanisms can occur in combination with a processive mode of action, meaning that the substrate is not released after successful cleavage but slides through the active site for the next cleavage event to occur. This precludes full understanding of nature's chitinolytic machineries.

Serratia marcescens is an efficient biological degrader of chitin and one of the most extensively studied chitinolytic bacteria. When grown on chitin as a sole carbon source, *S. marcescens* produces three chitinases (ChiA, ChiB, and ChiC), a chitin-binding protein (CBP21) lacking chitinase activity, and a hexosaminidase which further degrades the major end product of the chitinases, GlcNAc2 in the culture supernatant as the major proteins (Suzuki *et al.* 1998). All three chitinases belong to the family 18 of glycosyl hydrolases (Henrissat *et al.* 1995), which possess an eight-fold α/β barrel catalytic domain with approximately six sugar subsites (Perrakis *et al.* 1994). The three *S. marcescens* chitinases have different chitin-binding domains, which also called carbohydrate-binding modules (CBMs): ChiA contains a fibronectin type III (FnIII)-like CBM; ChiB contains a family 5 CBM and ChiC contains a family 12 and an FnIII-like CBM (Horn *et al.* 2006). It is conceivable that the primary role of these CBMs is to potentiate catalytic activity by disrupting the substrate, rather than simply to promote enzyme-substrate binding. Chitin hydrolysis by family 18 chitinases takes place through a substrate-assisted mechanism which involves the N-acetyl group of the -1 sugar and which leads to retention of the anomeric configuration (van Aalten *et al.* 2001). Thus, family 18 chitinases have an almost absolute preference for GlcNAc in their -1 substrate, whereas other subsites show less stringency in this respect to bind productively (Sorbotten *et al.* 2005).

Both ChiA and ChiB contain a substrate-binding domain which extents the substrate-binding cleft on the side where the non-reducing end of the substrate binds in ChiA (Henrissat *et al.* 1995) and on the side where the reducing end of the substrate binds in ChiB (van Aalten *et al.* 2000). In ChiA the deep substrate-binding cleft seems rather accessible, whereas the substrate-binding cleft of ChiB is relatively closed, giving the cleft a more tunnel-like character (van Aalten *et al.* 2001). On the basis of structural characteristics and enzymological work it has been suggested that ChiA and ChiB are exo-chitinases which degrade chitin chains from opposite ends, ChiA from the reducing end and ChiB from the non-reducing end (Horn *et al.* 2006). ChiC consists of a catalytic domain and two putative chitin-binding domains, which are located C-terminally in the sequence (Suzuki *et al.* 1999). The catalytic domain of ChiC lacks the $\alpha+\beta$ domain, suggesting that ChiC has a much more open substrate-binding groove, as observed in the crystal structure of the family 18 endo-chitinase hevamine (Terwisscha van Scheltinga *et al.* 1994). ChiC often occurs in two forms in cultures of *S. marcescens*: the complete protein, sometimes called ChiC1, and a proteolytically truncated variant, called ChiC2, which lacks the two putative chitin-binding domains (Suzuki *et al.* 1999). ChiC is a non-processive endo-chitinase, which cuts chitin randomly in the chains, thus releasing long fragments and new chain ends that are substrates for exo-chitinases (Horn *et al.* 2006). Therefore, the three chitinases have different and complementary activities (endo- vs. exo-) and directionalities, which can explain the synergistic effects that are observed for certain combinations of the enzymes. Synergistic effects could be due to one enzyme increasing substrate accessibility for other enzymes by hitherto unknown disruptive mechanisms involving the CBMs.

The recent finding that CBP21 produced by *S. marcescens* potentiates chitinase action by disrupting the structure of the chitin substrate points to another possible explanation for synergistic effects. Chitin-binding proteins without chitinase activity have been isolated from various sources including plants and microorganisms. All plant chitin-binding proteins sequenced to date contain a common structural motif with several conserved amino acid residues (Raikhel *et al.* 1993). In microorganisms, several chitin-binding membrane proteins were reported in *Vibrio harvayi* and *Vibrio cholerae* which mediate attachment of bacterial cells to chitin-containing substrates (Montgomery *et al.* 1993; Kirn *et al.* 2005). Attachment of bacterial cells to chitin-containing substrate is especially important in aquatic environments (Yu *et al.* 1991). Suzuki *et al.* first identified that a gram-negative bacterium, *S. marcescens*, produces a chitin-binding protein, CBP21 in the culture supernatant, which enhances chitin accessibility to bacterial chitinases and cloned the corresponding gene (Suzuki *et al.* 2002a). While chitinases such as ChiA, ChiB, and ChiC also contain auxiliary CBMs, most microorganisms containing chitinases gene also contain a gene encoding for a homologue of CBP21, that is, a non-catalytic chitin binding protein CBP21 is a 18 kDa protein with a structure consisting of a three-stranded and a four-stranded β-sheet that forms a β-sandwich (Vaaje-Kolstad *et al.* 2005b). The gene for CBP21 (cbp) is located 1.5kb downstream of one of the chitinase genes (ChiB) in *S. marcescens*, and the CBP21 protein is produced along with ChiB and the other two chitinases, ChiA and ChiC, suggesting coordinate regulation of the production of CBP21 and all chitinases (Suzuki *et al.* 2002a). In addition to three chitinases with apparent different roles and capabilities, CBP21 is essential for efficient chitin degradation by *S. marcescens* (Vaaje-Kolstad *et al.* 2005a). Through the sequence similarity, CBP21 is classified as a part of the CBM (carbohydrate-binding module) family 33 and identified to possess a chitin- binding domain (Hu *et al.* 1996). Within the 80 CBP sequences present in the Pfam database, 19 other bacterial CBPs cluster within CBP21. A multiple alignment of bacterial CBPs revealed that 40 highly conserved residues (>90% conserved) and a cluster of conserved, mainly hydrophilic, residues exist at the surface of CBP21 molecule (Vaaje-Kolstad *et al.* 2005b). It should be noted that carbohydrate-binding proteins and domains with the typical hydrophobic grooves/surfaces bind single carbohydrate chains/oligosaccharides, whereas CBP21 only binds to insoluble chitin (Suzuki *et al.* 1998), which is to an ordered array of multiple chains. The roles of six of these conserved surface-exposed residues (Tyr-54, Glu-55, Glu-60, His-114, Asp-182, and Asn-185) are analyzed to contribute to bind to the insoluble crystalline substrate and increase substrate accessibility (Vaaje-Kolstad *et al.* 2005b). Importantly, the majority of chitinase-producing pathogenic microorganisms contain a gene encoding for the homologue of the *cbp21* gene, suggesting the presence of a potential binding ability of chitinase-producing pathogenic bacteria to chitin-like carbohydrate via CBP21 homologue (Vaaje-Kolstad *et al.* 2005a). Therefore, it is possible that CHI3L1 may bind to CBP on bacteria and this binding may subsequently enhance the adhesion and invasion of these bacteria to CECs. Our group is currently investigating the possibility by utilizing CBP21 overexpressing *E. coli* strain *in vitro*.

6. Potential Interaction between CBP and CHI3L1 in Bacterial Adhesion

CHI3L1 protein contains a lectin-like structure that has no chitinase activity but is able to interact with chitin and chitin-like molecules with high affinity through a preserved hydrophobic substrate binding cleft (Fusetti *et al.* 2003). However, the overall functions and specific ligands (or receptors) of CHI3L1 are still unknown (Fusetti *et al.* 2003). Our study provides an unexpected impact that CHI3L1 acts as an enhancer of bacterial/host interactions as indicated by the fact that adhesion and invasion of intracellular bacteria including *S. typhimurium* and AIEC LF82 are up-regulated significantly in the presence of CHI3L1. Similar to classic chitinase, the binding ability of CHI3L1 to chitin or chitin-like carbohydrate is preserved (Fusetti *et al.* 2003), therefore, the CHI3L1 may bind bacteria and fungi directly or indirectly to attract these organisms to the surface of CECs.

We recently proposed a possibility that CHI3L1 expressed on CECs binds to bacterial CBP21 homologue via chitin or chitin-like molecule and this binding subsequently enhances the adhesion of bacteria to CECs. To identify the exact mechanism how CHI3L1 enhances the bacterial adhesion on CECs, we have utilized *S. marcescens* wild type (WT) and *S. marcescens* mutant form and genetically engineered CBP21-overexpressing non-pathogenic *E. coli. S. marcescens* is an efficient biological degrader of chitin and one of the most extensively studied chitinolytic bacteria. In addition to chitinases, *S. marcescens* and other chitin-degrading microorganisms secrete chitin-binding proteins (CBPs), which specifically bind to chitin and chitin-like glycan with high affinity (Suzuki *et al.* 1998; Zeltins *et al.*

1997). To analyze the biological function of CBP21, we performed adhesion and invasion assays of *S. marcescens* (Kawada *et al.* 2008). After transfected with the CHI3L1 vector to CECs, significantly increased number of *S. marcescens* bound and invaded into CECs as the same results as *S. typhimurium* and AIEC. These results suggest that CHI3L1 enhances the adhesion and invasion of *S. marcescens* to CECs. To analyze the function of CBP21 in bacterial adhesion to CECs, SW480 CEC line were infected with WT or CBP21 knockout strain of *S. marcescens*. Interestingly, CBP21 knockout strain of *S. marcescens* significantly decreased its adhesion but not invasion to CECs compared to WT of *S. marcescens*. Therefore, we next focused on analyzing the biological function of CBPs in bacterial adhesion to CECs.

To further examine the functional role of CBPs in bacterial adhesion to CECs, we generated WT CBP21-overexpressing BL21 strain of *E. coli* (non-pathogenic) and performed adhesion assays (Kawada *et al.* 2008). Interestingly, WT of CBP21-overexpressing *E. coli* (WT CBP21) significantly increased adhesion but not invasion to CECs, although control *E. coli* did not attach to CECs at all. Moreover, after transfected with the CHI3L1 vector to CECs, significantly increased number of WT CBP21-overexpressing *E. coli* bound to CECs. These results suggest that bacterial CBP21 is involved in the adhesion but not invasion of *E. coli* to CECs, and CHI3L1 enhances the effect of CBP21-mediated bacterial adhesion to CECs.

To examine our hypothesis that bacteria would bind to CECs through a CHI3L1/CBP complex, an effect of chitin on the adhesion efficiency of *S. marcescens* and WT CBP21 *E. coli* to CECs was determined (Kawada *et al.* 2008). One hour preincubation of the bacteria or CECs with chitin was followed by an adhesion assays. Not only *S. marcescens* but also WT CBP21-overexpressing non-pathogenic *E. coli* significantly decreased their binding to CECs after preincubation these bacteria with chitin compared to untreated control. In contrast, the adhesion of *S. marcescens* and WT CBP21-overexpressing *E. coli* to CECs were not inhibited when CECs instead of the bacteria were pretreated with chitin. These results suggest that chitin efficiently inhibits the adhesion of bacteria to CECs and bacterial CBP21 does not directly but indirectly bind to CHI3L1 on CECs. Therefore, the inhibitory effect of bacterial-epithelial interaction by chitin appears to be the results of existence of endogenous chitin-like molecule(s) on CECs, and the complex of endogenous chitin-like molecule(s) and CHI3L1 on CECs may indirectly interact the binding of CBP21 (on bacteria) and CHI3L1 (on CECs). As the candidate of endogenous chitin-like molecule, we speculate that endogenous polysaccharide (glycan) structure, like a GlcNAc monomer, on the surface of CECs would be the most probable components, and the complex is likely to bind with CBP-expressing bacteria.

The crystal structure of CBP21 shows that the roles of six highly conserved surface-exposed residues (Thy-54, Glu-55, Glu-60, His-114, Asp-182 and Asn-185) play an important role for binding between CBP21 and chitin (Vaaje-Kolstad *et al.* 2005b). To analyze the mechanisms by which parts of CBP21 binds to the complex of CHI3L1/endogenous chitin-like molecule on CECs, we generated mutants of these conserved surface residues to alanine (Y54A, E55A, E60A, H114A, D182A and N185A) of CBP21 in BL21 strain of *E. coli* (Kawada *et al.* 2008). As a result, genetically engineered *E. coli* with a single mutation of either Thy-54 or Glu-55 residue of CBP21 exhibited a significantly decreased binding to CECs compared to WT CBP21 *E. coli*. The binding ability was 74% reduced by the combined mutations of three amino acids of Thy-54, Glu-55 and Glu-60 (3MU) compared to WT CBP21 *E. coli*. It has been well analyzed by Vaaje-Kolstad *et al.* that the binding curve between CBP21 and chitin was decreased largely in point mutants of Y54A and E60A (Vaaje-Kolstad *et al.* 2005b). Thy-54 is the only one conserved aromatic residue that is fully exposed on the surface of CBP21 and is essential to generate the tyrosine ring for binding with chitin (Henrissat *et al.* 1995). These results suggest that bacterial CBP21 is actively involved in the adhesion of *E. coli* to CECs and residues in the conserved surface area, especially Thy-54 and Glu-55, are important for the ability of CBP21 to bind chitin or endogenous chitin-like molecule(s), which may express on CECs.

By utilizing the genetically engineered BL21 strain of *E. coli*, we next analyzed a biological significance of CHI3L1 in the interaction of bacterial CBP21 and CECs (Kawada *et al.* 2008). As the result, enhanced CHI3L1 expression in CECs significantly facilitated the adhesion of WT CBP21 *E. coli* to CECs, although the adhesion of control and 3MU CBP21 *E. coli* were not changes even after the CHI3L1 over-expression. To confirm whether enhanced adhesion of CBP21-overexpressing bacterial to CECs is mediated by CHI3L1, studies to inhibit CHI3L1 expression and function by utilizing anti-CHI3L1 Ab and CHI3L1 siRNA were performed. Pretreatment with anti-CHI3L1 Ab significantly reduced the WT CBP21 *E. coli* adhesion to the CECs, although the anti-CHI3L1 Ab treatment did not significantly suppress the adhesion of 3MU CBP21 *E. coli*. Inhibition of endogenously expressed CHI3L1 by CHI3L1 specific siRNA also showed significant fewer numbers of WT CBP21 *E. coli* attached on CECs. In contrast, 3MU CBP21 *E. coli* did not show significant reduction in the number of the adhered bacteria to CECs compared with control. Taken together, the series of these findings clearly show that CHI3L1 possesses the ability to facilitate the CBPs-expressing bacterial adhesion to CECs through the conserved amino acids region of bacterial CBP21 and presumably its homologues.

It has been elegantly described that a putative chitin-binding protein, GbpA (GlcNAc-binding protein A), which is produced by *Vibrio cholerae* significantly enhances the bacterial adhesion to both epithelial cell surfaces and chitin, most probably by a direct interaction with GlcNAc residues (Kirn *et al.* 2005). Interestingly, both GbpA and CBP21 belong to CBM (carbohydrate-binding module) family 33 and have multiple alignments in the sequences. Therefore, CBP21 may interact with GlcNAc residues on CECs same as GbpA. Structurally conserved chitin-binding properties exist in distantly related microorganisms, such as *Streptomyces*, *Serratia Yersinia* and *Vibrio*, suggesting a wide distribution of this type of CBPs in chitinolytic microorganisms (Suzuki *et al.* 1998). The series of our studies demonstrate that CHI3L1 plays a critical role in the adhesion of CBPs-expressing bacteria to CECs (Kawada *et al.* 2008). Specific adhesion of mucosa-associated bacteria in IBD and colon cancer might be enhanced by the inducible expression of mucosal glycoconjugates on CECs (Rhodes *et al.* 1996). In fact, CHI3L1 is expressed on CECs only under intestinal inflammatory conditions but not quiescent IBD or normal epithelial cells (Mizoguchi, 2006). We speculate that glycosylated-CHI3L1 is expressed on CECs under intestinal inflammatory conditions, and the attachment of CBPs-expressing bacteria to CECs is enhanced. The exact molecule expressing on CECs, which binds to GbpA is also still unknown, and further studies would be required to clarify the existence of mucosal glycoconjugates (presumably chitin-like glycans), which may generate complex formation with CHI3L1 on CECs under the inflammatory conditions. Our study demonstrates that CBP21 and its homologues may be required for the CHI3L1-mediated enhancement of potentially pathogenic bacterial adhesion to CECs through the conserved amino acid residues. CHI3L1 is involved in the enhancement of CBPs-expressing bacterial adhesion to CECs. Both CHI3L1 and CBPs may play an important as well as complicated role in the commensal bacterial adhesion to the CECs under the inflammatory conditions. Therefore, blockade of CHI3L1 and /or CBPs would be effective therapeutic strategy for treatment of IBD in the future. Insights into the microbial-host interrelationships are hampered by both the limited knowledge of the diversity and complexity of the microbial flora and the limitation of available tools to delineate these characteristics. Further analysis of microbiome may provide a foundation to achieve an understanding of the relevant, functional diversity of the flora in IBD

7. CHI3L1 and IBD

IBD is a group of chronic inflammatory disorders that affect individuals throughout life (Podolsky, 2002). The etiology and pathogenesis of IBD is still largely unknown. Colonic epithelial alteration and inflammatory mononuclear cell infiltrate appear to be important factors in the initiation and/or perpetuation of disease in both experimental colitis models and patients with UC and CD. Several studies have indicated that dysregulated host/enteric bacterial interactions are required for the development of chronic intestinal inflammation. Both CECs and the underlying immune cells play an important role in the immunoregulation of host/microbial interactions.

Vind et al. analyzed serum CHI3L1 level determined by ELISA in 164 patients with UC and 173 patients with CD (Vind *et al.* 2003). According the result, serum CHI3L1 is elevated in 40-50% of patients with UC and CD with active disease, and 30% of patients with clinically inactive CD (Vind *et al.* 2003). Interestingly, some of circulating CHI3L1 in CD patients seems to reflect ongoing fibrogenesis since serum level of CHI3L1 has been related to the degree of fibrosis in liver (Johansen *et al.* 2000). About 64% of CD patients with extra-intestinal manifestations (including arthralgia, erythema nodosum, uveritis, aphthous ulcerations and fistulas) had elevated serum CHI3L1, suggesting the strong relationship of serum CHI3L1 and presence of clinical complications in CD patients (Vind *et al.* 2003).

It was reported by Verheijden et al. that peripheral blood T cells from HLA class II-DR4 (DRB1*0401) positive RA patients strongly responses to CHI3L1 (Verheijden *et al.* 1997; Cope *et al.* 1999). Continuous study proved that immunological responses of the peripheral blood mononuclear cells (PBMC) response to CHI3L1 derived peptides in patients with inflammatory conditions including IBD (Vos *et al.* 2000a; Vos *et al.* 2000b). They generated 5 different peptides derived from CHI3L1 (75-87, 103-116, 259-271, 263-275, and 326-338), which are selected on the basis of a published DRB1*0401 motif on the basis of binding to DRB1*0401 (Verheijden *et al.* 1997). Only peptide 259-271 induces PBMC response in the RA patients group as compared to the healthy controls, and the immune response to peptide 259-271 correlates with disease activity in the RA patient group. In the IBD and osteoarthritis patient groups, two peptides 259-271 and 263-275 induced a PBMC response, which was significantly higher than in the healthy controls. Cope et al. also identified that antigen-specific T cells from DRB1*0401 transgenic mice produced significantly high amount of IFNγ and TNFα in response to CHI3L1 than did T cells from DRB1*0402 transgenic mice (Cope *et al.* 1999). These results strongly suggest that CHI3L1-derived peptides may be specific targets of the T-cell mediated inflammatory immune response in patients with arthritis and IBD. The structure of CHI3L1 shows that although region 266-275 maps onto the protein surface, residues 259-265 are buried in the vicinity of the chitin-binding groove (Fusetti *et al.* 2003).

In IBD patients, CHI3L1 is most likely produced by monocytes/macrophages and leukocytes in the inflamed intestine (Vind *et al.* 2003), however, our group also identified that CECs were major source of CHI3L1 in patients with UC and CD but not normal individuals detected by immunohistochemical staining with anti-CHI3L1 antibody (Mizoguchi, 2006). The CHI3L1 expression was more restricted to crypt and surface epithelium in active IBD patients. *In vitro* studies using several human colonic epithelial cell lines revealed that CHI3L1 was constitutively expressed in differentiated (e.g., Caco-2, SW480, T84) but not undifferentiated (e.g., COLO205, HT-29) cell lines. After stimulating with proinflammatory cytokines (TNFα, IL-1β, and IL-6), CHI3L1 expression in SW480 and T84 cells is significantly enhanced, suggesting several colitis-associated cytokines possess the ability to enhance the expression of CHI3L1 in CECs.

8. CHI3L1 and Cancer Development

Based on the dbEST database at the National Center for Biotechnology, CHI3L1 is expressed with different types of solid tumors including several types of adenocarcinoma, small cell lung carcinoma, glioblastoma, and melanoma (Johansen *et al.* 2006a). Interestingly, serum concentration of CHI3L1 (YKL-40) seems to have a high sensitivity for advanced cancer but not primary cancer (Johansen *et al.* 2006b). In addition, high serum CHI3L1 positively correlates with a poor prognosis in cancer patients (Dupont *et al.* 2004, Johansen et al. 2006b). For example, a multivariate Cox analysis including serum CHI3L1, serum carcinoembronic antigen (CEA), Dukes' stages, age and gender showed that a high CHI3L1 was an independent prognostic parameter for short survival of patients with colorectal cancer (Cintin *et al.* 1999). An elevated CHI3L1 level was also an independent prognostic parameter of short survival in patients with colon cancer (Cintin *et al.* 1999) , and metastatic prostate cancer (Brasso *et al.* 2006). Intensive immunohistochemical analyses of biopsies of colorectal cancer show that not all but some of cancers positively stained by anti-CHI3L1 antibody (Johansen *et al.* 2006b), suggesting CHI3L1 may reflect differences in the biology of various cancer cells within the same colon carcinoma.

The overall biological function of CHI3L1 in cancer is still not known clearly. It has been suggested that CHI3L1 may play a role in the proliferation and differentiation of cancer cells by protecting from undergoing apoptosis, stimulating migration and reorganization of vascular endothelial cells, and producing adhesion and migration factor(s) for vascular smooth muscle cells (Malinda *et al.* 1999; Nishikawa *et al.* 2003). Purified CHI3L1 protein acts synergistically with insulin-like growth factor-I in stimulating the growth of fibroblasts

(Recklies *et al.* 2002) and colon cancer (Mizoguchi *et al.* unpublished observation) in a concentration dependent manner. Ling et al showed that stimulation of human fibroblast or articular chondrocytes with IL-1 or TNFα in the presence of CHI3L1 resulted in a significant reduction of both p38 MAPK (mitogen-activated protein kinase) and SAPK/JNK (stress-activated protein kinase/Jun N-terminal kinase) phosphorylation, whereas activation of NF-κB (nuclear factor κB) is not altered (Ling *et al.* 2004). In addition, CHI3L1 suppresses the cytokine-induced secretion of matrix metalloproteinase (MMP)-1, MMP-3, and MMP-13, and the chemokine IL-8. All the suppressive and anti-catabolic effects of CHI3L1 are dependent on kinase activity, and treatment result in AKT-mediated serine/threonine phosphorylation of the apoptosis signal-regulator kinase, ASK1 in articular chondrocytes and skin fibroblasts (Ling *et al.* 2004). In contrast, CHI3L1 protein (purified from the MG63 human osteosarcoma cell line) significantly upregulate the activation of NF-κB pathway analyzed by Western blotting and immunocytochemistry of IkB-α degradation, and increased chemokine IL-8 and TNFα productions in SW480, HT29 and COLO-205 colon cancer cell line lines in a concentration dependent manner (Mizoguchi *et al.* unpublished observation). This result suggests that suppressive effect of CHI3L1 in pro-inflammatory cytokines production may be cell type dependent phenomenon. Cintin *et al.* reported one study of curatively operated 324 colorectal cancer patients who have elevated serum CHI3L1 levels during the follow-up after surgery (Cintin *et al.* 2002). Patients exhibiting elevated serum CHI3L1 had an increased hazard for death within the following six months compared to those patients with normal serum CHI3L1 level, suggesting serum concentration of CHI3L1 may be useful for the monitoring colon cancer patients after primary operation, adjuvant chemotherapy, and radiotherapy in order to identify the recurrence at the earliest time point. Since CHI3L1 may play an important role in cancer cells to proliferate, invade, and migrate into other organs, CHI3L1 could be attractive target in design of anticancer therapy in the future.

9. Anti-Inflammatory Effect of Anti-CHI3L1 Antibody in Colitis

Initially, polyclonal anti-CHI3L1 Ab was generated by Prof. Albert J.T. Millis's lab in 2003. Briefly, the Ab was prepared by multiple immunizations of rabbits using CHI3L1 antigen purified from vascular smooth muscle cell (VSMC) cultures which were derived from thoracic aortic explants of adult pigs (Nishikawa *et al.* 2003). The anti-serum was tittered by Western immunoassay and affinity purified via CHI3L1 protein. They identified that CHI3L1 (as low as 1 ng/ml) showed profound effects on VSMC migration but little or no effect on fibroblast migration.

CHI3L1 is a member of a novel family of chitinase-like lectins (Chi-lectins) and is able to bind chitooligosaccharides (oligomer of GlcNAc) with micromolar affinity (Houston *et al.* 2003). For analyzing the *in vivo* biological function of CHI3L1, we generated mouse/human CHI3L1 polyclonal antibody. For this purpose, we had initially analyzed for antigenesity and conserved sequences between mouse and human CHI3L1. We found that only four peptides sequences 27-40, 102-116, 141-155, and 327-341 are conserved between human and mouse and also show the antigenesity (maps onto the protein surface). Since 327-GYD-DQE-SVK-SKY-(Q/G)YL-341contains some negatively charged amino acids (1glutamic acids and 2 aspartic acid) and the sequence contains family 18 chitinase DXXDXDXE motif, which lies on strand β4 (Houston *et al.* 2003), we chose 327-341 peptide sequence for the immunization of the antibody. Rabbit were immunized with Keyhole limpet hemocyanin-conjugated mouse CHI3L1 peptide 325-339 by the glutaraldehyde-conjugation method. The peptide specific antibody was purified from the immune serum by affinity purification with the peptide 325-339 (Affinity BioReagents, Golden, CO). The specificity of this antibody was confirmed by Western blot analysis and enzyme-linked immunosorbent assay using human or mouse-purified CHI3L1/YKL-40. Our anti-CHI3L1 antibody showed similar reactivity as Prof. Millis's anti-CHI3L1 antibody detected by immunohistochemical analysis of colonic samples with IBD patients.

To assess the biological function of CHI3L1 *in vivo*, we intraperitoneally injected 1 mg/ml of house-made anti-CHI3L1 antibody or purified rabbit IgG (Bethyl Laboratories, Montgomery, TX) to the 4% dextran sulfate sodium (DSS)-induced colitis model. As previously described, DSS-induced colitis is one of the best models to analyze the effect of medicine, chemical compounds or antibodies during the recovery phase from the acute colitis (Kawada *et al.* 2007a). Administration of anti-CHI3L1 Ab intraperitoneally into the DSS-colitis model, significantly enhanced recovery from acute colitis as indicated by the improvement of body weight loss and clinical score compared with the control group that was administered normal rabbit serum (Mizoguchi, 2006; Kawada *et al.* 2007b). The anti-CHI3L1 Ab treated group showed significantly less epithelial damage, inflammatory cell infiltration, significantly lower bacterial loads in spleen, mesenteric lymph nodes and liver on DSS day 12 compared with the control group. This result strongly suggests that inhibition of CHI3L1 activity highly contributes to the suppression of DSS-induced colitis by reducing the internalization of luminal bacteria to colonic mucosa, and it is possible that anti-CHI3L1 Ab or CHI3L1 inhibitor may be a new and novel therapeutic strategy for IBD in the future.

Conclusion

Biological function of mammalian chitinases in gastrointerstinal tract is still unrevealed. One of the mammalian chitinase like proteins, CHI3L1 is a glycoprotein capable of binding to chitin and chitooligosaccharide without an enzymatic activity secreted by articular chondrocytes, synoviocytes, macrophages and CECs. Increased serum levels of CHI3L1 have been demonstrated in patients with active RA, giant cell arthritis, cancer, and IBD. The strong expression of CHI3L1 was observed in CECs and lamina propria macrophages in animal models of colitis as well as human IBD patients. CHI3L1 acts as a strong pathogenic mediator in acute and chronic colitis by enhancing the adhesion and invasion of chitin CBP-expressing intracellular bacteria to CECs. Inhibition of biological activities of CHI3L1 would shed a light for therapeutic strategy of chronic and persistent inflammatory conditions including RA and IBD in the near future.

Acknowledgment

The authors are grateful to Dr. Chun-Chuan Chen for helpful advices and discussions. We would also like to thank Ms. Sarah Murphy and Mayuko Segawa for their excellent secretarial assistance in preparing this chapter. This work has been supported by National Institute of Health (DK64289, DK74454, and DK43351) and grants from the Eli and Edythe L. Broad Medical Foundation and American Gastroenterological Association Foundation for Digestive Health and Nutrition to EM.

References

Arif S, Tree TI, Astill TP, Tremble JM, Bishop AJ, Dayan CM, Roep BO, Peakman M. Autoreactive T cell responses show proinflammatory polarization in diabetes but a regulatory phenotype in health. *J Clin Invest.* 2004; 113: 451-463.

Aronson NNJr, Halloran BA, Alexyev MF, Amable L, Madura JD, Pasupulati L, Worth C, Van Roey P. Family 18 chitinase-oligosaccharide substrate interaction: subsite preference and anomer selectivity of *Serratia marcescens* chitinase A. *Biochem J.* 2003; 376: 87-95.

Baeten D, Boots AM, Steenbakkers PG, Elewaut D, Bos E, VerheijdenGF, Berheijden G, Miltenburg AM, Rijnders AW, Veys EM, De Keyser F. Human cartilage gp-39+, CD16+ monocytes in peripheral blood and synovium: correlation with joint destruction in rheumatoid arthritis. *Arthritis Rheum.* 2000; 43: 1233-1243.

Bigg HF, Wait R, Rowan AD, Cawston TE. The mammalian chitinase-like lectin, YKL-40, binds specifically to type 1 collagen and modulates the rate of type 1 collagen fibril formation. *J Biol Chem.* 2006; 281: 21082-21095.

Boot RG, Blommaart EF, Swart E, Ghauharali-van der Vlugt K, Bijl N, Moe C, Place A, Aerts JM. Identification of a novel acidic mammalian chitinase distinct from chitotriosidase. *J Biol Chem.* 2001; 276: 6770-6778.

Boot RG, Renkema GH, Strijland A, van Zonneveld AJ, Aerts JM. Cloning of a cDNA encoding chitotriosidase, a human chitinase produced by macrophages. *J Biol Chem.* 1995; 270: 26252-26256.

Boudeau J, Glasser AL, Masseret E, Joly B, Darfeuille-Michaud A. Invasive ability of an *Escherichia coli* strain isolated from the ileal mucosa of a patient with Crohn's disease. *Infect Immun.* 1999; 67: 4499-4509.

Brasso K, Christensen IJ, Johansen JS, Teisner B, Garnero P, Price PA, Iversen P. Prognostic value of PINP, bone alkaline phosphatase, CTX-1, and YKL-40 in patients with metastatic prostate carcinoma. *Prostate* 2006; 66: 503-513.

Brurberg MB, Nes IF, Eijsink VG. Comparative studies of chitinases A and B from *Serratia marcescens. Microbiology* 1996; 142: 1581-1589.

Chupp GL, Lee CG, Jarjour N, Shim YM, Holm CT, He S, Dziura JD, Reed J, Coyle AJ, Kiener P, Cullen M, Grandsaigne M, Dombret MC, Aubier M, Pretolani M, Elias JA. A chitinase-like protein in the lung and circulation of patients with severe asthma. *N Engl J Med.* 2007; 357: 2016-2027.

Cintin C, Johansen JS, Christensen IJ, Price PA, Sorensen S, Nielsen HJ. High serum YKL-40 level after surgery for colorectal carcinoma is related to short survival. *Cancer* 2002; 95: 267-274.

Cintin C, Johansen JS, Christensen IJ, Price PA, Sorencen S, Nielsen HJ. Serum YKL-40 and colorectal cancer. *Br J Cancer* 1999; 79: 1494-1499.

Collinge DB, Kragh KM, Mikkelsen JD, Nielsen KK, Rasmussen U, Vad K. Plant chitinases. *Plant J.* 1993; 3: 31-40.

Cope AP, Patel SD, Hall F, Congia M, Hubers HA, Verheijden GF, Boots AM, Menon R, Trucco M, Rijnders AW, Sonderstrup G. T cell responses to a human cartilage autoantigen in the context of rheumatoid arthritis-associated and nonassociated HLA-DR4 alleles. *Arthritis Rheum.* 1999; 42: 1497-1507.

Debono M, Gordee RS. Antibiotics that inhibit fungal cell wall development. *Annu Rev Microbiol.* 1994; 48: 471-497.

Dupont J, Tanwar MK, Thaler HT, Fleisher M, Kauff N, Hensley ML, Sabbatini P, Anderson S, Aghajanian C, Holland EC, Spriggs DR. Early detection and prognosis of ovarian cancer using serum YKL-40. *J Clin Oncol.* 2004; 22: 3330-3339.

Fukamizo T, Sasaki C, Schelp E, Bortone K, Robertus JD. Kinetic properties of chitinase-1 from the fungal pathogen *Coccidioides immitis*. *Biochemistry* 2001; 40: 2448-2454.

Fusetti F, Pijning T, Kalk KH, Bos E, Dijkstra BW. Crystal structure and carbohydrate-binding properties of the human cartilage glycoprotein-39. *J Biol Chem.* 2003; 278: 37753-37760.

Gianfrancesco F, Musumeci S. The evolutionary conservation of the human chitotriosidase gene in rodents and primates. *Cytogenet Genome Res.* 2004, 105: 54-56.

Gooday GW. Aggressive and defensive roles for chitinases. *EXS* 1999, 87 : 157-169.

Hakala BE, White C, Recklies AD. Human cartilage gp-39, a major secretory product of articular chondrocytes and synovial cells, is a mammalian member of a chitinase protein family. *J Biol Chem.* 1993; 268: 25803-25810.

Henrissat B, Davies G. Structural and sequence-based classification of glycoside hydrolases. *Curr Opin Struct Biol.* 1997; 7: 637-644.

Henrissat B, Romeu A. Families, superfamilies and subfamilies of glycosyl hydrolases. *Biochem J.* 1995; 311: 350-351.

Herrera-Estrella A, Chet I. Chitinases in biological control. *EXS* 1999, 87:171-184, 1999.

Hollak CE, van Weely S, van Oers MH, Aerts JM. Marked elevation of plasma chitotriosidase activity. A novel hallmark of Gaucher disease. *J Clin Invest.* 1994; 93: 1288-1292.

Horn SJ, Sorbotten A, Synstad B, Sikorski P, Sorlie M, Varum KM, Eijsink VG. Endo/exo mechanism and processivity of family 18 chitinases produced by *Serratia marcescens*. *FEBS J.* 2006; 273: 491-503.

Houston DR, Recklies AD, Krupa JC, van Aalten DM. Structure and ligand-induced conformational change of the 39-kDa glycoprotein from human articular chondrocytes. *J Biol Chem.* 2003; 278: 30206-30212.

Hu B, Trinh K, Figueira WF, Price PA. Isolation and sequence of a novel human chondrocyte protein related to mammalian members of the chitinase protein family. *J Biol Chem.* 1996; 271: 19415-19420.

Johansen JS. Studies on serum YKL-40 as a biomarker in diseases with inflammation, tissue remodeling, fibrosis and cancer. *Dan Med Bull.* 2006a; 53: 172-209.

Johansen JS, Christoffersen P, Moller S, Price PA, Henriksen JH, Garbarsch C, Bendtsen F. Serum YKL-40 is increased in patients with hepatic fibrosis. *J Hepatol.* 2000; 32: 911-920.

Johansen JS, Drivsholm L, Price PA, Christensen IJ. High serum YKL-40 level in patients with small cell lung cancer is related to early death. *Lung cancer* 2004; 46: 333-340.

Johansen JS, Jensen BV, Roslind A, Nielsen D, Price PA. Serum YKL-40, a new prognostic biomarker in cancer patients? *Cancer Epidermiol. Biomarkers Prev.* 2006b; 15: 194-202.

Johansen JS, Stoltenberg M, Hansen M, Florescu A, Horslev-Petersen K, Lorenzen I, Price PA. Serum YKL-40 concentrations in patients with rheumatoid arthritis: relation to disease activity. *Rheumatology* 1999; 38: 618-626.

Johansen JS, Williamson MK, Rice JS, Price PA. Identification of proteins secreted by human osteoblastic cells in culture. *J Bone Miner Res.* 1992; 7: 501-512.

Kawada M, Arihiro A, Mizoguchi E. Insights from advances in research of chemically induced experimental models of human inflammatory bowel disease. *World J Gastroenterol.* 2007a; 13: 5581-5593.

Kawada M, Hachiya Y, Arihiro A, Mizoguchi E. Role of mammalian chitinases in inflammatory conditions. *Keio J Med.* 2007b; 56: 21-27.

Kawada M, Chen CC, Arihiro A, Nagatani K, Watanabe T, Mizoguchi E. Chitinase 3-like-1 enhances bacterial adhesion to colonic epithelial cells through the interaction with bacterial chitin-binding protein. *Lab Invest.* 2008; 88: 883-895.

Kelleher TB, Mehta SH, Bhaskar R, Sulkowski M, Astemborski J, Thomas DL, Moore RE, Afdhal NH. Prediction of hepatic fibrosis in HIV/HCV co-infected patients using serum fibrosis markers: the SHASTA index. *J Hepatol.* 2005; 43: 78-84.

Kirkpatrick RB, Emery JG, Connor JR, Dodds R, Lysko PG, Rosenberg M. Induction and expression of human cartilage glycoprotein 39 in rheumatoid inflammatory and peripheral blood monocyte-derived macrophages. *Exp Cell Res.* 1997; 237: 46-54.

Kirn TJ, Jude BA, Taylor RK. A colonization factor links *Vibrio cholerae* environmental survival and human infection. *Nature* 2005; 438: 863-866.

Kzhyshkowska J, Mamidi S, Gratchev A, Kremmer E, Schmuttermaier C, Krusell L, Haus G, Utikal J, Schledzewski K, Scholtze J, Goerdt S. Novel stabilin-1 interacting chitinase-like protein (SI-CLP) is up-regulated in alternatively activated macrophages and secreted via lysosomal pathway. *Blood* 2006; 107: 3221-3228.

Lee JY, Spicer AP. Hyaluronan: a multifunctional, megaDalton, stealth molecule. *Curr Opin Cell Biol.* 2000; 12: 581-586.

Ling H, Recklies AD. The chitinase 3-like protein human cartilage glycoprotein 39 inhibits cellular responses to the inflammatory cytokines interleukin-1 and tumor necrosis factor-alpha. *Biochem J.* 2004; 380: 651-659.

Malinda KM, Ponce L, Kleinman HK, Shackelton LM, Millis ATJ. Gp38k, a protein synthesized by vascular smooth muscle cells, stimulates directional migration of human umbilical vein endothelial cells. *Exp Cell Res.* 1999; 250: 168-173.

McManus MT. Small RNAs and immunity. *Immunity* 2004; 21: 747-756.

Mizoguchi E. Chitinase 3-like-1 exacerbates intestinal inflammation by enhancing bacterial adhesion and invasion in colonic epithelial cells. *Gastroenterol.* 2006; 130: 398-411.

Montgomery MT, Kirchman DL. Role of chitin-binding Proteins in the Specific Attachment of the Marine Bacterium *Vibrio harveyi* to chitin. *Appl. Environ Microbiol.* 1993; 59: 373-379.

Nishikawa KC, Millis AJ. gp38k (CHI3L1) is a novel adhesion and migration factor for vascular cells. *Exp Cell Res.* 2003; 287: 79-87.

Nordenbaek C, Johansen JS, Junker P, Borregaard N, Sorensen O, Price PA. YKL-40, a matrix protein of specific granules in neutrophils, is elevated in serum of patients with community-acquired pneumonia requiring hospitalization. *J Infect Dis.* 1999; 180: 1722-1726.

Nyirkos P, Golds EE. Human synovial cells secrete a 39 kDa protein similar to a bovine mammary protein expressed during the non-lactating period. *Biochem J.* 1990; 269: 265-268.

Ostergaard C, Johansen JS, Benfield T, Price PA, Lundgren JD. YKL-40 is elevated in cerebrospinal fluid from patients with purulent meningitis. *Clin Diagn Lab Immunol.* 2002;9:598-604.

Palli SR, Retnakaran A. Molecular and biochemical aspects of chitin synthesis inhibition. *EXS* 1999, 87: 85-98.

Peltomaa R, Paimela L, Harvey S, Helve T, Leirisalo-Repo M. Increased level of YKL-40 in sera from patients with early rheumatoid arthritis: a new marker for disease activity. *Rheumatol Int.* 2001; 20: 192-196.

Perrakis A, Tews I, Dauter Z, Oppenheim AB, Chet I, Wilson KS, Vorgias CE. Crystal structure of a bacterial chitinase at 2.3 A resolution. *Structure* 1994; 2: 1169-1180.

Podolsky DK. Inflammatory bowel disease. N Engl J Med. 2002; 347: 417-429.

Raikhel NV, Lee HI, Broekaert WF. Structure and function of Chitin-Binding Proteins. *Annu Rev Plant Physiol.* 1993; 44: 591-615.

Recklies AD, White C, Ling H. The chitinase 3-like protein human cartilage glycoprotein 39 (HC-gp39) stimulates proliferation of human connective-tissue cells and activates both extracellular signal-regulated kinase- and protein kinase B-mediated signaling pathways. *Biochem J.* 2002; 365: 119-126.

Reese TA, Liang HE, TagerAM, Luster AD, Van Rooijen N, Voehringer D, Locksley RM. Chitin induces accumulation in tissue of innate immune cells associated with allergy. *Nature* 2007; 447: 92-96.

Rehli M, Krause SW, Andreesen R. Molecular characterization of the gene for human cartilage gp-39 (CHI3L1), a member of the chitinase protein family and marker for late stages of macrophage differentiation. *Genomics* 1997; 43: 221-225.

Renkema GH, Boot RG, Au FL, Donker-Koopman WE, Strijland A, Muijsers AO, Hrebicek M, Aerts JM. Chitotriosidase, a chitinase, and the 39-kDa human cartilage glycoprotein, a chitin-binding lectin, are homologues of family 18 glycosyl hydrolases secreted by human macrophages. *Eur J Biochem.* 1998; 251: 504-509.

Renkema GH, Boot RG, Muijsers AO, Donker-Koopman WE, Aerts JM. Purification and characterization of human chitotriosidase, a novel member of the chitinase family of proteins. *J Biol Chem.* 1995; 270: 2198-2202.

Rhodes JM. Unifying hypothesis for inflammatory bowel disease and associated colon cancer: sticking the pieces together with sugar. *Lancet* 1996; 347: 40-44.

Sartor RB. Intestinal microflora in human and experimental inflammatory bowel disease. *Curr Opin Gastroenterol.* 2001; 17: 324-330.

Sartor RB. Therapeutic manipulation of the enteric microflora in inflammatory bowel disease: antibiotics, probiotics, and prebiotics. *Gastroenterol.* 2004; 126: 1620-1633.

Shackelton LM, Mann DM, Millis AJ. Identification of a 38-kDa heparin-binding glycoprotein (gp38k) in differentiating vascular smooth muscle cells as a member of a group of proteins associated with tissue remodeling. *J Biol Chem.* 1995; 270: 13076-13083.

Shahabuddin M, Toyoshima T, Aikawa M, Kaslow DC. Transmission-blocking activity of a chitinase inhibitor and activation of malarial parasite chitinase by mosquito protease. *Proc Natl Acad Sci USA.* 1993, 90: 4266-4270.

Sorbotten A, Horn SJ, Eijsink VG, Varum KM. Degradation of chitosans with chitinase B from *Serratia marcescens*. Production of chito-oligosaccharides and insight into enzyme processivity. *FEBS J.* 2005; 272: 538-549.

Suzuki K, Sugawara N, Suzuki M, Uchiyama T, Katouno F, Nikaidou N, Watanabe T. Chitinase A, B, and C1 of *Serratia marcescens* 2170 produced by recombinant *Escherichia coli*: enzymatic properties and synergism on chitin degradation. *Biosci Biotechnol Biochem.* 2002a; 66: 1075-1083.

Suzuki K, Suzuki M, Taiyoji M, Nikaidou N, Watanabe T. Chitin binding protein (CBP21) in the culture supernatant of *Serratia marcescens* 2170. *Biosci Biotechnol Biochem.* 1998; 62: 128-135.

Suzuki K, Taiyoji M, Sugawara N, Nikaidou N, Henrissat B, Watanabe T. The third chitinase gene (chiC) of *Serratia marcescens* 2170 and the relationship of its product to other bacterial chitinases. *Biochem J.* 1999; 343: 587-596.

Suzuki M, Fujimoto W, Goto M, Morimatsu M, Syuto B, Iwanaga T. Cellular expression of gut chitinase mRNA in the gastrointestinal tract of mice and chickens. *J Histochem Cytochem.* 2002b; 50: 1081-1089.

Swidsinski A, Ladhoff A, Pernthaler A, Swidsinski S, Loening-Baucke V, Ortner M, Weber J, Hoffmann U, Schreiber S, Dietel M, Lochs H. Mucosal flora in inflammatory bowel disease. *Gastroenterol.* 2002; 122: 44-54.

Terwisscha van Scheltinga AC, Kalk KH, Beintema JJ, Dijkstra BW. Crystal-structures of hevamine, a plant defence protein with chitinase and lysozyme activity, and its complex with an inhibitor. *Structure* 1994; 2, 1181-1189.

Tjoelker LW, Gosting L, Frey S, Hunter CL, Trong HL, Steiner B, Brammer H, Gray PW. Structural and functional definition of the human chitinase chitin-binding domain. *J Biol Chem.* 2000; 275: 514-520.

Tsark EC, Wang W, Teng YC, Arkfeld D, Dodge GR, Kovats S. Differential MHC class II-mediated presentation of rheumatoid arthritis autoantigens by human dendritic cells and macrophages. *J Immunol.* 2002; 169: 6625-6633.

Uchiyama T, Katouno F, Nikaidou N, Nonaka T, Sugiyama J, Watanabe T. Roles of the exposed aromatic residues in crystalline chitin hydrolysis by chitinase A from *Serratia marcescens* 2170. *J Biol Chem.* 2001; 276: 41343-41349.

Vaaje-Kolstad G, Horn SJ, van Aalten DM, Synstad B, Eijsink VG. The non-catalytic chitin-binding protein CBP21 from *Serratia marcescens* is essential for chitin degradation. *J Biol Chem.* 2005a; 280: 28492-28497.

Vaaje-Kolstad G, Houston DR, Riemen AH, Eijsink VG, van Aalten DM. Crystal structure and binding properties of the *Serratia marcescens* chitin-binding protein CBP21. *J Biol Chem.* 2005b; 280: 11313-11319.

van Aalten DM, Komander D, Synstad B, Gaseidnes S, Peter MG, Eijsink VG. Structural insights into the catalytic mechanism of a family 18 exo-chitinase. *Proc Natl Acad Sci USA*. 2001;98:8979-8984.

van Aalten DM, Synstad B, Brurberg MB, Hough E, Riise BW, Eijsink VG, Wierenga RK. Structure of a two-domain chitotriosidase from *Serratia marcescens* at 1.9-A resolution. *Proc Natl Acad Sci USA*. 2000; 97: 5842-5847.

van Bilsen JH, van Dongen H, Lard LR, van der Voort EI, Elferink DG, Bakker AM, Miltenburg AM, Huizinga TW, de Vries RR, Toes RE. Functional regulatory immune responses against human cartilage glycoprotein-39 in health vs. proinflammatory responses in rheumatoid arthritis. *Proc. Natl Acad Sci USA.* 2004; 101: 17180-17185.

Varki A. Does DG42 synthesize hyaluronan or chitin? : A controversy about oligosaccharides in vertebrate development. *Proc Natl Acad Sci USA.* 1996; 93: 4523-4525.

Verheijden GF, Rijnders AW, Bos E, Coenen-de Roo CJ, van Staveren CJ, Miltenburg AM, Meijerink JH, Elewaut D, de Keyser F, Veys E, Boots AM. Human cartilage glycoprotein-39 as a candidate autoantigen in rheumatoid arthritis. *Arthritis Rheum.* 1997; 40: 1115-1125.

Viglietta V, Baecher-Allan C, Weiner HL, Hafler DA. Loss of functional suppression by CD4+CD25+ regulatory T cells in patients with multiple sclerosis. *J Exp Med.* 2004; 199: 971-979.

Vind I, Johansen JS, Price PA, Munkholm P. Serum YKL-40, a potential new marker of disease activity in patients with inflammatory bowel disease. *Scand J Gastroenterol.* 2003; 38: 599-605.

Volck B, Price PA, Johansen JS, Sorensen O, Benfield TL, Nielsen HJ, Calafat J, Borregaard N. YKL-40, a mammalian member of the chitinase family, is a matrix protein of specific granules in human neutrophils. *Proc Assoc Am Physicians* 1998; 110: 351-360.

Vos K, Miltenburg AM, van Meijgaarden KE, van den Heuvel M, Elferink DG, van Galen PJ, van Hogezand RA, van Vliet-Daskalopoulou E, Ottenhoff TH, Breedveld FC, Boots AM, de Vries RR. Cellular immune response to human cartilage glycoprotein-39 (HC gp-39)-derived peptides in rheumatoid arthritis and other inflammatory conditions. *Rheumatology* 2000a; 39: 1326-1331.

Vos K, Steenbakkers P, Miltenburg AM, Bos E, van Den Heuvel MW, van Hogezand RA, de Vries RR, Breedveld FC, Boots AM. Raised human cartilage glycoprotein-39 plasma levels in patients with rheumatoid arthritis and other inflammatory conditions. *Ann Rheum Dis.* 2000b; 59: 544-548.

Watanabe T, Kimura K, Sumiya T, Nikaidou N, Suzuki K, Suzuki M, Taiyoji M, Ferrer S, Regue M. Genetic analysis of the chitinase system of *Serratia marcescens* 2170. *J Bacteriol.* 1997; 179: 7111-7117.

Watanabe T, Kobori K, Miyashita K, Fujii T, Sakai H, Uchida M, Tanaka H. Identification of glutamic acid 204 and aspartic acid 200 in chitinase A1 of Bacillus circulans WL-12 as essential residues for chitinase activity. *J Biol Chem.* 1993; 268: 18567-18572.

Yu C, Lee AM, Bassler BL, Roseman S. Chitin utilization by marine bacteria. A physiological function for bacterial adhesion to immobilized carbohydrates. *J Biol Chem.* 1991; 266: 24260-24267.

Zeltins A, Schrempf H. Specific interaction of the Streptomyces chitin-binding protein CHB1 with alpha-chitin – the role of individual tryptophan residues. *Eur J Biochem.* 1997; 246: 557-564.

Zhu Z, Zheng T, Homer RJ, Kim YK, Chen NY, Cohn L, Hamid Q, Elias JA. Acidic mammalian chitinase in asthmatic Th2 inflammation and IL-13 pathway activation. *Science* 2004; 304: 1678-1682.

In: Binomium Chitin-Chitinase: Recent Issues ISBN 978-1-60692-339-9
Editor: Salvatore Musumeci and Maurizio G. Paoletti

Chapter XXII

The Biochemical Significance of Allosamidins as Chitinase Inhibitors

Shohei Sakuda
Department of Applied Biological Chemistry, The University of Tokyo, Bunkyo-ku, Tokyo 113-8657, Japan
Phone: +81-3-5841-5133 Fax: +81-3-5841-8022 E-mail:asakuda@mail.ecc.u-tokyo.ac.jp

Abstract

Chitinases occur widely in nature with various physiological roles dictated by the producing organisms. Chitinase inhibitors are useful to investigate a physiological role of chitinase of each organism and have a potential as useful drugs, mainly as insecticides or anti-asthmatic agents. Among chitinase inhibitors, allosamidin, a *Streptomyces* metabolite, has been used in basic research most frequently. Its structure and effects on a variety of organisms including insects, yeasts, parasites and mammals have provided clues to elucidate chitinase enzymology and physiological roles.

1. Introduction

Chitinases are present not only in chitin-containing organisms such as insects or fungi but also in non-chitin-containing ones such as bacteria, plants or mammals. Therefore, there are a variety of physiological roles of chitinases. In the former organisms, chitin is an essential polysaccharide constituting insect cuticle or fungal cell wall. Therefore, chitinases may have an important role in insect moulting or hyphal growth of fungi. In the latter organisms, bacteria may produce chitinases to use chitin as a nutrient when they live in chitin-containing habitats such as soils. Chitinases of plants and mammals may be speculated to have a defensive function toward pathogens.

Therefore, basic studies on a chitinase in each organism to investigate its properties and physiological role are very important. In basic research on an enzyme, a specific inhibitor is very useful and can sometimes provide crucial evidence of enzyme functions such as the catalytic mechanism or the physiological role in its producing organism. In the case of chitinases, allosamidin, the first chitinase inhibitor, has been used in basic research concerning chitinases and chitin metabolism during the last two decades. Since chitinase inhibitors have a potential usefulness as specific insecticides or fungicides, screening works to obtain new inhibitors have been carried out. In this chapter, works on allosamidins and other chitinase inhibitors are reviewed.

2. Isolation and Chemistry of Allosamidins

2.1 Isolation of allosamidin

Allosamidin was found during a study on insect ecdysis inhibitors. Chitin synthase and chitinase are key enzymes for chitin metabolism. At the ecdysis stage of insects, it had been speculated that synthesis of new skin by chitin synthase and degradation of old skin by chitinase occurred in parallel. In early 1980s, inhibitors of chitin synthase such as nikkomycin had been available and shown to inhibit insect moulting (Cohen and Cashida, 1982), but no chitinase inhibitor was known to show if chitinase is essential for insect ecdysis. Thus, it was planned to obtain an insect chitinase inhibitor. Since chitinase enzymology had not been developed well at the time, a crude chitinase extracts from the midgut of silkworm were used in the enzyme reaction for bioassay (Sakuda *et al.* 1987a). By the screening using the enzyme bioassay system, allosamidin was isolated from the mycelial extracts of a soil bacterium belonging to *Streptomyces* species in 1986 (Sakuda *et al.* 1986). Allosamidin inhibited the moulting of silkworm by injection, demonstrating that chitinase was important for insect ecdysis and that a chitinase inhibitor was a potential means for insect control (Sakuda *et al.* 1987a).

2.2 Structure of allosamidin

Structure of allosamidin (**1** in Fig. 1) was very novel. It has a pseudotrisaccharide structure mimic to chitin (Sakuda *et al.* 1986) (Sakuda *et al.* 1987b) (Sakuda *et al.* 1988). Two units of *N*-acetyl-D-allosamine and one unit of an aminocyclitol derivative, allosamizoline (**8**), constitute the allosamidin molecule. *N*-Acetyl-D-allosamine is a C-3 epimer of *N*-acetyl-D-glucosamine and is known only in allosamidin among known natural products. Allosamizoline has a cyclopentanoid structure which is highly oxygenated and fused with a dimethylaminooxazoline ring. The dimethylamino moiety has strong basicity and the quaternary carbon on the oxazoline ring is highly electrophilic. A six membered ring is often observed in a pseudosaccharide moiety of glycosidase inhibitors such as acarbose or nojirimycin. The cyclopentane ring of allosamizoline was the first case of a five membered ring which interacts with an active center of an enzyme as mentioned later. Trehazolin, a trehalase inhibitor, is known to have a cyclopentane ring similar to that of allosamidin (Ando *et al.* 1991). A lot of synthetic works of allosamidin have been reported and excellent reviews have been published in 1999 (Berecibar *et al.* 1999) and 2005 (Anderses *et al.* 2005).

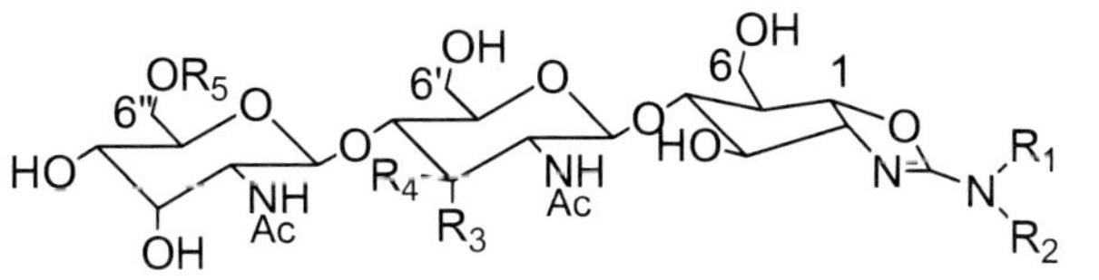

allosamidin (**1**) $R_1 = R_2 = CH_3$, $R_3 = OH$, $R_4 = H$, $R_5 = H$
methylallosamidin (**2**) $R_1 = R_2 = CH_3$, $R_3 = OH$, $R_4 = H$, $R_5 = CH_3$
demethylallosamidin (**3**) $R_1 = CH_3$, $R_2 = H$, $R_3 = OH$, $R_4 = H$, $R_5 = H$
glucoallosamidin A (**4**) $R_1 = R_2 = CH_3$, $R_3 = H$, $R_4 = OH$, $R_5 = CH_3$
glucoallosamidin B (**5**) $R_1 = CH_3$, $R_2 = H$, $R_3 = H$, $R_4 = OH$, $R_5 = CH_3$
methyl-*N*-demethylallosamidin (**6**) $R_1 = CH_3$, $R_2 = H$, $R_3 = OH$, $R_4 = H$, $R_5 = CH_3$
didemethylallosamidin (**7**) $R_1 = R_2 = H$, $R_3 = OH$, $R_4 = H$, $R_5 = H$

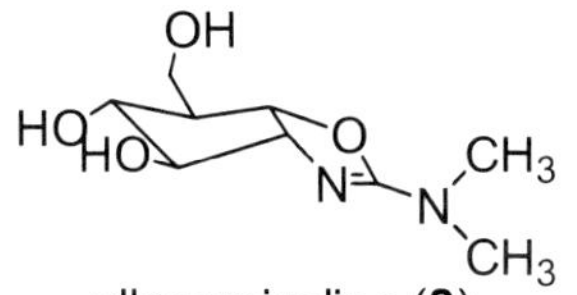

allosamizoline (**8**)

Figure 1. Structures of natural allosamidins and allosamizoline.

2.3 Isolation of allosamidin congeners

Until now, seven allosamidin congeners are isolated as natural products (Fig. 1). All allosamidins are *Streptomyces* metabolites. Methylallosamidin (**2**), 6"-*O*-methyl derivative of allosamidin, was isolated by work on insect chitinase inhibitors (Sakuda *et al.* 1987b). Demethylallosamidin (**3**), *N*-demethyl derivative of allosamidin, was found by the screening search for specific chitinase inhibitors of yeast, *Saccharomyces cerevisiae* (Sakuda *et al.* 1990). Glucoallosamidins A and B (**4**, **5**), and methyl-*N*-demethylallosamidin (**6**) were isolated during the work on inhibitors of *Candida albicans* chitinase (Nishimoto *et al.* 1991). Didemethylallosamidin (**7**), a *N,N*-didemethyl derivative, was found during the study on biosynthetic pathway of allosamidin (Zhou *et al.* 1993). Among these allosamidins, *N*-demethyl derivatives are important because they show much stronger inhibitory activity than allosamidin toward some chitinases as mentioned later.

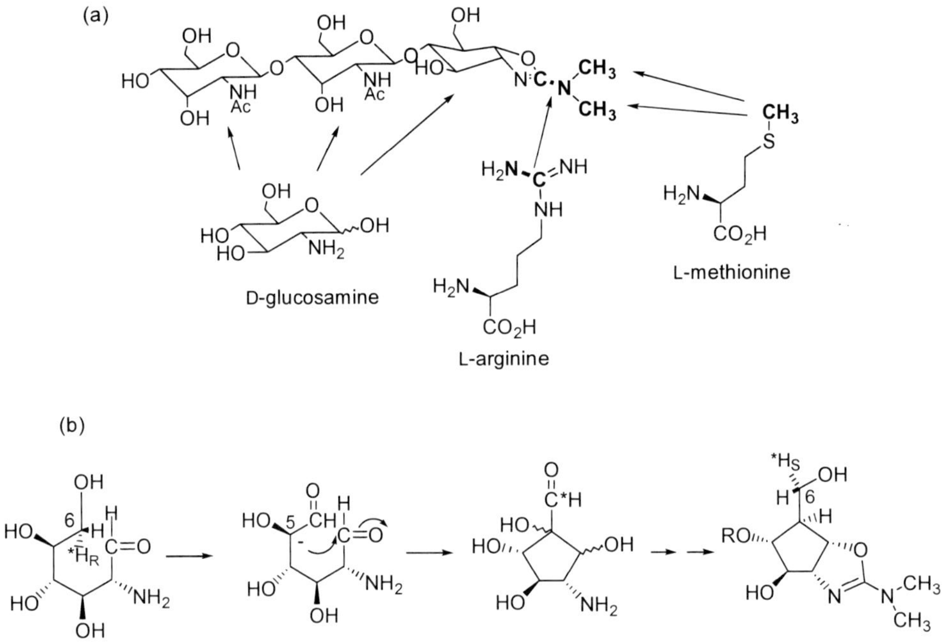

Figure 2. Biosynthesis of allosamidin.
(a) Origins of carbon and nitrogen atoms. (b) Plausible mechanism for formation of the cyclopentane ring of allosamizoline.

2.4 Biosynthesis of allosamidins

The unique structure of allosamidin indicated that some novel reactions were involved in its biosynthesis. Incorporation experiments of ^{13}C- and ^{15}N- labeled precursors revealed biosynthetic origins of carbon and nitrogen atoms of allosamidin molecule (Fig. 2a) (Zhou *et al.* 1992). Each carbon skeleton of allosamine and cyclopentane ring of allosamizoline is derived from D-glucosamine. Each nitrogen atom at C-2 of both the moieties also comes from the amino group of D-glucosamine. The aminooxazoline moiety was shown to originate from a carbon atom and one of the two nitrogen atoms of the guanidino group of L-arginine without cleavage of the C-N bond, and dimethyl group was shown to come from methionine. This biosynthesis of the aminooxazoline moiety of allosamizoline was different from that of trehazolin, which was shown to originate from the intact guanidino group of L-arginine (Sugiyama *et al.* 2002).

Conversion experiment with ^{14}C-labelled demethylallosamidin showed that allosamidin was produced by methylation of demethylallosamidin (Zhou *et al.* 1993). By a similar experiment with ^{14}C-labelled didemethylallosamidin, however, didemethylallosamidin was shown not to be converted to allosamidin or demethylallosamidin, suggesting that the first *N*-methyl group of the dimethylamino group may be introduced before the cyclization of the aminooxazoline moiety. In the case of didemethylallosamidin biosynthesis, an oxazoline ring would be formed before the methylation. To obtain a clue to reveal allosamidin biosynthetic pathway, a biosynthetic intermediate was searched among the metabolites of allosamidin-producing *Streptomyces*. While no biosynthetic intermediate was found, a disulfide compound consisting of cysteine, glucosamine and inositol was isolated (Sakuda *et al.* 1994) and was named mycothiol later when it was rediscovered as a co-factor widely distributed in gram-positive bacteria (Misset-Smits *et al.* 1997)

A cyclopentanoid skeleton biosynthesized from carbohydrate is uncommon in natural products and its biosynthetic mechanism is not clarified well, compared with a cyclohexanoid skeleton such as the case of inositol or sikimic acid. In the cyclopentane ring formation of allosamizoline, feeding experiments of ^{13}C-labelled glucosamine and glucose showed that the hydroxymethyl carbon of C-6 originates from C-6 of glucosamine and that C-C bond formation occurs between C-5 and C-1 of glucosamine. Further feeding experiments with a variety of specifically deuterium-labeled glucosamines suggested that the cyclization to form the cyclopentane ring may proceed *via* an intermediate with a 6-aldehyde group, which would undergo an aldol condensation on C-5 with C-1. The deuterium of (6*R*)-(6-$^{2}H_{1}$)-D-glucose was incorporated onto the *proS* position on C-6 of allosamizoline, but the deuterium of (6*S*)-(6-$^{2}H_{1}$)-D-glucose was not incorporated into allosamizoline. These experimental results indicated that overall inversion of stereochemistry of the C-6 methylene group occurred by stereospecific oxidation and reduction on C-6 during the formation of allosamizoline as shown in Fig. 2b (Sakuda *et al.* 2001).

3. Chitinase Inhibitory Activity of Allosamidins

3.1 Inhibitory activity of allosamidin

Chitinase inhibitory activity of allosamidin was first tested with insect chitinase. Allosamidin strongly inhibited the silkworm chitinase in a competitive way (Koga *et al.* 1987). Then, inhibitory activities of allosamidin toward a number of chitnases from a variety of organisms including fungi, bacteria, plants, parasites, nematodes, crustaceans, insects and mammals have been reported, which have been reviewed in 1993 (Sakuda *et al.* 1993) and 1999 (Spindler *et al.* 1999). Most of the chitinases tested were inhibited by allosamidin, but a small part of chitinases belonging to plants and *Streptomyces* were shown not to be inhibited by allosamidin (Koga *et al.* 1987) (Wang *et al.* 1993). The reason why allosamidin-sensitive and -insensitive chitinases are present in nature was unknown, but the data on amino acid and nucleotide sequences of various chitinases (Henrissat, 1991) indicate today that allosamidin inhibits family 18 chitinases, but not family 19.

3.2 Inhibitory activity of natural allosamidins and allosamidin derivatives

The inhibitory activities of natural allosamidins toward crude chitinases of silkworm (*Bombyx mori*), yeast (*Saccharomyces cerevisiae*) and fungus (*Trichoderma* sp.) were compared to each other by measuring IC_{50} values (Nishimoto *et al.* 1991). All allosamidins with the exception of didemethylallosamidin (**7**), showed similar strong inhibitory activity toward *Bombyx* and *Trichoderma* chitinases (IC_{50} = 0.05 – 0.1 μM for *Bombyx* chitinase, 1.3

– 2.6 μM for *Trichoderma* chitinase). The activity of didemethylallosamidin (**7**) was a little weaker (IC_{50} = 1.0 μM for *Bombyx* chitinase, 4.8 μM for *Trichoderma* chitinase). The inhibitory activity of allosamidins with *N*-dimethyl group (**1**, **2**, **4**) toward *Saccharomyces* chitinase (IC_{50} = 50 – 60 μM) was much weaker than that of allosamidins with *N*-monomethyl group (**3**, **5**, **6**) (IC_{50} = 0.5 – 0.8 μM). Inhibitory activity of didemethylallosamidin was medium (IC_{50} = 13 μM). Weaker inhibitory activity of didemethylallosamidin (**7**) than that of allosamidin or demethylallosamidin (**3**) was also shown against chitinases of a crustacean (*Artemia salina*), an insect (*Chironomus tentans*) and a bacterium (*Serratia marcescens*) (Spindler *et al.* 1997). Much stronger inhibitory activity of demethylallosamidin toward human chitotriosidase than that of allosamidin was reported (Rao *et al.* 2003).

Many allosamidin derivatives were prepared and their chitinase inhibitory activity was tested. A series of *N*-monoalkyl derivatives (**9**, **10**, **11**, **12** in Fig. 3) showed much weaker activities than that of allosamidin (Kinoshita *et al.* 1993). Inhibitory activities of a variety of 6- and 6"-*O*-acyl derivatives of allosamidin (**13**, **14**, **15**, **16**, **17**, **18** in Fig. 3) indicated that the 6"-*O*-acyl derivatives maintained relatively high activities. Synthetic analogues of allosamidin and demethylallosamidin with *N,N'*-diacetylchitobiosyl moiety (**19**, **20** in Fig. 3) maintained strong activity toward *Bombyx* and *Saccharomyces* chitinases, respectively, but they lost strong activity toward *Trichoderma* chitinase (Terayama *et al.* 1993) (Takahashi *et al.* 1994). Two pseudodisaccharides (**21**, **22** in Fig. 4) showed about the same inhibitory activities toward *Candida* and *Chironomus* chitinases as those of allosamidin (Nishimoto *et al.* 1991) (Blattner *et al.* 1997). Biotinylated and photoaffinity allosamidin derivatives (**23**, **24** in Fig. 4) (Sakuda and Sakurada, 1998, and Sakuda, unpublished data) maintaining strong inhibitory activity toward *Trichoderma* chitinase were prepared.

3.3 Mechanism of inhibition by allosamidin

Family 18 chitinases cleave chitin to yield the β configuration at C1 by a mechanism which leads to retention of the anomeric configuration after hydrolysis (Fig. 5). On the other hand, family 19 chitinases do it to yield α configuration at C1 by a mechanism leading to inversion of the anomeric configuration (Fig. 5). The novel structure of allosamizoline moiety of allosamidin contributed to elucidating the catalytic mechanism of family 18 chitinases. X-ray structure of hevamine, a family 18 plant chitinase complexed with allosamidin showed that allosamizoline moiety binds in the center of the catalytic site of the chitinase (van Scheltinga *et al.* 1995). This structure provided evidence for substrate assisted mechanism of the chitinase, in which allosamizoline structure is a mimic of a putative oxazolinium ion intermediate (Fig. 5) in the enzyme reaction (Tews *et al.* 1997). This catalytic model and allosamidin binding with family 18 chitinases were further examined by theoretical (Brameld *et al.* 1998) and NMR (Germer *et al.* 2002) studies and by X-ray crystallographic studies of the complexes of allosamidin with *Serratia marcescens* chitinases A (Papanikolau *et al.* 2003) and B (van Aalten *et al.* 2001), *Coccidioides immitis* chitinase (Bortone *et al.* 2002)

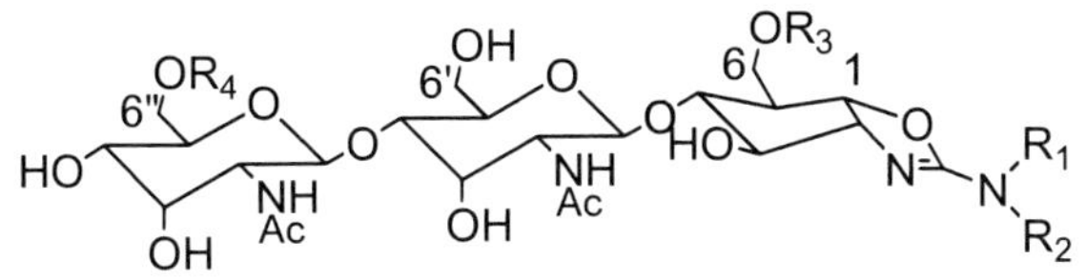

9 R_1 = ethyl, $R_2 = R_3 = R_4$ = H (**1/30**)*
10 R_1 = propyl, $R_2 = R_3 = R_4$ = H (**1/40**)
11 R_1 = butyl, $R_2 = R_3 = R_4$ = H (**1/150**)
12 R_1 = pentyl, $R_2 = R_3 = R_4$ = H (**1/150**)
13 $R_1 = R_2 = CH_3$, $R_3 = COC(CH_3)_3$, R_4 = H (**1/4**)
14 $R_1 = R_2 = CH_3$, $R_3 = CO(CH_2)_6CH_3$, R_4 = H (**1/20**)
15 $R_1 = R_2 = CH_3$, $R_3 = CO(OCH_2CH_2)_5CH_3$, R_4 = H (**1/25**)
16 $R_1 = R_2 = CH_3$, R_3 = H, $R_4 = COCH_2C_6H_5$ (**1/3**)
17 $R_1 = R_2 = CH_3$, R_3 = H, $R_4 = CO(CH_2)_6CH_3$ (**1/4**)
18 $R_1 = R_2 = CH_3$, R_3 = H, $R_4 = CO(OCH_2CH_2)_5CH_3$ (**1/5**)
* Values in parentheses indicate relative inhibitory activities toward *Bombyx* chtinase based on the activity of allosamidin.

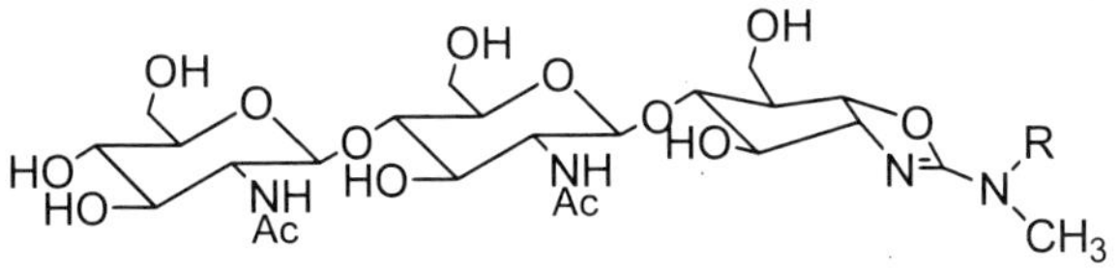

19 R = CH_3 (**1/5**)*
20 R = H (**3/5**) **
* Relative inhibitory activity toward *Bombyx* chitinase based on the activity of allosamidin.
** Relative inhibitory activity toward *Saccharomyces* chitinase based on the activity of demethylallosamidin.

Figure 3. Allosamidin derivatives and their chitinase inhibitory activities.

and human chitotriosidase (Rao *et al.* 2003). Recently, the thermodynamics of allosamidin binding to *S. marcescens* chitinase B was studied using isothermal titration calorimetry (Cederkvist *et al.* 2007), which theoretically explained pH dependency observed in inhibitory activity of allosamidin (Karasuda *et al.* 2004).

4. Styloguanidines, Diketopiperazines, Argifin, Argadin, and Miscellaneous Inhibitors

Until now, several natural products and synthetic compounds are known to inhibit chitinases. Styloguanidines (Kato *et al.* 1995) and a diketopiperazine of L-Arg and D-pro (Izumida *et al.* 1996) were isolated from marine organisms by the screening method with a chitin degrading bacterium *Schwanella* sp. Styloguanidine (**25** in Fig. 6), 3-bromostyloguanidine (**26**) and 2,3-dibromostyloguanidine (**27**) were obtained from an aqueous methanol extract of a sponge, *Stylotella aurantium*. They inhibited chitin degradation by the bacterium at a concentration of 2.5 μg/disk on the squid chitin plate and moulting of cyprid larvae of barnacles at 10 ppm *in vivo*. Cyclo(L-Arg-D-Pro) (**28** in Fig. 6) was isolated from the culture filtrate of a marine bacterium *Pseudomonas* sp. Inhibitory activity of the diketopiperazine toward *Schwanella* sp., *Bacillus* sp. and *Serratia marcescens* chitinases was weak (mM order of IC_{50} value), but it was shown to bind chtinase B of *S. marcescens* by mimicking the structure of the proposed reaction intermediate in family 18 chitinases above mentioned (Houston *et al.* 2004). Inhibitory activity and binding of several analogs of the diketopiperazine toward chitinase B were examined and it was shown that the cyclo(Gly-Pro) substructure is sufficient for binding.

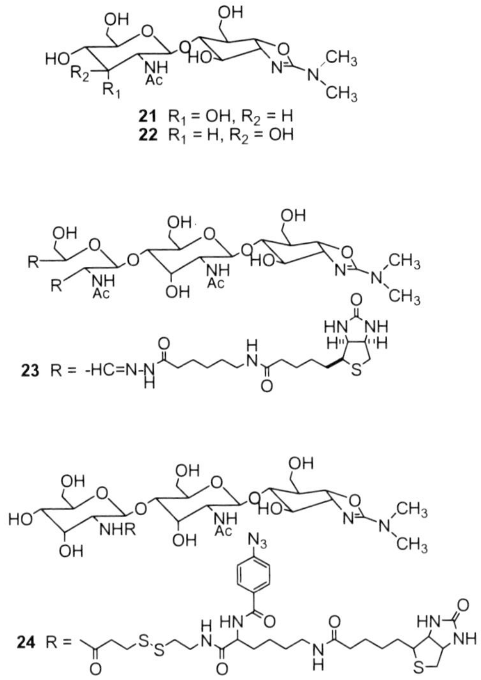

Figure 4. Pseudodisaccharides and allosamidin probes.

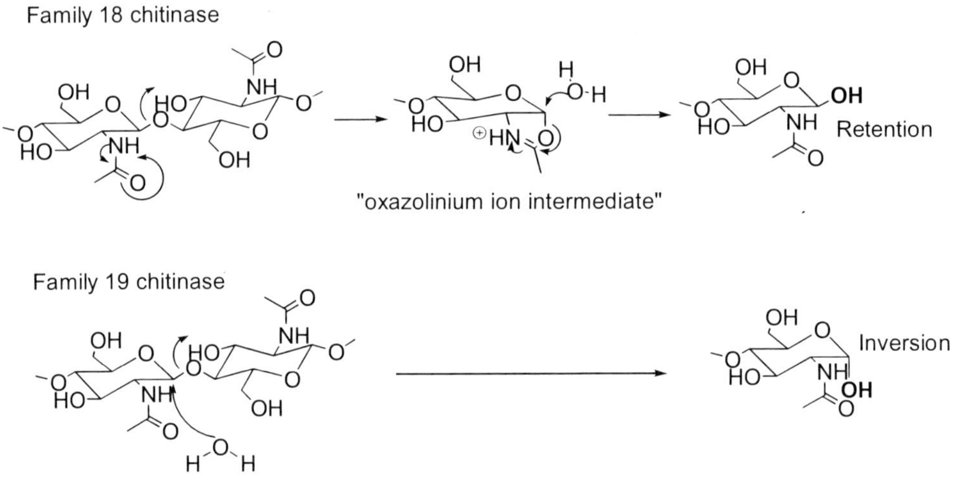

Figure 5. Catalytic mechanism of family 18 and 19 chitinases.

Argifin (**29** in Fig. 6) (Shiomi *et al.* 2000) and argadin (**30**) (Arai *et al.* 2000) were isolated from the culture broth of fungi, *Gliocladium* sp. and *Clonostachys* sp., respectively, as inhibitors of blowfly *Lucilia cuprina* chitinase. They have a unique cyclic pentapeptide structure containing an arginine residue whose ω-amino group is modified with *N*-methylcarbamoyl or acetyl group. Argadin showed much stronger inhibitory activity than argifin toward *L. cuprina* chitinase, *Serratia marcescens* chitinase B, *Aspergillus fumigatus* chitinase and human chitotriosidase with the IC_{50} or *Ki* values of 150 nM (IC_{50}), 20 nM (*Ki*), 500 nM (IC_{50}) and 13 nM (IC_{50}), respectively, by binding to the active site of these family 18 chitinases (Houston *et al.* 2002) (Rao *et al.* 2005). The inhibitory activity of argadin on *S. marcescens* chitinase B was stronger than that of allosamidin (*Ki* = 450 nM), whereas its activity on other three chitinases were weaker than that of allosamidin.

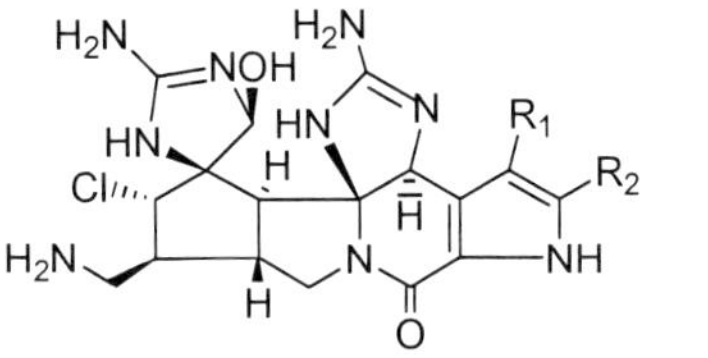

styloguanidine (**25**) $R_1 = R_2 = H$
3-bromostyloguanidine (**26**) R_1 = Br, R_2 = H
2,3-dibromostyloguanidine (**27**) $R_1 = R_2$ = Br

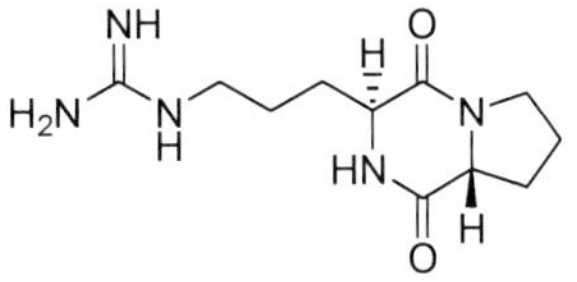

cyclo(L-Arg-D-Pro) (**28**)

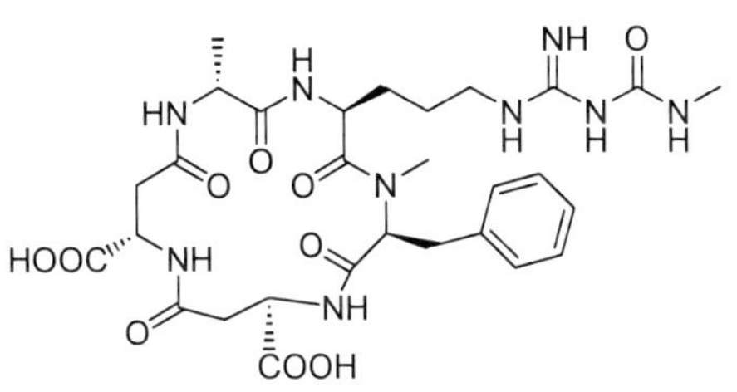

argifin (**29**)

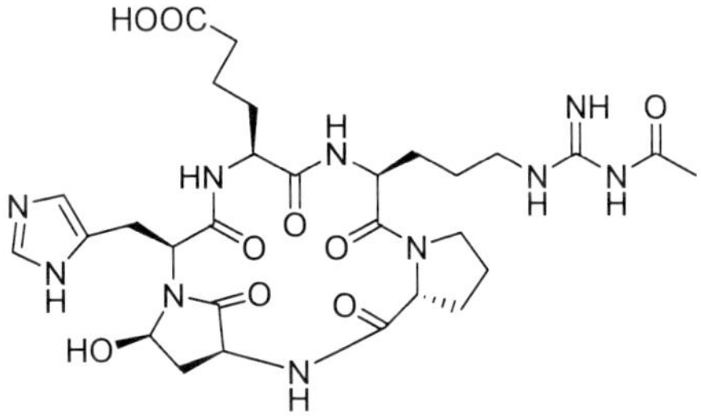

argadin (**30**)

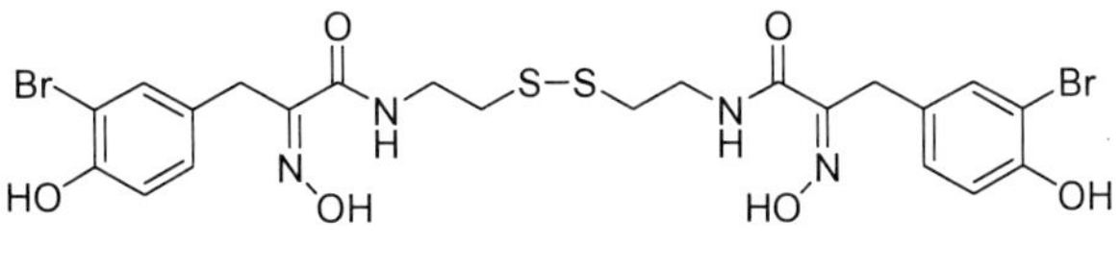

psammaplin A (**31**)

Figure 6. Structures of styloguanidines, cyclo(L-Arg-D-Pro), argifin, argadin and psammaplin A.

Psammaplin A (**31** in Fig. 6) was isolated from the marine sponge *Aplysinella rhax* by a screening (Tabudravu *et al.* 2002). It inhibited *Bacillus* sp. chitinase and *S. marcescens* chitinase B weakly by binding to near the active site of chitinase. However, its activity was not specific to chitinase because it showed a variety of biological activities including cytotoxic, antibacterial, antifungal and insecticidal ones.

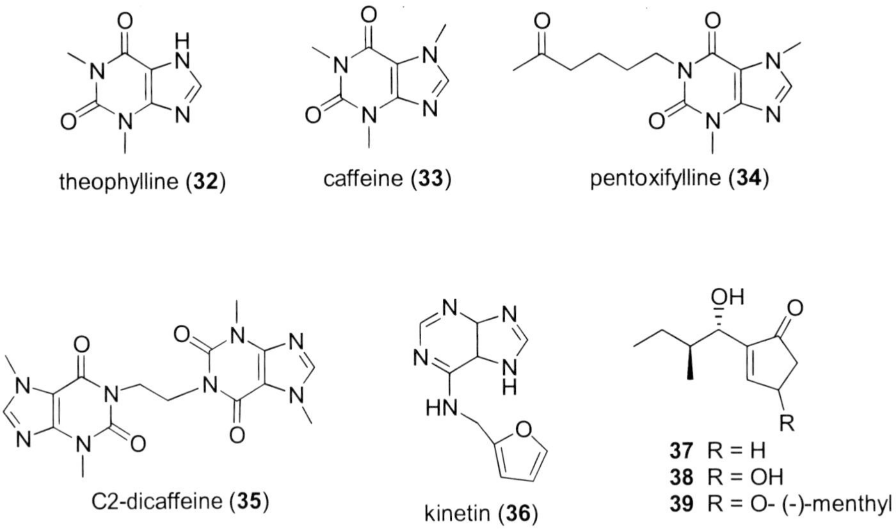

Figure 7. Structures of theophylline, caffeine, pentoxifylline, C2-dicaffeine, kinetin and butenolides.

Three methylxanthine derivatives (Fig. 7), theophylline (**32**), caffeine (**33**) and pentoxifylline (**34**), were found to inhibit *A. fumigatus* chitinase B1 by a high throughput screening (Rao *et al.* 2005). Their inhibitory activity on the chitinase was not strong (highest *Ki* value is 37 μM for pentoxifylline), but they were shown to bind the active center of the chitinase by mimicking the reaction intermediate similarly to the case of allosamidin. Based on the structure of pentoxifylline, virtual screening was carried out to obtain stronger inhibitors: in fact, a compound termed C2-dicaffeine (**35** in Fig. 7), was found to inhibit the fungal chitinase with *Ki* value around 2.8 μM (Schuttelkopf *et al.* 2006). C2-dicaffein inhibited *S. marcescens* chitinase and two mammalian chitinases (human chitotriosidase and mouse acidic mammalian chitinase), but its inhibitory activity on the three chitinases were weak. Recently, kinetin (**36** in Fig. 7) was found as a strong inhibitor for *Saccharomyces cerevisiae* chitinase (CTS1) (Hurtado-Guerrero *et al.* 2007).

A polysaccharide was isolated from the culture filtrate of a fungus, *Sphaeropsis* sp. as an inhibitor of the chitinase of common cutworm, *Spodoptera litura* (Nitoda *et al.* 2003a). The molecular mass of the polysaccharide was 16 kDa and *N*-acetylglucosamine, glucose and galactose were identified as major constituents of the polysaccharide. The polysaccharide showed strong inhibitory activity toward the insect chitinase with the IC_{50} value of 28 nM. Macromolecular insect chitinase inhibitors were found in culture broths of some fungal strains, but they are not well characterized (Nitoda *et al.* 2003b).

Weak chitinase inhibitory activity of some butenolides such as compounds **37** and **38** in Fig. 7 isolated from *Streptomyces antibioticus* was reported (Braum *et al.* 1995), but study on synthesis of them and their derivatives showed that natural products have no chitinase inhibitory activity and some *O*-menthyl derivatives such as compound **39** in Fig. 7 have weak inhibitory activity toward *S. marcescens* chitinases (Grossmann *et al.* 2005).

5. Biological Activities of Allosamidins

5.1 Biological activity on insects

Injection of allosamidin into the silkworm, *Bombyx mori*, completely inhibited the moulting from the last instar larva to pupa at the dose of 10 μg. At the time, new cuticle of pupa may be formed, but old cuticle of larva may not be degraded enough due to inactivation of chitinase, leading to failure of ecdysis, as shown in Fig. 8. Larval and pupal ecdysis of the common armyworm, *Leucania separate*, was also inhibited by application of allosamidin through injection (Sakuda *et al.* 1987a). Allosamidin could not affect the growth of these lepidopteran insects by oral or topical application, maybe due to its low permeability into skin of the insect. 6"-*O*-Acyl derivatives such as compound **18** in Fig. 3 showed relatively hydrophobic nature, but it did not show growth inhibitory activity toward silkworm larva by topical application. However, it was reported that allosamidin and two pseudodisaccharides (**21**, **22** in Fig. 4) showed inhibition of the moulting of the moth, *Tineola bisselliell*a, by oral application (Blattner *et al.* 1997). On the other hand, it was reported that allosamidin showed growth inhibition against the mite, *Tetranychus urticae*, larva of flies, *Musca domestica* (Somers *et al.* 1987) and *Lucilia cuprina* (Blattner *et al.* 1997), and the aphid, *Myzus persicae* (Saguez *et al.* 2006) by oral application. In insects, chitin is the main component not only of the skin but also of the peritrophic matrix in the midgut. It was reported that oral application of allosamidin to the mosquito, *Aedes aegypti*, led to the formation of an atypically thick peritrophic matrix, which delayed the digestive process of the mosquito (Filho *et al.* 2002). With respects of insecticidal activity of argifin and argadin, it was reported that they inhibited moulting of cockroach larvae upon injection.

Figure 8. Effect of allosamidin on ecdysis of the silkworm *Bombyx mori* from last instar to pupa. (a) Larval cuticle shedding is inhibited by injection of allosamidin (10 μg). (b) Normal pupa.

5.2 Biological activity on yeasts, fungi and bacteria

In yeasts, chitin is the main constituent of the primary septum between the mother and daughter cells at the budding cell stages. Yeast chitinases were deemed to have a role in the fission of the septum leading to cell separation. Addition of demethylallosamidin into the culture of *S. cerevisiae* caused clusterization of the yeast cells, which provided the first evidence of the role of yeast chitinase on cell separation (9). Demethylallosamidin did not show anti-yeast activity because the clustered cells grew normally. Similar phenomena were observed in another yeast, *Candida albicans*, and also in the fungi *Geotrichum candidum* (Yamanaka *et al.* 1994) and *Acremonium chrysogenum* (Sandor *et al.* 1998). Figures 9 and 10 shows abnormal cell clusters of *C. albicans* observed during the transition from mycelial form to yeast form cells and abnormal morphology of *G. candidum*, respectively, when demethylallosamidin was added to their culture media. In filamentous fungi, chitinases play an important role in living processes such as spore germination, hyphal growth and branching, or autolysis (Adams, 2004; Gooday *et al.* 1992). The fact that they have many genes encoding chitinases may support this speculation. For example, 18 chitinase genes are present in the genome of *Aspergillus nidulans*. Allosamidin or demethylallosamidin does not cause growth inhibition or morphological change of the fungus (Yamazaki *et al.* 2007) although disruption of one of the genes (*chiA*) affected germination frequency and hyphal growth (Takaya *et al.* 1998). Weak fungistatic effect of allosamidin on autolyzing *Penicillium chrysogenum* mycelia was reported (Sami *et al.* 2001), but critical effect of allosamidin or other chitinase inhibitors on fungal growth has not been observed. It is still unclear if a chitinase inhibitor is a potential antifungal agent.

It was reported that allosamidin inhibited the *in vivo* activity of the killer toxin of the yeast, *Kluyveromyces lactis*, by reducing action of the α-subunit of the toxin (Butler *et al.* 1991). The filamentous fungus *Trichoderma harzianum* is a mycoparasite against some plant pathogens such as *Rhizoctonia solani*. Addition of allosamidin during confrontation of the two fungi reduced inhibition of *R. solani* growth, indicating that chitinase is important for mycoparasitic interaction of *T. harzianum* with its host *R. solani* (Zelinger *et al.* 1999). The role of chitinase produced by *Bacillus thuringiensis* during pathogenesis in insects was demonstrated to be important by experiments with allosamidin (Sampson and Gooday, 1998). Allosamidin showed that chitinases produced by a chitinolytic dune soil β-subclass *Proteobacteria* were partially necessary for the bacterial invasion into hyphae of a host fungus such as *Chaetomium* or *Fusarium* sp. (Boer *et al.* 2001).

5.3 Biological activity on parasites

The encystment of *Entamoeba invadens* was markedly retarded by addition of allosamidin *in vivo* (Villagomez-Castro *et al.* 1992). During transmission of malaria ookinetes from humans to mosquito, the ookinetes chitinases may have a role for the penetration of ookinetes into the midgud epithelium of the mosquito by degrading the peritrophic matrix. Addition of allosamidin into a blood meal prevented oocyst formation in two mosquito models (*Aedes aegypti* infected with avian malaria parasite *Plasmodium gallinaceum*, and *Anopheles freeborni* infected with human malaria parasite *Plamodium falciparum*) (Shahabuddin *et al.* 1993; Vinetz *et al.* 1999), demonstrating that chitinase inhibitors may have a potential as malaria parasite transmission-blocking agents. Biotinylated allosamidin (**23**) was used for an experiment to detect the penetration of allosamidin into the cuticle of the filarial parasite *Onchocerca volvulus* (Wu *et al.* 2001).

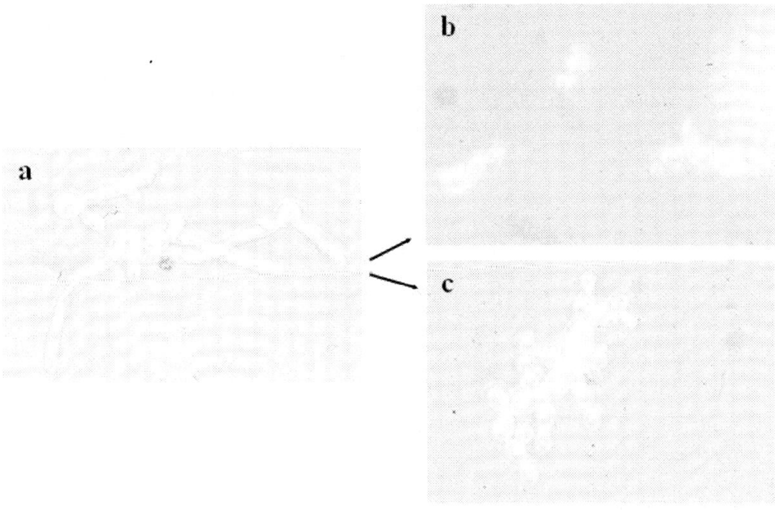

Figure 9. Effects of demethylallosamidin on *Candida albicans* during transition from mycelial form to yeast form cells.
(a) Mycelial form cells. (b) Yeast form cells converted from (a) without demethylallosamidin. (c) Clustered yeast form cells converted from (a) with demethylallosamidin (10 μg/mL). Calcofluor white M2R was added to stain cells.

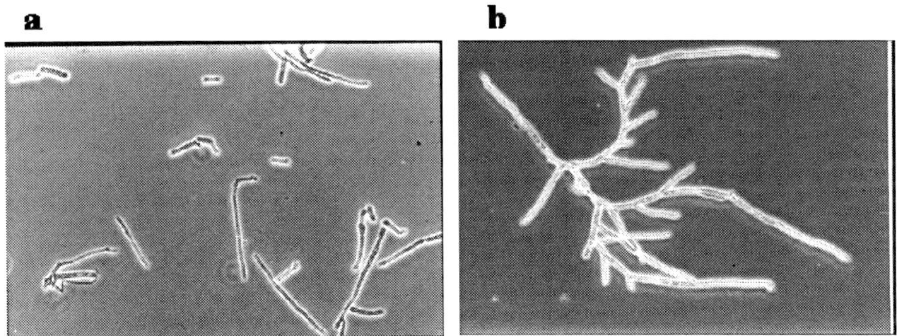

Figure 10. Effects of demethylallosamidin on morphology of *Geotrichum candidum*.
(a) Cultured without demethylallosamidin. (b) Cultured with demethylallosamdin (100 μg/mL).

5.4 Biological activity on asthma and inflammation

Although chitin is not present in mammals, two chitinases, chitotriosidase and acidic mammalian chitinase (AMCase), are present in human and mouse (Boot *et al.* 2001) (Boot *et al.* 2005). Physiological roles of the chitinases are not clear, but it may be possible to speculate that one of their roles is a defensive function against pathogens. AMCase has recently been associated with animal models of asthma. Bronchial asthma is a chronic inflammatory disease and T-helper-2 (Th2) cytokines are essential for generating asthmatic abnormalities. Among Th2 cytokines, IL-13 is now considered particularly critical. With respects to correlation between AMCase and asthma, it was reported that AMCase expression is upregulated in the response to allergen exposure or IL-13-induced inflammation in the lung (Homer *et al.* 2006) (Zhu *et al.* 2004). Inhibition of AMCase with anti-acidic mammalian chitinase sera lead to lower eosinophil counts and to reduce airway hyper-responsiveness in a murine model of asthma. Allosamidin suppressed allergen-induced airway eosinophilia in the asthma model and inhibition of AMCase by allosamidin was reported (Boot *et al.* 2001). Therefore, the effect of allosamidin on asthma may support the importance of AMCase in the mouse asthma. Although these observations suggest that AMCase acts as a proinflammatory mediator in IL-13 effector responses, it has been recently reported that transgenic mice overexpressing AMCase showed no signs of allergic inflammation (Reese *et al.* 2007). Thus, the precise role of AMCase in asthmatic responses is not clear enough and it may be necessary to reinvestigate the target molecule of allosamidin for its anti-asthmatic activity. Very recently, it was shown that allosamidin can reduce inflammatory signs observed in endotoxin-induced uveitis in rabbits (Bucolo *et al.* 2008).

5.5 Physiological activity on allosamidin-producing *Streptomyces*

The variety of secondary metabolisms present in microorganisms suggests that they developed them during evolution and each secondary metabolite might have a physiological role in its producer under some circumstances. The role of antibiotic production is presumable but has not been proved. It is entirely unknown why microorganisms produce many other compounds without antibiotic activity such as enzyme inhibitors. Allosamidin is a typical secondary metabolite of a microbe *Streptomyces* sp. It was recently found that allosamidin acts as a signal molecule for chitinase production in the producing *Streptomyces* strain itself. Allosamidin dramatically promoted chitinase production and growth of its producer, *Streptomyces* sp. AJ9463, in a chitin medium (Nakanishi *et al.* 2001) (Suzuki *et al.* 2006a). The chitinase whose production was enhanced by allosamidin was identified, and it was shown that two genes encoding proteins of a two-component regulatory system are present at the 5'-upstream region of the family 18 chitinase gene. Allosamidin enhanced the chitinase production through the two-component regulatory system (Suzuki *et al.* 2006b). Furthermore, a phenomenon in which allosamidin was released from the mycelia of its producer by responding to chitin was found (Suzuki *et al.* 2006a). The allosamidin effect on chitinase production among a variety of *Streptomyces* sp. was examined and it was shown that allosamidin can widely promote production of *Streptomyces* chitinases under control by the two-component system (Suzuki et al.. 2008), similarly to the case of strain AJ9463. Soil is rich in chitin-containing organisms such as fungi and insects. *Streptomyces* is thought to be a main microbe for the degradation of chitin in the soil. Therefore, the allosamidin's action on chitinase production and its generality suggest that allosamidin may affect chitin turnover and ecology in soil.

Conclusion

During the last two decades, many advances were made in basic research on chitinases. Allosamidins and chitinase inhibitors have been supporting them. The physiological roles of some chitinases are not well understood, however. Chitinase-like proteins deprived of chitinase activity occur widely in nature, but their physiological roles are not investigated; recently, we showed that allosamidin can tightly bind to a chitinase-like protein (Sakuda, unpublished data). Chitinase inhibitors including new ones may be useful for further basic and applied researches not only on chitinases but also on chitinase-like proteins.

References

van Aalten DMF, Komander D, Synstad B, Gaseidnes S, Peter MG, Eijsink VGH. Structural insights into the catalytic mechanism of a family 18 exo-chitinase. *Proc Natl Acad Sci* USA. 2001; 98: 8979-8984.

Adams DJ. Fungal cell wall chitinases and glucanases. Microbiol 2004; 150: 2029-2035.

Anderses OA, Dixon MJ, Eggleston IM, van Aalten DMF. Natural product family 18 chitinase inhibitors. *Nat Prod Rep.* 1005; 22: 563-579.

Ando O, Satake H, Itoi K, Sato A, Nakajima M, Takahashi S, Haruyama H, Ohkuma Y, Kinoshita T, Enokita R. Trehazolin, a new treharase inhibitor. *J Antibiot.* 1991; 44: 1165-1168.

Arai N, Shiomi K, Yamaguchi Y, Masuma R, Iwai Y, Turberg A, Kolbl H, Omura S. Argadin, a new chitinase inhibitor, produced by *Clonostachys* sp. FO-7314. *Che Pharm Bull.* 2000; 48: 1442-1446.

Berecibar A, Grandjean C, Siriwardena A. Synthesis and biological activity of natural aminocyclopentitol glycosidase inhibitors: mannostatins, trehazoline, allosamidins, and their analogues. *Chem Rev.* 1999; 99: 779-844.

Blattner R, Gerard PJ, Spindler-Barth M. Synthesis and biological activity of allosamidin and allosamidin analogues. *Pestic Sci.* 1997; 50: 312-318.

Boer WD, Gunnewiek PJAK, Kowalchuk GA, van Veen JA. Growth of chitinolytic dune soil β-subclass *Proteobacteria* in response to invading fungal hyphae. *Appl Environ Microbiol.* 2001; 67: 3358-3362.

Boot RG, Blommaart EFC, Swart E, Ghauharali-van der Vlugt K, Bijl N, Moe C, Place A, Aerts MF. Identification of a novel acidic mammalian chitinase distinct from chitotriosodase, *J Biol Chem.* 2001; 276: 6770-6778.

Boot RG, Bussink AP, Verhoek M, de Boer PAJ, Moorman AFM, Aerts JMFG. Marked Diffrences in tissue-specific expression of chitinases in mouse and man. *J Histochemistry & Cytochemistry.* 2005; 53:1283-1292.

Bortone K, Monzingo AF, Ernst S, Robertus JD. The structure of an allosamidin complex with the *Coccidioides immitis* chtinase defines a role for a second acid residue in substrate-assisted mechanism. *J Mol Biol.* 2002; 320: 293-302.

Brameld KA, Shrader WD, Imperiali B, Gddard III WA. Substrate assistance in the mechanism of family 18 chitinases: Theoretical studies of potential intermediates and inhibitors. *J Mol Biol* 1998; 280: 913-923.

Braum D, Pauli N, Sequin U, Zahner H. New butenolides from the photoconductivity screening of *Streptomyces antibioticus* (Waksman and Woodruff) Waksman and Henrici 1948. *FEMS Microbiol Lett.* 1995; 126: 37-42.

Bucolo C, Musumeci M, Maltese A, Drago F, Musumeci S. Effect of chitinase inhibitors on endotoxin-induced uveitis (EIU) in rabbits. *Pharmacol Res.* 2008; 57(3):247-52.

Butler AR, O'Donnell RW, Martin VJ, Gooday GW, Stark MJR. *Kluyveromyces lactis* toxin has an essential chitinase activity. *Eur J Biochem.* 1991; 199: 483-488.

Cederkvist FH, Saua SF, Karlsen V, Sakuda S, Eijsink VGH, Sorlie M. Thermodynamic analysis of allosamidin binding to a family 18 chitinase. *Biochemistry* 2007; 46: 12347-12354.

Cohen H, Cashida JE. Properties and inhibition of insect integumental chitin synthetase. *Pestic Biochem Physiol.* 1998; 17: 301-306.

Filho BPD, Lemos FJA, Secundino NFC, Pascoa V, Pereira ST, Pimenta PFP. Presence of chitinase and beta-*N*-acetyl glucosaminidase in the *Aedes aegypti* a chitinolytic system involving peritrophic matrix formation and degradation. *Insect Biochem Mol Biol.* 2002; 32: 1723-1729.

Germer A, Klod S, Peter MG, Kleinpeter E. NMR spectroscopic and theoretical study of the complexation of the inhibitor allosamidin in the binding pocket of the plant chitinase hevamine. *Mol Model* 2002; 8:231-236.

Gooday GW, Zhu WY, O'Donnell RW. What are the roles of chitinases in the growing fungus? *FEMS Microbiol Lett.* 1992; 100: 387-392.

Grossmann G, Jolivet B, Bornand M, Sequin U, Spindler KD. Synthesis and biological evaluation of some furanones as putative chitinase inhibitors. *Synthesis* 2005; 1543-1549.

Henrissat BA. Classification of glycosyl hydrolases based on amino acid sequence similarities. *Biochem J.* 1991; 280: 309-316.

Homer RJ, Zhu Z, Cohn L, Lee CG, White WI, Chen S, Elias, JA. Differential expressions of chitinases identify subsets of murine airway epithelial cells in allergic inflammation. *Am J Physiol Lung Cell Mol Physiol.* 2006; 291:L502-11.

Houston DR, Shiomi K, Arai N, Omura S, Peter MG, Turberg A, Synstad, B, Eijsink VGH, van Aalten DMF. High-resolution structures of a chitinase complexed with natural product cyclopentapeptide inhibitors: mimicry of carbohydrate substrate. *Proc Natl Acad Sci USA.* 2002; 99: 9127-9132.

Houston DR, Synstad B, Eijsink VGH, Stark MJR, Eggleston IM, van Aalten DMF. Structure-based exploration of cyclic dipeptide chitinase inhibitors. *J Med Chem.* 2004; 47: 5713-5720.

Hurtado-Guerrero R, van Aalten DMF. Structure of *Saccharomyces cerevisiae* chitinase 1 and screening-based discovery of potent inhibitors. *Chem Biol.* 2007; 14: 589-599.

Izumida H, Imamura N, Sano H. A novel chitinase inhibitor from a marine bacterium, *Pseudomonas* sp. *J Antibiot.* 1996: 49: 76-80.

Kato T, Shizuri Y, Izumida H, Yokoyama A, Endo M. Styloguanidines, new chitinase inhibitors from the marine sponge *Stylotella aurantium*. *Tetrahedron Lett.* 1995; 36: 2133-2136.

Karasuda S, Yamamoto K, Kono M, Sakuda S, Koga, D. Kinetic analysis of a chitinase from red sea bream, *Pagrus major*. *Biosci Biotechnol Biochem.* 2004; 68: 1338-1344.

Kinoshita M, Sakuda S, Yamada Y. Preparation of *N*-monoalkyl and *O*-acyl derivatives of allosamidin, and their chitinase inhibitory activities. *Biosci Biotech Biochem.* 1993; 57: 1699-1703.

Koga D, Isogai A, Sakuda S, Matsumoto S, Suzuki A, Kimura S, Ide, A. Specific inhibition of *Bombyx mori* chitinase by allosamidin. *Agric Biol Chem.* 1987; 51: 471-476.

Misset-Smits M, van Ophem PW, Sakuda S, Duine JA. Mycothiol, 1-*O*-(*N*-acetyl-L-cysteinyl)amido-2'-deoxy-α-D-glucopyranosyl)-D-*myo*-inositol, is the factor of NAD/factor-dependent formaldehyde dehydrogenase. *FEBS Lett.* 1997; 409: 221-222.

Nakanishi E, Okamoto S, Matsuura H, Nagasawa H, Sakuda S. Allosamidin, a chitinase inhibitor produced by *Streptomyces*, acts as an inducer of chitinase production in its producing strain. *Proc Japan Academy Ser. B* 2001; 77: 79-82.

Nishimoto Y, Sakuda S, Takayama S, Yamada, Y. Isolation and characterization of new allosamidins. *J Antibiot.* 1991; 44: 716-722.

Nitoda T, Usuki H, Kanzaki H. A potent insect chitinase inhibitor of fungal origin. *Z Naturforsch* 2003a; 58c: 891-894.

Nitoda T, Usuki H, Kurata A, Kanzaki H. Macromolecular insect chitinase inhibitors produced by fungi: screening and partial characterization. *J Pesticide Sci.* 2003; 28:33-36.

Papanikolau Y, Tavlas G, Vorgias CE, Petratos K. De novo purification scheme and crystallization conditions yield high-resolution structures of chitinase A and its complex with the inhibitor allosamidin. *Acta Cryst.* 2003; D59: 400-403.

Rao FV, Houston DR, Boot RG, Aerts JMFG, Hodkinson M, Adams DJ, Shiomi K, Omura S, van Aalten DMF. Specificity and affinity of natural product cyclopentapeptide inhibitors against *A. fumigatus*, human, and bacterial chitinases. *Chem Biol.* 2005; 12: 65-76.

Rao FV, Andersen OA, Vora KA, DeMartino JA, van Aalten DMF. Methylxanthine drugs are chitinase inhibitors: investigation of inhibition and binding modes. *Chem Biol.* 2005; 12: 973-980.

Rao FV, Houston DR., Boot RG, Aerts JMF, Sakuda S, van Aalten MF. Crystal structures of allosamidin derivatives in complex with human macrophage chitinase. *J Biol Chem.* 2003; 278: 20110-20116.

Reese TA, Liang HE, Tager AM, Luster AD, van Rooijen N, Voehringer D, Locksley RM. Chitin induces accumulation in tissue of innate immune cells associated with allergy. *Nature* 2007; 447:92-6.

Saguez J, Dubois F, Vincent C, Laberche JC, Sangwan-Norreel B, Giordanengo P. Differential aphicidal effects of chitinase inhibitors on the polyphagous homopteran *Myzus persicae* (Sulzer). *Pest Manag Sci.* 2006; 62: 1150-1154.

Sakuda S, Isogai A, Matsumoto S, Suzuki A, Koseki K. Structure of allosamidin, a novel insect chitinase inhibitor, produced by *Streptomyces* sp. *Tetrahedron Lett.* 1986; 27: 2475-2478.

Sakuda S, Isogai A, Matsumoto S, Suzuki, A. Search for microbial insect growth regulators II. Allosamidin. A novel insect chitinase inhibitor. *J Antibiot.* 1987a; 40: 296-300.

Sakuda S, Isogai A, Makita T, Matsumoto S, Koseki K, Kodama H, Suzuki, A. Structures of allosamidins, novel insect chitinase inhibitors, produced by actinomycetes. *Agric Biol Chem.* 1987b; 51: 3251-3259.

Sakuda S, Isogai A, Matsumoto, S, Suzuki A, Koseki K, Kodama H, Yamada, Y. Absolute configuration of allosamizoline, an aminocyclitol derivative of the chitinase inhibitor allosamidin. *Agric Biol Chem.* 1988; 52: 1615-1617.

Sakuda S, Nishimoto Y, Ohi M, Watanabe M, Takayama S, Isogai A, Yamada Y. Effects of demethylallosamidin, a potent yeast chitinase inhibitor, on the cell division of yeast. *Agric Biol Chem.* 1990; 54: 1333-1335.

Sakuda S, Isogai A, Suzuki A, Yamada Y. Chemistry and biochemistry of the chitinase inhibitors, allosamidins. *Actinomycetologica* 1993; 7: 50-57.

Sakuda S, Zhou ZY, Yamada Y. (1994). Structure of a novel disulfide of 2-(*N*-acetylcycteinyl)amino-2-deoxy-α-D-glucopyranosyl-*myo*-inositol produced by *Streptomyces* sp. *Biosci Biotech Biochem.* 1994; 58: 1347-1348.

Sakuda S, Sakurada M. Preparation of biotinylated allosamidins with strong chitinase inhibitory activities. *Bioorg Med Chem Lett.* 1998; 8: 2987-2990.

Sakuda S, Sugiyama Y, Zhou ZY, Takao H, Ikeda H, Kakinuma K, Yamada Y, Nagasawa H. Biosynthetic studies on the cyclopentane ring formation of allosamizoline, an aminocyclitol component of the chitinase inhibitor allosamidin. *J Org Chem.* 2001; 66: 3356-3361.

Sami L, Pusztahelyi T, Emri T, Varecza Z, Fekete A, Grallert A, Karanyi Z, Kiss L, Pocsi I. Autolysis and aging of *Penicillium chrysogenum* cultures under carbon starvation: Chitinase production and antifungal effect of allosamidin. *J Gen Appl Microbiol.* 2001; 47: 201-211.

Sampson MN, Gooday GW. Involvement of chitinases of *Bacillus thuringiensis* during pathogenesis in insects. *Microbiol.* 1998; 144: 2189-2194.

Sandor E, Pusztahelyi T, Karaffa L, Karanyi Z, Posci I, Biro S, Szentirmai A, Pocsi I. Allosamidin inhibits the fragmentation of *Acremonium chrysogenum* but does not influence the cephalosporin-C production of the fungus. *FEMS Microbiol Lett.* 1998; 164: 231-236.

van Scheltinga ACT, Armand S, Kalk KH, Isogai A, Henrissat B, Dijkstra BW. Stereochemistry of chitin hydrolysis by a plant chitinase/lysozyme and X-ray structure of a complex with allosamidin: Evidence for substrate assisted catalysis. *Biochemstry* 1995; 34: 15619-15623.

Schuttelkopf AW, Andersen OA, Rao FV, Allwood M, Lloyd C, Eggleston IM, van Aalten DMF. Screening-based discovery and structural dissection of a novel family 18 chitinase inhibitor. *J Biol Chem.* 2006; 281: 27278-27285.

Shahabuddin M, Toyoshima T, Aikawa M, Kaslow DC. Transmission-blocking activity of a chitinase inhibitor and activation of malarial parasite chitinase by mosquito protease. *Proc Natl Acad Sci USA.* 1993; 90: 4266-4270.

Shiomi K, Arai N, Iwai Y, Turberg A, Kolbl H, Omura S. Structure of argifin, a new chitinase inhibitor produced by *Gliocladium* sp. *Tetrahedron Lett.* 2000; 41: 2141-2143.

Somers PJB, Yao RC, Doolin LE, McGowan MJ, Fukuda DS, Mynderse JS. Methods for detection and quantitation of chitinase inhibitors in fermentation broths; isolation and insect life cycle effect of A82516. *J Antibiot.* 1987; 40: 1751-1756.

Spindler K D, Spindler-Barth M. (1999). Inhibitor of chitinases. In P. Jolles, & R. A. A. Muzzarelli (Eds.), *Chitin and Chitinases:* pp. 201-209. Basel: Birkhauser Verlag.

Spindler KD, Spindler-Barth M, Sakuda, S. Effect of demethylation on the chitinase inhibitory activity of allosamidin. *Arch Insect Biochem Physiol.* 1997; 36: 223-227.

Sugiyama Y, Nagsawa H, Suzuki A, Sakuda S. Biosynthesis of the trehalase inhibitor trehazolin. *J Antibiot.* 2002; 55: 263-269.

Suzuki S, Nakanishi E, Ohira T, Kawachi R, Nagasawa H, Sakuda S. Chitinase inhibitor allosamidin is a signal molecule for chitinase production in its producing *Streptomyces.* I. Analysis of the chitinase whose production is promoted by allosamidin and growth accelerating activity of allosamidin. *J Antibiot.* 2006; 59: 402-409.

Suzuki S, Nakanishi E, Ohira T, Kawachi R, Ohnishi Y, Horinouchi S, Nagasawa H, Sakuda S. Chitinase inhibitor allosamidin is a signal molecule for chitinase production in its producing *Streptomyces.* II. Mechanism for regulation of chitinase production by allosamidin through a two-component regulatory system. *J Antibiot.* 2006; 59: 410-417.

Suzuki S, Nakanishi E, Furihata K, Miyamoto K, Tsujibo H, Watanabe T, Ohnishi Y, Horinouchi S, Nagasawa H, Sakuda S. Chtinase inhibitor allosamidin promotes chitinase production of *Streptomyces* generally, *Int J Biol Macromol.* 2008; 43: 13-19.

Tabudravu JN, Eijsink VGH, Gooday GW, Jaspars M, Komander D, Legg M, Synstad B, van Aalten DMF. Psammaplin A, a chitinase inhibitor isolated from the Fijian marine sponge *Aplysinella rhax. Bioorg Med Chem.* 2002; 10: 1123-1128.

Takahashi S, Terayama H, Kuzuhara H, Sakuda S, Yamada Y. Preparation of a demetylallosamidin isomer having *N,N'*-diacetylchitobiosyl moiety and its potent inhibition against yeast chitinase. *Biosci Biotech Biochem.* 1994; 58:2301-2302.

Takaya N, Yamazaki D, Horiuchi H, Ohta A, Takagi M. Cloning and characterization of a chitinase-encoding gene (*chiA*) from *Aspergillus nidulans*, disruption of which decreases germination frequency and hyphal growth. *Biosci Biotechnol Biochem.* 1998; 62: 60-65.

Terayama H, Kuzuhara H, Takahashi S, Sakuda S, Yamada Y. Synthesis of a new allosamidin analog, *N,N'*-diacetyl-β-chitobiosyl allosamizoline, and its inhibitory activity against some chitinases. *Biosci Biotech Biochem.* 1993; 57: 2067-2069.

Tews I, van Scheltinga ACT, Perrakis A, Wilson KS, Dijkstra BW. Substrate-assisted catalysis unifies two families of chitinolytic enzymes. *J Am Chem Soc.* 1997; 119:7954-7959.

Villagomez-Castro JC, Calvo-Mendez C, Lopez-Romero E. Chtinase activity in encysting *Entamoeba invadens* and its inhibition by allosamidin. *Mol Biochem Parasitol.* 1992; 52: 53-62.

Vinetz JM, Dave SK, Specht CA, Brameld KA, Xu B, Hayward R, Fidock DA. The chitinase PfCHT1 from the human malaria parasite *Plasmodium falciparum* lacks proenzyme and chitin-binding domains and displays unique substrate preferences. *Proc Natl Acad Sci USA*. 1999; 96: 14061-14066.

Wang Q, Zhou, ZY, Sakuda S, Yamada Y. Purification of allosamidsin-sensitive and –insensitive chitinases produced by allosamidin-producing *Streptomyces*. *Biosci Biotech Biochem.* 1993; 57: 467-470.

Wu Y, Egerton G, Underwood AP, Sakuda S, Bianco AE. Expression and secretion of a larval-specific chtinase (family 18 glycosyl hydrolase) by the infective stages of the parasitic nematode, *Onchocerca volvulus*. *J Biol Chem.* 2001; 276: 42557-42564.

Yamanaka S, Tsuyoshi N, Kikuchi R, Takayama S, Sakuda S, Yamada Y. Effect of demethylallosamidin, a chitinase inhibitor, on morphology of fungus *Geotrichum candidum*. *J Gen Appl Microbiol.* 1994; 40: 171-174.

Yamazaki H, Yamazaki D, Takaya N, Takagi M, Ohta A, Horiuchi H. A chitinase gene, chiB, involved in the autolytic process of *Aspergillus nidulans*. *Curr Genet.* 2007; 51: 89-98.

Zeilinger S, Galhaup C, Payer K, Woo SL, Mach RL, Fekete C, Lorito M, Kubicek CP. Chtinase gene expression during mycoparasitic interaction of *Trichoderma harzianum* with its host. *Fungal Genet Biol.* 1999; 26: 131-140.

Zhou ZY, Sakuda S, Kinoshita M, Yamada, Y. Biosynthetic studies of allosamidin 2. Isolation of didemethylallosamidin, and conversion experiments of ^{14}C-labeled demethylallosamidin, didemethylallosamidin and their related compounds. *J Antibiot.* 1993; 46: 1582-1588.

Zhou ZY, Sakuda S, Yamada Y. Biosynthetic studies on the chitinase inhibitor, allosamidin. Origin of the carbon and nitrogen atoms. *J Chem Soc Perkin Trans. I* 1992: 1649-1652.

Zhu Z, Zheng T, Homer RJ, Kim YK, Chen NY, Cohn L, Hamid Q, Elias JA. Acidic mammalian chitinase in asthmatic Th2 inflammation and IL-13 pathway activation. *Science* 2004; 304:1678-1682.

In: Binomium Chitin-Chitinase: Recent Issues
Editor: Salvatore Musumeci and Maurizio G. Paoletti
ISBN 978-1-60692-339-9

Chapter XXIII

From Danger Signal to Messengers of Symbiosis: Recognition of Chitin and Chitin-Derived Chito-Oligosaccharides in Nature

Franco H. Falcone[17]
Associate Professor, Division of Molecular and Cellular Science, The School of Pharmacy, University of Nottingham, Boots Science Building, Science Road, Nottingham NG7 2RD, United Kingdom

Abstract

Chitin is a widespread carbohydrate polymer with unique biomechanical properties. In its crystalline form and many occurring additional modifications, chitin is a tough, resilient compound which is difficult to degrade, even by specialised enzymes. It is totally insoluble in water and plays an important part in the carbon and nitrogen cycles and as an energy source, particularly in the marine biosphere. Chitinolytic bacteria are endowed with a complex machinery enabling them to detect the presence of chitin, move chemotactically towards it following a gradient, attach to its surface via specialised pili, to release enzymes and accessory proteins for its degradation, and finally to import and metabolise its fragments. Binding of bacteria to chitin also has implications for human health, as pathogenic bacteria such as *Vibrio cholerae* are found predominantly in association with chitin-bearing copepods or other organisms, and thus can be filtered from water easily. Modified chitin-like substances (nod factors) released by Rhizobia (nitrogen fixing bacteria) are involved in the symbiotic relationship with legumes. For most plants, however, chitin detection signals an impending danger from a fungal pathogen. Considerable advances in the understanding of chitin sensing in plants have been achieved in the past few years, leading to the identification and cloning of several receptors containing extracellular LysM and intracytoplasmic Ser/Thr kinase domains

17 E-mail: franco.falcone@nottingham.ac.uk.

involved either in recognition of nod factors or in the detection of chitin fragments, the latter eliciting a complex immune response by the infected plant.

Very little is known regarding the interactions of chitin with the immune system of animals. Mammals do not contain chitin thus, similarly to what is seen in plants, chitin could constitute a 'danger' signal. Indeed, recent work has suggested that chitin is recognised by the mammalian immune system. This work has led the authors to postulate receptors for chitin e.g. on macrophages. Because chitin is very insoluble in aqueous solutions, sensing of chitin in most systems studied to date is mediated via recognition of its soluble degradation fragments $GlcNAc_n$. Size discrimination by the known receptors allows distinction of chitin-derived chito-oligomers ($GlcNAc_n$ with $n \geq 2$) from GlcNAc monomers, which could also be derived from the degradation of glycoproteins or glycolipids, and thus do not constitute a danger signal. Taken together, the examples seen in bacteria and plants point to the possibility that chitin-sensing pattern recognition receptors could also be found in higher animals such as mammals, and that chitin recognition could be mediated via interaction of chitin-oligosaccharides (of variable size) with these receptors.

1. Chitin is an Insoluble Polymer

Chitin is one of the most widespread carbohydrate polymers in Earth's biosphere. Its biomechanical properties make chitin a unique natural compound where rigidity, or toughness and flexibility, are required. It is widespread in the animal and fungal kingdom, where it is found e.g. in the cuticles of arthropods, the endoskeletons of cephalopods, in nematodes and in the fungal cell wall.

Chitin is a large, chemically stable biopolymer which is not readily soluble in aqueous solutions. A testament to its stability, chitin has been detected in old fossilised remains of insects dating back to the late Oligocene (about 25 million years ago) (Stankiewicz *et al.*, 1997). Commercial preparations of chitin are usually obtained from shrimp and crab shells, but as chitin is tightly associated with inorganic salts such as calcium carbonate, contains lipids including pigments, and is covalently crosslinked with proteins, the preparations have to undergo a three-step extraction and purification process, beginning with an acidic treatment in dilute hydrochloric acid for demineralisation.

Chitin occurs mainly in two different crystalline polymorphic forms (α and β) of which the α form, with antiparallel planes of strands held together by hydrogen bonds between the carbonyl and the amine groups and nonbonding interactions between the pyranose rings, is the most common, most rigid and most insoluble form. The β-form of chitin, which is characteristically found in squid pen and cuttlefish bone, undergoes fewer modifications and no mineralisation, and is therefore a superior source of clean, high molecular weight chitin with a higher degree of N-acetylation.

Due to its crystalline nature, chitin is insoluble in biological water-based media. For this reason, studies of chitinases since the 1950s have used colloidal chitin, rather than crystalline chitin (Muzzarelli 1999). Colloidal chitin is obtained by hydrolysing chitin in concentrated HCl at 40°C, and removing the excess acid by ion-exchange after hydrolysis. In this colloidal form, chitin can be more readily degraded by microbial chitinases (Muzzarelli 1999). Conversion to colloidal chitin has two major effects: it reduces its crystallinity, and results in a much larger exposed surface area per unit weight. This results in more rapid breakdown of chitin by exo- and endochitinases. Without this treatment, chitin breakdown would be much slower. Alternatively, chitinase studies have used fluorescently labelled chito-oligosaccharides, which are relatively small and soluble. While these assays are suitable for quantification of enzymatic activity, they do not reflect the complexity of chitin-chitinase interactions in the biological context (Eijsink *et al.* 2008). Finally, the modifications which chitin is known to undergo in many organisms, such as covalent links with proteins and the oxidation of diphenolic compounds in a process called sclerotisation, may further reduce its accessibility for chitinases (Ruiz-Herrera and Martínez-Espinosa, 1999). In fungi, chitin can also be found covalently linked with other carbohydrate polymers such as β-glucans and galactomannans (Ruiz-Herrera and Martínez-Espinosa, 1999).

Thus the physicochemical characteristics and covalent modifications of chitin raise the important question to which extent and at which rate chitin is accessible and can be degraded by exogenous chitinases in a biologically relevant context. A supermolecular structure of chitin in its natural forms has yet to be proposed, but it is certain to be a 2-phase system, containing areas of high crystallinity and amorphous areas, in which chitin is more accessible for degrading enzymes (Eijsink *et al.* 2008).

As chitin represents a major source of carbon, nitrogen and energy, particularly in the marine environment, and is a key component in fungi pathogenic for plants and animals, it is perhaps not surprising that sophisticated machineries for chitin sensing have evolved (see Fig 1). These mechanisms allow bacteria to locate and move towards a chitin containing source of energy, to attach to their surface, degrade chitin and metabolise its fragments (Bassler *et al.* 1989). In plants and perhaps also in animals, chitin sensing allows these organisms to detect and mount a protective immune response when encountering chitin-containing fungi and other pathogens. However, while the process of chitin recognition has been extensively studied in bacteria and is making steady progress in plant science, there is little understanding of these processes in higher animals, including humans.

This chapter will briefly review what is known about chitin sensing in bacteria and plants. Much work has been performed regarding chitin recognition in plants and bacteria; a detailed review is beyond the scope of this chapter. The aim is to highlight how little is known about the corresponding processes in mammals, and to explore how the corresponding knowledge obtained from other natural systems could be applied to the study of chitin recognition in higher animals (Figure 1).

2. Chitin Recognition in Bacteria

A large proportion of chitin is found in the aquatic environment. It is frequently quoted that $>10^{11}$ metric tons of chitin are produced each year, most of it in aquatic environments (Gooday 1990). While the presence of chitin is well known in the exoskeletons of crustaceans such as crabs, lobsters, crayfish and shrimps, or in internal structures such as the squid pen, it is also found in other widespread organisms including arrow worms, corals and, as shown only very recently, sponges (Ehrlich *et al.* 2007). Because chitin is the most abundant nitrogen-containing carbohydrate in the aquatic biosphere, it plays an important role in both the carbon and nitrogen cycles in the sea. The chitin derived from cast cuticles of molting and dying organisms such as zooplankton, is a major constituent of the so-called 'marine snow', a constant supply of organic detritus that sediments from upper layers to the sea bottom (Keyhani and Roseman, 1999). It is therefore not surprising to find that several marine bacteria are endowed with complex machinery enabling them to utilize chitin as a nutrient and energy source. Under laboratory conditions, many chitinolytic bacteria can grow on chitin as their only energy source. The importance of chitin and the role of chitinolytic bacteria in the aquatic environment have been recognised since the 1930s (Hock, 1940).

To date, the best studied bacterial species from this point of view is *Vibrio* (Family: *Vibrionaceae*), which is also the most widespread bacterial family in the marine environment (Colwell, 1984). This may reflect the special ability of these bacteria to utilise the enormous amounts of chitin produced yearly in the aquatic environment. Chitin utilisation by bacteria, as shown for *Vibrio furnissii*, is thought to occur in at least three distinct phases: chitin sensing and chemotaxis, attachment to chitinous surfaces, and chitin degradation (Keyhani and Roseman, 1999) (see Figure 2).

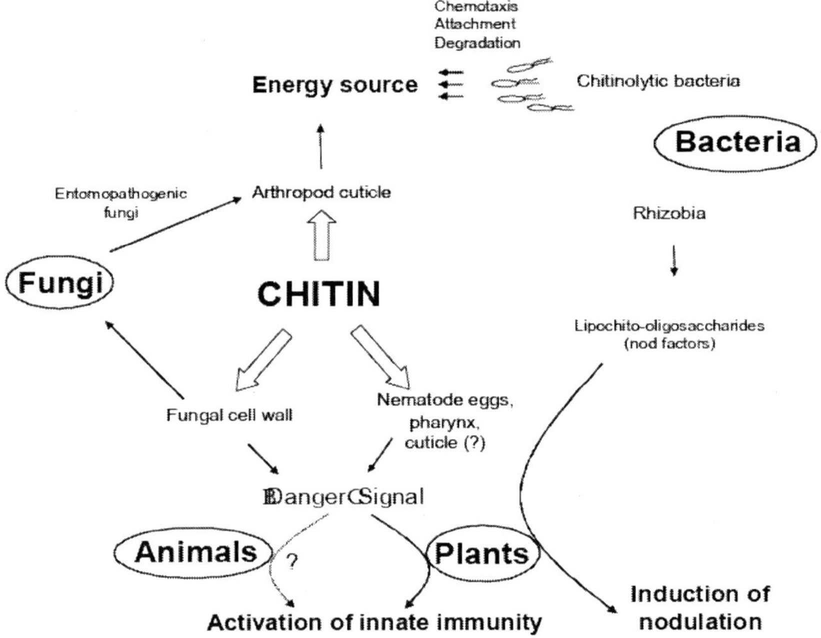

Figure 1. Schematic illustration of the roles of chitin in nature discussed in this chapter. Due to its biomechanical properties, chitin is a unique organic polymer found in many animals as well as in fungi. It is a precious source of carbon, nitrogen and energy in marine environments, where most of the chitin is produced and turned over. As a constituent of the fungal cell wall, and as it is not found in mammals or plants, chitin represents a 'danger' signal which can alert the immune system to the presence of a pathogen. Because chitin is very insoluble in aqueous solutions, sensing of chitin in most systems studied to date is mediated via recognition of its soluble degradation fragments $GlcNAc_n$. Size discrimination by the known receptors allows distinction of chitin-derived chito-oligomers ($GlcNAc_n$ with $n \geq 2$) from GlcNAc monomers, which could also be derived from the degradation of glycoproteins or glycolipids, and thus do not constitute a danger signal.

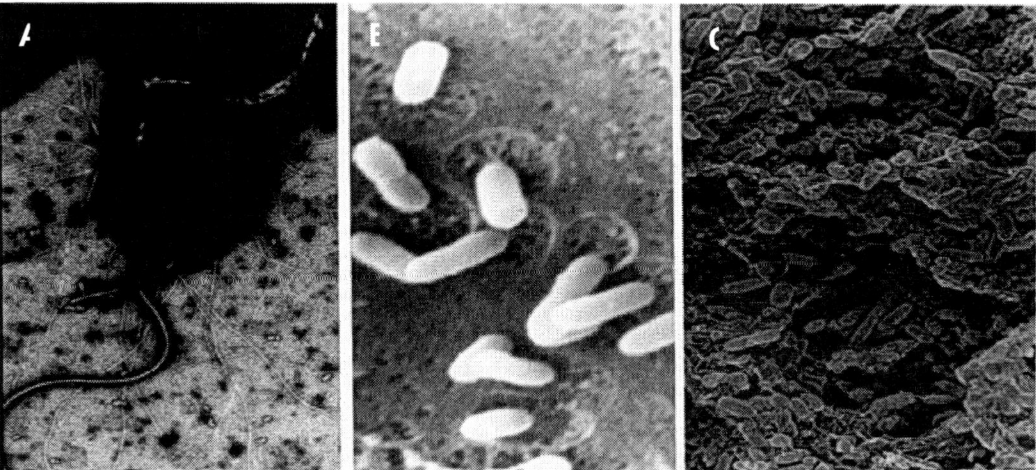

Figure 2. Electron micrographs of *Vibrio furnissii* (A) illustrating the two larger flagella used for movement, and numerous smaller fimbriae. These cells are attached to a chitinous substrate. (B) *V. furnissii* attached to chitinous substrate with clear signs of crater-like degradation, and (C) Bacterial association with a 'marine snow' aggregate particle. Courtesy of Dr. Keyhani, Dr. Roseman and Elsevier Publishers, from Keyhani and Roseman, 1999. Photograph in Panel A: Dr. Charles Yu, Johns Hopkins University, Panels B and C Dr. Kevin Carmen, Louisiana State University.

Sensing and Chemotaxis:

Because of its insolubility in water, chitin cannot diffuse and generate a gradient that can be sensed chemotactically. Therefore, encounters between motile planktonic bacteria and chitinous surfaces would appear to be a result of random collisions in the water column. However, as pointed out in the review article of Keyhani and Roseman (1999), free-living bacteria are relatively rare in the water column, as they tend to be associated with other organisms, such as microalgae, crustaceans or with marine snow, where their concentration is 2- to 5-fold orders of magnitude greater than in the surrounding water. This would make the chance of a random collision a relatively rare event. A mechanism for sensing chitin would therefore allow a more efficient detection by the bacteria.

Indeed, the work of Bassler and colleagues has demonstrated the existence of a sophisticated chemotactic mechanism which enables *Vibrio furnissii* to detect chitinous organisms (Bassler *et al.* 1989; Bassler *et al.* 1991). The bacterium displays low level constitutive chemotaxis towards the chitin monomer N-acetylglucosamine (GlcNAc). Cells grown in the presence of GlcNAc or its oligomers showed strongly enhanced migratory responses to the oligomers, due to induction processes. The authors also presented evidence for the induction of at least three inducible chemoreceptors, one which recognises GlcNAc, and two which recognise GlcNAc oligomers of different sizes with partially overlapping specificity. The bacteria also displayed chemotactic responses towards several aminoacids.

This work however raised the question of where the initial chitin oligomers come from. How do the first bacteria find and colonise chitinous organisms, and initiate the cascade leading to chitin degradation? A later publication by the same group of researchers suggested a possible answer to this conundrum. Yu and co-authors showed that *V. furnissii* are attracted to components released by injured or dying organisms, for example by the carbohydrate trehalose, which is a major component of crustacean or insect haemolymph (Yu *et al.* 1993). Thus, bacteria are able to locate and move towards dead or dying chitin-containing organisms independently of chitin, and as discussed below, can attach to them. The release of extracellular chitinases by the first colonising bacteria would release sufficient amounts of chito-oligosaccharides, acting as powerful inducers of chemotaxis and recruiting more bacteria to the chitinous growth substrate.

Adhesion/De-Adhesion

The next phase is the attachment, or adhesion of the bacteria to chitinous surfaces. This is mediated by chitin-specific lectins, some of which are – perhaps not surprisingly – induced by chitin and/or chito-oligosaccharides. Such chitin-specific lectins have been demonstrated e.g. in *Vibrio harveyi, V. parahaemolyticus* and *V. alginolyticus* (reviewed in Keyhani and Roseman, 1999). The process of adhesion was found to be dependent on the presence of other nutrients, resulting in what was called a 'nutrient sensorium'. The implication was that in the absence of essential nutrients such as phosphate or lactate, in other words, in conditions that would not support the survival of the bacterium, the organisms would detach (de-adhesion) in search of a more suitable, nutrient rich microenvironment (Keyhani and Roseman, 1999).

In more recent years, the induction of several chitinolytic enzymes in bacteria such as *Chromobacterium violaceum* (Chernin *et al.* 1998) or *Pseudomonas aeruginosa* (Folders *et al.* 2001) has been shown to be regulated genetically by an intercellular communication process called 'quorum sensing'. In Gram-negative bacteria, quorum sensing is typically mediated by N-acyl homoserine lactone molecules (AHLs). These compounds regulate many phenotypical characteristics such as the production of secondary metabolites, exoenzymes, chemoluminescence, and biofilm formation (Hardie and Heurlier, 2008). Biofilm formation is a critical factor, as biofilms provide protection from toxic compounds, thermal stress, predation and allow bacteria to reach higher numbers, thus making successful infection with pathogenic bacteria more likely and antibiotic therapy less effective (Huq *et al.* 2008).

The attachment of *Vibrio* to chitinous surfaces, as well as biofilm formation, has direct implications for human health. It has become clear since the mid-1970s that pathogenic bacteria, such as *Vibrio cholerae* and *Vibrio parahaemolyticus,* attach to chitinous surfaces, particularly to copepods (Nalin *et al.* 1979). Colwell an co-workers later showed how this characteristic behaviour related to the seasonal outbreaks of cholera in Bangladesh, which follow shortly after periodic algal bloom events and a corresponding explosive population increase of copepods and other zooplankton feeding on these algae (Huq *et al.* 1983). As most of the cholera bacteria are found on the zooplankton, rather than in the free water, most of the cholera pathogens (>99%) can be removed from contaminated water by employing a simple filtration technique in which sari cloth is folded over at least four times. This technique has been shown to reduce the incidence of cholera by almost 50% (Colwell *et al.* 2003).

Degradation

Once bacteria have located and adhered to the chitin substrate, the next step is its degradation (see Fig 3). This begins with the release of extracellular chitinases. Exochitinases (β-1,4-N-acetylglucosaminidases) cleave GlcNAc monomers from the nonreducing end. Endochitinases release fragments of various sizes initially, followed by chitobiose ($GlcNAc_2$) fragments (Eijsink *et al.* 2008). *Serratia marcescens* has been shown to express an accessory protein called CBP21 which binds chitin and is thought to speed up its degradation by partially opening the crystalline structure and increasing accessibility to endochitinases; similar proteins are found in many other chitinolytic organisms, including some insect viruses (Eijsink *et al.* 2008). The final products of chitin degradation are acetate, ammonia and fructose-6-phosphate, the latter of which can be shunted into the pentose-phosphate or the glycolytic pathway. The precise mechanisms and enzymes involved and their regulation are complex and beyond the scope of this chapter. Smaller chito-oligosaccharides can diffuse into the periplasm of Gram-negative bacteria through non-specific porins. In *Vibrio furnissii*, a specific chitoporin located in the outer membrane and induced by chito-oligosaccharides, but not by monomers, has been described (Keyhani *et al.* 2000).

Once the chito-oligosaccharides have reached the periplasm, they are degraded to mono- and disaccharides by specific β-N-acetylglucosaminidases and are taken up into the cell by specific transport systems, where they are further catabolised.

Taken together, due to the lack of solubility of chitin and the uneven distribution of nutrients in sea water, a complex machinery has to operate for chitinolytic bacteria to be thriving in the marine environment. This includes extracellular chitinases, chemotactic sensors for chitin oligomers, which act as powerful chemoattractants, chitin specific porins and other channels through the outer membrane, hydrolases located in the periplasm that generate GlcNAc monomers and dimers, specific transporters located in the cell membrane, and finally a series of catabolic enzymes (Li and Roseman, 2004). Through the combination of gene arrays and more traditional genetic studies, more genes involved in what has been called the chitinolytic cascade have been discovered (Meibom *et al.* 2004). New proteins such as a chitin-regulated Pilus (ChiRP) important in the colonisation of chitin by *Vibrio cholerae*, have been identified. This chitin-inducible Type IV pilus requires the cooperation of at least 12 different genes involved in its biogenesis (Meibom *et al.* 2004).

But how can such a complex response be co-ordinated? As shown in other bacterial metabolic pathways, specific key sensors can act as master switches, turning on or off the complex chitinolytic cascade machinery when needed. Li and Schoolnik have proposed a model which could operate in the case of *Vibrio* (Li and Roseman, 2004). In this model, two proteins called ChiS and CBP are the orchestrators of this process. When *Vibrio* is starving, the bacteria produce extracellular chitinases, which upon encountering chitin will liberate $(GlcNAc)_n$ oligomers and generate a gradient. These chito-oligomers will recruit more bacteria actively following the gradient, and will switch on the chitinolytic cascade, comprising of at least 50 different genes by derepressing the master switch ChiS. Once chitin is depleted, $(GlcNAc)_n$ levels will decrease, and the second regulator, a $(GlcNAc)_n$ high-affinity binding protein (CBP), will bind to ChiS and switch off the signal.

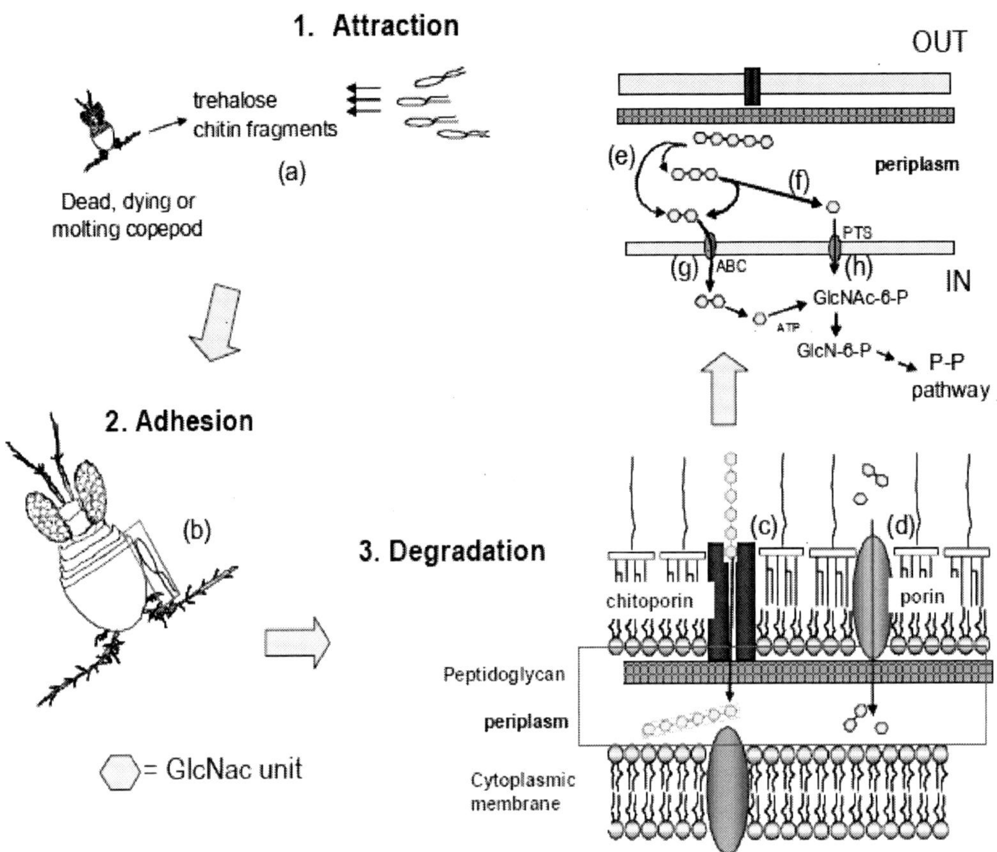

Figure 3: Initial encounters may be triggered by the release of e.g. trehalose from dying copepods, or by chitin fragments released by the action of extracellular bacterial chitinases such as ChiA (a). Chitinolytic marine bacteria follow the gradient and attach to the chitinous surface (b). This process may involve a chitin–inducible pilus (ChiRP) and chitin-specific lectins (chitovibrin). Chito-oligosaccharides are transported through the outer cell wall into the periplasm via a 40 kDa chitoporin (c), while GlcNAc and chitobiose can enter via a different porin.. The periplasm contains hydrolytic enzymes such as a chitodextrinase (d) and a β-N-acetyl-glucosaminidase (f) which break down the chito-oligomers into dimers or monomers. Chitobiose enters through an ABC-type transporter (g), while the monomer is transported into the bacterial cell cytosol via a PTS type transport system (h) resulting in its phosphorylation. Once in the cytosol, chitobiose is converted into two GlcNAc-6P via the action of an N,N'-diacetylchitobiose phosphorylase, a GlcNAc-1P-mutase and an ATP dependent GlcNAc specific kinase. GlcNAc-6-P is converted to Glc-6-P and Frc-6-P and can enter the pentose phosphate pathway or glycolysis.

It is advantageous for bacteria to recognise GlcNAc oligomers rather than the monomer as a trigger, because GlcNAc can be derived from non chitinous sources such as glycoproteins, while the GlcNAc oligomers are a better indicator of the presence of chitin (Li and Roseman, 2004).

3. Chitin Recognition in Plants

While chitin represents an important source of nutrients and energy for chitinovorous bacteria, for higher organisms such as plants, the presence of chitin is an indication of a potential threat (with one notable exception that will be discussed later), a 'danger' signal which alerts the plant to the presence of a fungal pathogen, a root knot nematode or a herbivorous insect. Such 'danger' signals are also known as 'pathogen associated molecular patterns' (PAMPs), and are seen as a central part of plant or animal innate immunity. PAMPs are usually molecules with fundamental intrinsic properties which the carrying organisms are not able to modify without loss of function. Examples of PAMPS are the endotoxin in the outer membrane of Gram-negative bacteria, peptidoglycan in the bacterial cell wall, or beta-glucans in the fungal cell wall. There are marked differences in how plants recognise pathogens in comparison with animal cells (Nurnberger *et al.* 2004; Parker, 2003; Zipfel and Felix, 2005), but some PAMPs, including LPS, peptidoglycan and bacterial flagellin, induce powerful innate responses in both organisms (Parker, 2003). The receptors involved however share no homology and are probably the result of convergent evolution, and animal cells appear to have much higher sensitivity for LPS (Zipfel and Felix, 2005).

PAMPs, when detected by plants, elicit the production of an array of innate immunity factors which include chitinases. The role of chitinases in plant immunity against fungi has been relatively well characterised, and many plant chitinases have also received interest because of their prominent allergenicity. The chitinase response will limit infection by attacking (in combination with other enzymes) the fungal cell wall, but the chitin fragments released by this interaction can also elicit the expression of further defence genes.

When encountering pathogens, after wounding, or under stress conditions, plants produce several inducible so-called pathogenesis-related proteins (PR); the functions and biological activities of these PRs are multiple, and their role is not just limited to defence against pathogens. These PR-proteins have been classified into 17 families designated PR-1 to PR-17, based on common types of proteins shared between plants (van Loon *et al.* 2006). Chitinases are found in PR-3, PR-4, PR-8, PR-11, and they are mostly endochitinases with the ability to cleave chitin in the fungal cell wall. However, these chitinases are substantially more effective in synergy with other enzymes such as beta-1,3-glucanases (found in PR-2) (Mauch *et al.* 1988). The target of these chitinases could also extend beyond fungal pathogens, e.g. including nematodes (e.g. in the nematode eggshell) or herbivorous insects (van Loon *et al.* 2006).

Is chitin a PAMP? Mounting an inducible response to chitin-bearing pathogens requires sensing of chitin or, in theory, of other components associated with chitin. The insolubility of crystalline chitin, and therefore the question of how its presence can be sensed, is also an issue in plants. It has been known for several years that plants quickly respond to fungal elicitors including chito-oligosaccharides, suggesting the existence of appropriate recognition systems. In plants such as *Arabidopsis*, barley, carrot, wheat and rice, chitin recognition has been shown to be mediated by chito-oligosaccharides with at least three N-acetylglucosamine units (Nurnberger *et al.* 2004; Kaku *et al.* 2006). Sensing of chitin is known to induce defensive responses such as the generation of reactive oxygen species, phytoalexin production, cell division stimulation and medium alkalinisation (Kasprzewska, 2003).

A putative receptor in rice, a plasma membrane glycoprotein called CEBiP with an extracellular, but no cytosolic domain, has been cloned. Engagement of this chitin receptor with chito-octaose ($GlnNAc_8$) induces upregulation of a series of immunity related proteins, including an endochitinase (Kaku *et al.* 2006). Another receptor, called LysM-RLK1 or CERK1 is essential for the chitin response in *Arabidopsis* (Miya *et al.* 2007; Wan *et al.* 2008). This receptor has three extracellular domains with the LysM motif, which is thought to bind to chito-octaose used as elicitor in these studies, as well as an intracytoplasmic Ser/Thr kinase domain (Wan *et al.* 2008). Wan found that as many as 890 different genes were either up- or downregulated after elicitation with chito-octaose, while only very few genes were changed in their expression when the receptor was mutagenised.

It has been shown that nematodes themselves can rapidly induce chitinase synthesis in infected roots of plants, although it is not known whether this induction is mediated via chitin or other nematode-derived molecules (Lambert, 1995; Williamson and Hussey, 1996). It is however clear that chitinases, derived either from the plants or from associated bacteria, play a beneficial role for the host, as they are known to rapidly destroy nematode eggs (Mercer *et al.* 1992). This is the rationale for the inclusion of chitinous waste (such as ground shrimp shells, etc.) in some commercial soil preparations, an additive which is thought to confer protection against fungal and nematode pathogens, probably by selectively enhancing the growth of chitinase-producing micro-organisms (Sarathchandra *et al.* 1996).

The evolutionary arms race between plant and pathogen has also resulted in interesting host-pathogen interactions by which the pathogen is able to circumvent the potentially lethal chitinase response of the host. One obvious way is to mask the PAMPs that elicit an immune response (Zipfel and Felix, 2005), but other strategies take advantage of subtle modifications in the structure of chitin. The hemibiotrophic fungal pathogen *Colletotrichum lindemuthianum*, which is the causative agent of a feared disease called 'bean anthracnose', exposes chitin during appressorium-mediated penetration of plant tissue. This is detected by the infected plant and results in the production of chitinases to combat the pathogen, which in return produces a chitin deacetylase which converts exposed chitin into chitosan (Blair *et al.* 2006). Chitosan lacks the N-Acetyl groups which are essential for cleavage by chitinases, allowing the fungus to escape the plant's chitinase immune response.

Chitin signalling also fulfils less 'baleful' roles in the plant-microbe relationship. Chitin oligomers, or more specifically, lipochito-oligosaccharides (Lerouge *et al.* 1990), are mediators of the symbiotic interaction between rhizobia and legumes. Structurally, these factors contain 3 to 5 β-1,4-linked GlcNAc units with covalent modifications including acylation with unsaturated fatty acids and sulphation.

These nod(ulation) factors (NF) are released by rhizobial bacteria upon sensing of flavonoids released by the roots. The NFs are sensed by receptors located on the root surface of legume plants. These receptors (NFR1 and NFR5 in legumes) contain an extracellular LysM domain which binds the NF and an intracellular portion with kinase activity. They belong to the large family of LysM domain-containing receptor-like kinases (LysM RLKs (Limpens *et al.* 2003). Binding of NFs induces receptor activation and signalling through the kinase domain, resulting in a root hair deformation and cell division (Stracke *et al.* 2002) necessary for the uptake of the bacterium. This occurs by forming an intracellular tube called the infection thread. The bacteria replicate and differentiate becoming bacteroids in nodules, where they acquire the ability to fix nitrogen via their nitrogenases, supplying the plant with NH_4^+ as a needed source of nitrogen, while receiving carbohydrate and protein nutrients from the plant in return.

Work performed with soybean has shown that additionally to the NFs, a second signal delivered by unmodified chito-oligosaccharides might be necessary for nodule initiation (Day *et al.* 2001). Interestingly, root knot nematodes appear to have hijacked these signalling systems for their own purpose, as secretions of these nematodes have recently been shown to induce a response similar to NFs in plant roots, including roothair waviness and branching (Weerasinghe *et al.* 2005).

Finally, chitinases are also induced in plants by herbivory (Inbar *et al.* 1998). These induced chitinases, when ingested by herbivorous insects, may lead to the disruption of the peritrophic membrane (Inbar *et al.* 1998), with parallels to the situation seen in the interaction of malaria-transmitting mosquitoes after a blood meal (see Malaria chapter XV by Musumeci) with human chitinases. This suggests that plants and mammals may use similar strategies targeting the chitin in the peritrophic membrane in their defence against insects.

4. Chitin Recognition in Animals – *Terra Incognita*

Much less is known regarding the recognition of chitin in higher animals. Insects have to rely solely on their innate immunity, as adaptive immunity is only found in vertebrates. Insects contain chitin as a constituent of their exoskeletons and possess the corresponding enzymatic machinery to achieve its synthesis, i.e. chitin synthases.

Chitin in the insect cuticle, covalently crosslinked with proteins, is the primary barrier for microbial pathogens, comparable with the function of mammalian skin, but with a rigidity that limits body growth. The development of insects thus proceeds through sequential molts, in which the entire exoskeleton is rebuilt to allow for an increase in body size. The rigid exoskeleton is shed, and replaced by a soft, expansible cuticle, known as procuticle. The outer part subsequently sclerotises resulting in an epicuticle and exocuticle, in which cuticular proteins are covalently linked with chitin. This process is linked with the production of endogenous chitinases and proteases to digest the old cuticles during the molting process. These chitinases usually possess a chitin-binding domain, which facilitates chitinase binding and efficient degradation of chitin (Arakane *et al.* 2003).

The process of molting (a.k.a. ecdysis) is highly regulated, but there is no evidence that its regulation involves chito-oligosaccharide recognition (Merzendorfer and Zimoch, 2003). In fact, it is known that insect molting is regulated by ecdysteroids, of which 20-hydroxy-ecdysone is the best known hormone. Ecdysone, or synthetic agonists such as tebufenozide (RH5992), induce chitinase synthesis (Zheng *et al.* 2003).

The full genome sequencing of *Drosophila*, *Anopheles* and *Tribolium* (red flour beetle) has revealed that each of these insects possesses a multitude (16-23) of different chitinases and chitinase-like genes, which can be divided in at least 5 different groups, and many of which are involved in different aspects of molting (Zhu *et al.* 2008). Gene knockdown experiments using RNA interference are beginning to unravel the relative contributions of chitinases and chitinase-like proteins to molting or their function in other aspects of insect development.

An unexplored question is whether insect chitinases, in addition to their roles in growth and development (molting), also play a role in insect immunity to pathogens, similarly to the protective role of chitinases in plants. Entomopathogenic fungi such as *Beauveria bassiana* produce powerful extracellular proteases and chitinases that degrade the external barrier and ultimately allow access of fungal hyphae to the nutrient-rich haemolymph (Fan *et al.* 2007). These fungi will also contain chitin in their cell wall, and this could be the target of an insect chitinase immune response. This would require the ability to sense the presence of a chitinous pathogen, raising the question of how the insect could discriminate between its own and pathogen-derived chitin. Interestingly, chitin sensors have been found in silkworms, and (LPS-free) chito-oligomers ranging from dimers to hexamers all induce the production of antimicrobial peptides (Furukawa *et al.* 1999). The authors unfortunately did not assess whether this also leads to chitinase production.

5. Does Chitin Sensing Also Occur in Mammals?

In contrast to other known small, soluble PAMPs, such as endotoxins or lipopeptides, it is not clear how polymeric, insoluble chitin can interact with and activate putative chitin-sensitive pattern recognition receptors (PRRs) in animals. A physical interaction between chitinous material and cellular receptors has to be postulated, but this will depend on an appropriate exposure or release of chitin by the pathogen.

A recently published model described the induction of an allergic response, with the recruitment of interleukin-4 (IL-4)-producing cells to the lung of mice after instillation of insoluble chitin (Reese *et al.* 2007) (see Figure 4). It is unclear how chitin is recognised in this setting. A possibility is that, as seen in bacteria and plants, chitin first needs to be degraded into smaller, soluble fragments by resident chitinases in order to be sensed by innate immunity. In this context, it is important to mention that both known mammalian chitinases, chitotriosidase (Boot *et al.* 1995) and acidic mammalian chitinase (AMCase) (Boot *et al.* 2001; Boot *et al.* 2005) are expressed constitutively, albeit at lower levels, in the healthy murine and human lung. Thus the initial release of soluble fragments by resident chitinases could initiate a cascade of events leading to the activation of innate immunity processes, culminating in the cellular events described by Reese.

The question whether chitin fragments can activate mammalian immune cells is relatively unexplored. Human neutrophils can migrate chemotactically towards N-acetylchitohexaose, and this is increased by prostaglandin E2, but the underlying mechanisms or putative receptors have not been characterised in detail (Tokoro *et al.* 1988). Intraperitoneal administration of different chito-oligosaccharides resulted in the recruitment of polymorphonuclear phagocytes and increased fungicidal activity *in vitro* (Suzuki *et al.* 1986). Work performed by the same (Suzuki *et al.* 1984) and others (Rementeria *et al.* 1997) has shown that chitin treatment can protect mice from systemic candidiasis, although the latter group also found that administration of a lower amount of chitin diminished long term survival after challenge with *Candida albicans*.

Reese/Locksley model: protective role of AMCase (infection model)

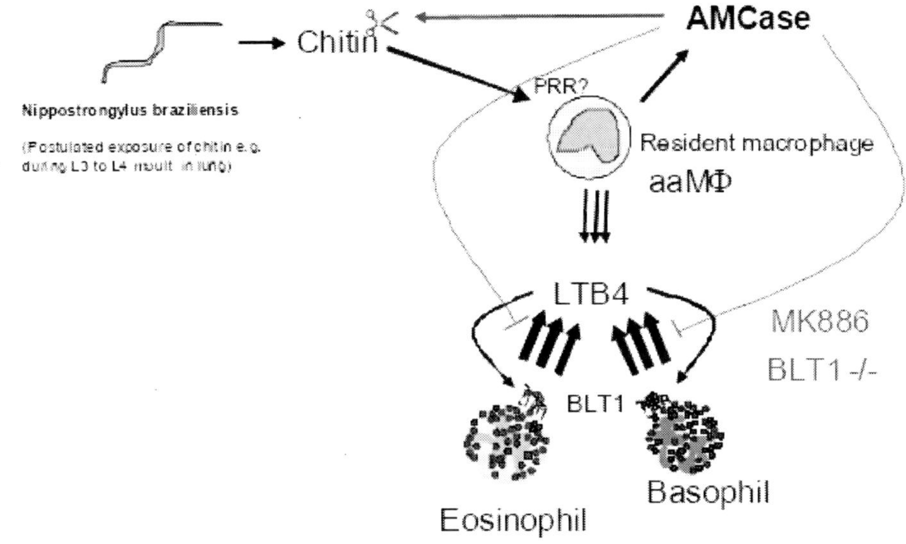

Figure 4. The Reese/Locksley model suggests a protective role for AMCase. In this model, exposure of chitin results in a direct induction of LTB4 from resident macrophages or recruited alternatively activated macrophages, which in turn results in a direct attraction of basophils and eosinophils via the BLT1 receptor. AMCase degrades chitin and ablates recruitment of eosinophils and basophils to the lung. Mice constitutively overexpressing AMCase in the lung do not show apparent abnormalities.

Overall, there is no data which could discriminate whether protection conferred by chitin treatment before infectious challenge is due to direct recognition of chitin or to the sensing of chito-oligosaccharides derived from chitinolysis by immune chitinases. Chitin may also be recognised in its insoluble form, as chitin particles of a certain size are known to exert a strong immune modulatory effect, which might be due to recognition via the mannose receptor (Shibata *et al.* 1997).

Taken together, the examples seen in bacteria and plants point to the possibility that chitin-sensing PRRs could also be found in higher animals such as mammals, and that chitin recognition could be mediated via interaction of chitin-oligosaccharides (of variable size) with these receptors. This lack of knowledge defines an exciting new area of investigation which is certain to yield new insights into the innate immune system of higher animals in the near future.

Acknowledgments

The author wishes to thank his colleagues Prof George Roberts (for the many interesting discussions about the properties of chitin) and Dr Marcos J. Alcocer (for making him think about innate immunity in plants).

References

Arakane Y, Zhu Q, Matsumiya M, Muthukrishnan S, Kramer KJ. Properties of catalytic, linker and chitin-binding domains of insect chitinase. *Insect Biochem Mol Biol.* 2003;33:631-48.

Bassler B, Gibbons P, Roseman S. Chemotaxis to chitin oligosaccharides by Vibrio furnissii, a chitinivorous marine bacterium. *Biochem Biophys Res Commun.* 1989;161:1172-6.

Bassler BL, Gibbons PJ, Yu C, Roseman S. Chitin utilization by marine bacteria. Chemotaxis to chitin oligosaccharides by Vibrio furnissii. *J Biol Chem.* 1991;266:24268-75.

Blair DE, Hekmat O, Schuttelkopf AW, Shrestha B, Tokuyasu K, Withers SG, van Aalten DM. Structure and mechanism of chitin deacetylase from the fungal pathogen Colletotrichum lindemuthianum. *Biochemistry* 2006;45:9416-26.

Boot RG, Renkema GH, Strijland A, van Zonneveld AJ, Aerts JM. Cloning of a cDNA encoding chitotriosidase, a human chitinase produced by macrophages. *J Biol Chem.* 1995;270:26252-6.

Boot RG, Blommaart EF, Swart E, Ghauharali-van der Vlugt K, Bijl N, Moe C, Place A, Aerts JM. Identification of a novel acidic mammalian chitinase distinct from chitotriosidase. *J Biol Chem.* 2001;276:6770-8.

Boot RG, Bussink AP, Verhoek M, de Boer PA, Moorman AF, Aerts JM. Marked Differences in Tissue-specific Expression of Chitinases in Mouse and Man. *J Histochem Cytochem.* 2005.

Colwell RR. Vibrios in the environment: John Wiley and Sons; 1984.

Colwell RR, Huq A, Islam MS, Aziz KM, Yunus M, Khan NH, Mahmud A, Sack RB, Nair GB, Chakraborty J, Sack DA, Russek-Cohen E. Reduction of cholera in Bangladeshi villages by simple filtration. *Proc Natl Acad Sci USA.* 2003;100:1051-5.

Chernin LS, Winson MK, Thompson JM, Haran S, Bycroft BW, Chet I, Williams P, Stewart GS. Chitinolytic activity in Chromobacterium violaceum: substrate analysis and regulation by quorum sensing. *J Bacteriol.* 1998;180:4435-41.

Day RB, Okada M, Ito Y, Tsukada K, Zaghouani H, Shibuya N, Stacey G. Binding site for chitin oligosaccharides in the soybean plasma membrane. Plant Physiol 2001;126:1162-73.

Ehrlich H, Krautter M, Hanke T, Simon P, Knieb C, Heinemann S, Worch H. First evidence of the presence of chitin in skeletons of marine sponges. Part II. Glass sponges (Hexactinellida: Porifera). *J Exp Zoolog B Mol Dev Evol.* 2007;308:473-83.

Eijsink VG, Vaaje-Kolstad G, Varum KM, Horn SJ. Towards new enzymes for biofuels: Lessons from chitinase research. *Trends Biotechnol.* 2008;26:228-35.

Fan Y, Fang W, Guo S, Pei X, Zhang Y, Xiao Y, Li D, Jin K, Bidochka MJ, Pei Y. Increased insect virulence in Beauveria bassiana strains overexpressing an engineered chitinase. *Appl Environ Microbiol.* 2007;73:295-302.

Folders J, Algra J, Roelofs MS, van Loon LC, Tommassen J, Bitter W. Characterization of Pseudomonas aeruginosa chitinase, a gradually secreted protein. *J Bacteriol.* 2001;183:7044-52.

Furukawa S, Taniai K, Yang J, Shono T, Yamakawa M. Induction of gene expression of antibacterial proteins by chitin oligomers in the silkworm, Bombyx mori. *Insect Mol Biol.* 1999;8:145-8.

Gooday GW. The ecology of chitin degradation. *Adv Microb Ecol.* 1990;11:387-430.

Hardie KR, Heurlier K. Establishing bacterial communities by 'word of mouth': LuxS and autoinducer 2 in biofilm development. *Nat Rev Microbiol.* 2008.

Hock CW. Decomposition of chitin by marine bacteria. *Biol Bull.* 1940;79:199-206.

Huq A, Small EB, West PA, Huq MI, Rahman R, Colwell RR. Ecological relationships between Vibrio cholerae and planktonic crustacean copepods. *Appl Environ Microbiol.* 1983;45:275-83.

Huq A, Whitehouse CA, Grim CJ, Alam M, Colwell RR. Biofilms in water, its role and impact in human disease transmission. *Curr Opin Biotechnol.* 2008.

Inbar M, Doostdar H, Sonoda RM, Leibee GL, Mayer RT. Elicitors of plant defensive systems reduce insect densities and disease incidence. *Journal of Chemical Ecology*, 1998 24 : 135-149.

Kaku H, Nishizawa Y, Ishii-Minami N, Akimoto-Tomiyama C, Dohmae N, Takio K, Minami E, Shibuya N. Plant cells recognize chitin fragments for defense signaling through a plasma membrane receptor. *Proc Natl Acad Sci USA.* 2006;103:11086-91.

Kasprzewska A. Plant chitinases--regulation and function. *Cell Mol Biol Lett.* 2003;8:809-24.

Keyhani NO, Roseman S. Physiological aspects of chitin catabolism in marine bacteria. *Biochim Biophys Acta* 1999;1473:108-22.

Keyhani NO, Li XB, Roseman S. Chitin catabolism in the marine bacterium Vibrio furnissii. Identification and molecular cloning of a chitoporin. *J Biol Chem.* 2000;275:33068-76.

Lambert KN. Isolation of Genes Induced Early in the Resistance Response to Meloidogyne javanica in Lycopersicon esculentum. [PhD Thesis]. Davis: University of California, Davis; 1995.

Lerouge P, Roche P, Faucher C, Maillet F, Truchet G, Prome JC, Dénarié J. Symbiotic host-specificity of Rhizobium meliloti is determined by a sulphated and acylated glucosamine oligosaccharide signal. *Nature* 1990;344(6268):781-4.

Limpens E, Franken C, Smit P, Willemse J, Bisseling T, Geurts R. LysM domain receptor kinases regulating rhizobial Nod factor-induced infection. *Science* 2003;302:630-3.

Li X, Roseman S. The chitinolytic cascade in Vibrios is regulated by chitin oligosaccharides and a two-component chitin catabolic sensor/kinase. *Proc Natl Acad Sci USA.* 2004;101:627-31.

Mauch F, Mauch-Mani B, Boller T. Antifungal Hydrolases in Pea Tissue : II. Inhibition of Fungal Growth by Combinations of Chitinase and beta-1,3-Glucanase. *Plant Physiol.* 1988;88:936-942.

Meibom KL, Li XB, Nielsen AT, Wu CY, Roseman S, Schoolnik GK. The Vibrio cholerae chitin utilization program. *Proc Natl Acad Sci USA.* 2004;101:2524-9.

Mercer CF, Greenwood DR, Grant JL. Effect of plant and microbial chitinases on the eggs and juveniles of Meloidogyne hapla Chitwood (Nematoda: Tylenchida). *Nematologica* 1992;38:227-236.

Merzendorfer H, Zimoch L. Chitin metabolism in insects: structure, function and regulation of chitin synthases and chitinases. *J Exp Biol.* 2003;206:4393-412.

Miya A, Albert P, Shinya T, Desaki Y, Ichimura K, Shirasu K, Narusaka Y, Kawakami N, Kaku H, Shibuya N. CERK1, a LysM receptor kinase, is essential for chitin elicitor signaling in Arabidopsis. *Proc Natl Acad Sci USA.* 2007;104:19613-8.

Muzzarelli RA. Native, industrial and fossil chitins. Exs 1999;87:1-6.

Nalin DR, Daya V, Reid A, Levine MM, Cisneros L. Adsorption and growth of Vibrio cholerae on chitin. *Infect Immun.* 1979;25:768-70.

Nurnberger T, Brunner F, Kemmerling B, Piater L. Innate immunity in plants and animals: striking similarities and obvious differences. *Immunol Rev.* 2004;198:249-66.

Parker JE. Plant recognition of microbial patterns. Trends Plant Sci 2003;8:245-7.

Reese TA, Liang HE, Tager AM, Luster AD, Van Rooijen N, Voehringer D, Locksley RM. Chitin induces accumulation in tissue of innate immune cells associated with allergy. *Nature* 2007;447(7140):92-6.

Rementeria A, Abaitua F, Garcia-Tobalina R, Hernando F, Ponton J, Sevilla MJ. Resistance to candidiasis and macrophage activity in chitin-treated mice. *FEMS Immunol Med Microbiol.* 1997;19:223-30.

Ruiz-Herrera J, Martínez-Espinosa AD. Chitin biosynthesis and structural organization in vivo. *Exs* 1999;87:39-53.

Sarathchandra SU, Watson, Cox NR, Di Menna ME, Brown JA, Burch G, Neville FJ. Effects of chitin amendment of soil on microorganisms, nematodes, and growth of white clover (Trifolium repens L.) and perennial ryegrass (Lolium perenne L.) *Biology and fertility of soils* 1996; 22: 221-226.

Shibata Y, Metzger WJ, Myrvik QN. Chitin particle-induced cell-mediated immunity is inhibited by soluble mannan: mannose receptor-mediated phagocytosis initiates IL-12 production. *J Immunol.* 1997;159:2462-7.

Stankiewicz BA, Briggs DEK, Evershed RP, Flannery MB, Wuttke M. Preservation of Chitin in 25-Million-Year-Old Fossils. *Science.* 1997;276:1541-1543.

Stracke S, Kistner C, Yoshida S, Mulder L, Sato S, Kaneko T, Tabata S, Sandal N, Stougaard J, Szczyglowski K, Parniske M. A plant receptor-like kinase required for both bacterial and fungal symbiosis. *Nature* 2002;417(6892):959-62.

Suzuki K, Okawa Y, Hashimoto K, Suzuki S, Suzuki M. Protecting effect of chitin and chitosan on experimentally induced murine candidiasis. *Microbiol Immunol.* 1984;28:903-12.

Suzuki K, Tokoro A, Okawa Y, Suzuki S, Suzuki M. Effect of N-acetylchito-oligosaccharides on activation of phagocytes. *Microbiol Immunol.* 1986;30:777-87.

Tokoro A, Suzuki K, Matsumoto T, Mikami T, Suzuki S, Suzuki M. Chemotactic response of human neutrophils to N-acetyl chitohexaose in vitro. *Microbiol Immunol.* 1988;32:387-95.

van Loon LC, Rep M, Pieterse CM. Significance of inducible defense-related proteins in infected plants. *Annu Rev Phytopathol.* 2006;44:135-62.

Wan J, Zhang XC, Neece D, Ramonell KM, Clough S, Kim SY, Stacey MG, Stacey G. A LysM receptor-like kinase plays a critical role in chitin signaling and fungal resistance in Arabidopsis. *Plant Cell.* 2008;20:471-81.

Weerasinghe RR, Bird DM, Allen NS. Root-knot nematodes and bacterial Nod factors elicit common signal transduction events in Lotus japonicus. *Proc Natl Acad Sci USA.* 2005;102:3147-52.

Williamson VM, Hussey RS. Nematode pathogenesis and resistance in plants. *Plant Cell.* 1996;8:1735-45.

Yu C, Bassler BL, Roseman S. Chemotaxis of the marine bacterium Vibrio furnissii to sugars. A potential mechanism for initiating the chitin catabolic cascade. *J Biol Chem.* 1993;268:9405-9.

Zheng YP, Retnakaran A, Krell PJ, Arif BM, Primavera M, Feng QL. Temporal, spatial and induced expression of chitinase in the spruce budworm, Choristoneura fumiferana. *J Insect Physiol.* 2003;49:241-7.

Zhu Q, Arakane Y, Beeman RW, Kramer KJ, Muthukrishnan S. Functional specialization among insect chitinase family genes revealed by RNA interference. *Proc Natl Acad Sci USA.* 2008;105:6650-5.

Zipfel C, Felix G. Plants and animals: a different taste for microbes? *Curr Opin Plant Biol.* 2005;8:353-60.

Index

A

B

C

D

E

F

G

H

I

J

K

L

M

N

O

P

Q

R

S

T

U

V

W

X

Y

Z